DISCOVERING THE UNIVERSE

DISCOVERING THE UNIVERSE

Second Edition

William J. Kaufmann, III

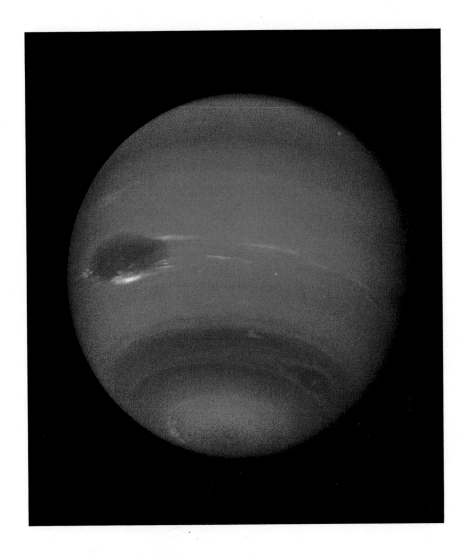

W. H. Freeman and Company
New York

To Wanda, with love

Front cover: This picture of Neptune was produced from images taken by the *Voyager 2* spacecraft through ultraviolet, violet, and green filters. The resulting "false color" photograph emphasizes details of the cloud structure and paints the clouds different colors associated with altitude. The highest clouds are colored white and pink, lower clouds are blue, and the lowest clouds are shaded green. The most prominent feature seen in this picture is the Great Dark Spot, a high-pressure storm system roughly the same size as Earth. (JPL; NASA)

Title page: Neptune's south pole was tilted slightly toward *Voyager 2* when the spacecraft took this picture. The planet's distinctive bluish color is caused by methane in its atmosphere. The Great Dark Spot is located 22° south of Neptune's equator. White, wispy clouds accompany the Great Dark Spot. A smaller dark spot, seen faintly in a dark blue band encircling the south pole, is at a latitude of 54° south. (JPL; NASA)

page vii: η Carinae Nebula (NOAO); **page viii:** Spacewalk (NASA); **page ix:** A solar prominence (NASA); **page x:** Dust lanes in M16 (Anglo-Australian Telescope Board © 1986); and **page xi:** A Seyfert galaxy, NGC 1566 (Anglo-Australian Telescope Board © 1987).

Library of Congress Cataloging-in-Publication Data
Kaufmann, William J.
   Discovering the universe / William J. Kaufmann III.—2nd ed.
     p.   cm.
   Bibliography: p.
   Includes index.
   ISBN 0-7167-2054-X
   1. Astronomy.   2. Cosmology.   I. Title.
QB43.2.K376   1989
520—dc20

89-16773
CIP

Illustration credits are listed on page 416

Copyright © 1987, 1990
by W. H. Freeman and Company

3 4 5 6 7 8 9 0   KP   9 9 8 7 6 5 4 3 2 1

Contents overview

Contents

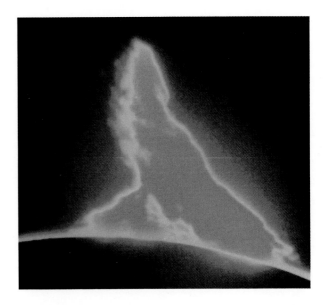

Preface

Each of us has probably had the experience of looking at the sky on a starry night, feeling overcome by its vastness, and wondering what kind of order governs it. It is easy to imagine the earliest people doing this; it is no wonder that astronomy is the oldest of sciences. Prehistoric ruins throughout the world bear witness to human preoccupation with the heavens.

An adventuresome spirit pervades astronomy, a sense that things wonderful and unimagined are yet to be discovered. For millenia people have been fascinated by topics that astronomers explore today: the creation of the universe, the formation of the Earth and other planets, the motions of the stars, the structure of space and time. Armed with the powers of observation, the laws of physics, and the resourcefulness of the human mind, astronomers survey alien worlds, follow the life cycles of stars, and probe the distant reaches of the cosmos.

In *Discovering the Universe* we join astronomers both past and present in their quest for knowledge about the cosmos. We will investigate phenomena and explore realms far removed from our daily experience. Indeed, many of the objects that astronomers study are too vast, distant, or intangible to sample directly; in fact, many of the phenomena observed today occurred very long ago. When viewed in the context of the evolution of the universe, even the dimensions of space and time take on new meaning.

This book was written with the conviction that the subjects astronomers concern themselves with can—and should—be understood by just about everyone. In today's increasingly technological world, we owe it to ourselves to stay informed as best we can. In this spirit, *Discovering the Universe* offers a broad view of astronomy to readers having little or no background in science or mathematics. The narrative is largely descriptive, and a minimum of equations and formulas appear.

As its title implies, this book is as much about the nature of scientific inquiry as it is about the physical universe itself. Of all the lessons to be derived from an introductory science course, perhaps none is more important than an understanding of how scientists reason. In this book I have tried to convey how astronomers have come to know what they know. By studying the methods that astronomers have used to explore and understand the universe, we sharpen our own reasoning skills and are better able to address other scientific issues.

Organization

This text was written for the one-term, descriptive astronomy course. It is considerably shorter and less rigorous than its parent text, *Universe*, which offers a different selection of detail and much more expansive coverage.

Having just nineteen succinct chapters, *Discovering the Universe* can be covered in a course as short as ten weeks.

The traditional Earth-outward organization of the text emphasizes how our understanding of the universe developed and invites the reader to share in the excitement of astronomical discovery. The first celestial objects we will examine are those that were observed by ancient astronomers. Moving outward from the planets to the stars and galaxies, we encounter a modern realm of observations, including those from outside the visible range and those made from space. These new observations in turn raise new questions, which draw the reader on to the limits of our universe and understanding.

The first five chapters introduce the foundations of astronomy, including descriptions of such naked-eye phenomena as eclipses and planetary motions and such basic tools as Kepler's laws and the optics of telescopes. A discussion of the formation of the solar system in Chapter 6 prepares the reader for the next four chapters, which cover the planets.

Chapter 11 introduces the Sun and sets the stage for stellar astronomy in Chapter 12. In Chapters 13 through 15, stellar evolution is described chronologically from birth to death. Molecular clouds, star clusters, nebulae, neutron stars, black holes, and various other phenomena are presented in the sequence in which they naturally occur in the life of a star, thus unifying the discussion of a wide variety of objects that astronomers find scattered about the universe.

A survey of the Milky Way introduces galactic astronomy in Chapter 16, followed by a chapter each on galaxies and quasars. Chapter 19, on cosmology, emphasizes exciting recent developments in our understanding of the physics of the early universe. An afterword addresses the question of whether intelligent life might exist elsewhere in the universe.

Major changes in this edition

The first step in planning the second edition of *Discovering the Universe* was to turn to a broad spectrum of instructors who had used the first edition, whose comments were most influential. Numerous suggestions that friends and colleagues have been sending me since the publication of the first edition also served to inspire this edition. Many of the improvements made in the second edition therefore reflect considerable classroom experience.

Two major organizational changes were made in this edition. The first involves the treatment of light. Virtually all topics relating to the physics of electromagnetic radiation, including spectroscopy and the Doppler effect, are now consolidated in one chapter near the beginning of the book. This chapter, entitled "The nature of light and matter," provides a complete and concise treatment of this critical topic, and increases the text's flexibility as well. For instance, an instructor who chooses to defer or sidestep coverage of the planets can go directly from the chapter on light (Chapter 5) to the first chapter on stellar astronomy (Chapter 11) without any loss of continuity.

The second significant organizational change in this edition is the placement of the chapter on the Sun, which now precedes the discussion of stars in general. As a result, the chapter on solar astronomy offers a comfortable transition between our studies of the planets and the stars.

Many of the people who commented on the first edition expressed the desire for expanded coverage of stellar and galactic astronomy. Chapters 12 through 14 of this edition therefore cover such topics as mass transfer in close binaries, Type I and II supernovae, and SN 1987A. Chapters on galactic and extragalactic astronomy (Chapters 16 through 18) now include extended discussions of the formation of spiral arms, the evolution of galaxies, and collisions between galaxies. The afterword on life in the universe is also new to this edition. Here we consider how life may be the result of ordinary chemical processes and how scientists determine whether a search for extraterrestrial life is worthwhile.

Many smaller, yet equally important revisions enhance the rest of the text. The text was, of course, updated throughout. Indeed, the latest results from the *Voyager 2* flyby of Neptune was added while the book was in press. The writing has become sharper and numerous color illustrations have been enhanced and added. With these changes, I have ventured to create a comprehensive yet entertaining text that gives a fair and accurate picture of the full scope of astronomy.

Pedagogical emphasis

I very much care that students find this book a pleasure to read and learn from. The field is dynamic, compelling, and relevant, and no book about it should be otherwise. I hope these pages will be turned with relish and that interest in astronomy will grow daily.

For those taking astronomy as a first science course, I would have the experience be as rewarding as possible. Ease in understanding will play a large role in this. To this end, I have tried to emphasize the central ideas around which astronomy (and indeed other sciences) revolves. Each chapter begins with a one-paragraph abstract that gives a clear idea of the chapter's contents. The chapter headings are given in the form of declarative sentences to highlight main concepts, and a formal summary outlines the essential facts addressed in each chapter. Each chapter concludes with a series of questions grouped by difficulty and content into three categories: review, advanced, and discussion. Answers to questions that require computation (marked with an asterisk) appear at the end of the book. Care has been taken to include questions whose answers require reasoning rather than rote memorization. Finally, a glossary is included that provides page references, so the reader may easily refer back to an appropriate discussion in the text if desired.

Ancillaries

It is a pleasure to announce the availability of the following outstanding ancillaries to the second edition of *Discovering the Universe:*

An *Instructor's Manual* prepared by Thomas H. Robertson at Ball State University contains chapter outlines, key points and teaching strategies, sample test questions, and suggestions for lecture topics. Included is a resource update of *Universe in the Classroom* by Andrew Fraknoi of the Astonomical Society of the Pacific, providing an invaluable bibliography of books, articles, audiovisual materials, and software.

A printed *Test Bank* and an *IBM PC Test Bank,* revised and expanded by T. Alan Clark at the University of Calgary, Alberta, Canada, are designed for a one-term course in basic astonomy. This two-disk set of multiple-choice questions, indexed by chapter and topic category, uses an upgraded, comprehensive question editing and handling program, QED, written by Uwe Oehler, University of Guelph, Ontario, Canada. Both the computerized test bank and the printed version are available free to adopters.

An attractive and useful set of *Overhead Transparencies* of full-color line diagrams from the text is available to adopters.

For more information and to request copies of these supplements, please contact:

Marketing Department
W. H. Freeman and Company
41 Madison Avenue
New York, NY 10010

Acknowledgments

I would like to begin by thanking my many friends and colleagues who offered suggestions, comments and corrections to the first edition. Many of the improvements in the new edition are a direct result of this unsolicited input. I am also deeply grateful to the following people who carefully reviewed the manuscript for the second edition:

Wyatt W. Anderson	University of Georgia
John W. Burns	Mt. San Antonio College
Barbara Bowman	University of California, Berkeley
Alexei Filippenko	University of California, Berkeley
John Friedman	University of Wisconsin, Milwaukee
Douglas J. Futuyama	State University of New York, Stony Brook
Hollis R. Johnson	Indiana University
Klaus Keil	The University of New Mexico
Steven L. Kipp	Mankato State University
Richard S. Marasso	Sierra College
Norman L. Markworth	Stephen F. Austin State University
Steve McMillan	Drexel University
Larry C. Oglesby	Pomona College
R. P. Olowin	St. Mary's College of California
Tobias Owen	State University of New York, Stony Brook
Hugh O. Peebles, Jr.	Lamar University
James G. Peters	San Francisco State University
Dale R. Snider	University of Wisconsin, Milwaukee
Thomas H. Robertson	Ball State University
Daniel Rothstein	Kent State University
David R. Wood	Wright State University

The reviewers of the preceding edition also deserve acknowledgment, as their advice has had an ongoing influence:

Jay Boleman	University of Central Florida
Gladwin Comes	Broward Community College
Robert J. Dukes	College of Charleston
Roger A. Freedman	University of California, Santa Barbara

Paul Helminger	University of South Alabama
Richard Henry	University of Oklahoma
Hal R. Jandorf	Moorpark College
Yong Hak Kim	Saddleback College
Terry Rettig	University of Notre Dame
Michael Stewart	San Antonio College
Takamasa Takahashi	St. Norbert College
Harley Thronson, Jr.	University of Wyoming
Louis Winkler	Pennsylvania State University

Thanks are due to many other people who have participated in the preparation of this book. Foremost among them is my developmental editor, Carol Pritchard-Martinez, who has worked closely with me since the first edition of *Universe* was conceived eight years ago. I also thank my editor, Jerry Lyons, and W. H. Freeman's president, Linda Chaput, for their support and encouragement during this project. It was once again a pleasure to work with Georgia Lee Hadler, my project editor; Ellen Cash, the production manager; Laurie Sulzburgh, the designer; Patricia Holtz, the illustration coordinator; and the staff of York Graphic Services, all of whom maintain very high standards and pride in their work. I wish also to acknowledge my copyeditor, Tom Szalkiewicz, and the marvelous airbrush artistry of George Kelvin.

Every effort has been made to make this book error-free. Nevertheless, some errors may have crept in. I would appreciate hearing from anyone who finds an error or who wishes to comment on the text. You may write to me care of the Physics Department at SDSU. I will respond personally to all correspondence.

William J. Kaufmann, III
Department of Physics
San Diego State University

1 Astronomy and the universe

The Horsehead and Orion Nebulae *New stars are forming in the clouds of interstellar gas and dust shown in this photograph, which covers an area of the sky approximately 4° × 6°. The gases glow because of the radiation emitted by newborn, massive stars. Clouds of interstellar dust block light, and so they appear as dark regions silhouetted against glowing background nebulosity. The two brightest stars near the top of the picture are in Orion's belt. The Horsehead Nebula, so named because of the shape of a silhouetted dust cloud, is in the upper left corner (also see Figure 13-1). The Orion Nebula, overexposed in this view, is at the bottom (also see Figure 13-9). Most of the nebulosity and many of the stars in this photograph are 1600 light years from Earth. (Royal Observatory, Edinburgh)*

Astronomy is the study of the universe. A brief preview here of the following chapters provides an outline of the scope and content of astronomy. We introduce important tools such as angular measure and powers-of-ten notation. We learn enough about the solar system, stars, nebulae, and galaxies to get a sense of where we will be going in this book. Above all, we learn that the universe is indeed comprehensible. Although some questions remain unanswered, there is no reason to think that any aspect of the physical universe is arbitrary or unexplainable.

Modern city dwellers usually pay little attention to the night sky. If they do look toward the heavens, they are likely to see little more than the Moon and a few of the brightest stars.

For the many generations who lived without electric lights and smog, however, the breathtaking panorama of the night sky was one of the central experiences of life. Thousands of stars are scattered from horizon to horizon, with the delicate mist of the Milky Way tracing a faerie path through the patterns of brighter stars. The Moon and the planets shift their positions from night to night against this glorious stellar background while the entire spectacle swings slowly overhead from east to west as the night progresses. Our ancestors learned to tell time and directions from these changing patterns in the sky. They mapped the stars into picture outlines of the most important legends and ideas in their cultures.

When we gaze at the stars, as did these earlier people, we find our thoughts turning to profound questions. How was the universe created? Where did the Earth, Moon, and Sun come from? What are the planets and stars made of? And how do we fit in—what is our place in the cosmic scope of space and time?

Speculation about the nature of the universe is one of the most ancient human endeavors. The study of the stars transcends all boundaries of culture, geography, and politics. The modern science of astronomy carries on an ancient tradition of observation and speculation, using the newest tools of technology and mathematics. In the most literal sense, astronomy is a universal subject—its subject is indeed the universe.

Astronomers use angles to denote the apparent sizes and positions of objects in the sky

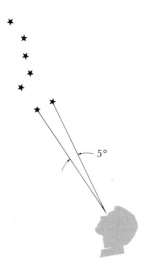

Figure 1-1 An angle An angle is the opening between two lines that meet at a point. The angular distance between the two stars at the front of the Big Dipper is 5°. For comparison, the angular diameter of the Moon is ½°.

Astronomers have inherited many useful concepts from antiquity. For example, ancient mathematicians invented angles and a system of angular measure that is still used to denote the positions and apparent sizes of objects in the sky.

An **angle** is the opening between two lines that meet at a point. **Angular measure** provides an exact description of the shape or "size" of an angle. The basic unit of angular measure is the **degree,** designated by the symbol °. A full circle is divided into 360°. A right angle measures 90°. As shown in Figure 1-1, the angle between the two "pointer stars" at the front of the Big Dipper is 5°.

Astronomy uses angular measure in a wide range of situations. For example, we can use an angle to describe how big an object appears in the sky. Imagine looking up at the full moon. The angle covered by the Moon is nearly ½°. We therefore say that the **angular diameter** or **angular size** of the Moon is ½°. Alternatively, astronomers say that the Moon **subtends** an angle of ½°. Ten full moons could fit side by side between the two pointer stars in the Big Dipper.

From everyday experience, we know that an object looks big when it is nearby, but small when it is far away. The angular size of an object therefore does not necessarily tell you anything about its actual physical size. In order to convert angular size to physical size, you also need to know the distance to the object. For instance, the fact that Moon's angular diameter is ½° does not tell you how big the Moon really is. But if you also happen to know that the distance to the Moon is 376,300 km, then it is possible to calculate that its physical diameter is 3476 km.

To talk about small angles, we subdivide the degree into 60 minutes of arc (abbreviated 60 arc min or 60′). A minute of arc is further subdivided into 60 arc seconds (abbreviated 60 arc sec or 60″). Thus,

$$1° = 60 \text{ arc min} = 60'$$

$$1' = 60 \text{ arc sec} = 60''$$

Your thumbnail held at arm's length subtends an angle of 1°, and a dime viewed from a distance of 1 mile has an angular diameter of about 2 arc seconds.

In the *Astronomical Almanac* for 1988, we can read that Venus had an angular diameter of 34.62 seconds of arc on May 3. That is a very convenient, precise statement of how big the planet appeared in the sky on that date.

Powers-of-ten notation is a useful shorthand system of writing numbers

Astronomy is a subject of extremes. As we examine alien environments, we find an astonishing range of conditions, from the oppressive, acid-drenched clouds of Venus to the near-perfect vacuum of interplanetary space. To describe these conditions accurately, we need a wide range of large and small numbers. To avoid such confusing terms as "a million billion billion," astronomers use a shorthand method called powers-of-ten notation. All of the cumbersome zeros that accompany a large number are consolidated into one term consisting of 10 followed by a superscript, or **exponent**. The exponent indicates how many zeros you would need to write out the long form of the number. Thus,

$$10^0 = 1$$

$$10^1 = 10$$

$$10^2 = 100$$

$$10^3 = 1000$$

$$10^4 = 10,000$$

and so forth.

The exponent tells you how many factors of 10 must be multiplied together to give the desired number. For example, 10,000 can be written as 10^4 (read "ten to the fourth") because $10^4 = 10 \times 10 \times 10 \times 10 = 10,000$.

With this notation, numbers are written as a figure between 1 and 10 multiplied by the appropriate power of 10. For example, the distance between the Earth and the Sun can be written as 1.5×10^8 km. After you are familiar with it, you will find this notation more convenient than using "150,000,000 kilometers" or "one hundred and fifty million kilometers."

This shorthand system can be extended to numbers less than one by using a minus sign in front of the exponent. A negative exponent tells you the location of the decimal point, as follows:

$$10^0 = 1$$
$$10^{-1} = 0.1$$
$$10^{-2} = 0.01$$
$$10^{-3} = 0.001$$
$$10^{-4} = 0.0001$$

Figure 1-2 *Examples of powers-of-ten*
At the center is the Taj Mahal, which adorns the 10-meter world easily in reach of our senses. Dimensions grow smaller to the left: first, crystalline skeletons of single-celled diatoms 10^{-4} meter (0.1 millimeter) in size and, far left, are tungsten atoms, 10^{-10} meter in diameter. On the right, looking across the Indian Ocean toward the South Pole, we see the curvature of the Earth, 10^7 meters in diameter. At the far right is a galaxy, 10^{21} meters in diameter and 10^{24} meters from Earth. (Courtesy of Scientific American, NASA, AAO)

and so forth.

For example, the diameter of a hydrogen atom is 1.1×10^{-8} cm. That is easier than saying "0.000000011 centimeters" or "eleven billionths of a centimeter." Thus, with both very large and very small numbers, the powers-of-ten notation bypasses all those awkward zeros in a convenient fashion. Figure 1-2 gives several examples of the use of powers-of-ten notation.

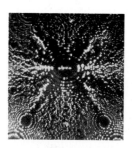    

Astronomers use the laws of physics to explore and understand the universe

We begin, in Chapter 2, by examining some of the astronomical observations and ideas of our ancestors. We shall see that the course of civilization has been dramatically affected by the realization that the universe is indeed comprehensible. This first glimpse of the power and potential of the human mind is one of the great gifts to come to us from ancient Greece. By observing the heavens and carefully thinking about what we see, we can figure out how the universe operates. For example, we shall see that the ancient Greeks measured the size of the Earth and understood eclipses.

Some people think astronomy deals with faraway places of no possible significance to life here on Earth, but nothing could be farther from the truth. For example, in Chapter 3 we see that the seventeenth century scientist Isaac Newton succeeded in describing why the planets orbit the Sun. Unhampered by air resistance or friction, the motions of the planets reveal to us some of the most fundamental laws of nature in their simplest forms. From Newton's work we obtained our first complete, coherent description of the behavior of the physical universe. The resulting body of knowledge, called **Newtonian mechanics,** speaks in concrete terms about force, mass, acceleration, momentum, and energy. It is no coincidence that the Industrial Revolution followed hard on the heels of this knowledge because Newtonian mechanics provided the theoretical and mathematical basis for the construction of machines, factories, buildings, and bridges.

Astronomers use Newtonian mechanics along with other physical principles (usually called the laws of physics) to interpret their observations and to understand the processes that occur in the universe. The laws of physics,

Figure 1-3 [left]   An astronaut on the Moon   *Humanity has taken its first small step out into the universe. As we explore distant worlds, we gain a new perspective on our own planet and a broadened understanding of our relationship to the cosmos. This photograph shows Apollo 11 astronaut Edwin Aldrin on July 21, 1969, at the first lunar landing site. (NASA)*

Figure 1-4 [right]   The Space Shuttle   *The Space Shuttle is used to transport telescopes and other astronomical equipment into orbit, far above the obscuring effects of the Earth's atmosphere. NASA's original plans to launch interplanetary probes from the Space Shuttle were set back by a disastrous explosion during a shuttle launch in 1986. This photograph shows the first Space Shuttle rising majestically from Cape Canaveral on its maiden voyage April 12, 1981. (NASA)*

particularly those involving optics and light, can also be used to develop new tools and techniques with which to examine and explore the universe.

In Chapter 4 we discuss the astronomer's most important tool, the telescope. Until recently, everything we knew about the distant universe was based on visible light. Astronomers would peer through telescopes, take photographs of what they saw, and analyze the starlight. Toward the end of the nineteenth century, however, scientists began discovering nonvisible forms of light such as X rays and gamma rays, radio waves and microwaves, and ultraviolet and infrared radiation. Visible light and radio waves succeed in penetrating the air we breathe, and so optical and radio observatories can operate on the Earth's surface. But the Earth's atmosphere is opaque to most other types of radiation. To detect these other forms of light from stars and galaxies, astronomers must place instruments in space (see Figures 1-3 and 1-4).

Astronomers recently have placed telescopes in orbit to detect nonvisible forms of light. Far above the obscuring effects of Earth's atmosphere, these astronomical instruments give us views of the universe vastly different from what our eyes can see. This new information is crucial to our understanding of such familiar objects as the Sun and gives us important clues about such exotic objects as neutron stars, pulsars, quasars, and black holes.

We complete our introduction to astronomy in Chapter 5 with a discussion of light. By understanding how objects emit radiation and how light interacts with matter, we acquire the skills to analyze and interpret the wealth of information coming to us from the stars and galaxies. In later chapters we shall see how astronomers use these skills to obtain fundamental information about stars, galaxies, and the evolution of the universe.

By exploring the planets, astronomers uncover clues about the formation of the solar system

This book presents the substance of modern astronomy in three segments, corresponding to three major steps out into the universe: the planets, the stars, and the galaxies. In Chapters 6 through 10, we explore the solar system, beginning with Earthlike worlds and moving outward toward the frigid

Figure 1-5 Jupiter, Io, and Europa
Jupiter is orbited by many moons, four of which are so large that they could qualify as planets in their own right. Two of these giant satellites are shown in this view: ruddy Io on the left, ice-bound Europa on the right. Closeup examination of these worlds has dramatically broadened our understanding of Earthlike planets. This photograph was taken by Voyager 1 in 1979, when the spacecraft was 20 million kilometers (12 million miles) from Jupiter's colorful cloudtops. (NASA)

depths of space where comets spend most of their time. Throughout this journey, we shall find that our discoveries are relevant to the quality of human life here on Earth. Until recently, our knowledge of such subjects as geology, geophysics, meteorology, and climatology was based on data from only one planet, Earth. With the advent of space exploration, however, we have a range of other planets with which to compare our own (see Figure 1-5). As a result, we are making important strides in understanding the creation and evolution of the Earth and the entire solar system. These investigations give us significant insight into the origin and extent of all our natural resources.

Astronomical distances are often measured in AUs, parsecs, or light years

In the second half of this book, when we turn our attention to the stars and galaxies, we find that some of our traditional measurement systems are cumbersome. It is fine to use kilometers (or miles) to give the diameters of craters on the Moon or the heights of volcanoes on Mars. But it is as awkward to use kilometers for distances to stars or galaxies as it would be to talk about the distance from New York to San Francisco in inches or millimeters. Astronomers have therefore invented new units of measure.

When discussing distance across the solar system, astronomers like to use a unit of length called the **astronomical unit** (abbreviated AU), which is the average distance between the Earth and the Sun:

$$1 \text{ AU} = 1.496 \times 10^8 \text{ km}$$

Thus, for example, the distance between the Sun and Jupiter is stated as 5.2 AU.

When talking about distances to the stars, astronomers use one of two different units of length. The **light year** (abbreviated ly) is the distance light travels in one year:

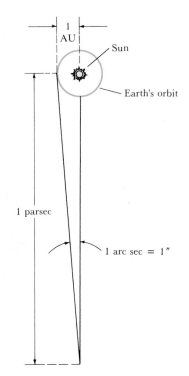

Figure 1-6 A parsec *The parsec, a unit of length commonly used by astronomers, is equal to 3.26 light years. The parsec is defined as the distance at which 1 AU subtends an angle of 1 second of arc.*

$$1 \text{ ly} = 9.46 \times 10^{12} \text{ km}$$

One light year is roughly equal to 6 trillion miles. The nearest star, Proxima Centauri, is 4.3 light years from Earth.

The second commonly used unit of length is the **parsec** (abbreviated pc). Imagine taking a journey far into space, beyond the orbit of Pluto. As you look back toward the Sun, the Earth's orbit subtends a small angle in the sky. One parsec is the distance at which 1 AU subtends an angle of one second of arc, as shown in Figure 1-6.

The parsec turns out to be longer than the light year. Specifically,

$$1 \text{ pc} = 3.09 \times 10^{13} \text{ km}$$
$$= 3.26 \text{ ly}$$

The distance to the nearest star, for instance, can be stated as 1.3 pc or as 4.3 light years. Whether one uses light years or parsecs is a matter of personal taste.

For even greater distances, astronomers commonly use **kiloparsecs** and **megaparsecs** (abbreviated kpc and Mpc), in which the prefixes simply mean "thousand" and "million," respectively:

$$1 \text{ kpc} = 10^3 \text{ pc}$$
$$1 \text{ Mpc} = 10^6 \text{ pc}$$

Thus, for example, the distance from Earth to the center of our Milky Way Galaxy is 9 kpc, and the rich cluster of galaxies in the constellation of Virgo is 20 Mpc away.

Some astronomers prefer to talk about thousands or millions of light years rather than kiloparsecs and megaparsecs. Once again, the choice is a matter of personal taste.

By studying stars and nebulae, astronomers discover how stars are born, grow old, and eventually die

We begin our study of the stars in Chapters 11 and 12 with our own star, the Sun. In the 1920s and 1930s, physicists figured out how the Sun shines. At its center, thermonuclear reactions convert hydrogen into helium. This violent process releases a vast amount of energy that eventually makes its way to the Sun's surface and escapes as sunlight. By 1950, physicists learned how to reproduce this thermonuclear reaction here on Earth. Hydrogen bombs operate on the same basic principles as those that govern the energy production at the Sun's center. Thermonuclear weapons stockpiled around the world have a profound effect on international politics and could dramatically influence the future of life on our planet. Once again, we see the surprising impact of astronomy on the course of civilization.

As we look deeper into space, we find star clusters and clouds of glowing gas, called **nebulae,** scattered across the sky. In Chapter 13 we learn that these beautiful objects tell us much about the lives of stars. We discover that stars are born in huge clouds of interstellar gas and dust such as the Orion Nebula shown in Figure 1-7. After billions of years, stars eventually die. Some end their lives with a spectacular detonation called a **supernova** that blows the star apart, producing objects such as the Crab Nebula seen in Figure 1-8.

Figure 1-7 [left] The Orion Nebula
This nebula (also called M42 or NGC 1976) is a fine example of a stellar "nursery." Intense radiation from hot, newly formed stars causes the surrounding gases to glow. Many of the stars embedded in this nebula are less than 1 million years old. The Orion Nebula is 1600 light years from Earth, and the distance across the nebula is about 23 light years. (U.S. Naval Observatory)

Figure 1-8 [right] The Crab Nebula
This nebula (also called M1 or NGC 1952) is a supernova remnant. A dying star exploded, and this "funeral shroud" was created by the gases that were blasted violently into space. These gases are still moving outward at about 1000 km/sec (roughly 2 million miles per hour). The Crab Nebula is 3600 light years from Earth, and the distance across the nebula is about 6 light years. (Lick Observatory)

During their death throes, stars return gas to interstellar space. We shall learn that this gas contains many heavy elements created by thermonuclear reactions in the stars' interiors. Interstellar space thus becomes enriched with chemicals that did not exist in earlier times, as a forest does when decaying leaves and logs enrich its soil for future generations of trees. The Sun and its planets were formed from enriched interstellar material. We therefore arrive at the surprising realization that virtually everything, including the atoms in our bodies, was originally created deep inside ancient, now-dead stars.

Chapters 14 and 15 discuss how dying stars can produce some of the strangest objects in the sky. The majority of dying stars become **white dwarfs,** which are very compact objects roughly the same size as the Earth. Some dead stars become **pulsars** from which we observe pulses of radio waves or **bursters** that emit powerful bursts of X rays. Massive dead stars become **black holes,** surrounded by incredibly powerful gravity from which nothing (not even light) can escape. Many of these bizarre stellar corpses have been discovered in recent years with Earth-orbiting telescopes that detect nonvisible light.

By observing galaxies, astronomers learn about the creation and fate of the universe

Stars are not spread uniformly across the universe but are grouped together in huge assemblages called **galaxies,** the largest individual objects in the universe. A typical large galaxy, like our own Milky Way, contains several hundred billion stars.

In Chapter 16 we tour the Milky Way Galaxy. We discover that our Galaxy has beautiful, arching spiral arms (like those of M83 in Figure 1-9) that are active sites of star formation. We are surprised to learn that the center of our Galaxy is emitting vast quantities of energy.

In Chapter 17 we explore other galaxies, and find that they come in a wide range of shapes and sizes. Some galaxies are quite small and contain only a few hundred million stars. Others are veritable monstrosities that devour neighboring galaxies in a process of "galactic cannibalism."

Some of the most intriguing galaxies appear to be in the throes of violent convulsions. The centers of these strange, distorted galaxies are often pow-

Figure 1-9 The galaxy M83
This spectacular galaxy (also called NGC 5236) contains about 200 billion stars. The galaxy's spiral arms are outlined by numerous nebulae that are the sites of active star formation. This galaxy has a diameter of about 35,000 light years and is at a distance of 12 million light years from Earth. (Courtesy of R. J. Dufour)

erful sources of X rays and radio waves. In many cases, it looks as though the entire galaxy is being blown apart.

Even more dramatic sources of energy are found still deeper in space. As described in Chapter 18, at distances of billions of light years from Earth we find the mysterious **quasars.** Although quasars look like stars (see Figure 1-10), they are probably the most distant and most luminous objects in the sky. A typical quasar shines with the brilliance of a hundred galaxies. We shall examine data suggesting that quasars draw their awesome energy from enormous black holes.

Finally, in Chapter 19, we turn to the most fundamental questions about the creation and fate of the universe. We shall see how the motions of the galaxies reveal that we live in an expanding universe. Extrapolating backward, we learn that the universe was probably born from an infinitely dense state roughly 15 billion years ago.

Most astronomers believe that the universe began with a cosmic explosion, called the **Big Bang,** that occurred throughout all space at the beginning of time. Many features of the universe today were dictated by events that happened during the Big Bang. We shall learn how astronomers are making significant progress in understanding these cosmic events. Indeed, we may be about to discover the origin of some of the most basic properties of the universe. We shall see, finally, how the motions of the most distant galaxies tell us the ultimate fate of the universe: whether it will expand forever or someday stop and collapse back on itself.

An underlying theme throughout this book is the idea that reality is rational. The universe is not a hodgepodge of unrelated things behaving in unpredictable ways. Rather, we find strong evidence for the existence of the fundamental laws that govern the nature and behavior of everything in the universe. This powerful unifying concept enables us to explore realms far removed from our earthly experience. Thus a scientist can do laboratory experiments to determine the properties of light or the behavior of atoms

Figure 1-10 The quasar 3C48
Quasars are the most distant and most luminous objects that astronomers have ever seen. At first glance, a quasar is easily mistaken for a faint star. This quasar is about 4.8 billion light years from Earth. (Palomar Observatory)

Figure 1-11 An Earth-orbiting industrial complex *A new generation of high-technology materials could easily be manufactured in space. Exotic alloys, foam metals, ultrapure semiconducting crystals, and rare vaccines are among the obvious practical applications of zero-gravity industry. This artist's conception shows an industrial space station under construction. (NASA)*

and then use this knowledge to discover the life cycles of stars and the structure of the universe.

These discoveries have a direct, profound influence on humanity. The past four centuries of civilization clearly show that major scientific advances sooner or later make their way into our lives. In the near future we can look forward to the benefits of space technology. Weightlessness and the near-perfect vacuum of space will enable us to manufacture a wide range of exceptional substances, from exotic alloys to ultrapure medicines (see Figure 1-11).

The dreams of Jules Verne and H. G. Wells pale in comparison to the reality of today. Ours is an age of exploration and discovery more profound than any since Columbus and Magellan set sail across uncharted seas. We have walked on the Moon, we have dug in the Martian soil, probed the poisonous clouds of Venus, and seen the craters on Mercury. We have discovered active volcanoes and barren ice fields on the satellites of Jupiter, and have visited the shimmering rings of Saturn. Never before has so much been revealed in such a short time.

As you explore the cosmos in the pages of this book, you might be tempted to look back at Earth and see yourself as living a brief and meaningless existence on a tiny rock orbiting an insignificant star. One of the great lessons of modern astronomy is the awesome power of the human mind to reach out, to explore, to observe, and to comprehend, thereby transcending the limitations of our bodies and the brevity of human life.

Summary

· The universe is comprehensible.

· The laws of physics govern the nature and behavior of everything in the universe.

· Observation of the heavens has led to discovery of some of the fundamental laws of nature. Important contributions to astronomical knowledge were made by many individuals in many cultures over the centuries.

· Angular measure and powers-of-ten notation are important tools for the study of astronomy.

> Angular measure can be used to describe the apparent size of an object in the sky.

> Powers-of-ten notation is a shorthand method of writing very large and very small numbers.

· Astronomers use the AU, the parsec, and the light year for conveniently measuring very large distances.

> The astronomical unit (AU) is commonly used to express distances across the solar system.

> Parsecs and light years are used to express distances to stars and galaxies.

· Study of the planets provides information about the origin, history, and evolution of the solar system. This study also gives us useful knowledge about the Earth's resources.

· Study of the stars and nebulae provides information about the origin and history of the Sun.

· Study of the galaxies provides information about the origin and history of the universe.

Review questions

1 With the aid of a diagram, explain what it means to say that the Moon subtends an angle of $\frac{1}{2}°$.

2 What is the relationship between degrees, minutes of arc, and seconds of arc?

* 3 Write the following numbers using the powers-of-ten notation: **(a)** ten million, **(b)** four hundred thousand, **(c)** six one-hundredths, **(d)** seventeen billion.

4 What is an AU?

5 What is a parsec?

6 What is the advantage to the astronomer of using the light year as a unit of distance?

Advanced questions

* 7 The speed of light is 3×10^{10} cm/sec. How long does it take light to get from the Sun to the Earth?

* 8 The diameter of the Sun is 1.4×10^{11} cm and the distance to the nearest star, Proxima Centauri, is 4.3 light years. If the Sun were reduced to the size of a basketball (about 30 cm in diameter), at what distance would Proxima Centauri be from the Sun on this reduced scale?

Discussion questions

9 How do astronomical observations and experiments differ from those of other sciences?

10 Discuss the meaning and justification of the assumption that "reality is rational."

11 Discuss the difference between basic or pure science and applied science. What practical applications are there for astronomy? What role should government play in funding each type of science?

For further reading

Asimov, I. *The Measure of the Universe.* Harper & Row, 1983.

Chaisson, E. *Cosmic Dawn.* Little, Brown, 1981.

Gore, R. "The Once and Future Universe." *National Geographic,* June 1983, p. 704.

Jastrow, R. *Red Giants and White Dwarfs,* 2nd ed. Norton, 1979.

King, I. "Man in the Universe." *Mercury,* November/December 1976, p. 7.

Morrison, P., and Morrison, P. *Powers of Ten.* Scientific American Books, 1982.

Seielstad, G. *Cosmic Ecology.* University of California Press, 1983.

2 Discovering the heavens

Circumpolar star trails This long exposure is aimed at the south celestial pole and shows the rotation of the sky. The foreground building is part of the Anglo-Australian Observatory, which houses one of the largest telescopes in the southern hemisphere. Many of the photographs in this book were taken with this telescope, whose primary mirror is 3.9 meters (12.8 feet) in diameter. During the exposure, someone carrying a flashlight walked along the dome's outside catwalk. Another flashlight made the wavy trail at ground level. (Anglo-Australian Observatory)

Ancient cultures made many important contributions to astronomy. Modern astronomers still use constellations described by the ancient Babylonians and Greeks. In this chapter, we learn to find our way around the sky, which is conveniently described as the celestial sphere. We discover that the seasons are related to the tilt of the Earth's axis of rotation and that the axis itself is slowly changing its orientation. We also learn about the phases of the Moon and their relationship to the Moon's motion about the Earth and the Earth's motion about the Sun. We see how ancient astronomers attempted to measure the size of the Earth and the distances from Earth to the Sun and the Moon. Eclipses of the Sun and the Moon played important roles in many of these early theories and measurements. We learn about the conditions under which eclipses occur and about their physical details.

The beauty of the star-filled night sky or the drama of an eclipse would suffice to make astronomy fascinating. But there are practical reasons as well for an interest in the universe. The ancient Greeks knew the connection between the seasons and the relative orientation of the Sun and the Earth. Many early seafaring cultures were aware that the tides are influenced by the position of the Moon.

Ancient civilizations placed great emphasis on careful astronomical observation. Hundreds of impressive monuments, such as Stonehenge (Figure 2-1), that dot the British Isles provide evidence of this preoccupation with astronomy. Alignments of the stones point to the rising and setting locations of the Sun and Moon at various times during the year. Similar astronomically oriented monuments are found in the Americas. The Medicine Wheel in Wyoming, constructed high atop a windswept plateau by the Plains Indians, has stones and markers aligned with the rising points of several bright stars and the Sun.

Architects of the Mayan city of Chichén Itzá on the Yucatán Peninsula of Mexico built an astronomical observatory, the Caracol, nearly a thousand years ago. The Caracol has a cylindrical tower that contains windows aligned with the northernmost and southernmost rising and setting points of both the Sun and the planet Venus. A similar four-story adobe building, probably constructed during the fourteenth century, is located at the Casa Grande site in Arizona. And in the ruined city of Tiahuanaco in Bolivia, ancient engineers built the Temple of the Sun with walls aligned north–south and east–west with an accuracy better than one degree. All of these structures bear witness to careful and patient astronomical observations by the people of many ancient civilizations.

Figure 2-1 Stonehenge This astronomical monument was constructed nearly 4000 years ago on Salisbury Plain in southern England. Originally, the monument consisted of 30 blocks of gray sandstone, each standing 4 meters (13½ feet) high, set in a circle 30 meters (97 feet) in diameter. These stones were topped with a continuous circle of smaller stones. Inside the circle are geometrical arrangements of other stones, most notably a horseshoe-shaped set of larger stones opening toward the northeast. (Courtesy of the British government)

Eighty-eight constellations cover the entire sky

The ancient roots of astronomy are most apparent in the **constellations.** Perhaps it was shepherds tending their flocks at night or priests studying the starry heavens who first imagined pictures among groupings of stars.

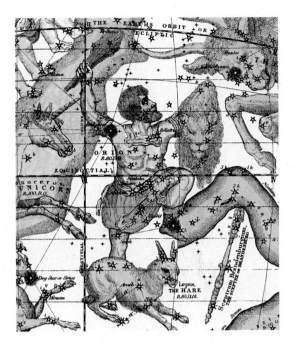

Figure 2-2 Orion Orion is a prominent winter constellation. From the United States, Orion is easily seen high above the southern horizon from December through March. Because of the time exposure—four minutes on Kodak Ektachrome—the colors of the stars are very noticeable. The fanciful drawing of Orion is from an 1835 star atlas. (Courtesy of Robert Mitchell and Janus Publications)

You may already be familiar with some of these patterns in the sky such as the Big Dipper, which is actually part of a large constellation called Ursa Major (the Great Bear). Many of these constellations, such as Orion in Figure 2-2, have names from ancient myths and legends. Although some star groupings vaguely resemble the figures they are supposed to represent, most do not.

Modern star charts divide the entire sky into 88 constellations. Some constellations cover very large areas (Ursa Major is one of the biggest), others small. The constellations are used to specify certain regions of the sky. Thus, for example, we might speak of "the galaxy M31 in Andromeda" much as we would refer to "the Ural Mountains in the Soviet Union."

The Earth rotates once every 24 hours (that is why we have day and night), and hence the constellations rise in the east and set in the west, as do the Sun and Moon. This daily, or **diurnal,** motion of the stars is apparent in time exposure photographs such as the picture on the first page of this chapter.

People who spend time outdoors at night are familiar with the diurnal motion of the constellations. Even if you are a city dweller, take the time to observe the basic facts of astronomy yourself. Go outdoors soon after dark, find a spot away from bright lights, and note the patterns of stars in the sky. A few hours later, check again. You will find that the entire pattern of stars (including the Moon, if it is visible) will have shifted its position. New constellations will have risen above the eastern horizon, while other constellations will have disappeared below the western horizon. If you check again just before dawn, you will find low in the western sky the stars that were just rising when the night began.

The constellations that you can see in the sky change slowly from one night to the next. This shift occurs as the Earth orbits the Sun, as shown in Figure 2-3. The Earth takes a full year to go once around the Sun, and thus

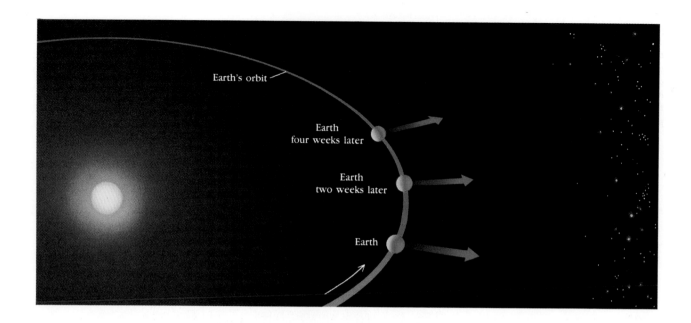

Figure 2-3 *Our changing view of the night sky* *As we orbit the Sun, the nighttime sky of the Earth gradually turns toward different parts of the sky. Thus, the constellations that we can see change slowly from one night to the next.*

the darkened, nighttime side of the Earth is gradually turned toward different parts of the heavens. Specifically, if you follow a particular star on successive evenings, you find that it rises approximately 4 minutes earlier each night. A set of star charts for the evening hours of selected months of the year is included at the end of this book.

It is often convenient to imagine that the stars are located on the celestial sphere

As you gaze at the heavens on a clear, dark night, you might think that you can see millions of stars. Actually, the unaided human eye can detect only about 6000 stars over the entire sky. At any one time, you can see roughly 3000 stars because only half of the sky is above the horizon.

Many ancient societies believed the Earth to be at the center of the universe. They also imagined that the stars were attached to the surface of a huge sphere centered on the Earth. This imaginary sphere, called the **celestial sphere,** is still a useful concept.

Of course, the stars actually are scattered at various distances from Earth. Many of the brightest stars you can see in the sky are 10 to 1000 light years away. These distances are so immense, however, that all the stars appear to be equally remote, fixed to a spherical backdrop. We can use this backdrop as a reference to specify the directions to objects in the sky.

As shown in Figure 2-4, the Earth is at the center of the celestial sphere. We can project key geographic features out into space to establish directions and bearings on the celestial sphere. If we project the Earth's equator onto the celestial sphere, we obtain the **celestial equator.** The celestial equator divides the sky into northern and southern hemispheres, just as the Earth's equator divides the Earth into two hemispheres.

We can also imagine extending the Earth's north and south poles out into space along the Earth's axis of rotation. This gives us the **north celestial pole** and the **south celestial pole,** also shown in Figure 2-4. With the celestial

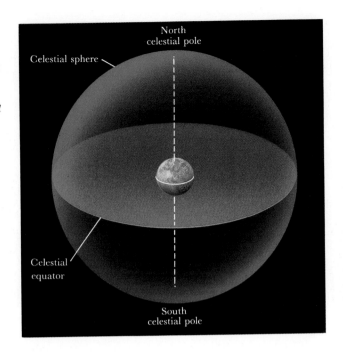

Figure 2-4 The celestial sphere
The celestial sphere is the apparent sphere of the sky. The celestial equator and pole are obtained by projecting the Earth's equator and axis of rotation out into space. The north celestial pole is therefore located directly over the Earth's north pole, while the south celestial pole is directly above the Earth's south pole.

Inside the figure:
North celestial pole
Celestial sphere
Celestial equator
South celestial pole

equator and poles as reference features, astronomers denote the position of an object in the sky much in the same way that longitude and latitude are used to specify a location on Earth.

The seasons are caused by the tilt of the Earth's axis of rotation

In addition to rotating on its axis every 24 hours, the Earth revolves around the Sun every $365\frac{1}{4}$ days. The seasonal changes we experience on Earth during a year result from the way the Earth's axis of rotation is tilted with respect to the plane of the Earth's orbit around the Sun.

The Earth's axis of rotation is not perpendicular to the plane of the Earth's orbit. Instead, the Earth's axis is tilted by $23\frac{1}{2}°$ away from the perpendicular, as shown in Figure 2-5. The Earth maintains this tilted orientation as it orbits the Sun. Thus, during part of the year the northern hemisphere is tilted toward the Sun and the southern hemisphere is tilted away, producing summer in the north and winter in the south. Half a year later the situation is reversed, with winter in the northern hemisphere (now tilted away from the Sun) while the southern hemisphere experiences summer. During March and September, when spring and fall begin, both hemispheres receive roughly equal amounts of illumination from the Sun.

Imagine that you could see the stars during the day, so that you could follow the Sun's apparent motion against the background constellations. As the Earth moves along its orbit, the Sun appears to shift its position gradually from day to day, tracing out a path in the sky, called the **ecliptic** (see Figure 2-6). The Sun takes one year to complete a trip around the ecliptic. There are about 365 days in a year and 360° in a circle, and so the Sun appears to move along the ecliptic at a rate of approximately 1° per day.

Because of the tilt of the Earth's axis of rotation, the ecliptic and the celestial equator are inclined to each other by $23\frac{1}{2}°$, as shown in Figure 2-6.

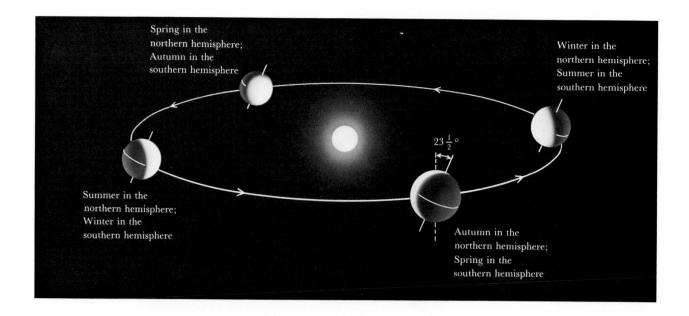

Spring in the
northern hemisphere;
Autumn in the
southern hemisphere

Winter in the
northern hemisphere;
Summer in the
southern hemisphere

$23\frac{1}{2}°$

Summer in the
northern hemisphere;
Winter in the
southern hemisphere

Autumn in the
northern hemisphere;
Spring in the
southern hemisphere

Figure 2-5 The seasons *The Earth's axis of rotation is inclined $23\frac{1}{2}°$ away from the perpendicular to the plane of the Earth's orbit. The Earth maintains this orientation (with its north pole aimed at the celestial north pole near the star called Polaris) throughout the year as the Earth orbits the Sun. Consequently, the amount of solar illumination and the number of daylight hours at any location on Earth varies in a regular fashion throughout the year.*

These two circles intersect at only two points, which are exactly opposite each other on the celestial sphere. Both points are called **equinoxes** (from the Latin words meaning "equal night") because, when the Sun appears at either point, daytime and nighttime are each 12 hours long at all locations on Earth.

The **vernal equinox** marks the beginning of spring in the northern hemisphere, as the Sun moves northward across the celestial equator about March 21. The **autumnal equinox** marks the moment when fall begins in the northern hemisphere (about September 22), as the Sun moves southward across the celestial equator.

Figure 2-6 Equinoxes and solstices
The ecliptic is the apparent annual path of the Sun on the celestial sphere. This path is inclined to the celestial equator by $23\frac{1}{2}°$ because of the tilt of the Earth's axis of rotation. The ecliptic and the celestial equator intersect at two points called the equinoxes. The northernmost point on the ecliptic is called the summer solstice. The corresponding southernmost point is called the winter solstice.

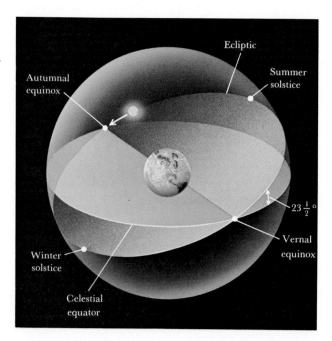

Ecliptic

Autumnal
equinox

Summer
solstice

$23\frac{1}{2}°$

Vernal
equinox

Winter
solstice

Celestial
equator

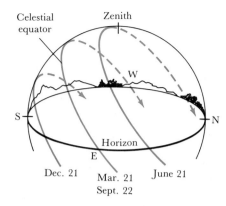

Figure 2-7 *The Sun's daily path* *On the first day of spring and the first day of fall, the Sun rises precisely in the east and sets precisely in the west. During summer, the Sun rises in the northeast and sets in the northwest. The maximum northerly excursion of the Sun occurs at the summer solstice. In the winter, the Sun rises in the southeast and sets in the southwest, with its maximum southerly excursion occurring at the winter solstice.*

We should always remember that the northern and southern hemispheres experience opposite seasons at any given point in time. For example, March 21 marks the beginning of autumn for people in Australia. The equinoxes were named during a time when virtually all astronomers lived north of the equator.

Between the vernal and autumnal equinoxes are two other significant locations along the ecliptic. The point on the ecliptic farthest north of the celestial equator is called the **summer solstice.** It is the location of the Sun at the moment summer begins in the northern hemisphere, about June 21. At the beginning of winter, about December 21, the Sun is farthest south of the celestial equator at a point called the **winter solstice.**

The Earth is round, and so people located at different latitudes view the celestial sphere from different orientations. For instance, a person at the Earth's north pole sees the north celestial pole directly overhead. A person on the equator sees both the celestial north and south poles on the horizon. At an intermediate latitude, such as in the United States, the celestial sphere is tilted as shown in Figure 2-7.

Seasonal changes in the Sun's daily path across the sky are diagrammed in Figure 2-7. On the first day of spring or fall (when the Sun is at one of the equinoxes), the Sun rises directly in the east and sets directly in the west. Daytime and nighttime are of equal duration. During the summer months, when the northern hemisphere is tilted toward the Sun, sunrise occurs in the northeast and sunset occurs in the northwest. The Sun spends more than 12 hours above the horizon and passes high in the sky at noontime. Incidentally, the point in the sky directly overhead is called the **zenith,** as shown in Figure 2-7. At the summer solstice, the Sun is as far north as it gets, giving the greatest number of daylight hours. During the winter months, when the northern hemisphere is tilted away from the Sun, sunrise occurs in the southeast. Daylight lasts for less than 12 hours as the Sun skims low over the southern horizon and sets in the southwest. Night is longest when the Sun is at the winter solstice.

Precession is a slow, conical motion of the Earth's axis of rotation

Ancient astronomers realized that the Moon orbits the Earth. They knew that the Moon takes roughly four weeks to go once around the Earth. Indeed, the word "month" comes from the same Old English root as the word "moon."

As seen from the Earth, the Moon is never far from the ecliptic. In other words, the Moon's path among the constellations is close to the Sun's path. The Moon's path remains within a band called the **zodiac** that extends about 8° on either side of the ecliptic. Thirteen constellations lie along the ecliptic, and the Moon is generally found in one of them. As the Moon moves along its orbit, it appears north of the celestial equator for about two weeks, and then it appears south of the celestial equator for about the next two weeks.

Both the Sun and the Moon exert a gravitational pull on the Earth. We will discuss gravity in greater detail in Chapter 3. For now, it is sufficient to realize that gravity is the universal attraction of matter for other matter.

The gravitational pull of the Sun and Moon affect the Earth's rotation because the Earth is not a perfect sphere. Our planet is slightly fatter, by about 43 kilometers (27 miles), across the equator than it is from pole to pole. The Earth is therefore said to have an "equatorial bulge." The gravita-

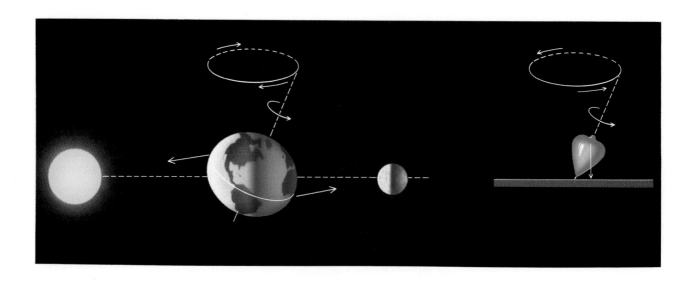

Figure 2-8 Precession *The gravitational pulls of the Moon and the Sun on the Earth's equatorial bulge cause the Earth to precess. As the Earth precesses, its axis of rotation slowly traces out a circle in the sky. The situation is analogous to that of a spinning top. As the top spins, the Earth's gravitational pull causes the top's axis of rotation to move in a circle.*

Figure 2-9 The path of the north celestial pole *As the Earth precesses, the north celestial pole slowly traces out a circle among the northern constellations in the sky. At the present time, the north celestial pole is near the moderately bright star Polaris, which serves as the "pole star."*

tional pull of the Moon and the Sun on this equatorial bulge gradually changes the orientation of the Earth's axis of rotation, producing a phenomenon called precession.

Imagine a spinning toy top, as illustrated in Figure 2-8. If the top were not spinning, gravity would pull it over on its side. But when it is spinning, the combined actions of gravity and rotation cause the top's axis of rotation to trace a circle—a motion called **precession.**

As the Sun and Moon move along the zodiac, each spends half the time north of the Earth's equatorial bulge and half the time south of it. The gravitational pull of the Sun and Moon tugging on the equatorial bulge tries to "straighten up" the Earth. In other words, as sketched in Figure 2-8, the gravity of the Sun and Moon tries to pull the Earth's axis of rotation toward a position perpendicular to the plane of the ecliptic. But the Earth is spinning. As with the toy top, the combined actions of gravity and rotation cause the Earth's axis to trace out a circle in the sky while remaining tilted about $23\frac{1}{2}°$ to the perpendicular.

The Earth's rate of precession is fairly slow. It takes 26,000 years for the north celestial pole to complete one full precessional circle around the sky (see Figure 2-9). At the present time, the Earth's axis of rotation points within 1° of the star Polaris. In 3000 BC, it was pointing near the star Thuban in the constellation of Draco (the dragon). In AD 14,000, the "pole star" will be Vega in Lyra (the harp). Of course, the south celestial pole executes a similar circle in the southern sky.

As the Earth's axis of rotation precesses, the Earth's equatorial plane also moves. Because the Earth's equatorial plane defines the location of the celestial equator in the sky, the celestial equator also precesses. The intersections of the celestial equator and the ecliptic define the equinoxes, and so these key locations in the sky also shift slowly from year to year. The entire phenomenon is often called the **precession of the equinoxes.** Today the vernal equinox is located in the constellation Pisces (the fishes). Two thousand years ago, it was in Aries (the ram). Around the year AD 2600, the vernal equinox will move into Aquarius (the water bearer).

Lunar phases are due to the Moon's orbital motion

Ancient astronomers knew that the Moon shines by reflected sunlight. They also understood that the Moon takes roughly four weeks to orbit the Earth. This knowledge was deduced from observations of the changing **phases of the Moon,** which occur as varying amounts of the Moon's illuminated hemisphere are exposed to observers on the Earth.

Figure 2-10 is an extraordinary photograph of both the Earth and Moon from space. The Moon orbits the Earth every 27.3 days, and so it takes approximately four weeks for the Moon to complete its cycle of phases. Figure 2-11 shows the Moon at several locations on its orbit (as viewed from above the Earth's north pole) along with the resulting phases.

The phase called **new moon** occurs when the unlit hemisphere of the Moon faces the Earth. We cannot see the Moon in this phase because it is in the same part of the sky as the Sun.

During the next seven days, more and more of the illuminated hemisphere of the Moon is progressively exposed to our Earth-based view. The term "waxing" means "growing larger," and so this phase is called **waxing crescent moon.** At **first quarter moon,** the angle between the Sun, Moon, and Earth is 90°, and so we see half of the Moon's illuminated hemisphere.

During the next week, still more of the illuminated hemisphere is seen from Earth, in a phase called **waxing gibbous moon.** When the Moon stands opposite the Sun in the sky, we see fully the illuminated hemisphere, producing the phase called **full moon.** Moonrise always occurs at sunset during full moon.

Over the subsequent two weeks, we see less and less of the illuminated hemisphere as the Moon continues along its orbit. The term "waning" means "growing smaller," and so this progression produces the phases called **waning gibbous moon, third quarter moon,** and **waning crescent moon,** as diagrammed in Figure 2-11.

Because the position of the Sun in the sky determines the local time, we can correlate the Moon's phase and location with the time of day. For example, during first quarter moon, the Moon is approximately 90° east of the Sun in the sky; hence, moonrise occurs approximately at noon. Figure 2-12 is a series of photographs showing the various phases of the moon.

It takes about one month for the Moon to complete an orbit around the Earth. However, astronomers are careful to distinguish between two types of months, depending on whether the Moon's motion is measured relative to the stars or to the Sun. Neither interval corresponds exactly to the months of our calendar.

The **sidereal month** is the time it takes for the Moon to complete one full orbit of the Earth, measured *with respect to the stars.* This interval is the Moons' true orbital period, equal to 27.3 days.

The **synodic month** is the time it takes for the Moon to complete one cycle of phases, as for example, from one new moon to the next. Consequently, the synodic month is measured *with respect to the Sun* and is equal to about 29.5 days.

Of course, the Earth is orbiting the Sun while the Moon is going through its phases. Thus, to get from one new moon to the next, the Moon must travel more than 360° along its orbit, as shown in Figure 2-13. The synodic month is thus approximately two days longer than the sidereal month.

The Moon stays in orbit about the Earth because of the gravitational attraction between these two bodies. However, the Sun also pulls on the

Figure 2-10 The Earth and the Moon
The Moon circles the Earth every 27.3 days at an average distance of 384,400 kilometers (238,900 miles). This picture was taken in 1977 from the Voyager 1 *spacecraft shortly after it was launched toward Jupiter and Saturn. (NASA)*

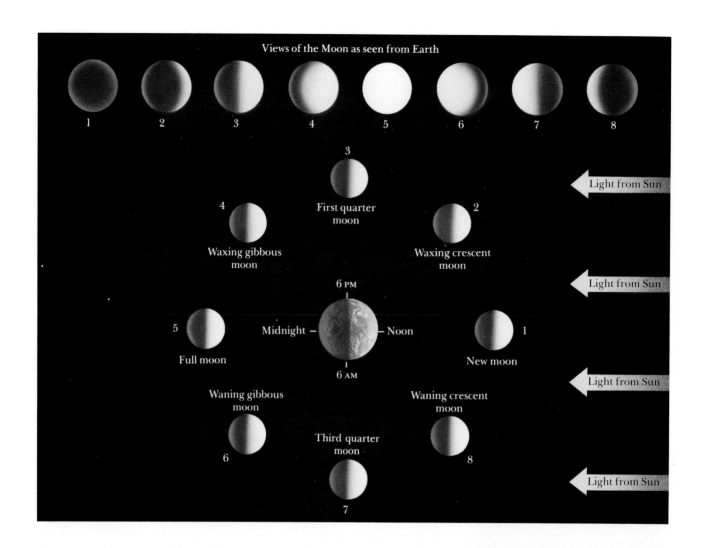

Views of the Moon as seen from Earth

1   2   3   4   5   6   7   8

3
First quarter
moon

4
Waxing gibbous
moon

2
Waxing crescent
moon

Light from Sun

6 PM

Midnight — Noon

Light from Sun

5
Full moon

1
New moon

6 AM

Waning gibbous
moon

Waning crescent
moon

6

Third quarter
moon

8

7

Light from Sun

Light from Sun

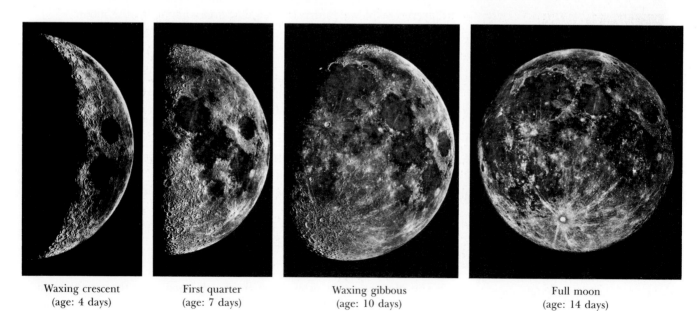

Waxing crescent
(age: 4 days)

First quarter
(age: 7 days)

Waxing gibbous
(age: 10 days)

Full moon
(age: 14 days)

Figure 2-11 [opposite, top] **The phases of the Moon** *This diagram shows the Moon at eight locations on its orbit (as viewed from above the Earth's north pole) along with the resulting lunar phases. Light from the Sun illuminates one half of the Moon, while the other half is dark. As the Moon orbits the Earth, we see varying amounts of the Moon's illuminated hemisphere. It takes 29½ days for the Moon to go through all its phases.*

Figure 2-12 [below, left and right] **The Moon's appearance** *The Moon always keeps the same side facing the Earth. Earth-based observers therefore always see the same craters and lunar mountains, regardless of the phase. (Lick Observatory)*

Figure 2-13 [above right] **The sidereal and synodic months** *The sidereal month is the time it takes the Moon to complete one revolution with respect to the background stars. However, because the Earth is constantly moving along its orbit about the Sun, the Moon must travel through more than 360° to get from one new moon to the next. Thus the synodic month is slightly longer than the sidereal month.*

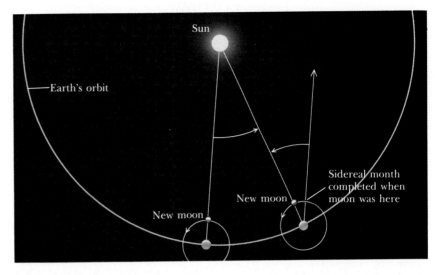

Moon, constantly altering the Moon's path. The final result is that both the sidereal and the synodic months are variable. The sidereal month (average length = 27^d 7^{hr} 43^{min} 11^{sec}) can vary by as much as seven hours. The synodic month (average length = 29^d 12^{hr} 44^{min} 3^{sec}) can vary by as much as half a day.

Ancient astronomers measured the size of the Earth and attempted to determine distances to the Sun and Moon

More than two thousand years ago, Greek astronomers were aware of the Earth's spherical shape. Eclipses of the Moon provided the convincing observations. During a **lunar eclipse,** the Moon passes through the Earth's shadow. Ancient astronomers noticed that the edge of the Earth's shadow is always circular. A sphere is the only shape that casts a circular shadow from any angle, and so the ancient astronomers concluded that the Earth is spherical.

Waning gibbous
(age: 20 days)

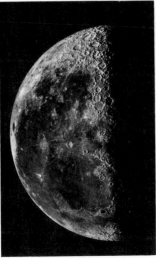

Third quarter
(age: 22 days)

Waning crescent
(age 26 days)

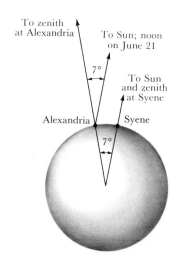

Figure 2-14 *Eratosthenes's method of determining the Earth's size Eratosthenes noticed that the Sun is about 7° south of the zenith at Alexandria when it is directly overhead at Syene. The angle is about one-fiftieth of a circle, and so the distance between Alexandria and Syene must be about one-fiftieth of the Earth's circumference.*

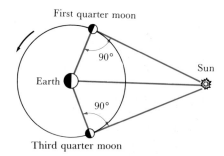

Figure 2-15 *Aristarchus's method of determining distances to the Sun and Moon Aristarchus knew that the Sun, Earth, and Moon form a right triangle at first quarter and third quarter phases. How he measured the acute angles in these right triangles is not known. Nevertheless, using Euclidean geometry, he argued that the Sun is about 20 times farther from the Earth than the Moon. The sizes of the Earth, the Moon, and the Moon's orbit are greatly exaggerated in this diagram.*

Around 200 BC, the Greek astronomer Eratosthenes devised a way to measure the circumference of the Earth. He was intrigued by reliable reports from the town of Syene in Egypt (near the modern Aswan) that the Sun shone directly down vertical wells on the first day of summer. Eratosthenes knew that the Sun never appeared at the zenith from his home in Alexandria, on the Mediterranean Sea almost due north of Syene. He measured the position of the Sun at local noon on the summer solstice in Alexandria and found it to be about 7° south of the zenith, as shown in Figure 2-14. This angle is one-fiftieth of a circle, and so Eratosthenes concluded that the distance from Alexandria to Syene was one-fiftieth of the Earth's circumference.

In Eratosthenes's day, the distance from Alexandria to Syene was said to be 5000 stades. (The stade was a Greek unit of length.) Therefore, Eratosthenes found the Earth's circumference to be 50 × 5000 stades = 250,000 stades. Unfortunately, no one today is sure of the exact length of the stade. One guess is that it was about $\frac{1}{6}$ kilometer, which would mean that Eratosthenes obtained a circumference of about 42,000 kilometers—within 5 percent of the modern value of about 40,000 kilometers.

Eratosthenes was only one of several brilliant astronomers to emerge from the distinguished Alexandrian school. Around 280 BC, one of the first Alexandrian astronomers, Aristarchus of Samos, devised a method of determining the relative distances to the Sun and Moon.

Aristarchus knew that the Sun, Moon, and Earth form a right triangle at the moment of first or third quarter moon, as diagrammed in Figure 2-15. For reasons lost in antiquity, he also argued that the angle between the Moon and the Sun is 3° less than a right angle (that is, 87°) at first and third quarter. Using the rules of Euclidean geometry, Aristarchus concluded that the Sun is about 20 times farther from us than the Moon is. We now know that the average distance to the Sun is about 390 times longer than the average distance to the Moon. Nevertheless, it is impressive that people were logically trying to measure distances across the solar system more than 2000 years ago.

Aristarchus also used lunar eclipses in an equally bold attempt to determine the relative sizes of the Earth, Moon, and Sun. Observing how long it takes for the Moon to move through the Earth's shadow, Aristarchus estimated that the diameter of the Earth is about three times the diameter of the Moon. To determine the diameter of the Sun, he simply pointed out that the Sun and the Moon have the same angular size in the sky, and so their diameters must be in the same ratio as their distances. In other words, because Aristarchus believed that the Sun is 20 times farther from the Earth than the Moon, he concluded that the Sun must be 20 times larger than the Moon. According to modern measurements, the Sun is about 400 times larger in diameter than the Moon.

You can now appreciate the significance of Eratosthenes's measurement of the Earth's circumference. The Greeks knew that the Earth's diameter is equal to its circumference divided by the constant called π (pi). Knowing the Earth's diameter, Alexandrian astronomers could calculate the diameters of the Sun and Moon as well as their distances from Earth. Although some of these ancient measurements are far from their modern values, our ancestors' achievements stand as an impressive exercise in observation and reasoning.

Eclipses occur only when the Sun and Moon are both on the line of nodes

Eclipses are among the most spectacular of nature's phenomena. In only a few minutes, the Sun can seemingly be blotted from the sky as broad daylight is suddenly steeped in eerie twilight, or the brilliant full moon on a cloud-free night can gradually grow dark and red.

A **lunar eclipse** occurs when the Moon passes through the Earth's shadow. This can happen only when the Sun, Earth, and Moon are in a straight line at full moon. A **solar eclipse** occurs when the Earth passes through the Moon's shadow. As seen from Earth, the Moon moves in front of the Sun. This can happen only when the Sun, Moon, and Earth are aligned at new moon.

New moon and full moon both occur at intervals of $29\frac{1}{2}$ days, but solar and lunar eclipses happen much less often. Eclipses occur infrequently because the Moon's orbit is tilted slightly (5°) out of the plane of the Earth's orbit, as shown in Figure 2-16. Because of this tilt, new moon and full moon usually occur when the Moon is either above or below the plane of the Earth's orbit. In such positions, a perfect alignment between the Sun, Moon, and Earth is not possible and an eclipse cannot occur.

The plane of the Earth's orbit and the plane of the Moon's orbit intersect along a line called the **line of nodes,** which passes through the Earth and is pointed in a particular direction in space, as shown in Figure 2-17. Eclipses

Figure 2-16   The line of nodes
The plane of the Moon's orbit is tilted slightly with respect to the plane of the Earth's orbit. These two planes intersect along a line called the line of nodes.

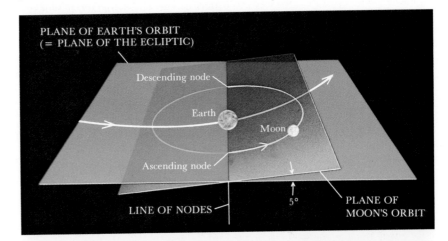

Figure 2-17   Conditions for eclipses
A solar eclipse occurs only if the Moon is very near the line of nodes at new moon. A lunar eclipse occurs only if the Moon is very near the line of nodes at full moon. When new moon or full moon phases occur away from the line of nodes, no eclipse is seen because the Moon and the Earth do not pass through each other's shadows.

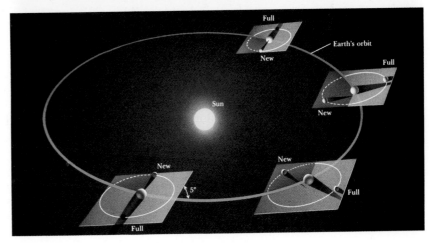

can occur only when both the Sun and Moon are on or very near the line of nodes, because only then do the Sun, Earth, and Moon lie along a straight line.

Knowing the orientation of the line of nodes is clearly important to anyone who wants to predict upcoming eclipses. However, predicting eclipses is complicated by the fact that the line of nodes gradually changes its direction in space. The constant gravitational pull of the Sun on the Moon causes the Moon's orbit gradually to shift its orientation in space. The resulting slow westward movement of the line of nodes is one of several details that astronomers must include in their calculations for times of upcoming eclipses.

At least two and not more than five solar eclipses occur each year. (The last year in which five solar eclipses occurred was 1935.) Lunar eclipses occur just about as frequently as solar eclipses, but the maximum number of eclipses (both solar and lunar) possible in a year is seven.

Lunar and solar eclipses can be either partial or total, depending on the alignment of the Sun, Earth, and Moon

During the few hours of a lunar eclipse, you can see the Earth's shadow move across the Moon. The Earth's shadow has two distinct parts, as diagrammed in Figure 2-18. The **umbra** is the darkest part of the shadow, from which no portion of the Sun's surface can be seen. In the **penumbra,** only part of the Sun's surface is blocked out.

Three kinds of lunar eclipses can occur, depending on exactly how the Moon travels through the Earth's shadow. First, the Moon might pass through only the Earth's penumbra, creating a **penumbral eclipse.** During this type of eclipse, none of the lunar surface is completely shaded by the Earth because only part of the Sun is covered by the Earth. At mid-eclipse, the Moon merely looks a little dimmer than usual from Earth and there is no "bite" taken out of the Moon by the Earth's umbra. The Moon still looks full, and thus it is easy to miss a penumbral eclipse.

Figure 2-18 The geometry of a lunar eclipse People on the nighttime side of the Earth see a lunar eclipse when the Moon moves through the Earth's shadow. The umbra is the darkest part of the shadow. In the penumbra, only part of the Sun is covered by the Earth.

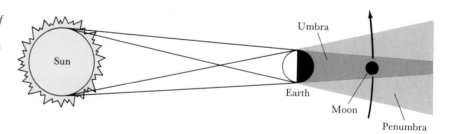

Most people notice a lunar eclipse only if the Moon passes into the Earth's umbra. During the umbral phase of such an eclipse, a "bite" seems to be taken out of the Moon. If the Moon's orbit is oriented so that only part of the lunar surface passes through the umbra, then we see a **partial eclipse.** When the Moon travels completely into the umbra, as sketched in Figure 2-19, we see a **total eclipse** of the Moon. The maximum duration of totality occurs when the Moon travels directly through the center of the umbra. The Moon's speed through the Earth's shadow is roughly 1 kilometer per second (2300 miles per hour), which means that totality can last for as much as 1 hour 42 minutes.

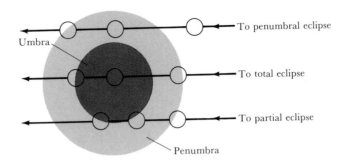

Figure 2-19 [right]   Various lunar eclipses   *This diagram shows the Earth's umbra and penumbra at the distance of the Moon's orbit. Different kinds of lunar eclipses are seen, depending on the Moon's path through the Earth's shadow.*

To penumbral eclipse

Umbra

To total eclipse

To partial eclipse

Penumbra

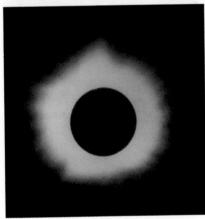

Figure 2-20 [top]   A total eclipse of the Moon   *This photograph was taken by an amateur astronomer during the lunar eclipse of September 6, 1979. Notice the distinctly reddish color of the Moon. (Courtesy of M. Harms)*

Figure 2-21 [bottom]   A total eclipse of the Sun   *During a total solar eclipse, the Moon completely covers the Sun's disk, and the solar corona can be seen. This halo of hot gases extends for thousands upon thousands of kilometers into space. Only the brightest, inner portions of the solar corona are seen in this photograph of the solar eclipse of March 7, 1970. (NASA)*

To illustrate the frequency of lunar eclipses, Table 2-1 lists all the total and partial eclipses from 1990 through 1994. Penumbral eclipses are not included in this listing.

Even during a total eclipse, the Moon does not completely disappear. A small amount of sunlight passing through the Earth's atmosphere is deflected into the Earth's umbra. Most of this deflected light is red, and so the darkened Moon glows faintly in reddish hues during totality, as shown in Figure 2-20.

As with the Earth's shadow, the darkest part of the Moon's shadow is also called the **umbra.** You must be within the Moon's umbra in order to see a **total solar eclipse,** for that is the only region from which the Moon completely covers the Sun. Because the Sun and Moon have nearly the same angular diameter as seen from Earth—about $\frac{1}{2}°$—the Moon "fits" over the Sun during a total solar eclipse. This is important to astronomers because the hot gases called the **solar corona** that surround the Sun can be photographed and studied in detail during the few precious moments when the eclipse is total (see Figure 2-21).

Surrounding the Moon's umbra is a region of partial shadow called the **penumbra.** During a solar eclipse, the Moon's penumbra extends over a large portion of the Earth's surface. When the observer is within the Moon's penumbra, the Sun's surface appears only partly covered by the Moon. This is called a **partial eclipse** of the Sun.

Table 2-1   Lunar eclipses 1990–1994

Date		Percentage eclipsed (100% = total)	Duration of totality (hr min)
1990	February 9	100	0 46
1990	August 6	68	——
1991	December 21	9	——
1992	June 15	69	——
1992	December 10	100	1 14
1993	June 4	100	1 38
1993	November 29	100	0 50
1994	May 25	28	——

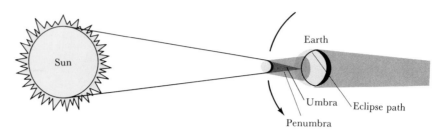

Figure 2-22 The geometry of a total solar eclipse During a total solar eclipse, the tip of the Moon's umbra traces an eclipse path across the Earth's surface. People inside the eclipse path see a total solar eclipse, whereas people inside the penumbra see only a partial eclipse.

Only the tip of the Moon's umbra reaches the Earth's surface, as shown in Figure 2-22. As the Earth turns, the tip of the umbra traces an **eclipse path** across its surface. Only those people inside this path are treated to the spectacle of a total solar eclipse. (Figure 2-23 shows the dark spot produced by the Moon's umbra on the Earth's surface.)

Figure 2-23 [right] The Moon's shadow on the Earth This photograph was taken from an Earth-orbiting satellite during the total eclipse of March 7, 1970. The Moon's umbra appears as a dark spot on the eastern coast of the United States. (NASA)

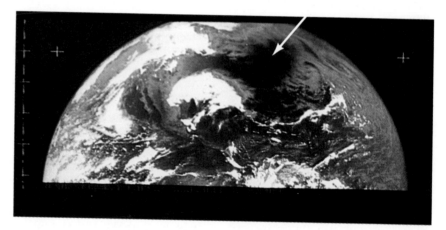

During a total eclipse, the Earth's rotation and the orbital motion of the Moon cause the umbra to race along the eclipse path at speeds in excess of 1700 kilometers per hour (1050 miles per hour). People along the eclipse path therefore observe totality for only a few moments. Totality never lasts for more than $7\frac{1}{2}$ minutes at any one location on the eclipse path. In a typical total solar eclipse, various factors such as the Earth–Moon distance limit the duration of totality to much less than the maximum $7\frac{1}{2}$ minutes.

The width of the eclipse path depends primarily on the Earth–Moon distance during an eclipse. The eclipse path is widest—up to 270 kilometers (170 miles)—if the Moon happens to be at the point in its orbit nearest the Earth. Usually it is much narrower. In fact, sometimes the Moon's umbra does not even reach down to the Earth's surface. If the alignment for a solar eclipse occurs when the Moon is farthest from the Earth, then the Moon's umbra falls short of the Earth, and no one sees a truly total eclipse. From the Earth's surface, the Moon appears too small to cover the Sun completely, and a thin ring of light is seen around the edge of the Moon at mid-eclipse. An eclipse of this type is called an **annular eclipse** (see Figure 2-24). The length of the Moon's umbra is nearly 5000 kilometers (3100 miles) shorter than the average distance between the Moon and the Earth's surface. Thus, the Moon's shadow often fails to reach the Earth, making annular eclipses more common than total eclipses.

Figure 2-24 An annular eclipse of the Sun This composite of six exposures taken at sunrise in Costa Rica shows the progress of an annular eclipse of the Sun that occurred on December 24, 1974. Note that at mid-eclipse the limb of the Sun is visible around the Moon. (Courtesy of D. di Cicco)

Table 2-2 Solar eclipses 1990–1994

Date		Area	Type	Notes
1990	January 26	Antarctic	Annular	
1990	July 22	Finland, Russia, Pacific	Total	Max. length 2 min 33 sec
1991	January 15/16	Australia, New Zealand, Pacific	Annular	
1991	July 11	Pacific, Mexico, Brazil	Total	Max. length 6 min 54 sec
1992	January 4/5	Central Pacific	Annular	
1992	June 30	South Atlantic	Total	Max. length 5 min 20 sec
1992	December 24	Arctic	Partial	84% eclipsed
1993	May 21	Arctic	Partial	74% eclipsed
1993	November 13	Antarctic	Partial	93% eclipsed
1994	May 10	Pacific, Mexico, United States, Canada	Annular	
1994	November 3	Peru, Brazil, South Atlantic	Total	Max. length 4 min 23 sec

To illustrate the frequency of solar eclipses, Table 2-2 lists all the total, annular, and partial eclipses from 1990 through 1994. All of the details of a solar eclipse are calculated well in advance and are published in reference books such as *The Astronomical Almanac*.

Ancient astronomers achieved limited ability to predict eclipses

A total solar eclipse is a dramatic event. The sky begins to darken, the air temperature falls, and the winds increase as the Moon's umbra races toward you. All nature responds: birds go to roost, flowers close their petals, and crickets begin to chirp as if evening had arrived. As totality approaches, the landscape is bathed in shimmering bands of light and dark as the last few rays of sunlight peek out from behind the edge of the Moon. And finally the corona blazes forth in a star-studded midday sky. It is an awesome sight.

In ancient times, the ability to predict eclipses must have seemed very desirable. Archaeological evidence suggests that astronomers in many civilizations struggled to predict eclipses, with varying degrees of success. The number and placement of certain holes in the ground around Stonehenge suggest some ability to predict eclipses more than 4000 years ago. One of three priceless manuscripts to survive the devastating Spanish conquests shows that Mayan astronomers in Mexico and Guatemala had a fairly reliable method of eclipse prediction. And there are many apocryphal stories such as the one about the great Greek astronomer Thales of Miletus, who is said to have predicted the eclipse of 585 BC, which occurred during the middle of a war. The sight was so awesome and unexpected that the soldiers put down their arms and declared peace.

In retrospect, it seems that ancient astronomers actually produced eclipse "warnings" of various degrees of reliability rather than actual predictions. Working with historical records, these astronomers generally searched for cycles and regularities from which to anticipate future eclipses.

Ancient astronomers, particularly those of Greece, gave humanity a new and powerful way of thinking about the world. They gave the first clear demonstration that the tools of logic, reason, and mathematics can be used to discover and understand the workings of the universe. This general approach to reality underlies all modern science. Aristarchus even went so far as to suggest that the motions of the planets in the sky could be simply explained if all the planets, including Earth, orbit the Sun. As we shall see in the next chapter, he was nearly 2000 years ahead of his time.

Summary

- It is convenient to imagine the stars fixed to the celestial sphere with the Earth at its center.

  The surface of the celestial sphere is divided into 88 regions, or constellations.

  The celestial sphere appears to rotate around the Earth once in each day and night; in fact, of course, it is the Earth that is rotating.

  The poles and equator of the celestial sphere are determined by extending the axis of rotation and the equatorial plane of the Earth to the celestial sphere.

- Earth's axis of rotation is tilted at an angle of $23\frac{1}{2}°$ from the perpendicular to the plane of the Earth's orbit. The seasons are caused by this tilt, which tips one hemisphere or the other toward the Sun at certain times of the year.

- Equinoxes and solstices are significant points along the Earth's orbit, determined by the relationship between the Sun's path on the celestial sphere (the ecliptic) and the celestial equator.

- The Earth's axis of rotation moves slowly in a conical fashion—a phenomenon called precession. Precession is caused by the gravitational pull of the Sun and Moon on the Earth's equatorial bulge.

- The phases of the Moon are caused by the relative positions of the Earth, Moon, and Sun.

- The Moon completes one orbit around the Earth with respect to the stars in a sidereal month averaging 27.3 days.

- The Moon completes one cycle of phases in a synodic month averaging 29.5 days.

- Ancient astronomers made great progress in determining the sizes and relative distances of the Earth, Moon, and Sun.

- A lunar eclipse occurs when the Sun and Moon are both on the line of nodes at full moon. A solar eclipse occurs when the Sun and Moon are both on the line of nodes at new moon. The line of nodes is the line where the planes of the Earth's orbit and the Moon's orbit intersect.

- Detailed knowledge of the Moon's orbit around the Earth is needed to predict eclipses.

- The shadow of an object has two parts: the umbra, where the light source is completely blocked, and the penumbra, where the light source is only partially obscured.

- Depending on the relative positions of the Sun, Moon, and Earth, lunar eclipses may be penumbral, partial, or total, and solar eclipses may be partial, annular, or total.

Review questions

1 Why is the ancient idea of the celestial sphere still a useful concept today?

* **2** At what location on Earth is the north celestial pole on the horizon?

3 Where do you have to be on the Earth in order to see the Sun at the zenith? If you stay at that location for a full year, on how many days will the Sun pass through the zenith?

* **4** Where do you have to be on Earth in order to see the south celestial pole directly overhead? What is the maximum possible elevation of the Sun above the horizon at that location? On what date is this maximum elevation observed?

* **5** At what point on the horizon does the vernal equinox rise?

6 Sketch a diagram of the relative positions of the Earth, Moon, Sun, and *Voyager 1* when the photograph in Figure 2-11 was taken.

7 Which type of eclipse, lunar or solar, do you think most people have seen? Why?

8 Is it possible for a total eclipse of the Sun to be followed three months later by a lunar eclipse? Why?

9 Can one ever observe an annular eclipse of the Moon? Why?

Advanced questions

10 Consult a star map of the southern hemisphere and determine which, if any, bright southern stars could some day become south celestial "pole" stars.

11 Using a star map, determine which, if any, bright stars could some day mark the location of the vernal equinox. Give the approximate years when that would happen.

12 During a lunar eclipse, does the Moon enter the Earth's shadow from the east or from the west? Explain why.

Discussion questions

13 Examine a list of the 88 constellations. Are there any that obviously date from modern times? Where are they located? Why do you suppose they do not have archaic names?

14 Describe the seasons if the Earth's axis of rotation were tilted 0° and 90° to its orbital plane.

15 Describe the cycle of lunar phases that would be observed if the Moon moved about the Earth in an orbit perpendicular to the plane of the Earth's orbit. Is it possible for solar and lunar eclipses to occur under these circumstances?

For further reading

Allen, R. *Star Names: Their Lore and Meaning*. 1899. Dover reprint, 1963.

Berry, R. *Discover the Stars*. Harmony/Crown, 1987.

Chartrand, M., and Wimmer, H. *Skyguide*. Western Publishing, 1982.

Cornell, J. *The First Stargazers*. Scribners, 1981.

Gallant, R. *The Constellations: How They Came to Be*. Four Winds Press, 1979.

Kals, W. *The Stargazer's Bible*. Doubleday, 1980.

Krupp, E. *Echoes of the Ancient Skies*. Harper & Row, 1983.

Menzel, D., and Pasachoff, J. "Solar Eclipse: Nature's Superspectacular." *National Geographic,* August 1970.

Ridpath, I., and Tirion, W. *Universe Guide to Stars and Planets*. Universe Books, 1985.

3 Gravitation and the motions of the planets

Apollo 11 leaving the Moon
The lunar module Eagle *returns from the Moon after completing the first successful manned lunar mission. This photograph was taken from the command module* Columbia *in which the astronauts returned to Earth. All of the orbital maneuvers to dock the* Eagle *and then to set* Columbia *on course for Earth were based on Newtonian mechanics and Newton's law of gravity. These same physical principles are used by astronomers to understand a wide range of phenomena from the motions of double stars to the rotation of the entire Galaxy. (NASA)*

Attempts to explain the motions of the planets led astronomers to an understanding of gravitation. Ancient astronomers believed that the heavens revolve around a stationary Earth. We begin this chapter with a look at the complicated assumptions needed to explain the detailed motions of the planets in such an Earth-centered view of the universe. We then discuss Copernicus's Sun-centered explanation of the universe, which provides simpler general explanations. We learn how Kepler's elliptical orbits for the planets improved and simplified the Sun-centered explanation and how Galileo's telescopic observations supported the idea that the Earth is one of the planets that all move around the Sun. Next, we see how Newton used Kepler's work to develop laws of gravity and motion, which provide a compact explanation and description of the motions of planets, comets, and satellites. Finally, we take a brief look at the new ideas about space, time, and gravitation introduced in this century by Einstein.

It is not obvious that the Earth moves around the Sun. Indeed, our daily experience strongly suggests that the opposite is true. The daily rising and setting of the Sun, Moon, and stars could lead us to believe that the entire cosmos revolves about us, with the Earth the center of the universe. That was just what most observers did believe for thousands of years.

Ancient astronomers invented geocentric cosmology to explain planetary motions

Ancient Greek astronomers were among the first to leave a written record of their attempts to explain just how the universe works. Their universe consisted of the Earth and those objects that can be seen with the naked eye: the Sun, the Moon, the stars, and five planets (Mercury, Venus, Mars, Jupiter, and Saturn). Most of these Greeks astronomers also assumed that the universe revolves about the Earth and thus they developed a **geocentric cosmology,** or Earth-centered theory of the universe.

The planets are usually quite noticeable in the night sky because they look like bright stars that do not twinkle. For instance, at maximum brilliancy, Venus is 16 times brighter than the brightest star and thus is readily recognizable. A careful observer also notices that the planets slowly shift their positions from night to night with respect to the background of the "fixed" stars in the constellations. Indeed, the word planet comes from a Greek term meaning "wanderer."

Observations of the planets' positions against the stars from night to night soon make it clear that the planets do not revolve uniformly about the Earth. Explaining the motions of the five known planets was one of the main challenges facing the astronomers of antiquity.

As seen from Earth, the planets move primarily across the constellations of the **zodiac.** As mentioned in Chapter 2, these constellations circle the sky in a band centered on the ecliptic. If you follow a planet as it travels across the zodiac from night to night, you find that the planet usually moves slowly eastward against the background stars. Occasionally, however, the planet seems to stop and then back up for several weeks or months. This occasional westward movement is called **retrograde** motion. These motions are much slower than the daily rotation of the sky caused by the Earth's rotation. Both direct and retrograde motion are best detected by mapping the position of a planet against the background stars from night to night over a long period of observation. Figure 3-1 shows direct and retrograde paths of Mars during 1988.

Greek astronomers devised various theories to account for retrograde motion and the loops that the planets trace out against the background stars. One of the most successful ideas was investigated by Hipparchus and elaborated by the last of the great Greek astronomers, Ptolemy, who lived in Alexandria during the second century AD. The basic concept is sketched in Figure 3-2. Each planet is assumed to revolve about a small circle called an **epicycle,** which in turn revolves about a larger circle called a **deferent,** which is approximately centered on the Earth. As viewed from Earth, the epicycle moves eastward along the deferent, and both circles rotate in the same direction (counterclockwise in Figure 3-2).

Most of the time, the motion of the planet on its epicycle adds to the eastward motion of the epicycle on the deferent. Thus the planet is seen to be in direct (eastward) motion against the background stars throughout most of the year. However, when the planet is on the part of its epicycle

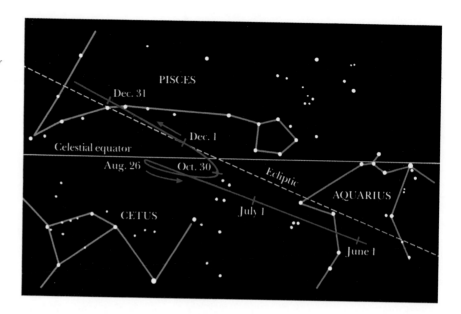

Figure 3-1   The path of Mars in 1988
During the summer and fall of 1988, Mars moved across the constellations of Aquarius and Pisces. From August 26 through October 30, Mars's motion was retrograde.

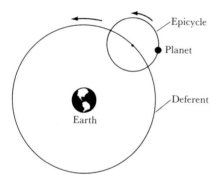

Figure 3-2   A geocentric explanation of planetary motion   *Each planet revolves about an epicycle, which in turn revolves about a deferent centered approximately on the Earth. As seen from Earth, the speed of the planet on the epicycle alternately adds to or subtracts from the speed of the epicycle on the deferent, thus producing alternating periods of direct and retrograde motion.*

nearest the Earth, its motion along the epicycle subtracts from the motion of the epicycle along the deferent. The planet thus appears to slow and then halt its usual eastward movement among the constellations, even seeming to go backward for a few weeks or months. This concept of epicycles and deferents provides a general explanation of the retrograde loops that the planets execute.

Using the wealth of astronomical data in the library at Alexandria, including records of planetary positions covering hundreds of years, Ptolemy deduced the sizes of the epicycles and deferents and the rates of revolution needed to produce the recorded paths of the planets. After years of arduous work, Ptolemy assembled his calculations in 13 volumes collectively called the *Almagest*, in which the positions and paths of the Sun, Moon, and planets were described with unprecedented accuracy. The *Almagest* became the astronomer's bible.

For over a thousand years, Ptolemy's cosmology endured as a useful description of the workings of the heavens. Eventually, however, things began going awry. Tiny errors and inaccuracies that were unnoticeable in Ptolemy's day compounded and multiplied over the years, especially with regard to precession. Fifteenth century astronomers made some cosmetic adjustments to the Ptolemaic system. However, the system became less and less satisfactory as more complicated and arbitrary details were added to keep it consistent with the observed motions of the planets.

Nicolaus Copernicus devised the first comprehensive heliocentric cosmology

Imagine driving on a freeway at high speed. As you pass a slowly moving car, it appears to move backward even though it is traveling in the same direction as your car. This sort of observation inspired the ancient Greek astronomer Aristarchus to suggest a more straightforward explanation of retrograde motion—one in which all the planets, including the Earth, revolve about the Sun. The retrograde motion of Mars occurs, for example,

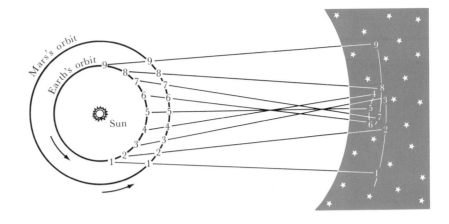

Figure 3-3 *A heliocentric explanation of planetary motion* *The Earth travels around the Sun more rapidly than does Mars. Consequently, as the Earth overtakes and passes this slower-moving planet, Mars appears to move backward for a few months.*

when the Earth overtakes and passes Mars, as shown in Figure 3-3. In Aristarchus's day, however, it seemed inconceivable that something as huge as the Earth could be in motion.

Almost 2000 years elapsed before someone had the insight and determination to work out the details of a heliocentric (Sun-centered) description of the motions of the planets. That person was a sixteenth century Polish lawyer, physician, mathematician, economist, monk, and artist named Nicolaus Copernicus. Copernicus's astronomical studies detailed the advantages of a straightforward Sun-centered cosmology over the cumbersome Earth-centered theory. This work instilled a revolutionary concept that pervades modern science: simplicity is a hallmark of correctness. Thus, nowadays, when a theory becomes unduly elaborate and complicated, scientists begin to suspect that the theory is probably wrong.

Copernicus realized that, with a heliocentric perspective, he could determine which planets are closer to the Sun than is the Earth and which are farther away. Because Mercury and Venus are always observed fairly near the Sun, Copernicus concluded that their orbits must be smaller than the Earth's. The other visible planets—Mars, Jupiter, and Saturn—can be seen in the middle of the night, when the Sun is far below the horizon, which can occur only if the Earth comes between the Sun and a planet. Copernicus concluded that the orbits of Mars, Jupiter, and Saturn are larger than the Earth's orbit.

It is often useful to specify various points on a planet's orbit, as in Figure 3-4. These points help us identify certain geometrical arrangements, or **configurations**, between the Earth, another planet, and the Sun. For example, when Mercury or Venus is between the Earth and the Sun, we say the planet is at **inferior conjunction**. When they are on the opposite side of the Sun, they are at **superior conjunction.**

The angle between the Sun and a planet as viewed from the Earth is called the planet's **elongation.** At **maximum eastern elongation,** Mercury or Venus is as far east of the Sun as it can be. At such times, the planet appears above the western horizon after sunset and is often called an "evening star." Similarly, at **maximum western elongation,** Mercury or Venus is as far west of the Sun as it can possibly be and rises before the Sun, gracing the eastern predawn sky as a "morning star."

When one of the outer planets is behind the Sun, we say that the planet is at **conjunction**. When it is exactly opposite the Sun in the sky, we say that it

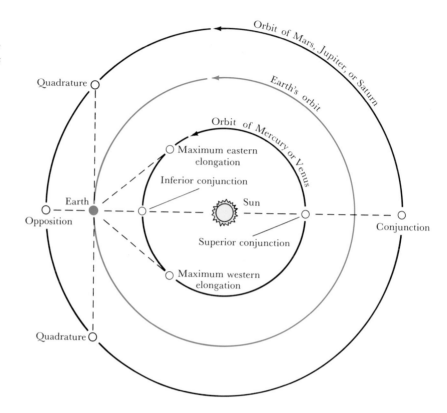

Figure 3-4 Planetary configurations
It is often useful to specify key points along a planet's orbit as shown in this diagram. These points identify specific geometric arrangements between the Earth, another planet, and the Sun.

Quadrature

Orbit of Mars, Jupiter, or Saturn

Earth's orbit

Orbit of Mercury or Venus

Maximum eastern elongation

Inferior conjunction

Quadrature

Earth

Sun

Opposition

Superior conjunction

Conjunction

Maximum western elongation

Quadrature

is at **opposition**. It is not difficult to determine when a planet happens to be located at one of the key positions in Figure 3-4. For example, when Mars is at opposition, it appears high in the sky at midnight.

Although it is easy to follow a planet as it moves from one configuration to another, these observations in themselves do not immediately provide relevant data about the planet's actual orbit around the Sun. Copernicus realized that the Earth, from which we make the observations, is also moving. He was therefore careful to distinguish between two characteristic time intervals, or **periods,** of each planet. The **synodic period** is the time that elapses between two successive identical configurations (as seen from the Earth)—from one opposition to the next, for example, or from one conjunction to the next. The **sidereal period** is the true orbital period of a planet, the time it takes the planet to complete one orbit of the Sun.

The synodic period of a planet can be determined by observing the sky, but the sidereal period must be calculated. Copernicus performed such calculations and obtained the results shown in Table 3-1.

Knowing the sidereal periods of the planets, Copernicus was able to devise a straightforward geometric method of determining the distances of the planets from the Sun. His answers turned out to be remarkably close to the modern values, as shown in Table 3-2.

From these two tables it is apparent that, the farther a planet is from the Sun, the longer the planet takes to travel around its orbit.

Copernicus compiled his ideas and calculations into a book entitled *De Revolutionibus Orbium Coelestium* ("On the Revolutions of the Celestial Spheres") that was published in 1543, just in time to reach Copernicus on his

Table 3-1 The synodic and sidereal periods of the planets

Planet	Synodic period	Sidereal period
Mercury	116 days	88 days
Venus	584 days	225 days
Earth	—	1.0 year
Mars	780 days	1.9 years
Jupiter	399 days	11.9 years
Saturn	378 days	29.5 years

Table 3-2 Average distances of the planets from the Sun (in astronomical units)

Planet	Copernicus	Modern
Mercury	0.38	0.39
Venus	0.72	0.72
Earth	1.00	1.00
Mars	1.52	1.52
Jupiter	5.22	5.20
Saturn	9.07	9.54

deathbed. Although he assumed that the Earth travels around the Sun along a circular path, he found that perfectly circular orbits cannot accurately describe the paths of the other planets. Copernicus had to add an epicycle to each planet to account for the slight variation in speed along its orbit. Thus, according to Copernicus, each planet revolves around a small epicycle, which in turn orbits the Sun along a circular path.

Placing the Sun at the center of the universe was a revolutionary proposal and one Renaissance astronomer, Tycho Brahe, tried to test Copernicus's ideas with detailed observations of the sky. Brahe spent his lifetime making accurate observations of the positions of the stars and planets, achieving an unprecedented level of precision. Brahe realized that these observations might reveal the motion of the Earth around the Sun.

We know that, when we walk from one place to another, nearby objects appear to shift their positions against the background of more distant objects. If Copernicus was correct, nearby stars should shift slightly against the background stars as the Earth orbits the Sun. In spite of Brahe's accurate observations, he could not detect any shifting of star positions. He therefore concluded that Copernicus was wrong.

Actually, the stars are so far away that Brahe's naked-eye observations could not have possibly detected the tiny shifting of star positions that has now been confirmed with telescopic observations. In spite of Tycho Brahe's failure to confirm Copernicus's ideas, his attempt to test them was quite revolutionary. At a time when heretics were burned at the stake for questioning dogma, Brahe was arguing that objective observations could be used to determine the validity or falsehood of an idea.

Brahe's astronomical records were destined to play an important role in the development of a heliocentric cosmology. Upon his death in 1601, many of his charts and books fell into the hands of his gifted assistant, Johannes Kepler.

Johannes Kepler proposed elliptical paths of the planets about the Sun

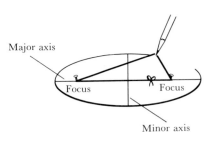

Figure 3-5 The construction of an ellipse An ellipse can be drawn with a pencil, a loop of string, and two thumbtacks, as shown in this diagram. If the string is kept taut, the pencil traces out an ellipse. The two thumbtacks are located at the two foci of the ellipse.

Until Johannes Kepler's time, at the beginning of the seventeenth century, astronomers had assumed that heavenly objects move in circles. The circle was considered perfect and the most harmonious of all geometric shapes. People of that time reasoned that since God is perfect, he wouldn't use any shape but a perfect one for the orbits of the planets. Kepler spent several years trying in vain to make circular motion fit Tycho Brahe's observations. Kepler's first major contribution to astronomy was the discovery that the planetary orbits are not circles.

Kepler first tried ovals, but he found they did not accurately describe the orbits of the planets about the Sun. Then he began working with a slightly different curve called an **ellipse.**

An ellipse can be drawn with a loop of string, two thumbtacks, and a pencil, as shown in Figure 3-5. Each thumbtack is at a **focus** (plural, **foci**). The longest diameter across an ellipse, called the **major axis,** passes through both foci. Half of that distance is called the **semimajor axis,** whose length is usually designated by the letter a.

To Kepler's delight, the ellipse turned out to be the curve he had been searching for. He published this discovery in 1609 in a book known today as *New Astronomy.* This important discovery is now called **Kepler's first law** and is stated as follows:

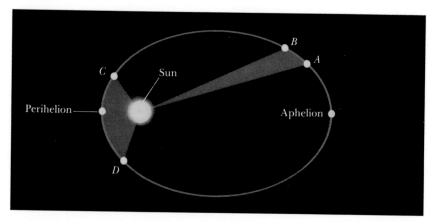

Figure 3-6 *Kepler's first and second laws* *According to Kepler's first two laws, every planet travels around the Sun along an elliptical orbit with the Sun at one focus in such a way that the line joining the planet and the Sun sweeps out equal areas in equal intervals of time.*

The orbit of a planet about the Sun is an ellipse with the Sun at one focus.

Kepler also realized that planets do not move at a uniform speed along their orbits. A planet moves most rapidly when it is nearest the Sun, at a point on its orbit called the **perihelion.** A planet moves most slowly when it is farthest from the Sun, at a point called the **aphelion.**

After much trial and error, Kepler discovered a way to describe how fast a planet moves along its orbit. This discovery, now called the **law of equal areas** or **Kepler's second law,** is illustrated in Figure 3-6. Suppose that it takes 30 days for a planet to go from point *A* to point *B*. During that time, the line joining the Sun and the planet sweeps out a nearly triangular area. Kepler discovered that the line joining the Sun and the planet sweeps out an equal area during any other 30-day interval. In other words, if the planet also takes a month to go from point *C* to point *D*, then the two shaded segments in Figure 3-6 are equal in area. Kepler's second law, also published in *New Astronomy,* can thus be stated:

A line joining a planet and the Sun sweeps out equal areas in equal intervals of time.

One of Kepler's later discoveries, published in 1619, stands out because of its impact on future developments in astronomy. Now called the **harmonic law** or **Kepler's third law,** it states a relationship between the sidereal period of a planet and the length of the semimajor axis of the planet's orbit:

The squares of the sidereal periods of the planets are proportional to the cubes of the semimajor axes of their orbits.

If a planet's sidereal period *P* is measured in years and the length of the semimajor axis *a* of its orbit is measured in astronomical units, then Kepler's third law is simply stated as

$$P^2 = a^3$$

The length of the semimajor axis is actually the average distance between a planet and the Sun. Using data from Tables 3-1 and 3-2, we can demon-

Table 3-3   A demonstration of Kepler's third law

Planet	Sidereal period P (in years)	Semimajor axis a (in AU)	P^2	a^3
Mercury	0.24	0.39	0.06	0.06
Venus	0.61	0.72	0.37	0.37
Earth	1.00	1.00	1.00	1.00
Mars	1.88	1.52	3.53	3.51
Jupiter	11.86	5.20	140.7	140.6
Saturn	29.46	9.54	867.9	868.3

strate Kepler's third law as shown in Table 3-3. This relationship can also be displayed on a graph as in Figure 3-7.

It is testimony to Kepler's genius that his three laws are rigorously obeyed in any situation where two objects orbit each other under the influence of their mutual gravitational attraction. Kepler's laws are obeyed not only by planets circling the Sun, but also by artificial satellites orbiting the Earth and by two stars revolving about each other in a double star system. Throughout this book, we shall see that Kepler's laws have a wide range of practical applications.

Figure 3-7   Kepler's third law   On this graph, the squares of the periods of the planets (P^2) are plotted against the cubes of the semimajor axes (a^3) of their orbits. The fact that the points fall along a straight line is verification of Kepler's discovery that $P^2 = a^3$.

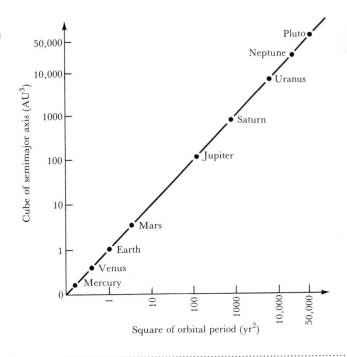

Galileo's discoveries with a telescope strongly supported a heliocentric cosmology

While Kepler was making rapid progress in central Europe, an Italian physicist was making observational discoveries equally dramatic. Galileo Galilei did not invent the telescope, but he was the first person to point one of the new devices toward the sky and publish his observations. He saw things that no one had ever dreamed of. He saw mountains on the Moon and sunspots on the Sun. He also discovered that Venus exhibits phases (see Figure 3-8).

Figure 3-8   The phases of Venus
This series of photographs shows how the appearance of Venus changes as it moves along its orbit. The number below each view is the angular diameter of the planet in seconds of arc. (New Mexico State University Observatory)

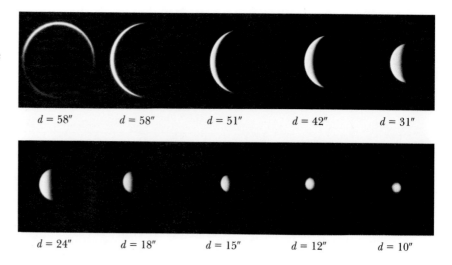

$d = 58''$ $d = 58''$ $d = 51''$ $d = 42''$ $d = 31''$

$d = 24''$ $d = 18''$ $d = 15''$ $d = 12''$ $d = 10''$

After only a few months of observation, Galileo noticed that the apparent size of Venus as seen through his telescope was related to the planet's phase. The planet appears smallest at gibbous phase and largest at crescent phase. There is also a correlation between the phases of Venus and the planet's angular distance from the Sun. These relationships clearly support the conclusion that Venus goes around the Sun (see Figure 3-9).

Figure 3-9 [right]   The changing appearance of Venus   *The phases of Venus are correlated with the planet's angular size and its angular distance from the Sun as sketched in this diagram. Galileo's observations of this correlation clearly supported the idea that Venus orbits the Sun.*

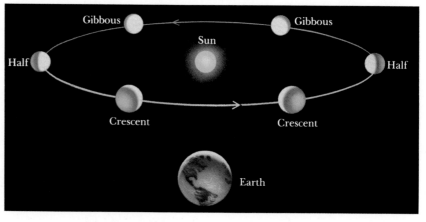

Figure 3-10 [above]   Jupiter and its largest moons   *This photograph, taken by an amateur astronomer with a small telescope, shows the four Galilean satellites alongside an overexposed image of Jupiter. Each satellite is bright enough to be seen with the unaided eye were it not overwhelmed by the glare of Jupiter. (Courtesy of C. Holmes)*

In 1610, Galileo also discovered four moons orbiting Jupiter (see Figure 3-10). He realized that they are orbiting Jupiter because they move back and forth from one side of the planet to the other. Confirming observations made by Jesuits in 1620 are shown in Figure 3-11. Astronomers soon demonstrated that these four moons obey Kepler's third law: the cube of a moon's average distance from Jupiter is proportional to the square of its orbital period about the planet. In honor of their discoverer, these four moons are today called the Galilean satellites.

These telescopic observations constituted the first fresh influx of fundamentally new astronomical data in almost 2000 years. In contradiction to prevailing opinions, these discoveries strongly suggested a heliocentric view of the universe. The Roman Catholic Church attacked his ideas because they

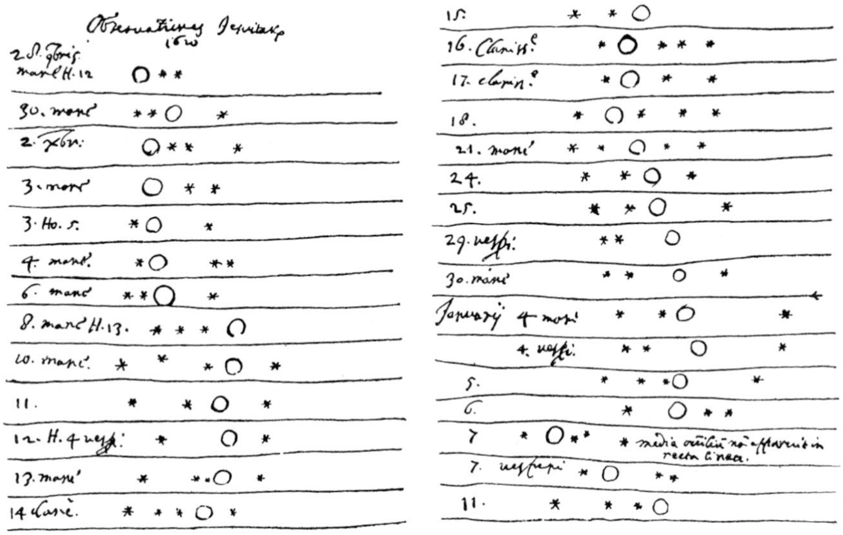

Figure 3-11 Early observations of Jupiter's moons In 1610, Galileo discovered four "stars" that move back and forth across Jupiter from one night to the next. He concluded that these are four moons that orbit Jupiter much as our Moon orbits the Earth. This drawing shows observations made in 1620. (Yerkes Observatory)

were not reconcilable with certain passages in the Bible or with the writings of Aristotle and Plato. Nevertheless, there was no turning back. Galileo was condemned to spend his latter years under house arrest "for vehement suspicion of heresy," but his revolutionary ideas soon inspired a sickly English boy born on Christmas day of 1642, less than a year after Galileo died. The boy's name was Isaac Newton.

Isaac Newton formulated a description of gravity that accounts for Kepler's laws and explains the motions of the planets

Until the mid-seventeenth century, virtually all mathematical astronomy had been entirely empirical, characterized by trial and error. From Ptolemy to Kepler, essentially the same approach had been used. Astronomers would work directly from data and observations, adjusting ideas and calculations until they finally came out with the right answers.

Isaac Newton introduced a new approach. He made three assumptions, now called **Newton's laws of motion,** about the nature of reality. Newton showed that Kepler's three laws follow logically from these laws of motion and from a formula for the force of gravity that he derived. Using this formula, Newton accurately described the observed orbits of the Moon, comets, and other objects in the solar system.

Newton's first law states that:

A body remains at rest, or moves in a straight line at a constant speed, unless acted upon by an outside force.

In other words, there must be an "outside" force acting on the planets or they would leave their curved orbits and move away from the Sun along straight-line paths at constant speeds. Since this does not happen, Newton concluded that the continuous action of this force confines the planets to their elliptical orbits.

Isaac Newton did not invent the idea of gravity (see Figure 3-12). An educated seventeenth century person had a vague appreciation of the fact that some force pulls things down to the ground. What Newton did was to

Copernicus
(1473–1543)

Kepler
(1571–1630)

Galileo
(1564–1642)

Figure 3-12 Major contributors to modern gravitational theory Each of these people is responsible for a major contribution to or breakthrough in our understanding of gravity. Because of their discoveries and insights, we now send spacecraft to the planets and probe the geometry of the universe.

Newton
(1642–1727)

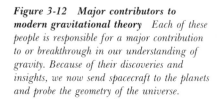

Einstein
(1879–1955)

give us a precise description of the action of gravity. Using his first law, Newton mathematically proved that the force acting on each of the planets is directed toward the Sun. This discovery led him to suspect that the force pulling a falling apple straight down to the ground is the same as the force the Sun and planets exert on one another.

Newton's second assumption, called Newton's second law, describes what a force does. Imagine an object floating in space. If you push on the object, it will begin to move. If you continue to push, the object's speed will continue to increase—in other words, the object will accelerate. **Acceleration** is the rate at which velocity changes. Newton's second law says:

The acceleration of an object is proportional to the force acting on the object.

Newton's final assumption, his third law, is the famous statement of action and reaction:

Whenever one body exerts a force on a second body, the second body exerts an equal and opposite force on the first body.

Newton then formulated a general statement that described the nature of the force—gravity—that keeps the planets in their orbits. Newton's **universal law of gravitation** states:

Two bodies attract each other with a force that is directly proportional to the product of their masses and inversely proportional to the square of the distance between them.

In other words, if two objects have masses m_1 and m_2 and are separated by a distance r, then the gravitational force F between these two masses is

$$F = G\frac{m_1 m_2}{r^2}$$

In this formula, G is a number called the *universal constant of gravitation*, whose value has been determined from laboratory experiments.

Using his law of gravity, Newton found that he could mathematically prove the validity of Kepler's three laws. For example, Kepler discovered by trial and error that $P^2 = a^3$. Newton demonstrated mathematically that this equation follows logically from his law of gravity anytime that two objects revolve about each other.

Newton also discovered new features of orbits around the Sun. For example, his equations soon led him to conclude that the orbit of an object around the Sun could be any one of a family of curves called conic sections. A **conic section** is any curve that you get by cutting a cone with a plane, as shown in Figure 3-13. You can get **circles** and **ellipses** by slicing all the way through the cone. By slicing the cone at a steep angle, you can get two "open" curves called **parabolas** and **hyperbolas.** Comets hurtling toward the Sun from the depths of space sometimes have parabolic orbits.

Figure 3-13 Conic sections A conic section is any one of a family of curves obtained by slicing a cone with a plane, as shown in this diagram. The orbit of one body about another can be any one of these curves: a circle, an ellipse, a parabola, or a hyperbola.

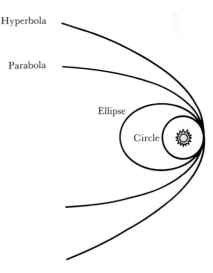

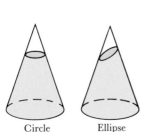

Circle Ellipse Parabola Hyperbola

Figure 3-14 Halley's comet Halley's comet orbits the Sun with an average period of about 76 years. During the twentieth century, the comet passed near the Sun twice—once in 1910 and again in 1986. This photograph shows how the comet looked in 1986. (Lumicon)

Figure 3-15 Uranus and Neptune
The existence of **(a)** Neptune (with one moon) was deduced from deviations in the predicted orbit of **(b)** Uranus (with three moons). This discovery was a major triumph for Newtonian mechanics. Both planets are shown along with some of their satellites. (Lick Observatory)

Newton's ideas and methods turned out to be incredibly successful in a wide range of situations. The orbits of the planets and their satellites could now be calculated with unprecedented precision. In addition, Newton's laws and mathematical techniques were capable of predicting new phenomena. For example, one of Newton's friends, Edmund Halley, was intrigued by historical records of a comet that was sighted about every 76 years. Using Newton's methods, Halley worked out the details of the comet's orbit and predicted its return in 1758. It was first sighted on Christmas night of that year, and to this day the comet bears Halley's name (see Figure 3-14).

Perhaps the most dramatic success of Newton's ideas was their role in the discovery of the eighth planet from the Sun. The seventh planet, Uranus, had been discovered by William Herschel in 1781 during a telescopic survey of the sky. Fifty years later, however, it was clear that Uranus was not following its predicted orbit. Two mathematicians, John Couch Adams in England and U. J. Leverrier in France, independently calculated that the deviations of Uranus from its orbit could be explained by the gravitational pull of a yet unknown, more distant planet. They each predicted that the planet would be found at a certain location in the constellation of Aquarius. A brief telescopic search on September 23, 1846, revealed Neptune less than 1° from the calculated position. Although sighted with a telescope (see Figure 3-15), Neptune was really discovered with pencil and paper.

Over the years, Newton's ideas were used in successful predictions and explanations of many phenomena, and Newtonian mechanics became the cornerstone of modern physical science. Even today, as we send astronauts to the Moon and probes to the outer planets, Newton's equations are used to calculate the orbits and trajectories of these spacecraft.

a

b

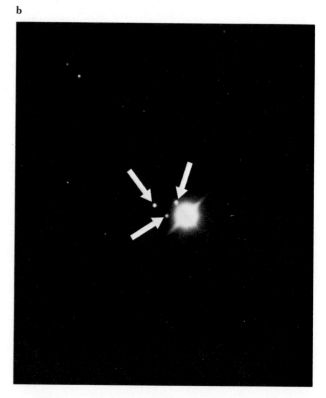

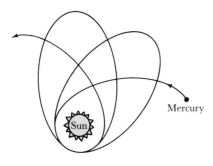

Figure 3-16 The advance of Mercury's perihelion As Mercury moves along its orbit, the orbit rotates. Consequently, Mercury traces out a rosette figure about the Sun, greatly exaggerated in this diagram. The excess rotation of Mercury's orbit (beyond that predicted by Newtonian mechanics) amounts to only 43 arc sec per century.

There was, however, one instance in which Newtonian mechanics was not quite in agreement with observations. During the mid-1800s, Leverrier pointed out that Mercury was not following its predicted orbit. As the planet moves along its elliptical orbit, the orbit itself slowly rotates (or precesses) as shown in Figure 3-16. Specifically, Mercury's perihelion shifts by an amount that cannot be explained with Newtonian mechanics. Although the effect is very small (the unexplained excess rotation of Mercury's major axis is only 43 arc sec per century), all attempts to account for this phenomenon met with failure. Nevertheless, Newton's laws proved so eminently successful in all other situations that most physicists and astronomers simply ignored this tiny discrepancy.

It is a testament to Isaac Newton's genius that his three laws were precisely the three basic ideas needed for a full understanding of the motions of the planets. In this way, Newton brought a new dimension of elegance and sophistication to our understanding of the workings of the universe. That is where things probably would have remained had it not been for the insight, vision, and genius of Albert Einstein.

Albert Einstein's theory states that gravity affects the shape of space and the flow of time

In the cosmologies of many ancient civilizations, the Earth and its inhabitants occupied a special place at the center of the universe. Copernicus's new cosmology made the Earth just one of a number of planets orbiting the Sun. Within less than a century, Kepler's accurate description of planetary orbits led directly to Newton's universal law of gravitation. Astronomers and philosophers then began to explore the possibility that even the Sun might occupy no special location in the universe—that our Sun could be merely one of many stars scattered through space.

Albert Einstein introduced an even more powerful perspective. He believed that the fundamental laws of the universe depend neither on our location in space nor on our motion through space. In other words, the laws of physics should be the same, whether we happen to be sitting on the Earth or moving through space at 95 percent of the speed of light. Inspired by this viewpoint, Einstein's endeavored to reformulate the laws of physics so that all observers would have equal status.

In 1916, Albert Einstein formulated a new theory of gravity, called the **general theory of relativity,** that is independent of one's location or motion. To achieve this goal, Einstein had to abandon the notion that rulers have fixed lengths and clocks always tick at an unchanging rate. The basic idea in general relativity is that the gravity of an object alters the properties of space and time around the object.

Einstein's description of gravity is radically different from Newton's. According to Newtonian mechanics, space is perfectly flat and extends infinitely far in all directions. Similarly, Newtonian clocks monotonously tick at an unchanging rate, never speeding up nor slowing down. In this rigid, unalterable framework of space and time, gravity is described as a "force" that acts at a distance. The planets literally pull on each other across empty space with a strength given by Newton's universal law of gravitation.

The general theory of relativity does not treat gravity as a force at all. Instead, gravity causes space to become curved and time to slow down. Far from any sources of gravity, space is flat and clocks tick at their normal rate. But around a massive object, space becomes curved as shown in Figure 3-17.

Figure 3-17 The gravitational curvature of space *According to Einstein's general theory of relativity, space becomes curved near a massive object such as the Sun. This diagram shows a two-dimensional analogy of the shape of space around a massive object.*

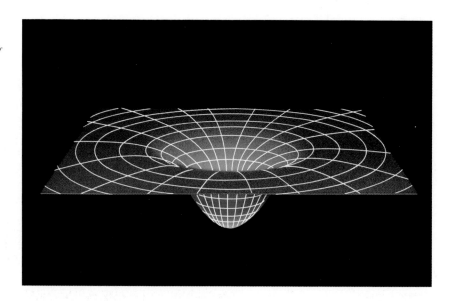

A planet or spacecraft passing near this massive object is deflected from a straight-line path because space itself is curved. The general theory of relativity predicts that only very strong gravitational fields have easily detectable effects upon space and time in their vicinity.

One of the first things Einstein did with his new theory was to calculate the orbits of the planets. With only one minor exception, general relativity theory gave almost exactly the same answers as Newtonian theory. Mercury is the only planet to pass so close to the Sun's strong gravitational field that the curvature of space produces an effect significantly different from that predicted by Newtonian mechanics. Indeed, throughout most of the solar system the curvature of space is so slight that Einstein's equations are essentially equivalent to Newtonian calculations. However, Einstein's theory did predict that Mercury's orbit should be a slowly rotating ellipse, thereby explaining a phenomenon that had frustrated astronomers for half a century.

To test the validity of his theory Einstein predicted new phenomena. He calculated that light rays passing near the surface of the Sun should appear to be deflected from their straight-line paths because the space through which they are moving is curved. In other words, gravity should bend light rays, an effect not predicted by Newtonian mechanics because light has no mass.

Figure 3-18 shows a beam of light from a star passing by the Sun and continuing down to the Earth. Because the light ray is bent, the star appears shifted from its usual location. At most, the position of a star is shifted by 1.75 arc sec for a light ray grazing the Sun's surface.

This prediction was first tested during a total solar eclipse in 1919. During the precious moments of totality, when the Moon blocked out the blinding solar disk, astronomers succeeded in photographing the stars around the Sun. Careful measurements revealed that the stars were shifted from their usual positions by an amount consistent with Einstein's theory. General relativity had passed another important test.

Einstein made a third prediction. Because gravity causes time to slow down, a clock on the first floor of a building should tick slightly more slowly than a clock on the roof, as shown in Figure 3-19. This prediction was tested

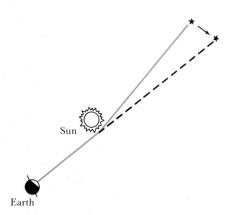

Figure 3-18 The gravitational deflection of light *Light rays passing near the Sun are deflected from their straight-line paths. By relativistic standards, the Sun's gravity is very weak. The maximum angle of deflection is only 1.75 arc sec for a light ray grazing the Sun's surface.*

Figure 3-19 The gravitational slowing of time A clock on the ground floor of a building is closer to the Earth than a clock at a higher elevation. Therefore, according to general relativity, the clock on the ground floor should tick more slowly than the clock on the roof.

in 1959 in a four story building at Harvard University using a radioactive substance as an accurate timepiece. Again Einstein was proven correct.

During the past several decades, these and similar tests have confirmed general relativity over and over again. This remarkable theory now stands as our most precise and complete description of gravity.

It is important to emphasize that Einstein did not prove Newton wrong. Rather, Einstein demonstrated that Newtonian gravitation is accurate only when applied to weak gravity. If extremely powerful gravitational fields (such as those of neutron stars and black holes) are involved, only a relativistic calculation will give correct answers.

Einstein's general theory of relativity evoked a sensation when it was first proposed. Newtonian gravitation had been overthrown by a radically different approach that worked better.

After the initial excitement, however, interest in general relativity rapidly waned. The complex mathematics of relativity seemed burdensome, and Newton's simpler approach gave reasonable precision in almost all conceivable circumstances. No one could imagine places where the curvature of space might have a major effect. Relativistic theory offered only a little more accuracy, as in the case of Mercury's orbit.

In recent years there has been a reawakening of interest in general relativity, primarily because of dramatic advances in our understanding of the evolution of stars. As we shall see in later chapters, we now have a reasonably complete picture of how stars are born, what happens to them as they mature, and where they go when they die. In particular, it appears that the most massive dying stars are doomed to collapse completely upon themselves, thereby producing some of the most bizarre objects in the universe—black holes. Gravity around one of these massive stellar corpses is so strong that it punches a hole in the fabric of space. Because of this intense gravitational effect, black holes can be described and discussed only in terms of general relativity.

As we set our sights on understanding the universe as a whole, we must once again turn to general relativity. All the matter in all the stars and galaxies is responsible for the overall curvature or shape of space. From the viewpoint of general relativity, it is therefore reasonable to ask about the actual shape of the universe. We shall see that the answer suggests the ultimate fate of the cosmos.

Summary

· Ancient astronomers believed that the Earth is at the center of the universe and invented a complex system of epicycles and deferents to explain the direct and retrograde motions of the planets.

· A heliocentric (Sun-centered) theory simplifies the general explanation of planetary motions. In a heliocentric system, the Earth is one of several planets that orbit the Sun.

· The sidereal period of a planet (measured with respect to the stars) is its true orbital period; its synodic period is measured with respect to the Sun as seen from the moving Earth.

· Ellipses describe the paths of the planets around the Sun much more accurately than do circles. Kepler's three laws give important details about elliptical orbits.

- The invention of the telescope led to new discoveries (phases of Venus, moons of Jupiter, and so forth) that supported a heliocentric view of the universe.

- Newton based his explanation of the universe on three assumptions, or laws of motion. Kepler's laws and extremely accurate descriptions of planetary motions can be deduced from Newton's laws and his universal law of gravitation.

- In general, the path of an object about the Sun is one of the curves called conic sections: a circle, an ellipse, a parabola, or a hyperbola.

- The general theory of relativity explains that gravity causes space to be curved and time to slow down.

- Although Newtonian mechanics accurately describes and predicts numerous phenomena, Einstein's general theory of relativity is more accurate where extremely intense gravitational fields are involved.

Review questions

1 How did ancient astronomers differentiate between a star and a planet?

2 With the aid of a drawing, show how an epicycle can explain the retrograde motion of a planet in a geocentric cosmology. With a second drawing, show how a heliocentric cosmology explains the same retrograde motion.

3 At what configuration (superior conjunction, greatest eastern elongation, etc.) would it be best to observe Mercury or Venus with an Earth-based telescope? At what configuration would it be best to observe Mars, Jupiter, or Saturn? Explain your answers.

4 In your own words, state Kepler's three laws of planetary motion.

5 In what ways did the astronomical observations of Galileo support a heliocentric cosmology?

6 How did Newton's approach to understanding planetary motions differ from that of his predecessors?

7 What are conic sections and in what way are they related to the orbits of objects in the solar system?

8 Why is the discovery of Neptune considered a major confirmation of Newtonian mechanics?

Advanced questions

9 Is it possible for an object in the solar system to have a synodic period of exactly one year? Explain your answer.

*10 A line joining the Sun and an asteroid was found to sweep out 5.2 square AUs of space in 1983. How much area was swept out in 1984? In five years?

*11 A comet moves in a highly elongated orbit about the Sun with a period of 1000 years. What is the comet's average distance from the Sun? What is the farthest it can get from the Sun?

*12 Is it possible for a planet's sidereal period to equal its synodic period? Would this planet be closer to the Sun than is the Earth or farther away? Is there a planet that nearly fits this description?

13 Look up the dates of various greatest eastern and western elongations for Mercury in a year of your choice. Does it take longer to go from eastern to western elongation or vice versa? Why do you suppose this is the case?

Discussion questions

14 Which planet would you expect to exhibit the greatest variation in apparent brightness as seen from Earth? Explain your answer.

15 Use two thumbtacks, a loop of string, and a pencil to draw several ellipses. Describe how the shape of the ellipse varies as you change the distance between the thumbtacks.

For further reading

Banville, J. *Kepler: A Novel*. Godine, 1981.

Christianson, G. *This Wild Abyss*. Free Press, 1978.

Cohen, O. "Newton's Discovery of Gravity." *Scientific American*, March 1981.

Durham, F., and Purrington, R. *Frame of the Universe*. Columbia University Press, 1983.

Gardner, M. *The Relativity Explosion*. Vintage, 1976.

Gingerich, O. "Copernicus and Tycho." *Scientific American*, December 1973.

Hoffman, B., and Dukas, H. *Albert Einstein: Creator and Rebel*. NAL Plume Books, 1972.

Kaufmann, W. *Relativity and Cosmology*. Harper & Row, 1977.

Kuhn, T. *The Copernican Revolution*. Harvard University Press, 1957.

Rogers, E. *Astronomy for the Inquiring Mind*. Princeton University Press, 1960.

Wilson, C. "How Did Kepler Discover His First Two Laws?" *Scientific American*, March 1972.

4 Light, optics, and telescopes

The Anglo-Australian Observatory
Astronomers prefer to build observatories far from city lights, usually on isolated mountain tops, where the air is dry, stable, and cloud-free. This aerial view shows the Anglo-Australian Observatory, about 400 km (250 miles) northwest of Sydney. The large dome in the background houses the largest telescope in Australia, the 3.9-meter Anglo-Australian Telescope. The foreground dome contains a Schmidt telescope operated by the Royal Observatory of Edinburgh. Many of the photographs in this textbook were taken with these telescopes. (Courtesy of D. Malin; AAT)

Until very recently, our knowledge about the universe came almost entirely from the visible light gathered by telescopes. In this chapter, we learn that light is a form of electromagnetic radiation and that many other forms of such radiation exist. We discuss the two major types of optical telescopes: refractors and reflectors. Then we learn about more exotic telescopes that detect radio waves and other electromagnetic radiation arriving at the Earth. Finally, we examine the orbiting observatories that are gathering a wide range of new information about the universe.

Telescopes have played a major role in revealing the universe since Galileo first saw craters on the Moon four centuries ago. The telescope remains the single most important tool of astronomy. With a telescope, we can see extremely faint objects in space far more clearly than we can with the naked eye.

Traditionally, telescopes detect visible light. Either lenses or mirrors bring light from a distant object to a focus where the resulting image is viewed or photographed. Recently, however, astronomers have built telescopes that detect nonvisible forms of light such as X rays and radio waves. To appreciate these developments and to understand how telescopes work, we must first learn something about the basic properties of light.

Light is electromagnetic radiation and is characterized by its wavelength

Galileo and Newton made important contributions to our modern understanding of light as well as to our theories of gravity and mechanics. In the early 1600s, Galileo made one of the first attempts to measure the speed of light. At night he stood on one hilltop while an assistant stood on another hilltop at a known distance, each holding a shuttered lantern. First Galileo opened the shutter of his lantern. As soon as the assistant saw the flash of light, he opened his own lantern. Using the rate of his pulse as a clock, Galileo attempted to measure the time that elapsed between opening his shutter and seeing the light from the assistant's lantern. By knowing this time and the distance between hilltops, he could then compute the speed at which light traveled to the distant hilltop and back again. He soon concluded that light travels so rapidly that slow human reactions and crude clocks, like his pulse, make it impossible to measure its speed.

The first successful measurement of the speed of light was made in 1675 by Olaus Roemer, a Danish astronomer, who carefully timed eclipses of Jupiter's satellites. Roemer discovered that the moment at which a satellite enters Jupiter's shadow seems to depend on the distance between the Earth and Jupiter. When the Earth–Jupiter distance is short (around the time of opposition), eclipses are observed slightly earlier than when Jupiter and the Earth are widely separated. Roemer correctly interpreted this phenomenon as the result of the time it takes light to travel across space. The greater the distance to Jupiter, the longer we must wait for the image of an eclipse to reach our eyes. From his timing measurements, Roemer concluded that it takes $16\frac{1}{2}$ minutes for light to traverse the diameter of the Earth's orbit (2 AU). In Roemer's day, no one knew the length of the astronomical unit and so he could not actually compute the speed of light in kilometers per second.

The first accurate laboratory experiments to measure the speed of light were performed in the mid-1800s, most of them using elaborate optical equipment. From these experiments, we now know that the speed of light in a vacuum is about 3×10^5 km/sec (186,000 miles per second).

A pioneering breakthrough in understanding light came from a simple experiment performed by Isaac Newton in the late 1600s. He passed a beam of sunlight through a glass prism that spread the light out into the colors of the rainbow as shown in Figure 4-1. This rainbow, called a **spectrum,** suggested to Newton that white light is actually a mixture of all colors. In Newton's day, most people erroneously believed that the colors were somehow added to the light by the glass. Newton disproved this idea by passing the spectrum through a second prism inverted with respect to the first. Only

Figure 4-1   A prism and a spectrum
When a beam of white light passes through a glass prism, the light is broken into a rainbow-colored band called a spectrum. The numbers on the right side of the spectrum indicate wavelengths as described in the text.

7000 Å
6000 Å
5000 Å
4000 Å

white light emerged from this second prism, and so the second prism reassembled the colors of the rainbow to give back the original beam of sunlight.

From his many experiments, Newton argued that light is composed of indetectably tiny particles of energy. As we shall see in the next chapter, the modern theory of quantum mechanics developed in the early 1900s also takes this viewpoint. Quantum theory explains that light is composed of tiny packets of energy called **photons.** In the mid-1600s, however, a rival explanation was proposed by the Dutch astronomer Christian Huygens, who suggested that light travels in the form of waves rather than particles. Scientists now realize that light posses both particlelike and wavelike characteristics.

The English physicist Thomas Young confirmed the wave nature of light in 1801. Young demonstrated that the shadows of objects in light of a single color are not crisp and sharp. Instead, the boundary between illuminated and shaded areas is overlaid with patterns of closely spaced dark and light bands. These patterns are similar to the patterns produced by water waves passing the edge of a reef or barrier in the ocean. Young showed that a wavelike description of light could explain the results of his experiment, but Newton's particle theory could not. Analyses of similar experiments soon produced overwhelming evidence for the wavelike behavior of light.

Further insight into the wave character of light came from calculations by the Scottish physicist James Clerk Maxwell in the 1860s. Maxwell succeeded in describing all the basic properties of electricity and magnetism in four equations. By combining these equations, Maxwell demonstrated that electrical and magnetic effects should travel through space in the form of waves. Furthermore, his calculations proved that these waves should travel with a speed of about 3×10^8 m/sec. Maxwell's suggestion that these waves do exist and are observed as light was soon confirmed by a variety of experiments. Light consists of vibrating electric and magnetic fields, as shown in Figure 4-2. Because of its electric and magnetic properties, light is called **electromagnetic radiation.** The distance between two successive wave crests is called the **wavelength** of the light (see Figure 4-2).

The wavelength of visible light is extremely small, less than a thousandth of a millimeter. To express these tiny distances conveniently, scientists use a unit of length called the angstrom (abbreviated Å), named after the Swedish physicist A. J. Ångström), where $1 \text{ Å} = 10^{-8}$ cm. Experiments demonstrated that visible light has wavelengths covering the range from about 4000 Å for violet light to about 7000 Å for red light. Intermediate colors of the rainbow fall between these wavelengths (see Figure 4-1).

Visible light includes a specific, narrow range of wavelengths. But Maxwell's equations place no restrictions on the wavelengths that electromag-

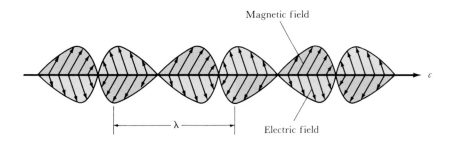
netic radiation can have. It therefore is possible that electromagnetic waves could exist with wavelengths both longer and shorter than the 4000–7000 Å range of visible light. Researchers began to look for invisible forms of light, forms to which the cells of the human retina do not respond.

The British astronomer William Herschel discovered radiation just beyond the red end of the visible spectrum. In 1888, the German physicist Heinrich Hertz succeeded in producing light having wavelengths longer than about 10 centimeters, now known as **radio waves.** In 1895, Wilhelm Röntgen invented a machine that produces light with a wavelength shorter than 100 Å, now known as **X rays.** Modern versions of Röntgen's machine are today found in medical and dental offices.

Over the years, radiation in many other wavelength ranges has been discovered. Visible light is only a tiny fraction of the full extent of possible wavelengths collectively called the **electromagnetic spectrum.** As shown in Figure 4-3, the electromagnetic spectrum stretches from the longest-wavelength radio waves to the shortest-wavelength **gamma rays.** For example, at wavelengths slightly longer than visible light, **infrared radiation** covers the range from about 7000 Å to 1 mm. From roughly 1 mm to 10 cm is the range of **microwaves,** beyond which is the domain of radio waves. Infrared and microwave wavelengths are sometimes measured in micrometers (μm), where 1 μm = 10^4 Å.

At wavelengths shorter than that of visible light, **ultraviolet radiation** extends from about 4000 Å down to 100 Å. Wavelengths between about 100 Å and 0.1 Å are X rays, beyond which is the domain of gamma rays.

Although these various types of electromagnetic radiation share many basic properties (for example, they all travel at the speed of light), they interact very differently with matter. Your body is transparent to X rays but not to visible light; your eyes respond to visible light but not to gamma rays; your radio detects radio waves but not ultraviolet light. Consequently, astronomers use fundamentally different kinds of telescopes to detect radiation in these various wavelength ranges. For example, a radio telescope that detects radio waves from space is quite different from either an X-ray telescope or an ordinary optical telescope. Because optical telescopes are the most common and familiar astronomical tool, we shall discuss them in detail before turning to the exotic instruments that reveal the nonvisible sky.

A refracting telescope uses a lens to concentrate incoming starlight at a focus

Although light travels at about 3×10^8 m/sec in a vacuum, it moves more slowly through a dense substance such as glass. The abrupt slowing of light entering a piece of glass is analogous to a person walking from a boardwalk onto a sandy beach: her pace suddenly slows as she steps from the smooth

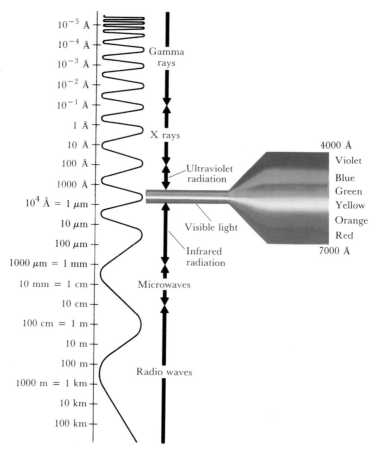

Figure 4-3 The electromagnetic spectrum *The full array of all types of electromagnetic radiation is called the electromagnetic spectrum. It extends from the shortest-wavelength gamma rays to the longest-wavelength radio waves. Visible light forms only a tiny portion of the full electromagnetic spectrum.*

pavement into the sand. And just as the person stepping back onto the boardwalk easily resumes her original pace, light exiting a piece of glass resumes its original speed.

Furthermore, a light ray is bent as it passes from one transparent medium into another at an angle oblique to the surface between the media. This phenomenon is called **refraction** and is caused by the change in the speed of light. Imagine driving a car from a smooth pavement onto a sandy beach. If the car approaches the beach at an angle, one of the front wheels is slowed by the sand before the other, causing the car to veer from its original direction.

To describe the refraction of a light ray entering a piece of glass, imagine drawing a perpendicular to the surface of the glass at the point where the light strikes the glass, as shown in Figure 4-4. As a light ray goes from a less-dense medium (such as air) into a more-dense medium (such as glass), the light is always bent toward the perpendicular direction, just as a car driving obliquely onto a beach veers toward the direction perpendicular to the pavement–sand boundary. Upon emerging from the other side of a piece of glass, light resumes its original high speed, and the light ray is bent away from the perpendicular direction. The exact amount of refraction depends on the speed of light in the glass, which depends on the chemical composition of the glass.

Because of the refracting property of glass, a convex lens—one that is fatter in the middle than at the edges—causes incoming light rays to con-

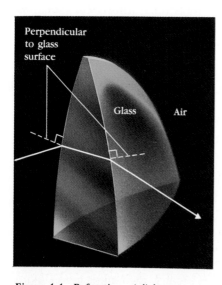

Figure 4-4 Refraction *A light ray entering a piece of glass is bent toward the perpendicular. As the ray leaves the piece of glass it is bent away from the perpendicular.*

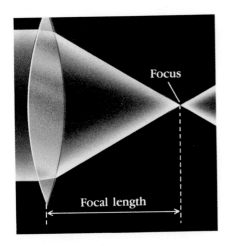

Figure 4-5 A convex lens *A convex lens causes parallel light rays to converge to a focus. The distance from the lens to the focus is the focal length of the lens.*

verge to a point called the **focus,** as shown in Figure 4-5. If the light source is extremely far away, then the incoming light rays are parallel, and they come to a focus at a specific distance from the lens called the **focal length** of the lens.

Stars and planets are so far away that light rays from these objects are essentially parallel. Consequently, a lens always focuses light from an astronomical object as shown in Figure 4-5. An image of the object is formed at the focus, and a second lens can be used to magnify and examine this image. Such an arrangement of two lenses is called a **refracting telescope,** or **refractor** (see Figure 4-6). The large-diameter, long-focal-length lens at the front of the telescope is the **objective lens.** The smaller, short-focal-length lens at the rear of the telescope is the **eyepiece lens.**

The **magnification,** or **magnifying power,** of a refracting telescope is equal to the focal length of the objective lens divided by the focal length of the eyepiece lens. For example, if the objective of a telescope has a focal length of 100 cm and the eyepiece has a focal length of $\frac{1}{2}$ cm, then the magnifying power of the telescope is 200 (usually written as $200\times$).

If you build a telescope using only the instructions given so far, you will probably be disappointed with the results. You will see stars surrounded by fuzzy, rainbow-colored halos. This optical defect, called **chromatic aberration,** exists because a lens bends different colors of light through different angles, just as a prism does (recall Figure 4-1).

By adding small amounts of various chemicals to a vat of molten glass, an optician can manufacture different kinds of glass. The speed of light in glass depends on the chemical composition of the glass—a fact that opticians use to correct for chromatic aberration. Specifically, a thin lens can be mounted just behind the main objective lens of a telescope as diagrammed in Figure 4-7. By carefully choosing different kinds of glass for these two lenses, the optician can ensure that different colors of light come to a focus at the same point.

Chromatic aberration is the most severe of a host of optical problems that must be solved in designing a high-quality refracting telescope. Nineteenth century master opticians devoted their lives to overcoming these problems, and several magnificent refractors were constructed in the late 1800s. The largest refracting telescope, completed in 1897, is located at the Yerkes

Figure 4-6 A refracting telescope
A refracting telescope consists of a large, long-focal-length objective lens and a small, short-focal-length eyepiece lens that magnifies the image formed at the focus of the objective lens.

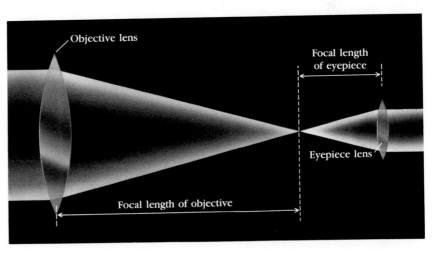

Figure 4-7 [right] Chromatic aberration
A single lens suffers from a defect called chromatic aberration in which different colors of light have slightly different focal lengths. This problem is corrected by adding a second lens made of a kind of glass that is different from that of the first lens.

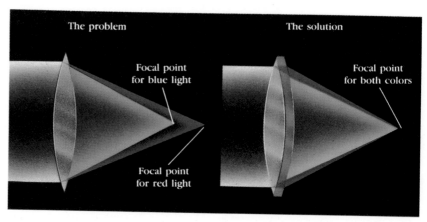

Figure 4-8 A large refracting telescope
This giant refracting telescope was built in the late 1800s and is housed at Yerkes Observatory near Chicago. The objective lens is 102 cm (40 in.) in diameter, and the telescope tube is 19½ m (63¼ ft) long. (Yerkes Observatory)

Observatory near Chicago (see Figure 4-8) and has an objective lens 102 cm (40 in.) in diameter. The second largest refracting telescope is located at Lick Observatory near San Jose, California. This refractor has an objective lens whose diameter is 91 cm (36 in.). All refractors have extremely long focal lengths. For example, the Yerkes refractor has a focal length of 19.35 m (63½ ft).

Few major new refracting telescopes have been constructed in the twentieth century. There are many reasons for the modern astronomer's lack of interest in this type of telescope. First, because faint light must pass readily through the objective lens, the glass from which the lens is made must be totally free of defects such as bubbles that frequently form when molten glass is poured into a mold. Consequently, the glass for the lens is extremely expensive. Second, glass is opaque to certain kinds of light. Even visible light is dimmed substantially in passing through the thick slab of glass at the front of a refractor, and ultraviolet radiation is largely absorbed by the glass lens. Third, it is impossible to produce a lens that is completely corrected to eliminate chromatic aberration. Fourth, it is difficult to support such a heavy lens without blocking the path of light into the telescope. All of these problems can be avoided by using mirrors instead of lenses.

A reflecting telescope uses a mirror to concentrate incoming starlight at a focus

Reflection can be described very simply. To understand reflection, imagine drawing a perpendicular to a mirror's surface at the point where a light ray strikes the mirror, as shown in Figure 4-9. The angle between the arriving (incident) light ray and the perpendicular is always equal to the angle between the reflected ray and the perpendicular. Knowing this, Isaac Newton realized that a concave mirror will cause parallel light rays to converge to a focus as shown in Figure 4-10. The distance between the reflecting surface and the focus is called the **focal length** of the mirror.

An image of a distant object is formed at the focus of a concave mirror. In order to view the image, Newton placed a small, flat mirror at a 45° angle in front of the focal point as sketched in Figure 4-11a. This secondary mirror deflects the light rays to one side of the **reflecting telescope,** or **reflector,** where the astronomer can place an eyepiece lens to magnify the image. A telescope having this optical design is appropriately called a **Newtonian reflector.** The magnifying power of such a reflecting telescope is calculated in

Figure 4-9 [left] **Reflection** *The angle at which a beam of light strikes a mirror (the angle of incidence* i*) is always equal to the angle at which the beam is reflected from the mirror (the angle of reflection* r*).*

Figure 4-10 [right] **A concave mirror** *A concave mirror causes parallel light rays to converge to a focus. The distance between the mirror and the focus is the focal length of the mirror.*

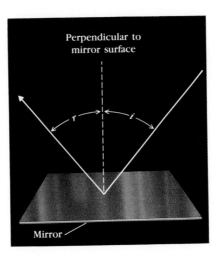

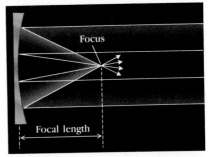

the same way as for a refractor: the focal length of the primary mirror is divided by the focal length of the eyepiece.

Useful modifications of Newton's original design have since been made. The primary mirrors of many major reflectors are so large that the astronomer can actually sit at the undeflected focal point, directly in front of the primary mirror. This arrangement is called a **prime focus** (see in Figure 4-11*b*). The "observing cage," in which the astronomer rides, blocks only a small fraction of the incoming starlight.

Figure 4-11 **Reflecting telescopes** *Four of the most popular optical designs for reflecting telescopes:* (a) *Newtonian focus,* (b) *prime focus,* (c) *Cassegrain focus, and* (d) *coudé focus.*

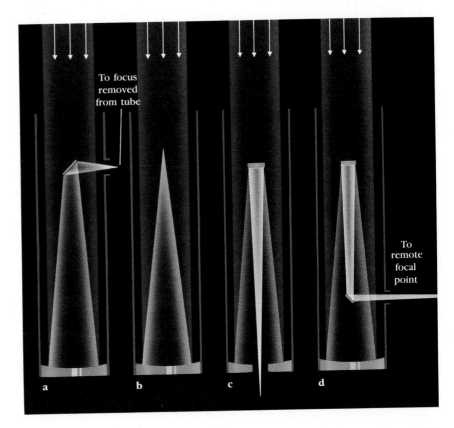

Figure 4-12   The 4-meter telescope at Cerro Tololo   This telescope is located on a mountaintop near Santiago, Chile. Its twin is at the Kitt Peak Observatory in Arizona. Both telescopes have been in operation since the early 1970s. (NOAO)

Another popular optical design, called a **Cassegrain focus,** has the advantage of placing the focal point at a convenient and accessible location. A hole is drilled directly through the center of the primary mirror. A convex secondary mirror placed in front of the original focal point is used to reflect the light rays back through the hole (see Figure 4-11*c*).

Alternatively, a series of mirrors can be used to channel the light rays away from the telescope to a remote focal point. Heavy optical equipment that could not be mounted directly on the telescope is located at the resulting **coudé focus** (see Figure 4-11*d*).

To make a reflector, an optician grinds and polishes a large slab of glass into the appropriate concave shape. The glass is then coated with silver or aluminum or a similar highly reflective substance. Defects inside the glass such as bubbles or flecks of dirt do not detract from the telescope's effectiveness, as they would in the objective lens of a refracting telescope.

Furthermore, reflection is not affected by the wavelength of the light. Therefore, the image formed by a mirror does not suffer from chromatic aberration. Finally, the mirror can be fully supported by braces on its back, so that a large and heavy mirror can be mounted without much danger of breakage or shape distortion.

Eleven reflectors exist with primary mirrors measuring at least 3 m in diameter. The largest, located in the Soviet Union, has a mirror 6 m (19.7 ft) in diameter. The second largest is at the Palomar Observatory in southern California, with a mirror 5 m (16.4 ft) in diameter. In the early 1970s, a matching pair of telescopes was built in Arizona and Chile. Both have mirrors 4 m (13.1 ft) in diameter. These two telescopes (see Figure 4-12) allow astronomers to observe the entire sky with essentially the same instrument.

Astronomers strongly prefer large telescopes. A large mirror intercepts and focuses more starlight than does a small mirror (see Figure 4-13). A large mirror therefore produces brighter images and detects fainter stars than does a small mirror. The **light-gathering power** of a telescope is directly related to the area of the telescope's primary mirror. For example, the 200-inch mirror at Palomar Observatory has four times the area of the 100-inch mirror at Mt. Wilson Observatory. Therefore, the Palomar telescope has four times the light-gathering power of the Mt. Wilson telescope.

A large telescope also increases the sharpness of the image and the extent to which fine details can be distinguished. This property is called **resolving power.** With low resolving power, star images are fuzzy and blurred together. With high resolving power, images are sharp and crisp.

The resolving power of a telescope is measured as the angular distance between two adjacent stars whose images can just barely be distinguished under ideal observing conditions. Large modern telescopes—such as those at Palomar, Kitt Peak, and Cerro Tololo—are calculated to have resolving powers better than 0.1 arc sec. In practice, however, this exceptionally high resolving power is never achieved. Turbulence and impurities in the air cause star images to jiggle around, or twinkle. Even when photographed through the largest telescopes, a star looks like a tiny circular blob rather than a pinpoint of light.

The angular diameter of a star's image is called the **seeing disk** and is a realistic measure of the best possible resolution that can be achieved. The size of the seeing disk varies from one observatory site to another. At Palomar and Kitt Peak, the seeing disk is roughly 1 arc sec. The best conditions in the world (seeing disk = $\frac{1}{4}$ arc sec) have been reported at the observatory

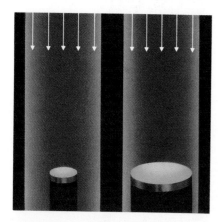

Figure 4-13   Light-gathering power
A large mirror intercepts more starlight than does a small mirror. Large mirrors therefore produce brighter images than small mirrors.

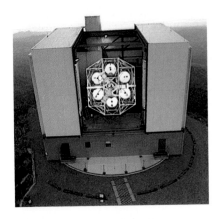

Figure 4-14 The Multiple-Mirror Telescope *This aerial photograph shows the six 1.8-m mirrors that together constitute the first multiple-mirror telescope. The total area of the six mirrors is equal to one 4.5-m mirror. (MMT)*

on top of Mauna Kea, a 14,000-foot volcano on the island of Hawaii. Only in the vacuum of outer space could the theoretical resolving power of a large telescope be achieved.

Significant engineering problems are associated with building large reflectors. Very large mirrors (more than about 4 m in diameter) are slabs of glass so heavy that the mirror's shape changes slightly as the telescope is turned toward different parts of the sky. The mirror actually sags under its own weight, thereby detracting from the sharpness of the focus and the quality of the resulting image. New techniques for building thin, light-weight mirrors should help alleviate this problem.

Another approach is to mount several smaller mirrors together and aim them at the same focal point. The Multiple-Mirror Telescope atop Mt. Hopkins in Arizona has six mirrors, each measuring 1.8 m (6 ft) in diameter, mounted together as shown in Figure 4-14. The total light-gathering power of this arrangement is equivalent to one $4\frac{1}{2}$-m mirror. The MMT, as it is called, thus really ranks as the third largest telescope in the world.

The MMT has proved so successful that astronomers around the world are now planning and building even larger multiple-mirror telescopes. For example, the Keck telescope will use 36 individually controlled hexagonal mirrors mounted together as shown in Figure 4-15. The result is equivalent to one mirror 10 m (32.8 ft) in diameter. This enormous telescope is under construction at the summit of Mauna Kea and will be in operation in the early 1990s.

Figure 4-15 The 10-meter Keck telescope *This model shows the design of a giant telescope currently under construction at the summit of Mauna Kea on the island of Hawaii. Thirty-six hexagonal mirrors, each measuring 1.8 m (5.9 ft) across, will together have the effect of one mirror 10 m in diameter. (California Institute of Technology)*

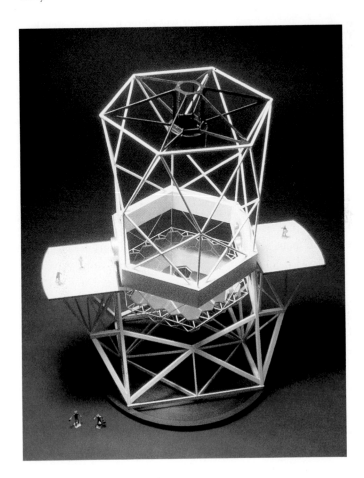

Electronic devices are often used to record the image at a telescope's focus

The invention of photography during the nineteenth century was a boon for astronomy. By taking a long exposure with a camera mounted at the focus of a telescope, an astronomer could record extremely faint features that could not be seen by looking through the telescope. Today, astronomical photography continues to reveal details in galaxies, star clusters, and nebulae.

Astronomers have long realized, however, that a photographic plate is not a very efficient light detector. Only 1 out of every 20 photons striking a photographic plate succeeds in triggering the chemical reaction in the photographic emulsion that is needed to produce an image. Thus, roughly 95 percent of the light falling onto a photographic plate is wasted.

The most sensitive and efficient light detector currently available to astronomers is the newly invented **charge-coupled device** (**CCD**). A CCD is a thin piece of silicon (see Figure 4-16), not unlike the silicon chips used in pocket calculators. The silicon wafer of a CCD is divided into an array of small light-sensitive squares called picture elements or, more commonly, **pixels**. For example, the latest CCDs have 640,000 pixels arranged in 800 columns by 800 rows in an area roughly the size of a postage stamp. When an image from a telescope is focused on the CCD, an electric charge builds up in each pixel in proportion to the intensity of the light falling on that pixel. When the exposure is finished, the amount of the charge on each pixel is read into a computer. From the computer, the image can be transferred onto ordinary photographic film or to a television monitor. Over certain wavelength ranges, nearly 75 percent of the photons falling on a CCD can be recorded.

Figure 4-17 shows one photograph and two CCD images of the same region of the sky, both taken with the same telescope. Notice that many details visible in the CCD images are totally absent in the ordinary photograph. In fact, the CCD pictures in Figure 4-17 show some of the faintest stars and galaxies ever recorded. Because of their extraordinary sensitivity and their use in conjunction with computers, CCDs are playing an increasingly important role in astronomy.

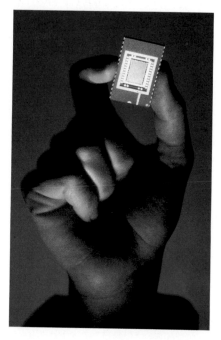

Figure 4-16 A charge-coupled device (CCD) *This tiny silicon rectangle contains 163,840 light-sensitive electric circuits that store images. At the end of each exposure, additional circuits in the silicon chip control the transfer and readout of the data to a waiting computer. (Smithsonian Institution Astrophysical Observatory)*

A radio telescope uses a large concave dish to reflect radio waves to a focus

Until recently, all information that astronomers could gather about the universe was based on ordinary visible light. With the discovery of nonvisible electromagnetic radiation, scientists began to wonder if objects in the universe might also emit radio waves, X rays, and infrared and ultraviolet radiation.

The first evidence of nonvisible radiation from outer space came from the work of a young radio engineer, Karl Jansky of Bell Telephone Laboratories. Using long antennas, Jansky was investigating the sources of radio static that affects short-wavelength radiotelephone communication. In 1932, he realized that a certain kind of radio noise is strongest when the constellation of Sagittarius is high in the sky. The center of our Galaxy is located in the direction of Sagittarius, and Jansky concluded that he was detecting radio waves from elsewhere in the Galaxy.

Astronomers were not quick to pursue this line of research. Only one person, Grote Reber (an electronics engineer living in Illinois), pursued the matter. In 1936, Reber built the first radio telescope in his backyard to map

a

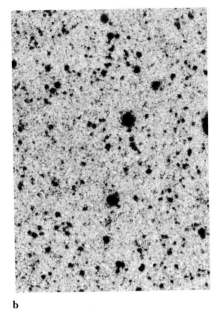
b

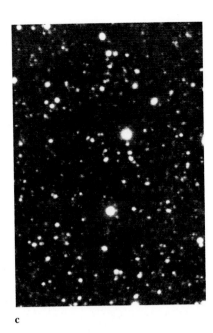
c

Figure 4-17 Ordinary photography versus a CCD image *These three views of the same part of the sky, each taken with the 4-m telescope shown in Figure 4-12, demonstrate the superiority of CCDs to ordinary photographic plates. (a) A negative print (black stars and white sky) of a 45-min exposure on a photographic plate. (b) The sum of fifteen 500-sec CCD images. Notice that many faint stars and galaxies virtually invisible in the ordinary photograph can be clearly seen in this CCD image. (c) This color view was produced by combining a series of CCD images taken through colored filters. The total exposure time was 6 hr. (Courtesy of P. Seitzer, NOAO)*

Figure 4-18 A radio telescope *The dish of this radio telescope is 45.2 m (140 ft) in diameter. It is one of several large instruments at the National Radio Astronomy Observatory near Green Bank, West Virginia. (NRAO)*

radio emission from the Milky Way. His design was modeled after an ordinary reflecting telescope, with a concave "dish" (reflecting antenna) measuring 9.1 m in diameter. The radio receiver at the focal point of the metal dish was tuned to a wavelength of 1.85 m.

By 1944, when Reber completed his map of the Milky Way, astronomers had begun to take notice of these developments. Shortly after World War II, radio telescopes began to spring up around the world. Radio observatories are today as common as major optical observatories.

Like Reber's prototype, the standard radio telescope has a large concave dish (see Figure 4-18). A small antenna tuned to the desired wavelength is located at the focus. The incoming signal is relayed to amplifiers and recording instruments, which are typically located in a room at the base of the telescope's pier.

a

b

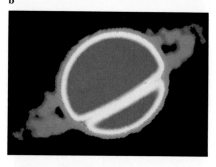

Figure 4-20   Optical and radio views of Saturn (a) This picture was taken by a camera on board a spacecraft as it approached Saturn. Sunlight reflecting from the planet's cloudtops and rings is responsible for this view. (b) This picture, taken by the VLA, shows radio emission from Saturn at a wavelength of 2 cm. Color is used to represent the intensity of radio emission. (NASA; NRAO)

Astronomers were at first not enthusiastic about detecting radio noise from space, in part because of the low resolving power of early radio telescopes. The resolving power of any telescope decreases with an increase in wavelength. In other words, the longer the wavelength, the fuzzier the picture. Because radio radiation has very long wavelengths, astronomers thought that radio telescopes could produce only blurry, indistinct views.

Very large radio telescopes can produce somewhat sharper radio images: the bigger the dish, the better the resolving power. For this reason, most modern radio telescopes have dishes more than 100 ft in diameter.

There is, however, another clever way to achieve high resolution. Two radio telescopes separated by many kilometers can be hooked together with electrical wires. This technique is called **interferometry** because the incoming radio signals are made to "interfere" or blend together, so that the combined signal is sharp and clear. The result is very impressive: the effective resolving power is equivalent to that of one gigantic dish with a diameter equal to the distance between the two telescopes.

Interferometry techniques were exploited for the first time in the late 1940s, and gave astronomers their first detailed views of radio objects in the sky. Radio telescopes separated by thousands of kilometers were linked together to give resolving power much higher than that of optical telescopes. This technique is called **very-long-baseline interferometry (VLBI).** The best possible resolution would be obtained by two telescopes on opposite sides of the Earth. In that case, features as small as 0.00001 arc sec could be distinguished at radio wavelengths—a resolution 100,000 times better than the sharpest pictures from ordinary optical telescopes.

One of the finest systems of radio telescopes began operating in 1980 in the desert near Socorro, New Mexico. Called the Very Large Array (VLA), it consists of 27 concave dishes, each 26 m (85 ft) in diameter. The 27 telescopes are arranged along the arms of a gigantic Y that covers an area 27 km (17 miles) in diameter. Only a portion of the VLA is shown in Figure 4-19. This system produces radio views of the sky with resolutions comparable to that of the very best optical telescopes.

Radio astronomers often use "false color" to display their radio views of astronomical objects. An example is shown in Figure 4-20. The most intense radio emission is shown in red, the least intense in blue. Intermediate colors of the rainbow represent intermediate levels of radio intensity. Black indicates that there is no detectable radio radiation. Astronomers working at other nonvisible wavelength ranges also frequently use false-color techniques to display views obtained from their instruments.

Telescopes in orbit around the Earth detect radiation that does not penetrate the atmosphere

As the success of radio astronomy began to mount, astronomers started exploring the possibility of making observations at other nonvisible wavelengths. Unfortunately, the Earth's atmosphere is opaque to many wavelengths. Very little radiation other than visible light and radio waves manages to penetrate the air we breathe.

Water vapor is the main absorber of infrared radiation from space. Infrared observatories are therefore located at sites of low humidity. For example, the 14,000-ft summit of Mauna Kea on Hawaii is exceptionally dry, and infrared observations are the primary function of NASA's 3-m telescope there.

Another possibility is to take a telescope up in an airplane. That is the basic idea behind the Kuiper Airborne Observatory (KAO) shown in Figure 4-21. The airplane carries a 1-m reflecting telescope to an altitude of 12 km (40,000 ft), placing the observatory above 99 percent of the atmospheric water vapor.

A telescope in Earth orbit offers the best arrangement. In 1983, the Infrared Astronomical Satellite (IRAS) was launched into a 900-km-high polar orbit. This satellite was designed around a small reflecting telescope that surveyed the entire sky at infrared wavelengths.

The European Space Agency (ESO) plans to launch an infrared telescope in the early 1990s. Later that same decade, NASA plans to launch SIRTF (Space Infrared Telescope Facility) into a circular orbit 10,000 km above the Earth (see Figure 4-22). Both telescopes will spend several years making observations at infrared wavelengths that never penetrate the Earth's atmosphere.

Figure 4-21 The Kuiper Airborne Observatory This C-141 jet airplane carries a 1-m reflecting telescope specifically designed for infrared observations. The observing portal (through which the telescope is aimed) can be seen on the fuselage just in front of the wing. (NASA)

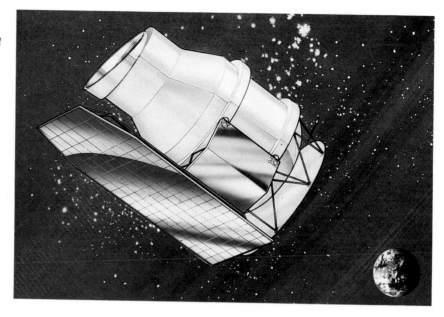

Figure 4-22 [right] The Space Infrared Telescope Facility (SIRTF) If all goes as planned, a rocket will carry this 5-ton infrared telescope into a high, circular orbit around the Earth in 1998. The telescope will have a mirror 95 cm (37 in.) in diameter and will be designed to operate for at least five years. (NASA)

Figure 4-23 [below] Orion as seen in ultraviolet, infrared, and visible wavelengths An ultraviolet view **(a)** of the constellation of Orion was obtained during a brief rocket flight on December 5, 1975. The 100-sec exposure covers the wavelength range 1250–2000 Å. The "false color" view **(b)** from IRAS uses color to display specific ranges of infrared wavelengths: red indicates long-wavelength radiation; green indicates intermediate-wavelength radiation; and blue shows short-wavelength radiation. For comparison, an ordinary optical photograph **(c)** and a star chart **(d)** are included. (Courtesy of George R. Carruthers, NRL; NASA; R. C. Mitchell, Central Washington University)

The best ultraviolet observations are also made from space. During the early 1970s, Apollo and Skylab astronauts carried small ultraviolet telescopes above the Earth's atmosphere to give us some of our first views of the ultraviolet sky. Small rockets have also been used to place ultraviolet cameras briefly above the Earth's atmosphere. A typical view is shown in Figure 4-23, along with a corresponding infrared view from IRAS, a view in visible light, and a star chart.

Some of the finest ultraviolet astronomy has been accomplished by the International Ultraviolet Explorer (IUE), which was launched in 1978. The satellite (see Figure 4-24) is built around a Cassegrain telescope with a 45-cm (18-in.) mirror and a total focal length of 6.74 m (22 ft). Observations cover the range from 1160 to 3200 Å.

a

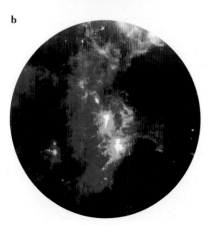

b

c

d

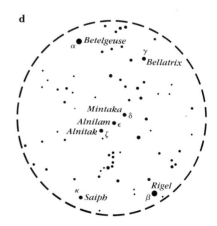

Figure 4-24 The International Ultraviolet Explorer (IUE) *Since its launch in 1978, this 671-kg satellite has produced superb observations in the far-ultraviolet. The dark blue panels at the midsection of the satellite are solar-cell arrays that provide electric power for the radio transmitters and other electronic equipment. (NASA)*

For decades, astronomers have dreamed of having a major observatory in space. Although satellites like the IRAS and IUE give excellent views of selected wavelength regions, astronomers are tantalized by the prospects of one very large telescope that could be operated at any wavelength—from the infrared through the visible range and out into the far-ultraviolet. This is the mission of the Hubble Space Telescope (see Figure 4-25) to be carried aloft by the Space Shuttle in 1990. This instrument will dominate astronomy for the rest of the twentieth century.

The primary mirror of the Hubble Space Telescope has a diameter of 2.4 m (7.8 ft). The resulting light-gathering power is so great and the sky in space is so dark that the telescope will be able to detect stars 40 times dimmer than those that can be photographed from Earth-based observatories where ambient light is scattered by the atmosphere.

Figure 4-25 The Hubble Space Telescope (HST) *The Space Shuttle will be used to place this 2.4-m Hubble Space Telescope into Earth orbit in 1990. During its anticipated 15-year lifetime, the telescope will be used to study the heavens over wavelengths from the infrared through the ultraviolet. (NASA)*

Figure 4-26 The Advanced X-ray Astrophysics Facility (AXAF) *If all goes as planned, this X-ray telescope will be placed into Earth orbit in the mid-1990s. Twelve cylindrical mirrors—the largest being about a meter long and 1¼ meters in diameter—will focus incoming X rays to produce images and observations of a wide variety of objects. (Courtesy of TRW)*

Because its images will not be degraded by the atmosphere, the actual resolution of the Hubble Space Telescope will be very close to the theoretical value. For example, when operated along with an auxiliary instrument called the faint-object camera, the telescope will have a resolution of 0.02 arc sec. That is a fiftyfold improvement over the best conditions at observatories such as Palomar, where the seeing disk is 1 arc sec.

The extraordinary light-gathering and resolving powers of the Hubble Space Telescope will permit astronomers to make observations that were unthinkable only a few years ago. For example, one observing program will have the Hubble Space Telescope search for new planets around other stars. Many familiar nebulae, star clusters, and galaxies whose photographs are used throughout this book will be seen with unprecedented clarity and brilliance. Currently unsuspected details and subtleties should be revealed.

Perhaps the greatest surprises will come from the discovery of totally new objects in space. Ever since Galileo turned his telescope toward the skies and saw four moons orbiting Jupiter, each new generation of astronomical instruments has disclosed previously unimagined objects and processes, often more bizarre than the strangest science fiction. It happened twice in the 1960s, when radio telescopes found quasars and pulsars, and again, in the 1970s, when X-ray telescopes detected bursters. With serendipity so commonplace, many astronomers predict that it will happen again in the 1990s.

Neither X rays nor gamma rays penetrate the Earth's atmosphere, and so observations at these extremely short wavelengths also must be done from space. Astronomers got their first quick look at the X-ray sky with brief rocket flights during the late 1940s. Several small satellites launched during the early 1970s viewed the entire X-ray and gamma-ray sky, revealing hundreds of previously unknown sources, including at least one black hole.

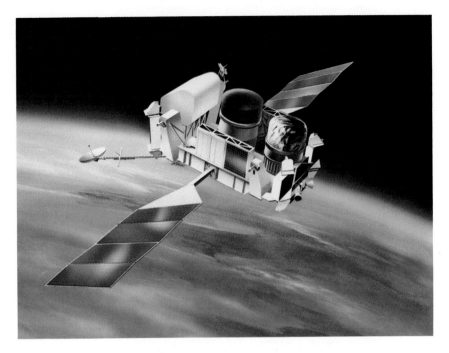

Figure 4-27 The Gamma Ray Observatory (GRO) *The best views of the high-energy gamma-ray sky will come from this satellite, which is scheduled for launch in 1990. The 15-ton satellite will be placed into Earth orbit by the Space Shuttle. (D. Kniffen, NASA; TRW)*

Although heroic in their day, these preliminary efforts pale in comparison to the detailed views and results from three huge satellites launched between 1977 and 1979. Called High Energy Astrophysical Observatories (HEAO), these satellites each carried an array of X-ray and gamma-ray detectors. Thousands of sources were discovered all across the sky. The second satellite in this series was especially successful in producing high-quality X-ray images of a wide range of exotic objects. This satellite was called the Einstein Observatory and was launched near the hundredth anniversary of Albert Einstein's birth. X-ray views from this observatory appear throughout this book.

A logical successor to the Einstein Observatory is AXAF, the Advanced X-Ray Astrophysics Facility, which is still in the planning stages at NASA. Like the Hubble Space Telescope, AXAF will be long-lived, adaptable, and will be controlled by astronomers on the ground. If this project receives funding in the near future, AXAF could be launched in the mid-1990s (see Figure 4-26).

Electromagnetic radiation with the shortest wavelengths and the most energy are gamma rays. Astronomers have had only their first tantalizing glimpses in this exotic region of the electromagnetic spectrum, and many hopes and expectations lie with the Gamma Ray Observatory (GRO) to be launched in 1990 (see Figure 4-27). It will carry four instruments capable of performing a variety of observations, ranging from the detection of objects that emit brief bursts of gamma rays to an all-sky survey of long-lived sources.

The advantages and benefits of these Earth-orbiting observatories cannot be overemphasized. We are no longer limited to the narrow ranges of wavelengths that manage to leak through our shimmering, hazy atmosphere (see Figure 4-28). For the first time we are really seeing the universe.

a Radio

b Infrared

c Visible

d Ultraviolet

e X ray

Figure 4-28   The entire sky at five wavelength ranges   These five views show the entire sky at **(a)** radio, **(b)** infrared, **(c)** visible, **(d)** ultraviolet, and **(e)** X-ray wavelengths. The Milky Way stretches horizontally across each picture. (Max Planck Institut für Radioastronomie; Jet Propulsion Laboratory; Griffith Observatory; Royal Observatory; E. Boldt, NASA)

Summary

· Electromagnetic radiation consists of vibrating electric and magnetic fields that carry energy through space at the speed of light (3×10^8 m/sec).

· Visible light, radio waves, microwaves, infrared and ultraviolet radiation, X rays, and gamma rays are all forms of electromagnetic radiation.

· Visible light forms only a small portion of the electromagnetic spectrum.

· The wavelength of light is associated with its color; wavelengths of visible light range from about 4000 Å for violet light to 7000 Å for red light.

· Infrared radiation, microwaves, and radio waves have wavelengths larger than those of visible light. Ultraviolet radiation, X rays, and gamma rays have wavelengths shorter than those of visible light.

· Refracting telescopes (refractors) produce images by bending light rays as they pass through glass lenses.

   Chromatic aberration is an optical defect whereby light of different wavelengths fail to come to a common focus.

· Reflecting telescopes (reflectors) produce images by reflecting light rays from concave mirrors to a focus point.

   Reflectors are not subject to many of the problems that limit the usefulness of refractors.

· Radio telescopes have large reflecting antennas (dishes) that are used to focus radio waves.

· Very sharp radio images are produced with combinations or arrays of radio telescopes with a technique called interferometry.

· The Earth's atmosphere is transparent to visible light and radio waves arriving from space, but it absorbs much of the radiation at other wavelengths. For observations at other wavelengths, astronomers depend upon telescopes carried above the atmosphere by high-altitude airplanes, rockets, or satellites.

· Satellite-based observatories are giving us a wealth of new information about the universe, permitting coordinated observation of the sky at all wavelengths.

Review questions

1   Give everyday examples of the phenomena of refraction and reflection.

2   What is chromatic aberration? What kinds of telescopes suffer from this defect, and how can it be corrected?

3   What is meant by the resolving power of a telescope?

4   What are the principal advantages of a reflector over a refractor?

5   What limits the resolving power of the 5-meter telescope at Palomar?

6   Why can radio astronomers observe at any time during the day whereas optical astronomers are mostly limited to nighttime observing?

7   Why must X-ray telescopes be placed above the Earth's atmosphere?

8   Why will very large telescopes of the future make use of multiple mirrors?

Advanced questions

9 Quite often, advertisements appear for telescopes that extol their magnifying abilities. Is this a good criterion for evaluating telescopes? Explain your answer.

10 This book was published at about the same time the Hubble Space Telescope was to be launched. Using whatever news sources you wish, investigate the current status of the HST. When was it launched? Are all its components working? Are there any problems? Have any observations been reported or pictures sent back?

Discussion questions

11 Discuss the advantages and disadvantages of using a small telescope in Earth orbit versus a large telescope on a mountain top.

12 If you were in charge of selecting a site for a new observatory, what factors would you consider important?

For further reading

Bok, B. "The Promise of the Space Telescope." *Mercury,* May/June 1983, p. 60.

Cohen, M. *In Quest of Telescopes.* Sky Publishing Corp. and Cambridge University Press, 1980.

Henbest, N., and Marten, M. *The New Astronomy.* Cambridge University Press, 1983.

Janesick, J., and Blouke, M. "Sky on a Chip: The Fabulous CCD." *Sky & Telescope,* September 1987, p. 238.

Learner, R. *Astronomy Through the Telescope.* Van Nostrand Reinhold, 1981.

Robinson, L. "Monster Telescopes for the 1990s." *Sky & Telescope,* May 1987, p. 495.

Schild, R. "Coloring the Electronic Sky." *Sky & Telescope,* February 1988, p. 144.

Smith, D. "From Aerobee to AXAF." *Sky & Telescope,* June 1987, p. 606.

Spradley, J. "The First True Radio Telescope." *Sky & Telescope,* July 1988, p. 28.

5

The nature of light and matter

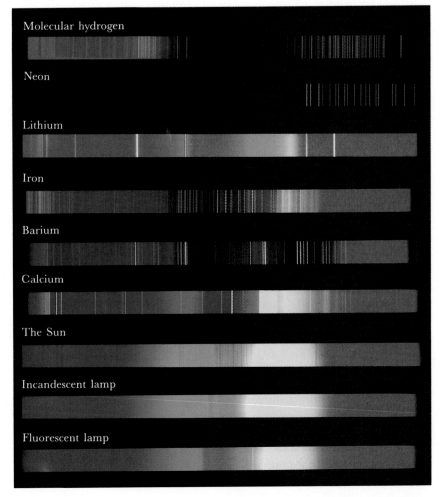

Spectra and spectral lines *When an electric spark passes through a gas, atoms of the gas emit radiation at certain wavelengths, and so a spectrum of that gas contains bright spectral lines. This photograph displays the spectra of various elements. Each element has a unique and distinctive pattern of spectral lines. Spectra of the Sun and two common, household lamps are also shown. The spectrum of an incandescent light bulb is quite like the continuous spectrum of a blackbody. (Courtesy of Bauch & Lomb)*

Our knowledge about the planets, stars, and galaxies comes primarily from analyzing the light they emit. We must therefore understand the nature of electromagnetic radiation and how it interacts with matter before turning to the study of the universe. This chapter examines the basic properties of light and the structure of atoms. We find that light has both wavelike and particlelike properties. We discover that the electromagnetic radiation emitted by an object is affected by the temperature of the object. We then examine the structure of atoms and see how each type of atom produces its own unique set of spectral lines. We also learn about spectrographs, machines that astronomers use to analyze starlight and the chemical composition of remote worlds.

Most of our knowledge about the universe comes from studying light arriving at the Earth from distant objects. When we speak of "light" from stars and galaxies, we mean the full **electromagnetic spectrum** (see Figure 4-3). The human eye responds only to a narrow band of wavelengths that we call visible light. Astronomers also detect gamma rays, X rays, ultraviolet and infrared radiation, microwaves, and radio waves emitted by astronomical objects.

Recall from Chapter 4 that electromagnetic radiation consists of oscillating electric and magnetic fields (see Figure 4-2) that move through empty space at a speed of nearly 300,000 km/sec (186,000 miles/sec). The distance between adjacent crests of an electromagnetic wave is called the **wavelength.**

All objects emit electromagnetic radiation. Stars like the Sun emit radiation primarily at visible wavelengths (from 4000 Å to 7000 Å). During the late 1800s, physicists discovered how the intensity and wavelengths of light emitted by an object depend on the object's temperature. Before we learn about these important discoveries, we need to know how scientists denote temperature.

Temperatures are usually given on the Celsius or Kelvin temperature scales

There are three temperature scales in common use today (see Figure 5-1). Temperatures are expressed throughout most of the world in degrees Celsius (°C). The **Celsius temperature scale** is based on the behavior of water, which freezes at 0°C and boils at 100°C at sea level. This scale, once known as the centigrade scale, was renamed in honor of the Swedish astronomer Anders Celsius, who proposed it in 1742.

For many purposes, scientists prefer to express temperatures in a unit called the **kelvin** (K), named after the British physicist William Thomson (Lord Kelvin), who made many important contributions to our knowledge about heat and temperature. On the Kelvin temperature scale, water freezes at 273 K and boils at 373 K. Because water must be heated through a change of either 100 K or 100°C to go from the freezing point to the boiling point, the "size" of a kelvin is the same as the "size" of a degree Celsius. A temperature expressed in kelvins is always equal to the temperature in degrees Celsius *plus* 273. Scientists usually prefer the Kelvin scale because of its physical interpretation of the meaning of temperature.

All substances are made of tiny **atoms** (typical atomic diameters are about 10^{-8} cm) that are constantly in motion. The temperature of a substance is directly related to the average speed of its atoms. If something is hot, its atoms are moving at high speeds. If a substance is cold, its atoms are moving much more slowly.

The coldest possible temperature is the temperature at which atoms move as slowly as possible (they can never quite stop completely). The minimum possible temperature, called **absolute zero,** is the starting point for the Kelvin scale. Absolute zero is 0 K, or −273°C.

In the United States, many people still use the archaic Fahrenheit scale, expressing temperatures in **degrees Fahrenheit** (°F). When the German physicist Gabriel Fahrenheit introduced this scale in the early 1700s, he intended 0°F to represent the coldest temperature then achievable (with a mixture of ice and saltwater) and 100°F to represent the temperature of a healthy human body. On the Fahrenheit scale, water freezes at 32°F and boils at 212°F.

Kelvin	Celsius	Fahrenheit	
1336	1063	1945	Melting point of gold
1234	961	1761	Melting point of silver
718	445	833	Boiling point of sulfur
373	100	212	Boiling point of water
273	0	32	Freezing point of water
90	−183	−297	Boiling point of oxygen
0	−273	−460	Absolute zero

Figure 5-1   Temperature scales
Three temperature scales are in common use: Kelvin, Celsius, and Fahrenheit. Scientists usually prefer the Kelvin scale because its starting point (0 K) is the coldest possible temperature—absolute zero.

Table 5-1   Various temperatures on different scales

	Kelvin (K)	Celcius (°C)	Fahrenheit (°F)
Absolute zero	0	−273	−460
Liquid helium boils	4	−269	−452
Liquid oxygen boils	90	−183	−297
Water freezes	273	0	32
"Room temperature"	295	22	72
Water boils	373	100	212
Sulfur boils	718	445	833
Iron melts	1808	1535	2795
Iron boils	3273	3000	5432
Carbon boils	5100	4827	8721

Table 5-1 shows various temperatures expressed on the Kelvin, Celsius, and Fahrenheit scales.

Many scientists who study the solar system express temperatures (such as the temperatures on the surfaces of the planets) using the Celsius scale. Temperatures relating to stars, galaxies, and other astronomical phenomena are, however, usually given in kelvins.

Any object emits electromagnetic radiation with intensity and wavelengths related to its temperature

The total amount of energy radiated by an object depends upon its temperature. The hotter the object, the more energy it emits as electromagnetic radiation. The dominant wavelength of the emitted radiation also depends upon the temperature. A cool object emits most of the energy at long wavelengths, whereas a hotter object emits most of the energy at shorter wavelengths.

These basic phenomena are familiar to anyone who has watched a welder or blacksmith heat a bar of iron (as shown in Figure 5-2). As it starts to heat up, the bar begins to glow with a deep red color. As the temperature rises, the bar begins to give off a brighter, reddish-orange light. At still higher

a

b
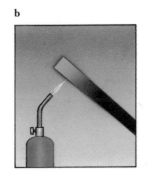

c

Figure 5-2   Heating a bar of iron
*This sequence of drawings (**a**→**b**→**c**) shows the changing appearance of a heated bar of iron as its temperature rises. As the temperature increases, the amount of energy radiated by the bar increases. The color of the bar also changes because, as the temperature goes up, the dominant wavelength of light emitted by the bar decreases. These effects are described by the Stefan–Boltzmann law and Wien's law.*

temperatures, it shines with a brilliant, yellowish-white light. If the bar could be prevented from melting and vaporizing, at very high temperatures it would emit a dazzling, blue-white light.

In 1879, the Austrian physicist Josef Stefan summarized the results of his experiments on this phenomenon by stating that *an object emits energy at a rate proportional to the fourth power of the object's temperature (measured in kelvins)*. In other words, if you double the temperature of an object (for example, from 500 to 1000 K), the energy emitted from the object's surface each second increases by a factor of $2^4 = 16$. If you triple the temperature (for instance, from 500 to 1500 K), the rate of energy emission increases by a factor of $3^4 = 81$.

Five years after Stefan announced this law, another Austrian physicist, Ludwig Boltzmann, showed how it could be derived mathematically from basic assumptions about atoms and molecules. The law today is commonly known as the **Stefan–Boltzmann law.**

The Stefan–Boltzmann law can be stated as a simple equation. If E is the energy emitted from each square centimeter of an object's surface each second, then

$$E = \sigma T^4$$

where T is the temperature of the object in kelvins and σ ("sigma") is a number called the Stefan–Boltzmann constant. The Stefan–Boltzmann law demonstrates that any object above absolute zero emits radiation.

Boltzmann showed that this law is best obeyed by an idealized object called a **blackbody.** A blackbody does not reflect any light; it instead absorbs all radiation falling on it. Since it does not reflect any light, the radiation that a blackbody does emit is entirely the result of its temperature. Ordinary objects are not perfect blackbodies—we can see them because they reflect some light—and thus the amount of energy they emit is slightly different from the amount calculated from the Stefan–Boltzmann law. Stars, however, do behave like blackbodies, and so astronomers can use the Stefan–Boltzmann law to relate a star's brightness to its surface temperature.

Any object emits radiation over a wide range of wavelengths, but there is always a particular wavelength (λ_{max}) at which the emission of energy is strongest. This dominant wavelength gives a glowing hot object its characteristic color.

In 1893, the German physicist Wilhelm Wien discovered that *the dominant wavelength (λ_{max}) of radiation emitted by a blackbody is inversely proportional to its temperature*. In other words, the hotter an object, the shorter the dominant wavelength of the electromagnetic radiation it emits. This relationship is today called **Wien's law.**

Wien's law can also be stated as a simple equation. If λ_{max} is measured in centimeters, then

$$\lambda_{max} = \frac{0.29}{T}$$

where T is the temperature of the blackbody measured in kelvins.

From the Stefan–Boltzmann law, we see that any object with a temperature above absolute zero (0 K) emits some electromagnetic radiation. From

Wien's law, we find that a very cold object with a temperature of only a few kelvins emits primarily microwaves. An object at "room temperature" (about 300 K) emits primarily infrared radiation. One with a temperature of a few thousand kelvins emits mostly visible light. Something with a temperature of a few million kelvins emits most of its radiation in the X-ray wavelengths.

Wien's law is very useful in computing the surface temperatures of stars because it does not require knowledge of the star's size or brightness; all we need to know is the dominant wavelength of the star's electromagnetic radiation.

Studies of blackbody radiation led to the discovery that light has particlelike properties

The Stefan–Boltzmann law and Wien's law describe only two basic properties of **blackbody radiation,** the electromagnetic radiation emitted by a hypothetical blackbody. A more complete picture is given by **blackbody curves** such as those in Figure 5-3. These curves show the relationship between the wavelength and the intensity of light emitted by a blackbody at a given temperature.

The total area under a blackbody curve is proportional to the energy emitted; the wavelength corresponding to the peak of the curve is the dominant wavelength λ_{max}. Note that the blackbody curves clearly illustrate both of the laws we have discussed: a cool object has a low curve that peaks at a long wavelength, and a hotter object has a much higher curve that peaks at a shorter wavelength.

Figure 5-4 shows how the intensity of sunlight varies with wavelength. Note that the peak of the curve is at a wavelength of about 0.5 μm = 5000 Å,

Figure 5-3 Blackbody curves
Three representative blackbody curves are shown here. Each curve shows the intensity of radiation at every wavelength emitted by a blackbody at a particular temperature. On this graph, wavelength is measured in micrometers (μm) where 1μm = 10,000 Å. The range of wavelengths of visible light is indicated.

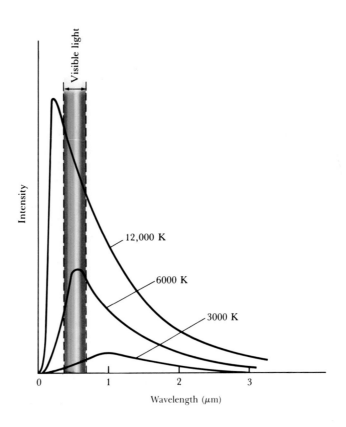

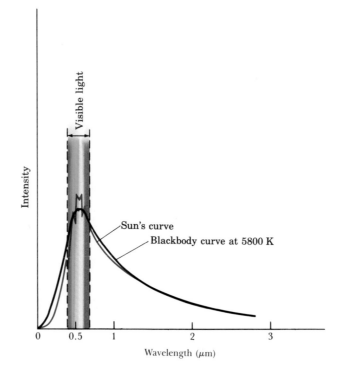

Figure 5-4 *The Sun as a blackbody*
This graph compares the intensity of sunlight over a wide range of wavelengths with the intensity of radiation from a blackbody at a temperature of 5800 K. Measurements of the Sun's intensity were made above the Earth's atmosphere. The Sun mimics a blackbody remarkably well.

which is nearly in the middle of the wavelength range for visible light. The blackbody curve for a temperature of 5800 K is also plotted in Figure 5-4. Note how closely the observed intensity curve for the Sun matches the blackbody curve. The close correlation between the observed intensity curves for most stars and the idealized blackbody curves is the reason why astronomers are justified in using the laws of blackbody radiation.

Despite these achievements, nineteenth century physicists were unable to explain the characteristic shape of blackbody curves. A breakthrough came in 1900, when the German physicist Max Planck discovered that he could derive a mathematical formula for the blackbody curves provided he assumed that electromagnetic radiation is emitted in separate packets of energy. This important assumption was verified in 1905 by Albert Einstein, who described light as consisting of particlelike packets called **photons.** *The energy carried by a photon of light is inversely proportional to its wavelength.* In other words, long-wavelength photons (such as radio waves and microwaves) carry little energy, whereas short-wavelength photons (like X rays and gamma rays) carry much more energy. This relationship between the energy of a photon and its wavelength is called **Planck's law.**

The energy of a photon is usually expressed in **electron volts (eV).** An electron volt is a tiny amount of energy, especially compared to the common unit of kilowatt-hours in which household electric energy is usually measured (a 1000-watt electric appliance consumes 1 kilowatt-hour of energy in 1 hour). Visible photons each carry between 2 and 3 electron volts, whereas 1 kilowatt-hour equals 2.25×10^{25} eV. Table 5-2 gives the energy carried by photons of various wavelengths.

Together, Planck's law and Wien's law relate the temperature of an object to the energy of the photons that it emits, as in Table 5-2. A cool object emits

Table 5-2 Some properties of electromagnetic radiation

	Wavelength (cm)	Photon energy (eV)	Blackbody temperature (K)
Radio	>10	$>10^{-5}$	<0.03
Microwave	10 to 0.01	10^{-5} to 10^{-2}	0.03 to 30
Infrared	0.01 to 7×10^{-5}	0.01 to 2	30 to 4100
Visible	7×10^{-5} to 4×10^{-5}	2 to 3	4100 to 7300
Ultraviolet	4×10^{-5} to 10^{-7}	3 to 10^{3}	7300 to 3×10^{6}
X ray	10^{-7} to 10^{-9}	10^{3} to 10^{5}	3×10^{6} to 3×10^{8}
Gamma ray	$<10^{-9}$	$>10^{5}$	$>3 \times 10^{8}$

NOTE: The symbol $>$ means "greater than." The symbol $<$ means "less than."

primarily long-wavelength photons that carry little energy, and a hot object gives off mostly short-wavelength photons that carry much more energy. In later chapters, we shall find these ideas very useful in understanding how stars of various temperatures interact with gas and dust in space.

Each chemical element produces its own unique set of spectral lines

Figure 5-5 The Sun's spectrum
Numerous spectral lines are seen in this photograph of the Sun's spectrum. The spectrum is so long that it had to be cut into convenient segments to fit on this page. (NOAO)

In 1814, the German master optician Joseph von Fraunhofer repeated Newton's classic experiment of shining a beam of sunlight through a prism (recall Figure 4-1), but he subjected the resulting rainbow-colored spectrum to intense magnification. Fraunhofer discovered that the solar spectrum contains hundreds of fine dark lines, which became known as **spectral lines.** Fraunhofer counted over 600 such lines, and today we know more than 20,000. Hundreds of spectral lines are visible in the photograph of the Sun's spectrum shown in Figure 5-5.

Half a century later, chemists discovered that they could produce spectral lines in laboratory experiments. Certain substances had long been identified

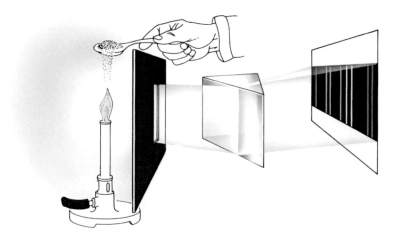

Figure 5-6   The Kirchhoff–Bunsen experiment   *In the mid-1850s, Kirchhoff and Bunsen discovered that when a chemical substance is heated and vaporized the resulting spectrum exhibits a series of bright spectral lines. In addition, they found that each chemical element produces its own characteristic pattern of spectral lines.*

by the distinctive colors emitted when they were sprinkled into a flame. About 1857, the German chemist Robert Bunsen invented a special gas burner that produces a clean, colorless flame. This Bunsen burner became very useful in the analysis of substances because its flame had no color of its own to confuse with the color produced by a substance sprinkled into it.

Bunsen's colleague, Prussian-born physicist Gustav Kirchhoff, suggested that light from the colored flames could best be studied by passing it through a prism (see Figure 5-6). They promptly discovered that the spectrum from a flame consists of a pattern of thin bright spectral lines against a dark background. They next found that *each chemical element produces its own characteristic pattern of spectral lines.* Thus was born in 1859 the technique of **spectral analysis,** the identification of chemical substances by their unique patterns of spectral lines.

A chemical **element** is a fundamental substance because it cannot be broken down into more basic chemicals. By the mid-1800s, chemists had conclusively identified such familiar elements as hydrogen, oxygen, carbon, iron, gold, silver, and so forth. Spectral analysis promptly lead to the discovery of additional elements, many of which are quite rare.

After Bunsen and Kirchhoff had recorded the prominent spectral lines of all the known elements, they soon began to discover other spectral lines in mineral samples. In 1860, for instance, they found a new line in the blue portion of the spectrum of mineral water. After chemically isolating the previously unknown element responsible for the line, they named it cesium (from the Latin *caesium,* "gray-blue"). The next year, a new spectral line in the red portion of the spectrum of a mineral sample led them to the discovery of the element rubidium (from *rubidium,* "red").

During a solar eclipse in 1868, astronomers found a new spectral line in light coming from the upper surface of the Sun while the main body of the Sun was hidden by the Moon. This line was attributed to a new element that was named helium (from the Greek *helios,* "Sun"). Helium was not actually discovered on the Earth until 1895, when it was identified in gases obtained from a uranium mineral.

A list of the chemical elements is most conveniently displayed in the form of a **periodic table** (see Figure 5-7). Each element is assigned a unique **atomic number** and the elements are arranged in the periodic table in that sequence. With a few exceptions, this sequence also corresponds to increas-

Figure 5-7 The periodic table of the elements *The periodic table is a convenient listing of the elements, arranged according to their weights and chemical properties.*

1 H																	2 He
3 Li	4 Be											5 B	6 C	7 N	8 O	9 F	10 Ne
11 Na	12 Mg											13 Al	14 Si	15 P	16 S	17 Cl	18 A
19 K	20 Ca	21 Sc	22 Ti	23 V	24 Cr	25 Mn	26 Fe	27 Co	28 Ni	29 Cu	30 Zn	31 Ga	32 Ge	33 As	34 Se	35 Br	36 Kr
37 Rb	38 Sr	39 Y	40 Zr	41 Nb	42 Mo	43 Tc	44 Ru	45 Rh	46 Pd	47 Ag	48 Cd	49 In	50 Sn	51 Sb	52 Te	53 I	54 Xe
55 Cs	56 Ba	57 La	72 Hf	73 Ta	74 W	75 Re	76 Os	77 Ir	78 Pt	79 Au	80 Hg	81 Tl	82 Pb	83 Bi	84 Po	85 At	86 Rn
87 Fr	88 Ra	89 Ac	104	105	106												

58 Ce	59 Pr	60 Nd	61 Pm	62 Sm	63 Eu	64 Gd	65 Tb	66 Dy	67 Ho	68 Er	69 Tm	70 Yb	71 Lu
90 Th	91 Pa	92 U	93 Np	94 Pu	95 Am	96 Cm	97 Bk	98 Cf	99 Es	100 Fm	101 Md	102 No	103 Lr

ing average mass of the atoms of the elements. Thus hydrogen (the symbol H) with atomic number 1 is the lightest element. Iron (Fe) has atomic number 26 and is a relatively heavy element. All the elements in a single vertical column of the periodic table have similar chemical properties. For example, the elements in the far right column are all gases at Earth-surface temperature and pressure, and they all tend to be very reluctant to react chemically with other elements.

In addition to nearly 100 naturally occurring elements, Figure 5-7 lists several artificially produced elements. All of them are heavier than uranium (the symbol U) and are highly radioactive, which means that they change into lighter elements within a short time after being created in the laboratory.

A spectrograph is an optical device that records a spectrum

Figure 5-8 Iron in the Sun's atmosphere *The upper spectrum is a portion of the Sun's spectrum from 4200 to 4300 Å. Numerous dark spectral lines are visible. The lower spectrum is a corresponding portion of the spectrum of vaporized iron. Several bright spectral lines can be seen against the black background. The fact that the iron lines coincide with some of the solar lines proves that there is some iron (albeit a very tiny amount) in the Sun's atmosphere. (Mount Wilson and Las Campanas Observatories)*

In the early 1800s, the French philosopher Auguste Comte argued that humanity would never know the nature and composition of the stars. Because they are so far beyond our earthly grasp, it seemed these remote celestial worlds would forever remain unfathomable and mysterious.

This pessimistic view was cast off when Bunsen and Kirchhoff discovered that each chemical element produces its own unique pattern of spectral lines. This discovery gave scientists the ability to determine the chemical composition of a remote astronomical object by identifying the spectral lines in its spectrum. For example, Figure 5-8 shows a portion of the Sun's spectrum along with the spectrum of iron. No other chemical can mimic iron's particular pattern of spectral lines at these wavelengths. It is iron's own distinctive "fingerprint." Since the spectral lines of iron appear in the Sun's spectrum, we can safely conclude the Sun's atmosphere contains some vaporized iron.

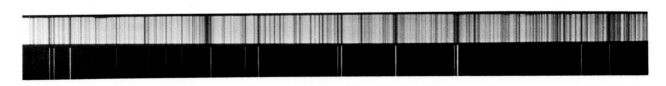

Figure 5-9 A prism spectrograph
This optical device uses a prism to break up the light of an object into a spectrum. Lenses focus that spectrum onto a photographic plate.

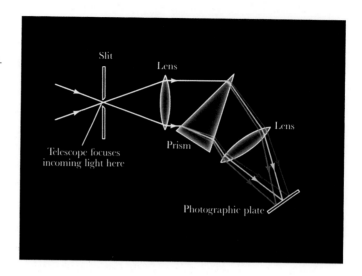

Bunsen and Kirchhoff collaborated in the design and construction of the first **spectroscope,** a device consisting of a prism and several lenses by which a spectrum could be magnified and examined. After the invention of photography, scientists preferred to produce a permanent photographic record of spectra. A device for photographing a spectrum is called a **spectrograph.** These machines have become one of the astronomer's most important tools, second only perhaps to the telescope itself.

In its basic form, a spectrograph consists of a slit, two lenses, and a prism arranged to focus the spectrum of an astronomical object onto a small photographic plate, as shown in Figure 5-9. This optical device typically is mounted at the focal point of a telescope, and the image of the object to be examined is focused on the slit. After the spectrum of a star or galaxy has been photographed, the exposed portion of the photographic plate is covered and light from a known source (usually an electric spark that vaporizes a small amount of iron) is focused on the slit. This exposes a "comparison spectrum" above and below the spectrum of the object being examined, as in Figure 5-10. The wavelengths of the spectral lines in the comparison spectrum are already known from laboratory experiments. These lines can therefore be used to identify and measure the wavelengths of the lines in the spectrum of the star or galaxy under study.

There are drawbacks to this old-fashioned spectrograph. A prism does not disperse the colors of the rainbow evenly. A prism compresses the red part of the spectrum, but spreads out the blue and violet portion. In addi-

Figure 5-10 A spectrogram
The photographic record of a spectrum is called a spectrogram. This spectrogram shows the spectrum of a star. Numerous dark lines can be seen against the brighter background of the spectrum. Above and below the star's spectrum are bright lines produced by an iron arc. These bright lines are the comparison spectrum. (Palomar Observatory)

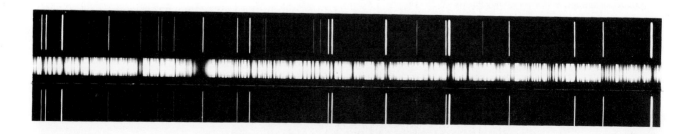

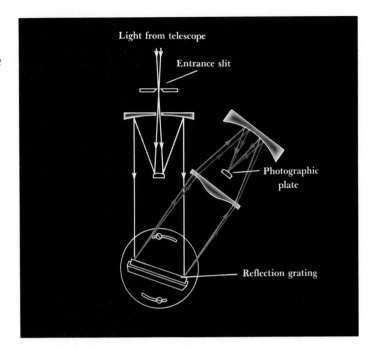

Figure 5-11 A grating spectrograph
This optical device uses a grating to break up the light of an object into a spectrum. An arrangement of lenses and mirrors focuses that spectrum onto a photographic plate.

Light from telescope

Entrance slit

Photographic plate

Reflection grating

tion, the blue and violet wavelengths must pass through more glass than the red wavelengths (examine Figure 4-1), resulting in an uneven absorption of light across the spectrum. Indeed, a glass prism is opaque to near-ultraviolet wavelengths.

A better device for breaking starlight into the colors of the rainbow is a **diffraction grating,** a piece of glass on which thousands of closely spaced lines are cut. Some of the finest diffraction gratings have as many as 10,000 lines per centimeter. These lines are usually cut by drawing a diamond back and forth across the piece of glass. The spacing of the lines must be very regular. Light rays leaving various parts of the grating interfere with each other so as to produce a spectrum. Figure 5-11 shows the design of a modern grating spectrograph.

In recent years, the television and electronics industries have produced a variety of light-sensitive devices that are significantly better than photographic film for recording spectra. For instance, many observatories now record spectra with a charge-coupled device, or CCD (see Figure 4-16). As described in Chapter 4, a CCD is a silicon chip approximately the size of a postage stamp divided into thousands of tiny squares. When light falls on one of these tiny squares (usually called a "pixel"), an electric charge builds up in proportion to the amount of light. When the exposure is finished, electronic equipment measures how much charge has accumulated in each pixel. The final result is a graph on which light intensity is plotted against wavelength. Dark lines in the rainbow-colored spectrum appear as depressions or valleys, while bright lines in the spectrum appear as peaks. For example, Figure 5-12 shows a CCD spectrum of Pluto. The prominent feature around 9000 Å is caused by methane, clearly indicating the presence of that substance (probably as methane ice because of Pluto's low temperature) on that planet's surface.

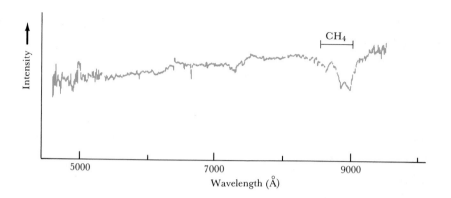

Figure 5-12 *A CCD spectrum of Pluto*
This graph shows the spectrum of Pluto as recorded by a charge-coupled device (CCD). The prominent absorption near 9000 Å is caused by methane (CH₄) and is conclusive evidence of the presence of that substance on the planet. (From observations by J. Apt, N. P. Carleton, and C. D. Mackay)

Spectral lines are bright or dark depending on conditions in spectrum's source

During his pioneering experiments with spectra, Kirchhoff noticed that sometimes he would see dark spectral lines, called **absorption lines,** among the colors of the rainbow. But a different experiment might give bright spectral lines, called **emission lines,** against an otherwise dark background. By the early 1860s, Kirchhoff's studies had progressed sufficiently for him to formulate three statements that describe the conditions under which various types of spectra are observed. These statements, called **Kirchhoff's laws,** are

Law 1   A hot object or a hot, dense gas produces a **continuous spectrum**—a complete rainbow of colors without any spectral lines.

Law 2   A hot, rarefied gas produces an **emission line spectrum**—a series of bright spectral lines against a dark background.

Law 3   A cool gas in front of a continuous source of light produces an **absorption line spectrum**—a series of dark spectral lines among the colors of the rainbow.

Figure 5-13 schematically illustrates Kirchhoff's three laws. The bright lines in the emission spectrum of a particular gas occur at exactly the same positions (wavelengths) as the dark lines in the absorption spectrum of that gas.

Kirchhoff's laws tell us that a gas absorbs light at specific wavelengths from white light passing through it, producing dark absorption lines among the colors of the continuous spectrum. The gas then gives up this energy by emitting light at precisely these same wavelengths in all directions, so that an observer at an oblique angle (without a light source in the background) sees bright emission lines.

Why do the atoms of a gas absorb light only at particular wavelengths? Why do they then emit light only at these same wavelengths? Traditional theories of electromagnetism could not answer these questions. The answers came early in the twentieth century with the development of quantum mechanics and nuclear physics.

An atom consists of a small, dense nucleus surrounded by electrons

The first important clue about the internal structure of atoms came from an experiment conducted in 1910 by Ernest Rutherford, a gifted chemist and physicist from New Zealand. Rutherford and his colleagues at the University

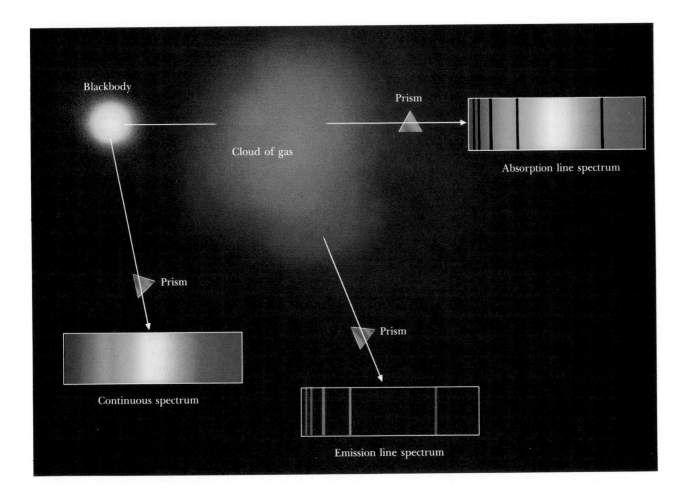

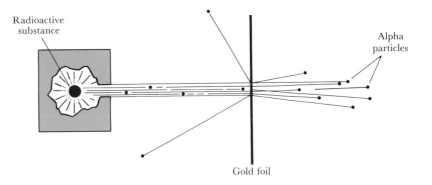

Figure 5-13   Kirchhoff's laws of spectral analysis *This schematic diagram summarizes Kirchhoff's laws. A hot, glowing object emits a continuous spectrum. If this source of light is viewed through a cool gas, dark absorption lines appear in the resulting spectrum. When the gas is viewed against a cold, dark background, bright emission lines are seen in the spectrum.*

of Manchester in England were investigating the recently discovered phenomenon of radioactivity. Certain radioactive elements such as uranium and radium were known to emit particles. One type of particle, called an alpha (α) particle, is quite massive (comparable to four hydrogen atoms) and is emitted from a radioactive substance with considerable speed.

In one series of experiments, Rutherford and his associates were using alpha particles as projectiles to probe the structure of solid matter. They directed a beam of alpha particles against a thin sheet of metal (see Figure 5-14). Almost all the alpha particles passed through the metal sheet with

Figure 5-14   Rutherford's experiment *Alpha particles from a radioactive source are directed against a thin metal foil. Most of the particles pass through the foil with very little deflection. Occasionally, however, a particle recoils dramatically, indicating that it has collided with the massive nucleus of an atom. This experiment provided the first evidence that atoms have nuclei.*

little or no deflection. To the surprise of the experimenters, however, an occasional alpha particle bounced back from the metal sheet as though it had struck something very dense. Rutherford later remarked, "It was almost as incredible as if you fired a 15-inch shell at a piece of tissue paper and it came back and hit you."

Rutherford was quick to realize the implications of this experiment. Most of the mass of an atom is concentrated in a compact, massive lump of matter that occupies only a small part of the atom's volume. The majority of the alpha particles pass freely through the nearly empty space that makes up most of the atom, but a very few particles happen to strike the dense mass at the center and bounce back.

Rutherford proposed a new model for the structure of an atom. According to this model, a massive, positively charged **nucleus** at the center of the atom is orbited by tiny, negatively charged electrons (see Figure 5-15). Rutherford concluded that at least 99.98 percent of the mass of an atom is concentrated in a nucleus whose diameter is only about one ten-thousandth the diameter of the atom.

We know today that the nucleus of an atom contains two types of particles: protons and neutrons. A proton has almost the same mass as a neutron and each is about 2000 times more massive than an electron. A proton has a positive electric charge and a neutron has no electric charge. The electrical attraction between the positively charged protons and the negatively charged electrons holds an atom together.

Normally, the number of electrons orbiting an atom is equal to the number of protons in the nucleus, making the atom electrically neutral. Furthermore, the number of protons in an atom's nucleus equals the atomic number for that particular element. Thus, a hydrogen nucleus has one proton, a helium nucleus has two, and so forth, up to uranium with 92 protons in its nucleus.

The number of protons in an atom's nucleus determines what element that atom is. The number of neutrons in the nuclei of the atoms of the same

Figure 5-15 Rutherford's model of the atom Electrons orbit the atom's nucleus, which contains most of the atom's mass. The nucleus contains two types of particles: protons and neutrons.

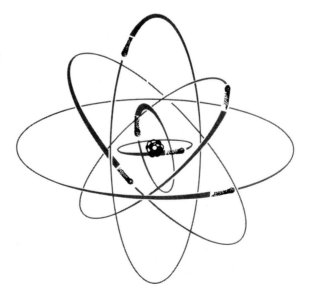

element may, however, vary. For example, oxygen, the eighth element on the periodic table, with an atomic number of 8, has exactly eight protons, but may have eight, nine, or ten neutrons. These three slightly different kinds of oxygen are called **isotopes.** The isotope with eight neutrons is by far the most abundant variety.

Spectral lines are produced when an electron jumps from one energy level to another within an atom

The task of reconciling Rutherford's atomic model with the observations of spectral analysis was undertaken by the young Danish physicist Niels Bohr, who joined Rutherford's group at Manchester in 1911.

Bohr began by trying to understand the structure of hydrogen, the simplest and lightest of the elements. A hydrogen atom consists of a single electron and a single proton. Hydrogen has a simple spectrum consisting of a pattern of lines that begins at 6563.1 Å and ends at 3645.6 Å. The first spectral line is called H_α, the second H_β, the third H_γ, and so forth, ending with H_∞ at 3645.6 Å. The closer you get to 3645.6 Å, the more spectral lines you see.

The regularity in this spectral pattern had been described mathematically in 1885 by Johann Jacob Balmer, an elderly Swiss schoolteacher. Balmer used trial and error to discover a formula with which the wavelengths of the hydrogen lines could be calculated. Since his formula is successful, the spectral lines of hydrogen at visible wavelengths are today called **Balmer lines** and the entire pattern from H_α to H_∞ is called the **Balmer series.** The spectrum of the star HD 193182 shown in Figure 5-16 exhibits more than two dozen Balmer lines from H_{13} through H_{40}.

Bohr realized that, if he were to succeed in understanding the structure of the hydrogen atom, he should be able to derive mathematically Balmer's formula from basic laws of physics. He began by assuming that the electron in a hydrogen atom orbits the nucleus only in certain specific orbits. As shown in Figure 5-17, it is customary to label these orbits $n = 1, n = 2, n = 3$, and so on. They are called the Bohr orbits.

Bohr argued that for an electron to jump from one orbit to another, the hydrogen atom must gain or lose a specific amount of energy. An electron jump from an inner orbit to an outer orbit would require energy. Conversely, an electron jump from an outer orbit to an inner one would release energy.

The energy gained or released by the atom when the electron jumps from one orbit to another is the difference in energy between these two orbits. According to Planck and Einstein, the packet of energy gained or released is a photon whose energy is inversely proportional to its wavelength. Using these ideas, Bohr mathematically derived the formula that Balmer had discovered by trial and error. Furthermore, Bohr's discovery elucidated the meaning of the Balmer series: all the Balmer lines are

Figure 5-16 Balmer lines in the spectrum of a star *This portion of the spectrum of a star called HD 193182 shows nearly two dozen Balmer lines. The series converges at 3645.6 Å, just to the left of H_{40}. This star's spectrum also contains the first 12 Balmer lines (H_α through H_{12}), but they are not visible in this particular spectrogram. (Mount Wilson and Las Campanas Observatories)*

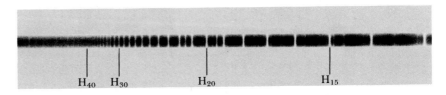

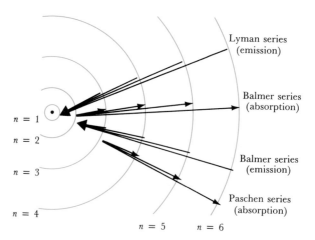

Figure 5-17 *The Bohr model of the hydrogen atom* An electron circles the nucleus in allowed orbits n = 1, 2, 3, and so on. A photon is absorbed by the atom as the electron jumps from an inner orbit to an outer orbit. A photon is emitted by the atom as the electron falls down to a low orbit. These absorptions and emissions of photons occur only at specific wavelengths, thereby producing characteristic patterns of lines in the hydrogen spectrum.

produced by electron transitions between the second Bohr orbit ($n = 2$) and higher orbits ($n = 3, 4, 5$, and so forth).

In addition to giving the wavelengths of the Balmer series, Bohr's formula correctly predicts the wavelengths of other series of spectral lines that occur at nonvisible wavelengths. For example, spectral lines occur because of electron transitions between the lowest Bohr orbit ($n = 1$) and all higher orbits ($n = 2, 3, 4, \ldots$). These transitions produce the **Lyman series,** which is entirely in the ultraviolet. This pattern of spectral lines begins with L_α ("Lyman alpha") at 1215 Å and converges on L_∞ at 912 Å. At infrared wavelengths is the **Paschen series,** which arises out of transitions to and from the third Bohr orbit ($n = 3$). It begins with P_α ("Paschen alpha") at 18,751 Å and converges on P_∞ at 8206 Å. Additional series exist at still longer wavelengths.

Bohr's ideas help explain Kirchhoff's laws. Each spectral line corresponds to one specific transition between the orbits of the atoms of a particular element. An absorption line is created when an electron jumps from an inner orbit to an outer orbit, extracting the required photon from an outside source of energy such as the continuous spectrum of a hot, glowing object. An emission line is produced when the electron falls back down to a lower orbit and gives up a photon.

Today's vision of the atoms owes much to the Bohr model, but it is different in certain ways. The modern picture is based on **quantum mechanics,** a branch of physics dealing with photons and subatomic particles that was developed during the 1920s. As a result of this work, physicists no longer imagine that electrons move in specific orbits about the nucleus. Instead, electrons are now said to occupy certain allowed **energy levels** in the atom. An extremely useful way of displaying the structure of an atom is with an **energy-level diagram,** such as that shown in Figure 5-18 for hydrogen. The lowest energy level, called the **ground state,** corresponds to the $n = 1$ Bohr orbit. An electron can jump from the ground state up to the $n = 2$ level only if the atom absorbs a Lyman-alpha photon of wavelength 1216 Å. As noted earlier, the energy of a photon is usually expressed in electron volts (eV). The Lyman-alpha photon has an energy of 10.19 eV, and so the energy level $n = 2$ is shown on the diagram as having an energy 10.19 eV above the energy of the ground state, conventionally assigned a value of 0 eV. Similarly, the $n = 3$ level is 12.07 eV above the ground state, and so forth up to the $n = \infty$ level at 13.6 eV. If the atom absorbs a photon of an energy greater

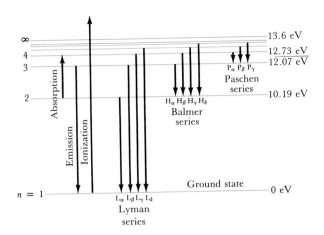

Figure 5-18 ***The energy-level diagram of hydrogen*** *The structure of the hydrogen atom is conveniently displayed in a diagram showing the allowed energy levels above the ground state. A variety of electron jumps, or transitions, are shown—including those that produce some of the most prominent lines in the hydrogen spectrum.*

than 13.6 eV, an electron from the ground state will be knocked completely out of the atom. This process, in which high-energy photons knock electrons out of atoms, is called **ionization.** In general, an atom that has been stripped of one or more electrons is called an **ion.**

With the work of people like Planck, Einstein, Rutherford, and Bohr, the interchange between astronomy and physics came full circle. Modern physics was born when Newton had set out to understand the motions of the planets. Two and a half centuries later, physicists in their laboratories discovered the basic properties of light and the structures of atoms. As we shall see in later chapters, the fruits of their labors have important astronomical applications.

The wavelength of a spectral line is affected by the relative motion between the source and the observer

Christian Doppler, a professor of mathematics in Prague, pointed out in 1842 that wavelength is affected by motion. As shown in Figure 5-19, light waves from an approaching light source are compressed. The circles represent waves emitted from consecutive positions as the source moves along. Because each successive wave is emitted from a position slightly closer to you, you see a shorter wavelength than you would if the source were stationary. All the spectral lines in the spectrum of an approaching source are shifted toward the short-wavelength (blue) end of the spectrum. This phenomenon is called a **blueshift.**

Conversely, light waves from a receding source are stretched out. You see a longer wavelength than you would if the source were stationary. All the spectral lines in the spectrum of a receding source are shifted toward the longer-wavelength (red) end of the spectrum, producing a **redshift.** In general, the effect of relative motion on wavelength is called the **Doppler effect.**

Suppose that λ_0 is the wavelength of a particular spectral line from a source that is not moving. It is the wavelength that you might look up in a reference book or determine in a laboratory experiment for this spectral line. If the source is moving, this particular spectral line is shifted to a different wavelength (λ). The size of the wavelength shift is usually written as $\Delta\lambda$, where $\Delta\lambda = \lambda - \lambda_0$. Thus $\Delta\lambda$ is the difference between the wavelength listed in reference books and the wavelength that you actually observe in the spectrum of a star or galaxy.

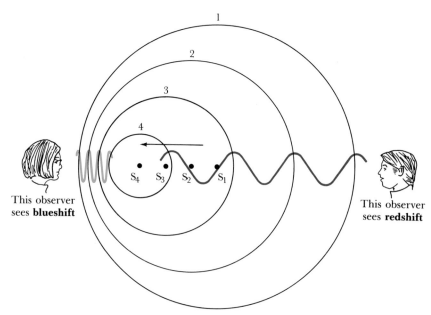

Figure 5-19 The Doppler effect
This diagram shows why wavelength is affected by motion between the light source and an observer. A source of light is moving toward the left. The four circles (numbered 1 through 4) indicate the location of light waves that were emitted by the moving source when it was at points S_1 through S_4, respectively. Note that the waves are compressed in front of the source, but stretched out behind the source. Consequently, wavelengths appear shortened (blueshifted) if the source is moving toward the observer. Wavelengths appear lengthened (redshifted) if the source is moving away from the observer. Motion perpendicular to an observer's line of sight does not affect wavelength.

Christian Doppler proved that the wavelength shift $\Delta\lambda$ is governed by the simple equation

$$\frac{\Delta\lambda}{\lambda_0} = \frac{v}{c}$$

where v is the speed of the source measured along the line of sight between the source and the observer. As usual, c is the speed of light (3×10^{10} cm/sec).

For example, the spectral lines of hydrogen appear in the spectrum of the bright star Vega as shown in Figure 5-20. The prominent hydrogen line H_α has a normal wavelength of 6562.85 Å, but in Vega's spectrum this line is located at 6562.55 Å. According to Doppler's formula, this wavelength shift of 0.30 Å corresponds to a speed of 13.7 km/sec. Since the wavelength of the H_α line has been shortened, Vega is coming towards us at that speed.

A speed determined in this fashion is called a **radial velocity,** because v is the component of the star's motion parallel to our line of sight, or along the "radius" drawn from Earth to the star. Of course, a sizable fraction of a star's motion may be perpendicular to our line of sight. This transverse movement, called **proper motion,** does not affect wavelengths.

The Doppler effect is a powerful tool that astronomers use in a wide range of situations to discover the motions of objects in space. For instance, careful spectroscopic studies of the Sun's surface reveal the speed of hot

Figure 5-20 The spectrum of Vega
Several Balmer lines are seen in this photograph of the spectrum of Vega, the brightest star in Lyra (the harp). The fact that all the spectral lines are shifted slightly toward the blue side of the spectrum indicates that Vega is approaching us. (NOAO)

gases as they rise and fall. Similarly, Doppler shift measurements of stars in double star systems give crucial data about the speeds of the stars along their orbits. Indeed, by measuring the redshifts of distant galaxies, we can determine the rate at which the entire universe is expanding. In later chapters we shall refer to the Doppler effect whenever we need to convert an observed wavelength shift into a speed.

Summary

- Electromagnetic radiation (including visible light) has wavelike properties and travels through empty space at the constant speed $c = 3 \times 10^8$ m/sec.

- The Stefan–Boltzmann law relates the temperature of an object to the rate at which it radiates energy.

- Wien's law relates the temperature of an object to the dominant wavelength at which it radiates energy.

- A blackbody is a hypothetical object that is a perfect absorber of electromagnetic radiation at all wavelengths. Since a blackbody does not reflect any light from outside sources, the radiation that it does emit depends only on its temperature.

  Stars closely approximate the behavior of blackbodies.

- The intensities of radiation emitted at various wavelengths by a blackbody at a given temperature are shown by a blackbody curve.

- When Planck assumed that energy is emitted or absorbed in discrete packets—later called photons—it became possible to explain the shapes of blackbody curves.

- Planck's law relates the energy of a photon to its wavelength. Kirchhoff's three laws of spectral analysis describe the conditions under which absorption lines, emission lines, or a continuous spectrum can be observed.

- Spectroscopy, the study of spectra, provides information about the chemical composition of remote astronomical objects.

- Spectral lines serve as distinctive "fingerprints" for detecting elements and chemical compounds in a light source.

- An atom consists of a small, dense nucleus (composed of protons and neutrons) surrounded by electrons that occupy only certain allowed energy levels.

- The spectral lines of a particular element correspond to the various electron transitions between allowed energy levels in the atoms of the element.

  When an electron jumps from one energy level to another, a photon of the appropriate energy (and hence a specific wavelength) is absorbed or emitted by the atom.

- The spectrum of hydrogen at visible wavelengths consists of the Balmer series, which arises from electron transitions between the second energy level of the hydrogen atom and higher levels.

- The process in which an atom absorbs an energetic photon and loses an electron entirely is called ionization.

- The spectral lines of an approaching light source are shifted toward short wavelengths (a blueshift); the spectral lines of a receding light source are shifted toward long wavelengths (a redshift).

- The Doppler effect explains that the size of a wavelength shift is proportional to the radial velocity between the light source and the observer.

Review questions

1 What is a blackbody? What does it mean to say that a star behaves like a blackbody? If stars behave like blackbodies, why are they not black?

2 What is the Stefan–Boltzmann law? Why do you suppose that astronomers are interested in it?

3 What is Wien's law? Of what use might it be to astronomers?

4 Using Wien's law and the Stefan–Boltzmann law, explain the color and brightness changes that are observed as the temperature of a hot, glowing object increases.

5 Describe the experimental evidence that supported the Bohr model of the atom.

6 Explain how the spectrum of hydrogen is related to the structure of the hydrogen atom.

7 Explain why the Doppler effect tells us only about the motion directly along the line of sight between a light source and an observer.

Advanced questions

* 8 Approximately how many times around the world could a beam of light travel in one second?

* 9 Imagine a star the same size as the Sun, but with a surface temperature twice that of the Sun. At what wavelength does that star emit most of its radiation? How many times brighter than the Sun is that star?

*10 Imagine a star the same size as the Sun with a surface temperature of 2900 K. Suppose both the Sun and this star were located at the same distance from you. Which would be brighter? By how much?

11 Describe how the spectrum of a helium atom might appear if one of its two electrons were stripped off.

*12 Imagine driving down a street toward a traffic light. How fast would you have to go so that the red light would appear green?

Discussion questions

13 Compare chemical identification based on spectral line patterns with the identification of people by line patterns in their fingerprints.

14 Suppose you look up at the night sky and observe some of the brightest stars with your naked eye. Is there any way of telling which stars are hot and which are cool? Explain.

For further reading

Cline, B. *Men Who Made a New Physics*. Signet, 1965.

Gingerich, O. "Unlocking the Chemical Secrets of the Cosmos." *Sky & Telescope*, July 1981, p. 13.

Rublowsky, J. *Light*. Basic Books, 1964.

Sobel, M. *Light*. University of Chicago Press, 1987.

van Heel, A., and Velzel, C. *What Is Light?* McGraw-Hill, 1968.

Weymann, R. "Extending the Visible Frontier: New Tools of the Optical Astronomer." *Mercury*, September/October 1975, p. 2.

6 Our solar system

Nebulae in Sagittarius *Planets are probably forming along with new stars in these nebulae in Sagittarius. The type of planet that forms at a particular distance from a star depends on such conditions as the temperature there and the substances— rock fragments, ice crystals, and gases— at that distance. In the solar system, planets composed primarily of rock formed near the Sun, but its heat drove off ices and gases. Far from the Sun, where temperatures are low, planets retained volatile substances, resulting in worlds composed primarily of gas. (Royal Observatory, Edinburgh)*

Modern instruments and space probes have given us a rich body of information about the solar system. We begin this chapter with a survey of the major physical characteristics of the planets. We find that they fall into two distinct classes: the inner (terrestrial) planets and the outer (Jovian) planets, with Pluto an oddity not easily classified. We also see that our Moon and six other giant satellites are large enough to be considered planets in their own right. We then discuss the materials of which the planets are composed and we learn that these substances were left over from stars that existed before the Sun was born. We outline current theories about the origin of the solar system that explain that the formation of the planets must have involved the collisions between many chunks of rock and ice that orbited the protosun 4.5 billion years ago. Finally, we explore the possibility that our Moon was formed from material torn from the Earth during one such collision with a planet-sized rock.

Looking up at the heavens and wondering about the nature and origin of the Sun, Moon, stars, and planets is a universal human experience. Unlike our ancestors, however, we possess a wealth of information. Especially within the past few decades, telescopic observations and interplanetary spacecraft have given us vast quantities of data from which to formulate and test theories. The Sun, planets, moons, asteroids, comets, and meteoroids that make up our tiny niche in the universe are finally accessible. Many of these objects are exceedingly ancient and contain records of the cosmic events that created our solar system.

In addition to gleaning information from the objects that orbit the Sun, we can observe active star formation occurring elsewhere in our Galaxy. Stars and planetary systems are now being formed in many beautiful nebulae scattered across the heavens. A general understanding of star creation coupled with knowledge of the Sun and its satellites gives us a reasonably comprehensive picture of how the solar system was created. Many details still need to be worked out, but the overall scenario seems remarkably sound. For the first time, we can truly appreciate what is unique and what is commonplace about our world. We have begun to fathom our connection with the rest of the cosmos.

The planets are classified as terrestrial or Jovian by their physical attributes

A brief overview of the solar system distinguishes two classes of planets. First, notice the striking dichotomy in the orbits of the planets as shown in Figure 6-1. The orbits of the four inner planets (Mercury, Venus, Earth, and Mars) are crowded close to the Sun. In contrast, the orbits of the next four

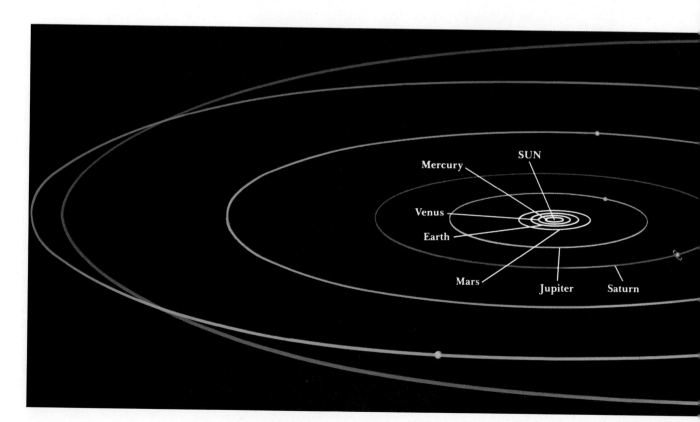

planets (Jupiter, Saturn, Uranus, and Neptune) are widely spaced at great distances from the Sun.

As you might expect, the range of surface temperatures that each planet experiences is related to its distance from the Sun. The four inner planets are quite warm. For example, noontime temperatures on Mercury climb to 600 K (=327°C = 621°F). At noon in the middle of summer on Mars, it is sometimes as warm as 300 K (=27°C = 81°F). Of course, the outer planets, which receive much less solar radiation, are much cooler. Typical temperatures range from about 150 K (=−123°C = −189°F) in Jupiter's cloudtops to 63 K (=−210°C = −346°F) on Neptune. Temperature plays a major role in determining whether substances exist as solids, liquids, or gases, thus profoundly affecting the appearance of the planets.

Most of the planets' orbits are nearly circular. As discussed in Chapter 3, Kepler discovered that these orbits are actually ellipses. Astronomers denote the elongation of an ellipse by its **eccentricity** (the eccentricity of a circle is zero). Most planets have orbital eccentricities that are very close to zero. The notable exception is Pluto, with an orbital eccentricity of 0.25. Because of its highly noncircular orbit, Pluto is sometimes closer to the Sun than Neptune is.

The planetary orbits all lie in nearly the same plane. In other words, the orbits of the planets are inclined at only small angles to the plane of the ecliptic. Again, however, Pluto is a notable exception. The plane of Pluto's orbit is tilted at 17° to the plane of the Earth's orbit (see Table 6-1).

As we compare the physical properties of the planets, we again find that they fall naturally into two groups—the four inner planets and the four

Figure 6-1   The solar system   *This scale drawing shows the distribution of planetary orbits around the Sun. Four inner planets are crowded close to the Sun, with five outer planets orbiting at much greater distances from the Sun.*

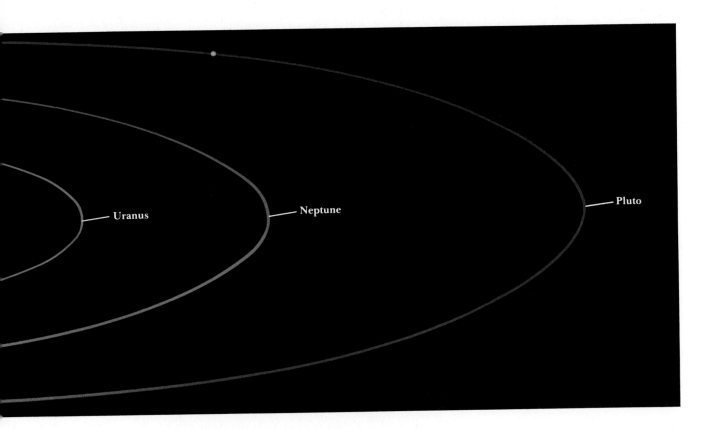

Uranus

Neptune

Pluto

Table 6-1 Orbital characteristics of the planets

	Mean distance from Sun		Orbital (sidereal) period (years)	Eccentricity	Inclination to the ecliptic (degrees)
	(AU)	(10^6 km)			
Mercury	0.39	58	0.24	0.206	7.0
Venus	0.72	108	0.62	0.007	3.4
Earth	1.00	150	1.00	0.017	0.0
Mars	1.52	228	1.88	0.093	1.8
Jupiter	5.20	778	11.86	0.048	1.3
Saturn	9.54	1427	29.46	0.056	2.5
Uranus	19.18	2870	84.01	0.047	0.8
Neptune	30.06	4497	164.79	0.009	1.8
Pluto	39.44	5900	247.7	0.250	17.2

outer planets—with Pluto once more an exception. For instance, let's examine the important properties of size, mass, and density.

The four inner planets are quite small. The Earth, with a diameter of 12,760 km (7930 miles) is the largest. In sharp contrast, the four outer planets are much larger. First place goes to Jupiter, whose equatorial diameter is 143,800 km (89,400 miles). Pluto, again the anomaly, is even smaller than the inner planets, despite its position as the outermost planet. Its diameter is only about 2300 km (1400 miles), roughly two-thirds the size of our Moon. Figure 6-2 shows the Sun and the planets drawn to the same scale. The diameters of the planets are given in Table 6-2.

A planet's mass is most easily determined if that planet has a satellite. The satellite's orbit obeys Newtonian mechanics, which relate the orbit to the

Figure 6-2 The Sun and the planets
This drawing shows the nine planets in front of the disk of the Sun, with all 10 bodies drawn to the same scale. The four planets that have orbits nearest the Sun (Mercury, Venus, Earth, and Mars) are small and made of rock. The next four planets (Jupiter, Saturn, Uranus, and Neptune) are large and composed primarily of gas.

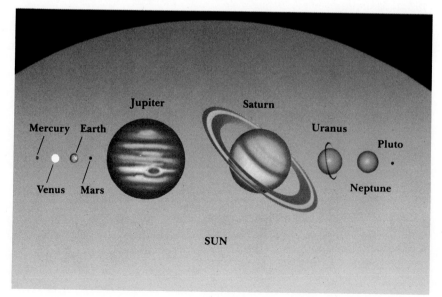

Table 6-2 Physical characteristics of the planets

	Diameter		Mass		Average density (g/cm³)
	(km)	(Earth = 1)	(gm)	(Earth = 1)	
Mercury	4,880	0.38	3.3×10^{26}	0.06	5.4
Venus	12,100	0.95	4.9×10^{27}	0.82	5.2
Earth	12,760	1.00	6.0×10^{27}	1.00	5.5
Mars	6,800	0.53	6.4×10^{26}	0.11	3.9
Jupiter	143,800	11.27	1.9×10^{30}	317.89	1.3
Saturn	120,000	9.44	5.7×10^{29}	95.15	0.7
Uranus	51,800	4.06	8.7×10^{28}	14.54	1.2
Neptune	49,500	3.88	1.0×10^{29}	17.23	1.7
Pluto	2,300	0.18	10^{25}	0.002	1.8

Figure 6-3 A terrestrial planet *Mars is a typical terrestrial planet. Its diameter is only 6800 km and it has an average density of 3.9 g/cm³, indicating that it is composed of rock. Volcanoes, canyons, and craters can be seen in this photograph taken by the* Viking 2 *spacecraft in 1976. (NASA)*

planet's mass. Astronomers can observe the satellite and measure its orbital period and semimajor axis. With this information, they can calculate the planet's mass.

If the planet does not have a satellite, astronomers can track a comet or spacecraft as it passes near the planet. The planet's gravity, which is directly related to its mass, deflects the path of the comet or spacecraft. By measuring the size of this deflection and using Newtonian mechanics, astronomers can determine the planet's mass. The four inner planets have low masses, the next four planets have substantially larger masses. Again, Jupiter is first, with a mass 318 times greater than the mass of the Earth. The masses of the planets are given in Table 6-2.

Average density (mass divided by volume) is a physical property that can often be used to deduce important information about the composition of an object. Scientists commonly express average density in grams per cubic centimeter, as in Table 6-2. The four inner planets have large average densities. The average density of the Earth, 5.5 g/cm³, may not seem large until you know that the average density of a typical rock is only about 3 g/cm³ and the average density of water is 1 g/cm³. The Earth must therefore contain a large amount of material that is more dense than rock. This information provides our first clue that some of the smaller planets, like the Earth, have iron cores.

In sharp contrast, the outer planets have very low densities. Indeed, Saturn is less dense than water. This strongly suggests that the giant outer planets are composed primarily of such light elements as hydrogen and helium. Again, Pluto is an exception. Although it is even smaller than the dense inner planets, its average density seems to be much like those of the giant outer planets. Table 6-2 lists the average densities of the planets.

These differences in size, mass, and density lead us to consider the four inner and four outer planets as two distinct groups. The four inner planets are called **terrestrial planets** because they resemble the Earth. They are small and dense, with craters, canyons, and volcanoes being common on their hard, rocky surfaces (see Figure 6-3). The four outer planets are called

Figure 6-4   A Jovian planet   Jupiter is the largest of the Jovian planets. Its equatorial diameter is 143,800 km. Its average density is only 1.3 g/cm³, indicating that the planet is primarily composed of light elements. Two of Jupiter's moons, Io and Europa, are seen in this view taken by the Voyager 1 spacecraft in 1979. Each of these satellites is approximately the same size as Earth's moon. (NASA)

Figure 6-5   A comet   The solid part of a comet is chunk of ice roughly 10 km in diameter. When a comet passes near the Sun, solar radiation vaporizes some of the comet's ices and the resulting gases form a tail millions of kilometers long. This photograph shows a comet that was seen in January 1974. (NASA)

Jovian planets because they resemble Jupiter. Vast swirling cloud formations dominate the appearance of these enormous gaseous spheres (see Figure 6-4). The solid cores of these planets are probably not much bigger than the Earth, buried beneath atmospheres that are tens of thousands of kilometers thick.

Although it is usually called a planet, Pluto clearly is an oddity. Its physical properties are not typical of either the terrestrial or the Jovian planets. Some astronomers speculate that Pluto is a moon that escaped from Neptune. Many of the satellites of the outer planets are composed of ice whose density is about 1 g/cm³. This is also the density of Pluto; hence Pluto may be a large ball of ice.

In addition to the nine planets, many smaller objects orbit the Sun. Between the orbits of Mars and Jupiter are thousands of small rocks (typical diameters about 40 km) called **asteroids.** The largest asteroid, Ceres, has a diameter of about 750 km. Quite far from the Sun, well beyond the orbit of Pluto, are chunks of ice called **comets.** Many comets have highly elongated orbits that occasionally bring them close to the Sun. When this happens, the Sun's radiation vaporizes some of the comet's ices, thereby producing a long flowing tail (see Figure 6-5).

Astronomers believe that asteroids and comets are debris left over from the formation of the solar system. In the inner regions of the solar system, rocky fragments have been able to endure continuous exposure to the Sun's heat. Far from the Sun, chunks of ice have survived for billions of years. Thus debris in the solar system naturally divides into two families (asteroids and comets) arranged according to distance from the Sun just as the two categories of planets (terrestrial and Jovian) are.

Seven large satellites resemble terrestrial planets

All the planets except Mercury and Venus have satellites. More than 40 satellites are known (Jupiter, Saturn, and Uranus each have at least 15) and certainly dozens of others await discovery. The known satellites fall into two distinct categories. There are seven giant satellites (listed in Table 6-3) that are roughly comparable to Mercury in size. The other satellites are much smaller, with diameters of less than 2000 km.

The recent interplanetary missions of *Voyager 1* and *Voyager 2* have revealed many fascinating characteristics of the giant satellites (see Figure 6-6). For example, Jupiter's satellite Io is one of the most geologically active worlds in the solar system, with numerous volcanoes belching sulfur-rich compounds. Saturn's largest satellite, Titan, has an atmosphere nearly twice as dense as Earth's.

Table 6-3   The seven giant satellites

Satellite name	Parent planet	Diameter (km)	Average density (g/cm³)
Moon	Earth	3480	3.3
Io	Jupiter	3630	3.6
Europa	Jupiter	3130	3.0
Ganymede	Jupiter	5280	1.9
Callisto	Jupiter	4820	1.8
Titan	Saturn	5120	1.9
Triton	Neptune	2720	2.0

Figure 6-6 Giant satellites *Jupiter's four largest satellites are shown here along with our Moon and Saturn's largest moon, Titan. All six worlds are reproduced to the same scale. Io has numerous active volcanoes and Europa has a smooth icy surface. Both are roughly the same size as our Moon. Ganymede and Callisto are covered with a 1000-km thick layer of ice. They are roughly the same size as Mercury. Titan has a thick methane-rich atmosphere. The remaining large satellite of the solar system, Triton, is not shown.*
(S. P. Meszaros; NASA)

Some scientists like to classify the seven giant satellites as planets, even though they do not orbit the Sun independently. Their hard surfaces (of either rock or ice) would place them in the terrestrial category. This expanded definition of a terrestrial planet gives us 10 Earthlike worlds to compare with our own planet.

The relative abundance of the elements is the result of cosmic processes

Some elements are very common, others are quite rare. Hydrogen is by far the most abundant substance, making up nearly three-quarters of the mass of all the stars and galaxies in the universe.

Helium is the second most abundant element. Together hydrogen and helium account for 98 percent of the mass of all the material in the universe—leaving only 2 percent for all the other elements combined.

There is a good reason for this overwhelming abundance of hydrogen and helium. Most astronomers believe that the universe began roughly 20 billion years ago with a violent event called the Big Bang. Only the lightest elements—hydrogen, helium, and a tiny amount of lithium—emerged from the enormously high temperatures following this cosmic event. All the other elements were manufactured deep inside stars at later times. In Chapters 13 and 14, we shall explore the processes by which elements are created at a star's center. For the moment, let us simply note that if it were not for the nuclear reactions inside stars, there would be no heavy elements in the universe today.

Near the ends of their lives, stars cast much of their matter out into space. This process can be a comparatively gentle one in which a star's outer layers are gradually expelled. Figure 6-7 shows a star that is losing material in this fashion. Alternatively, a star may end its life with a spectacular detonation called a supernova explosion, which blows the star apart. Either way, the interstellar gases in the galaxy become enriched with heavy elements dredged up from the dying star's interior, where they were created. New

Figure 6-7 A mass-loss star *This star is shedding material rapidly. The nebulosity around the star is caused by starlight reflected from dust grains. These grains may have condensed from the material cast off by the star. (Anglo-Australian Observatory)*

Table 6-4 Solar abundances of the most common elements

Atomic number	Symbol	Element	Relative abundance
1	H	Hydrogen	10^{12}
2	He	Helium	6×10^{10}
6	C	Carbon	4×10^{8}
7	N	Nitrogen	9×10^{7}
8	O	Oxygen	7×10^{8}
10	Ne	Neon	4×10^{7}
12	Mg	Magnesium	4×10^{7}
14	Si	Silicon	5×10^{7}
16	S	Sulfur	2×10^{7}
26	Fe	Iron	3×10^{7}

Figure 6-8 A dusty region of star formation *These young stars in the constellation of Orion are still surrounded by much of the gas and dust from which they formed. The bluish wispy nebulosity is caused by starlight reflecting off of abundant interstellar dust grains. These grains are made of heavy elements produced by earlier generations of stars. (Anglo-Australian Observatory)*

stars that form out of this enriched material thus have an ample supply of heavy elements from which to develop a system of planets, satellites, asteroids, and comets.

Our solar system is made of matter created in stars that existed billions of years ago. The Sun is a fairly young star, only 5 billion years old. All the heavy elements in our solar system were created and cast off by ancient stars during the first 10 billion years of our Galaxy's existence. We are literally made of star dust (see Figure 6-8).

Stars create the different elements in different amounts. For example, the elements carbon, oxygen, silicon, and iron are readily produced in a star's interior, whereas gold is created only under special circumstances. Thus, gold is rare in the solar system but carbon is not.

A convenient way to express the abundance of the various elements is to say how many atoms of a particular element are found for every trillion (that is, 10^{12}) hydrogen atoms in space. For example, for every trillion hydrogen atoms there are about 60 billion (6×10^{10}) helium atoms. From chemical analysis of Earth rocks, Moon rocks, and meteorites, scientists have been able to determine the relative abundances of the elements in our part of the galaxy. The most abundant elements are listed in Table 6-4.

In addition to these ten very common elements, five others are moderately abundant: sodium, aluminum, argon, calcium, and nickel. These elements have abundances in the range of 10^{6} to 10^{7} relative to the standard trillion hydrogen atoms. All other elements are much rarer. For instance, for every trillion hydrogen atoms in the solar system there are only six atoms of gold.

The planets were formed by the accumulation of material in the solar nebula during the birth of the Sun

Hydrogen and helium are the most common elements, but they are not plentiful on the inner four terrestrial planets. Warmth from the nearby Sun causes temperatures on the four inner planets to be comparatively high. The higher the temperature of a gas, the greater the speed of its atoms. The lightweight atoms of hydrogen and helium move swiftly enough to escape from the relatively weak gravity of these small planets.

A different situation prevails on the four Jovian planets. Far from the Sun, temperatures are low and the atoms of hydrogen and helium move slowly. The relatively strong gravity of the massive Jovian planets easily prevents the lightest gases from escaping into space. In fact, the Jovian planets are composed primarily of hydrogen and helium.

Temperature must also have been an important factor in determining conditions inside the vast cloud of gas and dust, called the **solar nebula,** out of which the solar system formed. It is thus extremely useful to categorize the abundant elements and their common compounds according to their behavior at various temperatures.

First, hydrogen and helium are gaseous except at extremely low temperatures and extraordinarily high pressures. Second, rock-forming compounds of iron and silicon are solids except at temperatures exceeding 1000 K. Finally, there is an intermediate class of common chemicals such as water, carbon dioxide, methane, and ammonia. At low temperatures (typically below 200 to 300 K), these substances solidify into solids called ices. At somewhat higher temperatures, they can exist as liquids or gases. In Table 6-5, the common planet-forming substances are listed according to these three broad classifications.

By observing the process of star formation elsewhere in our galaxy, astronomers can deduce the conditions that probably led to the formation of our solar system. For instance, Figure 6-9 shows a disk of material surrounding a star. Planets may still be forming in this disk of debris left over from the birth of that star.

Just before the birth of the Sun, atoms in the solar nebula were so widely spaced that no substance could exist as a liquid. Matter in this vast cloud existed either as a gas or as tiny grains of dust and ice. Astronomers find it useful to speak of **condensation temperature** to specify whether a substance is a solid or a gas under these conditions of extremely low pressure. Above its condensation temperature, a substance is a gas; below its condensation temperature, the substance solidifies into tiny specks of dust or snowflakes.

Rock-forming substances have very high condensation temperatures, typically in the range of 1300 to 1600 K. The ices have condensation temperatures in the range of 100 to 300 K. The condensation temperatures of hydrogen and helium are so near absolute zero that they always existed as gases during the creation of the solar system.

Initially, before the Sun was formed, the solar nebula must have been quite cold. Temperatures throughout the cloud were probably less than 50 K, which is below the condensation temperatures of all common substances

Table 6-5 Common planet-forming substances

Gas	Ice	Rock
Hydrogen (H)	Water (H_2O)	Iron (Fe)
Helium (He)	Methane (CH_4)	Iron Sulfide (FeS)
Neon (Ne)	Ammonia (NH_3)	Olivine ($(Mg,Fe)SiO_4$)
	Carbon Dioxide (CO_2)	Pyroxene ($CaMgSi_2O_6$)

Figure 6-9 *A circumstellar disk of matter* This computer-enhanced photograph shows a disk of material orbiting the star called β Pictoris. This star is located behind a small circular mask at the center of the picture that was needed to block the star's light, which would have otherwise overwhelmed light from the disk. The disk, seen nearly edge on, is believed to be very young, possibly no more than a few hundred million years old. (University of Arizona and JPL)

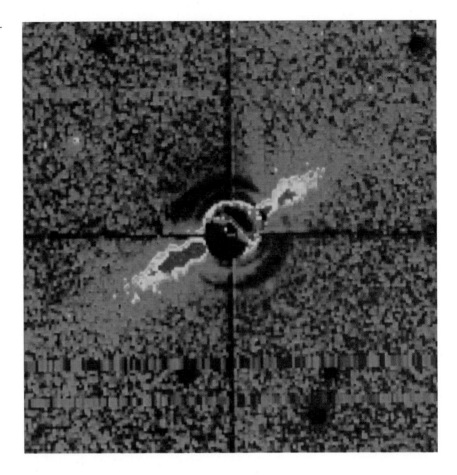

except hydrogen and helium. Snowflakes and ice-coated dust grains were scattered abundantly across the solar nebula, which had a diameter of at least 100 AU and a mass roughly two or three times the mass of the Sun.

The gravitational pull of the particles on one another caused them to begin a general drift toward the center of the solar nebula. The density and pressure at the center of the solar nebula began to increase, producing a concentration of matter called the **protosun.** Because of gravitational contraction, temperatures deep inside the solar nebula began to climb. The solar nebula also probably had some overall rotation, or **angular momentum,** as it is properly called. As the solar nebula contracted toward its gravitational center, it was transformed from a shapeless cloud into a rotating flattened disk that was warm at the center and cold at the edges. Astronomers feel confident in describing a flattened, disk-shaped solar nebula because all the planets today have orbits that are very nearly in the same plane.

Temperatures around the newly created protosun soon climbed to 2000 K while temperatures in the outermost regions of the solar nebula remained at less than 50 K. Figure 6-10 shows a probable temperature distribution throughout the solar nebula at this preliminary stage in the formation of the solar system. All the common icy substances in the inner regions of the solar nebula were vaporized by these high temperatures. Only the rocky substances remained solid, which is why the four inner planets are composed primarily of dense, rocky material. In contrast, snowflakes and ice-coated dust grains were able to survive in the cooler, outer portions of the solar

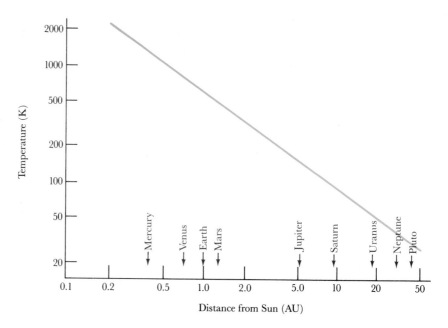

Figure 6-10 Temperature distribution in the solar nebula This graph shows how the temperature probably varied across the solar nebula as the planets were forming. For the inner planets, temperatures ranged from roughly 1200 K at Mercury to about 500 K at the orbit of Mars. Beyond Jupiter, temperatures were everywhere less than 200 K. (Based on calculations by J. S. Lewis)

nebula. That is why the Jovian planets have abundant quantities of methane and ammonia in their extensive low-density atmospheres. In addition, many of the satellites of the Jovian planets are either partially or almost entirely composed of ices.

The terrestrial planets formed out of rocky material, whereas the Jovian planets also incorporated vast amounts of ices and gases

The formation of the four inner planets was dominated by the fusing together of solid, rocky particles. Initially, dust grains collided and stuck together to form objects called **planetesimals,** with diameters of about 100 km. This stage of planetesimal formation probably took a few million years.

During the next stage, gravitational attraction between the planetesimals caused them to collide and coalesce into still larger objects called **protoplanets.** This accumulation of material is called **accretion.** In recent years, astronomers have used detailed computer simulation to improve our understanding of accretion. The computer is programmed to simulate a large number of planetesimals circling a hypothetical newborn sun along orbits dictated by Newtonian mechanics. Such studies show that accretion may continue for roughly 100 million years and will typically form about half a dozen planets.

A particularly successful computer simulation is summarized in Figure 6-11. The calculations begin with 100 planetesimals each having a mass of 1.2×10^{26} grams. This procedure ensures that the total mass (1.2×10^{28} grams) equals the mass of the four terrestrial planets (Mercury through Mars) plus their satellites. The initial orbits of these planetesimals are inclined to each other by angles less than 5°, to simulate a thin layer of asteroidlike objects orbiting the protosun.

In this simulation, after an elapsed time of 30 million years the 100 original planetesimals have coalesced into 22 protoplanets. After 79 million years, 11 larger protoplanets remain. Nearly another 100 million years

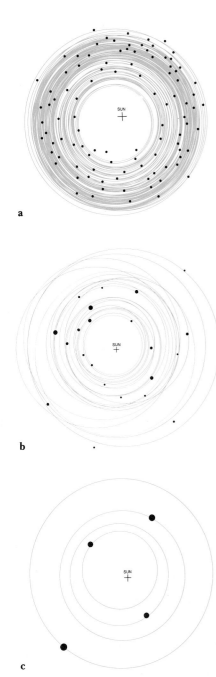

Figure 6-11 Accretion of the terrestrial planets *These three drawings show the results of a computer simulation of the formation of the inner planets. (a) The simulation begins with 100 planetesimals. (b) After 30 million years, these planetesimals have coalesced into 22 protoplanets. (c) This final view is for an elapsed time of 441 million years, but the formation of the inner planets is essentially complete after only 150 million years. (Adapted from a computer simulation by George W. Wetherill)*

elapse before the total number of growing protoplanets is reduced to six. Figure 6-11*c* shows four planets following nearly circular orbits after a total elapsed time of 441 million years. This exercise does not exactly reproduce the inner solar system because the fourth planet from the Sun ends up being the most massive. In fact, Earth is nine times more massive than Mars. Nevertheless, agreement with most characteristics of the inner solar system is very striking.

The material from which the inner protoplanets accreted was rich in mineral-forming elements having high condensation temperatures. Iron, silicon, magnesium, and sulfur were particularly abundant, followed closely by aluminum, calcium, and nickel. Energy released by the violent impact of planetesimals with the growing protoplanets—as well as the decay of radioactive elements—melted all this rocky material. The terrestrial planets thus began their existence as spheres of molten rock. During this time, denser iron-rich minerals sank to the center of the planets, forcing the less-dense silicon-rich minerals to the surface. That is why planets such as the Earth have dense iron cores surrounded by less-dense rock.

The formation of the outer planets probably proceeded slightly differently from that of the inner planets. Recall that the original solar nebula was a rotating disk of snowflakes and ice-coated dust particles. Analyses of the behavior of the outer regions of such a disk have shown that it would probably break up into rings around the developing protosun, separated by regions of space fairly free of matter. The material in each ring would then coalesce to form a large protoplanet. This protoplanet would be largely gaseous because the accumulation of the small particles would release enough energy to vaporize the abundant ices. These massive protoplanets would then sweep up vast amounts of gaseous hydrogen and helium as they moved along their orbits. The final result is four huge planets, each with an enormously thick atmosphere surrounding an Earth-sized core of rocky material. Figure 6-12 summarizes this story of the formation of the outer solar system.

Some of the details of this story about the formation of the planets will probably be revised in coming decades as we gather more evidence from spacecraft and orbiting observatories. However, most of the new discoveries in the past few decades have confirmed the broad outlines of this scenario, and astronomers are fairly confident that the solar system did form in the fashion depicted here. Of course, this discussion presents conclusions drawn from many different kinds of evidence, much of which we will discuss in later chapters.

During the millions of years while the planets were forming, temperatures and pressures at the center of the contracting protosun continued to climb. Finally, temperatures at the center of the protosun reached 8 million degrees, hot enough to ignite thermonuclear reactions, and the Sun was born. As we shall see in Chapter 13, detailed calculations have demonstrated that Sunlike stars take approximately 100 million years to form from the prestellar nebula. As a result, the Sun must have become a full-fledged star at roughly the same time the accretion of the inner protoplanets was complete.

A newborn star adjusts somewhat violently to the onset of thermonuclear reactions at its core, and any surrounding material is expelled into space (see Figure 6-13). In fact, many young stars experience an episode of mass loss during which the stars' tenuous outer layers are vigorously ejected. This

Figure 6-12   Formation of the solar system This series of sketches shows major stages in the birth of the solar system spanning 100 million years. Terrestrial planets accrete from rocky material in the warm, central regions of the solar nebula. Meanwhile, the huge, gaseous Jovian planets form in the cold outer regions. **(a)** The solar nebula in its initial stages. **(b)** The early solar system after 50 million years. **(c)** Planetary formation (nearly complete) after 100 million years.

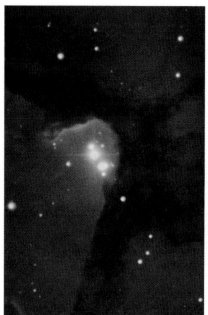

Figure 6-13   Newly formed stars A full-fledged star is born when thermonuclear reactions ignite at its center. This ignition is often accompanied by an outpouring of particles and radiation from the star's surface that sweeps the surrounding space clean of dust and gas from which the star had formed. This photograph shows dusty material being blown away from newly created stars at the center of the so-called Trifid Nebula in the constellation of Sagittarius. (Anglo-Australian Observatory)

brief burst of mass loss is called a **T Tauri wind** after the star in Taurus (the bull) where it was first identified. The T Tauri wind that heralded the birth of the Sun swept the solar system clean of excess gases, preventing further accretion. Many small rocks were left behind to pelt the planets over the next half billion years. For all practical purposes, however, the formation of the solar system was complete by the time the T Tauri wind began to blow.

The Sun today continues to lose matter gradually in a mild fashion. This on-going, gentle mass loss, called the **solar wind,** consists of high-speed protons and electrons leaking away from the Sun's outer layers.

Our Moon may have formed from material torn from the Earth during a collision with a protoplanet

Collisions between objects dominated the early history of the solar system. As we have seen, the formation of the planets involved smaller objects colliding and fusing together to build larger objects. Some of these collisions must have been quite spectacular. Astronomers have recently come to appreciate how important such collisions were in shaping and modifying the solar system.

The Moon's surface bears the scars of numerous impacts dating mostly from the final stages of the formation of the solar system (see Figure 6-14). These scars, called **craters,** are also seen on other planets and satellites that do not have appreciable atmospheres or geological activity to erase these features. Indeed, the Moon's airless environment has preserved important information about the early history of the solar system. A closeup view of a heavily cratered region of the lunar surface is seen in Figure 6-15.

Radioactive dating of moon rocks brought back by the Apollo astronauts demonstrates that the rate of impacts declined dramatically about 3.5 billion

Figure 6-14 Our Moon This photograph taken by astronauts in 1972 shows thousands of craters, most of them produced by impacts of rocky debris left over from the formation of the solar system. Age-dating of moon rocks brought back by the astronauts demonstrate that the Moon is about 4.5 billion years old. Most of the lunar craters were formed when the Moon was less than a billion years old. (NASA)

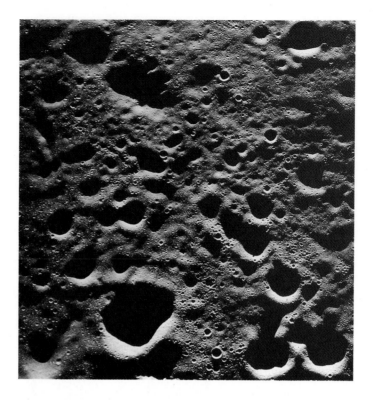

Figure 6-15 Lunar craters This closeup photograph of the Moon's surface was taken from lunar orbit by astronauts in 1969. The area shown measures about 30 km on a side. The airless environment of the Moon has preserved numerous craters that were formed 3.5 to 4.5 billion years ago. (NASA)

years ago. Since that time, impact cratering has proceeded at a very low rate that continues to the present. In other words, most of the craters on the Moon and planets were formed during the first billion years of the solar system's history as the young planets swept up rocky debris left over from the solar nebula.

One of the important goals of the manned lunar landings of the late 1960s and early 1970s was to obtain clues about the Moon's formation. Prior to the Apollo program, there were three main theories about the origin of the Moon. One theory said that the Moon was pulled out from a rapidly rotating protoearth, perhaps while it was still entirely molten. A second theory suggested that the Moon was formed elsewhere in the solar system and then captured into orbit about the Earth by the force of gravity. A third theory proposed that many small rocks may have been gravitationally captured by the young Earth and these rocks eventually collided and fused together to form the Moon.

Moon rocks brought back by the Apollo astronauts changed our thinking. As we shall see in Chapter 9, moon rocks resemble material from deep within the Earth. In addition, moon rocks are bone dry, whereas Earth rocks contain small amounts of volatile substances like water. In other words, moon rocks resemble deep-seated Earth rocks that have been baked at a high temperature to drive off substances that evaporate or vaporize easily. None of the three pre-Apollo theories seemed to account adequately for these important characteristics of moon rocks.

In the mid-1980s, scientists devised a new theory inspired by the fact that violent collisions between sizable protoplanets must have occurred during the final stages of the formation of the solar system. This new hypothesis, called the **collisional ejection theory,** posits that a large asteroidlike object

Figure 6-16 An asteroid hitting the Earth
This simulation, performed on a supercomputer, shows the effects of a Mars-sized object hitting the Earth. The mass of the object is one-tenth of the Earth's mass, and its speed at impact is 12 km/sec (27,000 miles/hour). Color is used to display density (refer to color scale) and the four frames show the situation at the moment of impact (t = 0) and at 10, 20, and 30 minutes after impact. A huge plume of vaporized rock is ejected by the impact. The Moon may have formed from debris ejected by such a collision. (Courtesy of M. Kipp and J. Meloch)

Color-coded
density scale
(g/cm^3)

13
11
9
7
5
3
1
.01

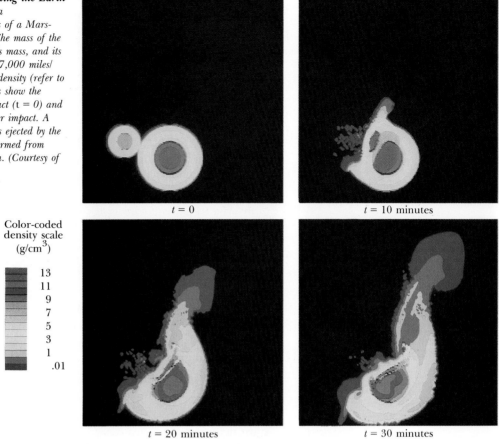

$t = 0$

$t = 10$ minutes

$t = 20$ minutes

$t = 30$ minutes

struck the Earth 4.5 billion years ago, shortly after our planet had formed. The impacting object may have been as large as Mars. A supercomputer simulation of this cataclysm is shown in Figure 6-16. Energy released during the collision produces a huge plume of vaporized rock that squirts out from the point of impact. After the ejected material cools, the Earth is left with a ring of debris that soon coalesces to form the Moon.

The collisional ejection theory is in agreement with many of the known facts about the Moon. For example, the vaporization of ejected rock during impact would have driven off volatile elements and water, leaving material that strongly resembles moon rocks. Indeed, the impact of an asteroid large enough to create the Moon could have also tipped the Earth's rotation axis relative to the Moon's orbit as we observe it today.

Astronomers and geologists have only recently come to appreciate the important role of collision between objects in shaping the planets. As we explore the solar system in the next four chapters, we shall see that craters cover the surfaces of many worlds. Of course, many impacts must have also scarred the Earth, but most of these craters have long since been erased by rain, wind, and snow. Nevertheless, we shall see that a large asteroid impact 63 million years ago may have killed off the dinosaurs and thus changed the course of life on the Earth.

Summary

· The four inner planets of the solar system share many characteristics and are distinctly different from the four giant outer planets.

· The four inner (terrestrial) planets are relatively small, have high average densities, and are composed primarily of rock.

· The giant outer (Jovian) planets have large diameters, low densities, and are primarily composed of hydrogen and helium.

· Pluto, the outermost planet, resembles the icy satellites that orbit the outermost planets.

· Hydrogen and helium, the lightest elements, were formed shortly after the creation of the universe; the heavier elements were produced much later in the centers of stars and cast into space when the stars died.

  By mass, 98 percent of the matter in the universe is hydrogen and helium.

· Planet-forming substances can be classified as gas, ice, and rock depending on their condensation temperature.

· The solar system formed from a disk-shaped cloud of hydrogen and helium, called the solar nebula, that also contained ice and dust particles.

  The inner planets formed through the accretion of dust particles into planetesimals and then into larger protoplanets.

  The outer planets probably formed through a similar accretion involving ice-coated dust and large amounts of gas that formed huge protoplanets.

· The Sun formed at the center of the solar nebula. After about 100 million years, temperatures at the protosun's center were high enough to ignite thermonuclear reactions.

· When the protosun became a star, excess gas and dust were vigorously blown away from the Sun, ending the process of planet formation.

· For nearly a billion years after the formation of the Sun, impacts of asteroidlike objects on the young planets dominated the early history of the solar system.

  Our Moon may have been created by the shattering impact of a Mars-sized object on the Earth 4.5 billion years ago.

Review questions

1 Name three differences between terrestrial and Jovian planets.

2 On what grounds can certain satellites be classified as terrestrial planets?

3 Why do you suppose there are no Jovian planets near the Sun?

4 What orbital properties of Pluto make it somewhat different from the other planets?

5 If hydrogen and helium account for 98 percent of the mass of all the material in the universe, why aren't the Earth and Moon composed primarily of these two gases?

* 6 Why are water (H_2O), methane (CH_4), and ammonia (NH_3) comparatively abundant substances?

7 Why is it reasonable to suppose that the solar system was formed during a relatively short interval?

8   Would you expect very old stars to possess planetary systems? If so, what types of planets would they have? Explain.

9   Why are craters common on the Moon but rare on the Earth?

Advanced questions

10   How does studying other planets and moons help us better understand the Earth?

*11   Suppose there were a planet having roughly the same mass as the Earth but located at 50 AU from the Sun. What do you think this planet would be made of? On the basis of this speculation, assume a reasonable density for this planet and calculate its diameter. How many times bigger (or smaller) than the Earth is it?

Discussion questions

12   Propose an explanation for the fact that the Jovian planets are orbited by terrestrial-like satellites.

13   Suppose a planetary system is now forming around some protostar in the sky. How might this system compare with our own solar system?

For further reading

Beatty, J. "The Making of a Better Moon." *Sky & Telescope*, December 1986, p. 558.

Beatty, J., et al., eds. *The New Solar System*, 2nd ed. Sky Publishing and Cambridge University Press, 1982.

Chapman, C. *Planets of Rock and Ice*. Scribners, 1982.

Frazier, K. *The Solar System*. Time-Life Books, 1985.

Gore, R. "Between Fire and Ice: The Planets." *National Geographic*, January 1985, p. 4.

Kaufmann, W. *Planets and Moons*. W. H. Freeman and Company, 1979.

Lewis, J. "The Chemistry of the Solar System." *Scientific American*, September 1975.

Murray, B., ed. *The Planets*. W. H. Freeman and Company, 1983.

Reeves, H. "The Origin of the Solar System." *Mercury*, March/April 1977, p. 7.

7 The major terrestrial planets

Earth

Mars

Venus

The three largest terrestrial planets *The Earth and Venus have nearly the same mass, size, and surface gravity. However, Venus's surface is perpetually shrouded by a dense cloud cover containing poisonous, corrosive gases. Mars is only about half the size and has only one-tenth the mass of Earth. The thin, dry Martian atmosphere does not protect the planet from the Sun's ultraviolet radiation. Both the Venusian and Martian environments are hostile to life-forms that thrive on Earth. (NASA)*

We begin our study of the solar system by examining our own planet and the two terrestrial worlds that most closely resemble the Earth: Venus and Mars. We explore these worlds as alien visitors might, first descending into the planet's atmosphere, observing its surface features, and then probing the planet's interior. On Earth we find an atmosphere profoundly affected by the presence of life. Mars's atmosphere is very sparse, but on Venus we find oppressive, hot, poisonous gases. We also discover that the Earth's interior is hot and partly molten because of decaying radioactive elements. We learn that the transfer of this heat outward from the Earth's interior to its crust profoundly affects our planet's surface features. Earth's crust is divided into huge plates that jostle each other, producing mountain ranges and earthquakes. On Mars, such plate movement ceased long ago as this small planet rapidly cooled, but Venus has two continentlike features. Finally, we see that molten iron in the Earth's interior produces a magnetic field that shields our planet from the solar wind, whereas Venus and Mars have no such protection.

Of all the planets that orbit the Sun, we are most familiar with the Earth. We walk on its surface, we drink its water, and we breathe its air. Naturally, we know more about the Earth than about any other object in the universe. Sometimes, however, this very familiarity makes it difficult for us to view the Earth in its place as a member of the solar system. Enlightening comparisons have come from explorations of our nearest planetary neighbors, Venus and Mars. What we have learned about both planets gives us perspective on how various conditions affect the evolution of Earthlike worlds. Venus shows us what the Earth might be like if it were nearer the Sun, and Mars shows us what Earth might be like if it were less massive.

Earth's atmosphere is a mixture of nitrogen and oxygen, whereas the atmospheres of Venus and Mars are nearly pure carbon dioxide

Imagine an alien spacecraft approaching the inner solar system. The spacecraft's occupants might first come upon Mars with its thin atmosphere and barren desert landscapes. Closer to the Sun, they would find Venus with its shroud of corrosive clouds hiding a forbiddingly hot surface. Between these two relatively unpromising planets, they would see Earth with its ever-changing ballet of delicate white cloud formations contrasting against the darker browns and blues of the continents and oceans (see Figure 7-1). Perhaps it is only our own bias that leads us to guess that the aliens would focus on the Earth as the most interesting and inviting of the terrestrial planets.

Figure 7-1   The Earth   Astronauts often report that Earth is the most invitingly beautiful object visible from their spacecraft. Our blue-and-white world is the largest of the terrestrial planets. This photograph was taken in 1969 by Apollo 11 *astronauts on the way to the first manned landing on Moon. Most of Africa and portions of Europe can be seen. (NASA)*

Table 7-1   The chemical composition of three planetary atmospheres

	Venus	Earth	Mars
Nitrogen	3%	77%	3%
Oxygen	Almost zero	21%	Almost zero
Carbon dioxide	97%	Almost zero	95%
Other gases	Almost zero	2%	2%

At first, alien visitors might be puzzled to find that Earth's atmosphere is very different from that of its neighbors. The atmospheres of both Venus and Mars consist almost entirely of carbon dioxide, whereas only 0.03 percent of the Earth's atmosphere is carbon dioxide. Earth's atmosphere is predominantly a 4-to-1 mixture of nitrogen and oxygen, but these two gases are found only in very small amounts on Venus and Mars, as Table 7-1 shows. The data for Venus and Mars in the table come from measurements by Soviet and American spacecraft that landed on these planets in the 1970s.

Why do Earth's planetary neighbors possess atmospheres of nearly pure carbon dioxide when Earth's own atmosphere is almost entirely nitrogen and oxygen? From study of the Earth's geological records, we know that living organisms have been active here for at least the past 3 billion years and that such biological processes as photosynthesis are largely responsible for the present chemical composition of Earth's atmosphere. As far as we know, a planetary atmosphere will contain large amounts of oxygen only as a result of biological activity. So aliens might single out Earth for investigation because its oxygen indicates the possibility of life.

Nearing Earth's surface, aliens would find that the temperature of our atmosphere varies with altitude in a complicated way, as shown in Figure 7-2. A graph like this can be used to deduce fundamental information about a planet's atmosphere because the temperature variations indicate how sunlight is absorbed differently at various altitudes.

Seventy-five percent of the mass of Earth's atmosphere lies below an altitude of 11 km (roughly 7 miles, or 36,000 ft) in a layer called the **troposphere.** All Earth's weather—clouds, rain, sleet, and snow—occur in this

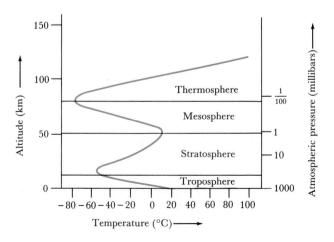

Figure 7-2   Temperature profile of Earth's atmosphere   *The atmospheric temperature varies with altitude because of the way sunlight interacts with atoms and ions at different elevations.*

lowest layer. Commercial jets generally fly at the top of the troposphere to minimize buffeting and jostling.

Above the troposphere is a region called the **stratosphere,** which covers a range in altitude from 11 to 50 km (7 to 30 miles) above the Earth's surface. Ozone (O_3) molecules in the stratosphere efficiently absorb solar ultraviolet rays, thus heating the air in this layer. Some scientists are concerned that certain chemicals (such as chlorofluorocarbons in aerosol cans) we have made and released into the air may destroy much of this ozone layer. If not absorbed by the ozone in the stratosphere, the ultraviolet radiation from the Sun would beat down on the Earth's surface. Because ultraviolet radiation breaks apart most of the delicate molecules that form living tissue, loss of the ozone layer could lead to the sterilization of our planet.

Above the stratosphere, atmospheric temperature declines with increasing altitude in the **mesosphere,** reaching a minimum temperature at an altitude of about 80 km (50 miles). This minimum marks the bottom of the **thermosphere,** above which the Sun's ultraviolet light strips atoms of one or more electrons. Ionized atoms and molecules of oxygen and nitrogen are the most prevalent elements. The stripped electrons easily reflect radio waves in a wide range of frequencies, which is why you can tune in a distant radio station: its transmissions bounce off this radioreflective layer.

At sea level on Earth, the air weights down with a pressure of 14.7 pounds per square inch. By definition, this pressure is called *one atmosphere* (1 atm). Because we are familiar with this pressure, atmospheric pressure on other planets is usefully expressed in atmospheres also. For example, the atmospheric pressure on the surface of Venus is 90 atm whereas the sparse Martian atmosphere exerts a pressure of only 0.01 atm. The atmospheric pressure above any planet decreases with increasing altitude because, the higher you go, the less atmosphere there is to weigh down upon you.

The surface of Venus is hidden beneath a very thick, highly reflective cloud cover

At first glance, Venus and Earth look like twins. They have almost the same mass, diameter, average density, and surface gravity. But Venus is exposed to a greater intensity of sunlight, which transforms its potentially Earthlike environment into an extremely hostile world.

Since Venus is closer to the Sun than is the Earth, we always find Venus near the Sun in the sky. Venus can most easily be seen above the western horizon after sunset (when it is called an evening star), or before sunrise above the eastern horizon (when it is called a morning star).

Venus is easy to identify because it is often one of the brightest objects in the night sky. Venus's cloud cover reflects 76 percent of the sunlight that falls on it. From the Earth, Venus can be 16 times brighter than the brightest star. Indeed, only the Sun and Moon outshine Venus at its greatest brilliancy.

Earth-based telescopic views of Venus reveal a thick, nearly featureless, unbroken layer of clouds. Figure 7-3 shows a closeup view of the clouds, as seen from a spacecraft orbiting the planet. This high reflective cloud cover is responsible for our long-standing ignorance of the planet's surface.

Spacecraft have descended into Venus's clouds and provided detailed information about the Venusian atmosphere. During the 1960s, while the Americans were busy landing astronauts on the Moon, Soviet scientists concentrated on building spacecraft that could survive a descent into the

Figure 7-3 **Venus** *Venus's thick cloud cover efficiently traps heat from the Sun, resulting in a surface temperature (480°C = 900°F) even hotter than that on Mercury. Unlike Earth's clouds, which are made of water droplets, Venus's clouds are very dry and contain droplets of concentrated sulfuric acid. This photograph was taken in 1979 by an American spacecraft in orbit about Venus. (NASA)*

Sunlight comes in

Infrared radiation cannot get back out

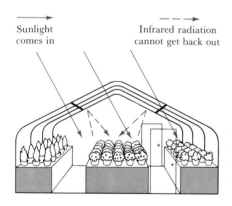

Figure 7-4 **The greenhouse effect** *Incoming sunlight easily penetrates the windows of a greenhouse and is absorbed by objects inside. These objects reradiate this energy at infrared wavelengths to which the glass windows are opaque. The trapped infrared radiation is absorbed by the air and objects, causing the temperature inside the greenhouse to rise.*

Venusian cloud cover. The task proved to be more frustrating than anyone had expected because the spacecraft would stop transmitting data before reaching the ground. Finally, in 1970, a Soviet spacecraft managed to transmit data for a few seconds directly from the Venusian surface. Soviet missions during the early 1970s measured a surface temperature of 750 K (= 480°C = 900°F) and a pressure of 90 atmospheres, which is equal to a crushing pressure of $\frac{2}{3}$ ton per square inch.

After some initial puzzlement, astronomers quickly realized there is a straightforward explanation for Venus's high surface temperature. Perhaps you have had the experience of parking your car in the sunshine on a warm summer day. You roll up the windows, lock the car, and go on an errand for a few hours. When you return, you are annoyed to discover that the interior of your automobile has become stiflingly hot, typically at least 20°C warmer than the outside air temperature.

What happened to make your car so warm? First, energy in the form of sunlight entered the car through the windows. The solar radiation was absorbed by the dashboard, the steering wheel, and the upholstery, causing their temperature to rise about 20°C. At this elevated temperature, the seats in your car reradiate the energy primarily at infrared wavelengths. Your car windows are opaque to these wavelengths. This energy therefore is trapped inside your car and absorbed by the air and interior surfaces. As more sunlight comes through the windows and is trapped, the temperature continues to rise.

A similar phenomenon occurs in Venus's atmosphere, and to a lesser extent in Earth's atmosphere. It is commonly called the **greenhouse effect** (see Figure 7-4). Sunlight enters the thick Venusian clouds, where it is absorbed and reradiated at infrared wavelengths that cannot get back out. This trapped radiation produces a high surface temperature.

While descending through the Venusian clouds, Soviet spacecraft measured atmospheric pressure and temperature. The results are shown in Figures 7-5 and 7-6. The pressure and temperature profiles of the Venusian atmosphere are simple: both decrease smoothly with increasing altitude. As we saw in Figure 7-2, Earth's atmosphere has a much more complicated relationship between temperature and altitude. In the chapters on Jupiter and Saturn, we shall see how pressure and temperature variation with altitude can have a profound effect on a planet's appearance.

Both Soviet and American spacecraft found the top of the Venusian clouds at an altitude of about 68 km (42 miles = 220,000 ft). However, the Russians also discovered the bottom of the clouds, at an elevation of about 31 km (19 miles = 100,000 ft) above the ground. Below this altitude, the Venusian atmosphere is remarkably clear.

Because Venus and Earth are so similar in size and mass, astronomers assumed that Venus's clouds were made of water vapor, like Earth's. This assumption was proved wrong in the 1960s, when scientists had great difficulty detecting any water in the Venusian atmosphere. The latest measurements demonstrate that Venus is extremely dry, with water making up far less than 1 percent of the clouds. If the clouds are dry, what are they made of?

Clues to the complex chemistry of the Venusian clouds came in bits and pieces in the 1960s when Earth-based observations and measurements from spacecraft demonstrated that the clouds efficiently absorb radiation at specific infrared and microwave wavelengths. Scientists soon proposed a surprising explanation for this absorption: droplets of sulfuric acid.

When sulfuric acid (H_2SO_4) is mixed with water (H_2O), the ions HSO_4^- and H_3O^+ are produced. These ions absorb infrared and microwave radiation at precisely the wavelengths Venus's clouds absorb it. Sulfuric acid is

Figure 7-5 [left]   Temperature in the Venusian atmosphere   *The temperature in Venus's atmosphere rises smoothly from a minimum of about 170 K (about −100°C, or −150°F) at an altitude of 100 km to a maximum of nearly 750 K (about 480°C, or 900°F) on the ground.*

Figure 7-6 [right]   Pressure in the Venusian atmosphere   *The pressure at the Venusian surface is a crushing 90 atm (1300 pounds per square inch). Above the surface, atmospheric pressure decreases smoothly with increasing altitude.*

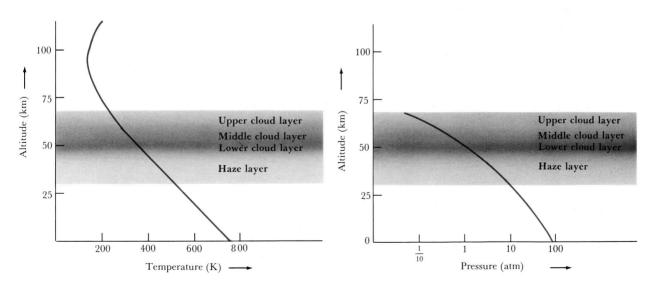

partly responsible for depleting the Venusian atmosphere of water by converting it into H_3O^+. The latest Soviet and American probes leave no doubt that the Venusian clouds are in fact made of droplets of concentrated sulfuric acid.

Venus appears yellowish or yellow-orange to the human eye, and data from spacecraft have indicated why: the upper clouds contain substantial amounts of sulfur dust. Over the temperature range of these upper clouds, sulfur is distinctly yellow or yellowish-orange. At lower elevations, large concentrations of sulfur compounds (especially SO_2, OCS, and H_2S) were found along with droplets of sulfuric acid. Because of the tremendous atmospheric pressure, the droplets do not fall as a rain; they are more or less permanently suspended in the clouds like an aerosol.

The sulfuric acid in Venus's atmosphere causes a number of chemical reactions. Reactions with fluorides and chlorides in surface rocks give rise to hydrofluoric acid (HF) and hydrochloric acid (HCl). Further reactions produce fluorosulfuric acid (HSO_3F), one of the most corrosive substances known to chemists, which can dissolve lead, tin, and most rocks. Indeed, the Venusian clouds are a cauldron of chemical reactions hostile to metals and other solid materials. No wonder Soviet and American spacecraft survive for only a few minutes after landing on Venus.

The Martian atmosphere is thin and dry

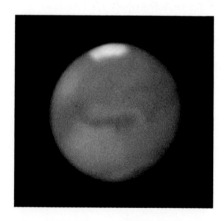

Figure 7-7 Mars viewed from the Earth
This high-quality, Earth-based photograph of Mars was taken in 1971, when the Earth–Mars distance was only 56 million kilometers (35 million miles). At that time, Mars presented a disk nearly 25 arc sec in angular diameter. Circumstances as favorable as these will not be repeated for the rest of the century. This photograph accurately portrays what you might typically see through a moderate-sized telescope under excellent observing conditions. Notice the prominent polar cap. (Courtesy of Stephen M. Larson)

Mars is the only planet whose surface features can be seen through Earth-based telescopes (see Figure 7-7). Even with the modest telescopes of the seventeenth century, astronomers made discoveries that hinted that Mars might be very Earthlike. For example, in 1666 the Italian astronomer Giovanni Domenico Cassini spent many nights carefully observing the motions of surface features on Mars and discovered that the planet's rotation period is 24 hours 37 minutes. Thus a day on Mars is only slightly longer than a day on Earth. A century later, the German-born English astronomer William Herschel determined the inclination of Mars's axis of rotation. Just as Earth's equatorial plane is tilted $23\frac{1}{2}°$ from the plane of its orbit, Mars's equator makes an angle of nearly 24° with its orbit. As a result, Mars experiences seasons much as Earth does.

Cassini also discovered that Mars has polar caps that are large during the Martian winter but shrink with the coming of summer. Early telescopic observers also reported seasonal color variations that seemed to indicate vegetation on the Martian surface, and some reported geometric patterns that might represent a planetwide system of artificial canals. Scientists speculated about the possibility of life on Mars, and science-fiction writers wove popular stories about invasions of Earth by hostile Martians.

Ironically, the first actual invasion came in 1976, when automated spacecraft from the Earth landed on the Martian surface (see Figure 7-8). These probes sent back pictures and data indicating that Mars is a barren, desolate world. The possibility of some form of Martian life has not been completely ruled out, but it now seems quite likely that Mars is as sterile and lifeless as the Moon.

The two American spacecraft that journeyed to Mars in 1976—*Viking 1* and *Viking 2*—made numerous measurements of conditions in the Martian atmosphere during their descents. Figure 7-9 shows the recorded atmospheric temperature plotted against altitude above the Martian surface. Also

Figure 7-8 Mars *Many of the major features of the Martian surface are seen in this photograph taken by* Viking 2 *as the spacecraft approached the planet in 1976. Clouds flank the western slopes of the huge volcano Olympus Mons near the top of the photograph. In the middle is the vast rift canyon called Valles Marineris. This canyon stretches nearly 4000 km along the planet's equator. At the bottom of the photograph, carbon dioxide snow lines the floor of the Argyre Basin and surrounding craters.* (NASA)

shown, for comparison, is the atmospheric temperature above the Earth's surface. Earth exhibits a maximum temperature at an altitude of 50 km because of the absorption of ultraviolet radiation by our ozone layer. The temperature profile of the Martian atmosphere exhibits relatively little variation with altitude, and so we conclude that Mars has no ozone layer. The absence of an ozone layer means that the Sun's ultraviolet radiation strikes

Figure 7-9 Temperature of the Martian atmosphere *Unlike Earth, Mars lacks an ozone layer in its atmosphere. Consequently, the temperature profile of the Martian atmosphere does not show a maximum at 50 km altitude as does the Earth's temperature profile.*

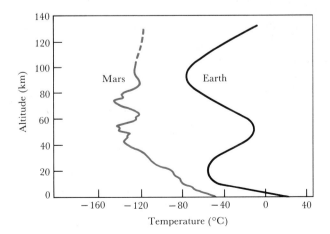

Figure 7-10   A panorama of the Viking 2 landing site   *This mosaic of three views shows about 180° of the Viking 2 landing site. Northwest is at the left, southeast on the right. The flat, featureless horizon is approximately 3 km (2 miles) away from the spacecraft. (NASA)*

the Martian surface directly. The sterilizing effect of ultraviolet light may be an important factor inhibiting the appearance of life on Mars.

The meteorological instruments on both Viking landers promptly confirmed that the surface atmospheric pressure on Mars is about 0.007 atm, as had been expected from previous flyby missions. The atmospheric pressure at sea level on Earth is 1 atm, and so the density of the Martian atmosphere is less than $\frac{1}{100}$ that of the Earth's atmosphere.

Direct chemical analysis of the Martian atmosphere by the Vikings' instruments reported a carbon dioxide abundance of 95 percent, with nitrogen at 2.7 percent, and argon 1.6 percent. The remaining fraction of a percent is mostly oxygen and carbon monoxide, with only a small amount of water vapor. If all the water vapor could somehow be squeezed out of the Martian atmosphere, it would not fill one of the five Great Lakes in North America.

In most of the pictures sent back from Mars, the sky has a distinctly pinkish-orange tint (see Figure 7-10). This coloration is thought to be caused by extremely fine-grained dust suspended in the Martian atmosphere. Indeed, Earth-based observers have often reported seeing planetwide dust storms that occasionally obscure the Martian surface features for several weeks at a time. Although the Martian atmosphere is very thin, its winds are sometimes strong enough to raise large amounts of fine dust particles high into the atmosphere.

After the Viking landers had been on the Martian surface for a few weeks, their data showed clearly that the atmospheric pressure at both landing sites was dropping steadily. Mars seemed to be rapidly losing its atmosphere, and some scientists joked that all the air would be gone in a few months. A straightforward explanation was, however, readily available: winter was coming to the southern hemisphere. At the Martian south pole, it was so cold that large amounts of carbon dioxide were solidifying out of the atmosphere, covering the ground with dry-ice snow.

In early 1977, when spring came to the southern hemisphere, the dry-ice snow rapidly evaporated, and the atmospheric pressure returned to prewinter levels. With the arrival of winter in the northern hemisphere, another decrease in atmospheric pressure was observed as dry-ice snow blanketed the northern latitudes (see Figure 7-11). The Martian surface clearly experiences extreme seasonal variations in temperature and atmospheric pressure.

Figure 7-11   Winter on Mars   *This picture, taken in May 1979, shows a thin frost layer that lasted for about a hundred days at the Viking 2 site. Freezing carbon dioxide adheres to water-ice crystals and dust grains in the atmosphere, causing them to fall to the ground. The sky is therefore not as pink as it was in the summertime view of Figure 7-10. (NASA)*

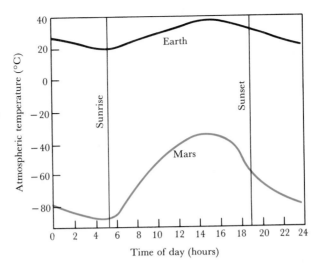

Figure 7-12   Daily temperature variation
The lower curve shows the daily temperature variation at the Viking 1 site. The upper curve shows the daily temperature variation at a desert site in California. The daily temperature range on Mars is about three times greater than that on Earth. The thin, dry Martian air does not retain heat as well as the Earth's atmosphere does.

Aside from an occasional dust storm, the weather on Mars would seem boring to someone accustomed to the temperate zones of the Earth. Atmospheric pressure varies with the seasons in a regular, predictable fashion. The atmospheric temperature varies with the time of day, also in a monotonously repetitive way. Actually, the daily temperature variation on Mars is quite similar to the temperature changes observable in a desert on Earth (see Figure 7-12). The warmest time of day occurs about two hours after local noon, and the coldest temperatures are recorded just before sunrise. The thin, dry Martian atmosphere is not capable of retaining much of the day's heat, however, and so the range of the daily temperature swings on Mars is larger than that on Earth.

The Earth has oceans and its crust consists of large plates whose motions produce mountain ranges, volcanoes, and earthquakes

Were aliens to land on Earth, they would find it radically different from either of its neighbors in yet another way: the Earth is very wet. Nearly 71 percent of the Earth's surface is covered with water. An alien space probe to Earth might send back thousands of photographs such as the one shown in Figure 7-13, which represents a random closeup view of Earth's surface. In

Figure 7-13   A typical closeup view of Earth's surface   *More than two-thirds of Earth's surface is covered with water. In contrast, there is no liquid water on Venus or Mars.*

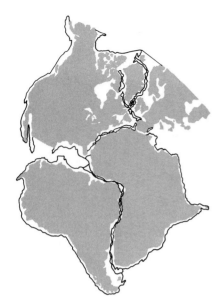

Figure 7-14 Comparing the continents
Africa, Europe, Greenland, and North and South America fit together as though they were once joined. The fit is especially convincing if the edges of the continental shelves (rather than today's shorelines) are used. (Adapted from P. M. Hurley)

contrast, Venus and Mars are extremely arid. No place on Earth is as dry as the surfaces of Venus and Mars. By Venusian or Martian standards, our Sahara desert is a veritable swamp.

One of the most important geological discoveries of the twentieth century is the realization that the surface of our planet is active and constantly changing. Geologists have learned that the Earth's crust is divided into huge **plates** that constantly jostle each other, producing earthquakes, volcanoes, and oceanic trenches.

The idea of huge, moving plates might occur to anyone who carefully examines a map of the Earth. You can see that, were it not for the Atlantic Ocean, South America would fit snugly against Africa. Indeed, the fit between land masses on either side of the Atlantic Ocean is remarkable (see Figure 7-14). This observation inspired people such as Alfred Wegener to propose the idea of "continental drift," suggesting that the continents on either side of the Atlantic Ocean have simply drifted apart. Wegener got the idea of drifting continents around 1910 while looking at a globe of the Earth. After much research, Wegener published the theory that there was originally a single, gigantic supercontinent he called Pangaea, which began to break up and drift apart some 200 million years ago. Pangaea first split into two smaller supercontinents called Laurasia and Gondwanaland. Gondwanaland later split into Africa and South America, while Laurasia divided to become North America and Eurasia.

Most geologists initially greeted Wegener's ideas with ridicule and scorn. Although it was generally accepted that the continents do "float" on denser material beneath them, few geologists could accept the idea that entire continents could move around the Earth at speeds that must be as great as several centimeters per year.

Then in the mid-1950s, scientists began discovering long mountain ranges on the ocean floors, such as the Mid-Atlantic Ridge (see Figure 7-15), which stretches all the way from Iceland to Antarctica. During the 1960s, careful examination of the ocean floor revealed that molten rock from the Earth's interior is oozing upward along the Mid-Atlantic Ridge, indicating a long chain of underwater volcanoes. Furthermore, accurate measurements revealed that the ocean floor is separating along the length of the mid-Atlantic Ridge, pushing South America and Africa apart at a speed of roughly 3 cm per year. This process of creating new crustal rock in a widening gap on the ocean floor is called **seafloor spreading.**

Seafloor spreading provided the mechanism that had been missing from Wegener's theory of continental drift. In the 1960s, geologists began to review Wegener's ideas and, with some embarrassment, found a great deal of new evidence to support them. This time, however, the emphasis was upon the motion of large plates of the crust (not simply the continents). The modern theory of crustal motion came to be known as **plate tectonics.** Geologists today realize that seismic activity indicates where plates are either colliding or separating. The boundaries between plates are clear when the locations of earthquakes are plotted on a map (see Figure 7-16).

Geologists believe that convective currents beneath the Earth's crust are responsible for the movement of the plates. Heat is transported from the Earth's interior toward its surface by **convection,** a process in which warm material rises while cool material sinks. Rock beneath the crust is not liquid, but it is hot enough to permit an oozing, plastic flow throughout a soft

Figure 7-15   The Mid-Atlantic Ridge
This artist's rendition shows the floor of the North Atlantic Ocean. The unusual mountain range in the middle of the ocean floor, called the Mid-Atlantic Ridge, is caused by lava seeping up from the Earth's interior along a rift that extends from Iceland to Antarctica. (Courtesy of Marie Tharp and Bruce Heezen)

Figure 7-16   The major plates The boundaries of major plates are the scenes of violent seismic and geologic activity. Most earthquakes occur where plates separate or collide. Plate boundaries therefore are easily identified by plotting the locations of earthquakes on a map.

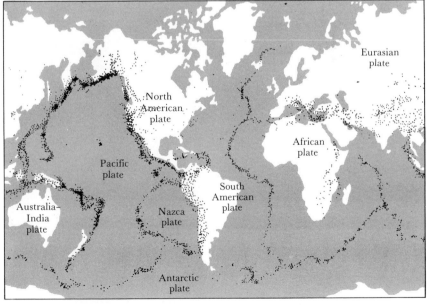

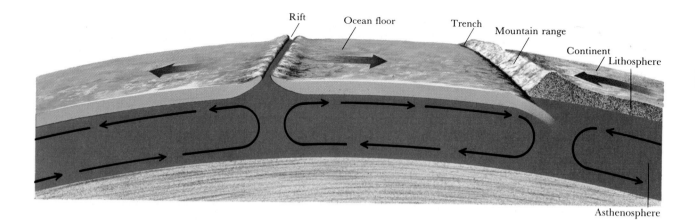

Figure 7-17 labels: Rift, Ocean floor, Trench, Mountain range, Continent, Lithosphere, Asthenosphere

Figure 7-17 The mechanism of plate tectonics *Convection currents in the soft upper layer of the Earth's interior (called the asthenosphere) are responsible for pushing around rigid, low-density plates that constitute the Earth's crust (also called the lithosphere). New crust is formed in oceanic rifts, where lava oozes upward between separating plates. Mountain ranges and deep oceanic trenches are formed where plates collide.*

region called the **asthenosphere.** As sketched in Figure 7-17, hot material (magma) seeps upward along **oceanic rifts** where plates are separating. Cool material sinks back down along **subduction zones** where plates are colliding. Above the plastic asthenosphere is a low-density, rigid layer called the **lithosphere,** divided into plates that simply ride on the convection currents of the asthenosphere.

The regions where plates meet are the sites of some of the most impressive geological activity on our planet (see Figure 7-18). Great mountain ranges, such as those along the western coasts of North and South America, are thrust up by ongoing collisions with the plates of the ocean floor (see also Figure 7-19). Subduction zones, where old crust is pushed back down into the Earth, are typically the locations of deep oceanic trenches such as those off the coasts of Japan and Chile.

Figure 7-18 The separation of two plates *The plates that carry Egypt and Saudi Arabia are moving apart, leaving the trench that contains the Red Sea. In this view taken by astronauts in 1966, Saudi Arabia is on the left and Egypt (with the Nile River) on the right. (NASA)*

Figure 7-19 *The collision of two plates* *The plates that carry India and China are colliding. The Himalaya Mountains have been thrust upward as a result of this collision. In this photograph taken by astronauts in 1968, India is on the left, Tibet on the right, and Mt. Everest is one of the snow-covered peaks near the center. (NASA)*

Mars has huge volcanoes and canyons while Venus has two continents, but neither planet shows significant tectonic activity

The convection process associated with seafloor spreading is the primary way in which heat is transported outward from the Earth's interior to the crust. Here again our planet differs from its neighbors. Venus and Mars apparently transport heat outward to their surfaces predominantly by a mechanism called **hot-spot volcanism.** Hot areas deep inside Venus and Mars squirt molten lava up through the crust. In the absence of any tectonic activity to move the crust around, millions of years of eruptions eventually build enormous volcanoes.

Hot-spot volcanism also occurs on Earth. The Hawaiian Islands, which are in the middle of the Pacific plate, are a fine example. There is a hot spot in the Earth's interior beneath Hawaii that continuously pumps lava up through the crust. However, the Pacific plate is moving northwest at the rate of several centimeters per year. New volcanoes are created over the hot spot as older volcanoes move away from the magma source, become extinct, and eventually erode and disappear beneath the ocean. In fact, the Hawaiian Islands are the most recent additions to a long chain of extinct volcanoes that stretches all the way back to Japan.

Martian volcanoes were first photographed by a spacecraft that was placed into orbit about Mars in 1971. The largest Martian volcano, Olympus Mons, rises 24 km (15 miles) above the surrounding plains—nearly three times as high as Mount Everest (see Figure 7-20). The highest volcano on Earth is Mauna Loa in the Hawaiian Islands, whose summit is only 8 km above the ocean floor. The huge size of Olympus Mons strongly suggests a lack of plate tectonics on Mars. A hot spot in Mars's interior kept pumping lava upward through the same vent for millions of years, producing one giant volcano rather than a chain of smaller volcanoes.

The base of Olympus Mons is ringed with cliffs and covers an area as big as the state of Missouri. At the volcano's summit are several overlapping

Figure 7-20   Olympus Mons
This photograph looks straight down on the largest volcano on Mars. Olympus Mons is 2½ times as tall as Mount Everest, and its cliff-ringed base measures nearly 600 km (370 miles) in diameter. (NASA)

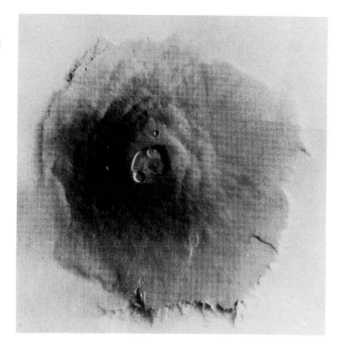

volcanic craters, forming a **caldera** large enough to contain the state of Rhode Island (see Figure 7-21).

Olympus Mons is one of several very large volcanoes centered just north of the Martian equator and clustered together on a huge bulge that covers an area 2500 km in diameter. Ground levels over this vast dome-shaped region are typically 5 to 6 km higher that the average ground level for the rest of the planet. Another major, though less dramatic grouping of volcanoes is located nearly on the opposite side of the planet. None of these volcanoes seem to be active today.

Figure 7-21   The Olympus caldera
This view of the summit of Olympus Mons is based on a mosaic of six pictures taken by one of the Viking spacecraft. The caldera consists of overlapping, collapsed volcanic craters and measures roughly 70 km across. The volcano itself is wreathed in midmorning clouds that formed from ice (H$_2$O) brought upslope by cool air currents. The cloudtops are about 8 km below the volcano's peak. (NASA)

For reasons we do not yet understand, most of the volcanoes on Mars are in the northern hemisphere, whereas craters are mostly found in the southern hemisphere. Between the two hemispheres is a vast canyon called Valles Marineris running roughly parallel to the Martian equator (see Figures 7-22 and 7-23).

Valles Marineris stretches 4000 km, beginning with heavily fractured terrain in the west and ending with ancient cratered terrain in the east. If this canyon were located on Earth, it could stretch all the way from New York to Los Angeles. Many geologists suspect that Valles Marineris is a fracture in the Martian crust caused by internal stresses, not unlike the rift valleys on Earth that result from plate tectonics. One such valley is the Red Sea (recall Figure 7-18), which is 3000 km long.

Current theories suggest that plate tectonics did operate on the young Mars but that this process long ago ceased there. A small planet loses its internal heat much more rapidly than a large one. Mars's diameter is only

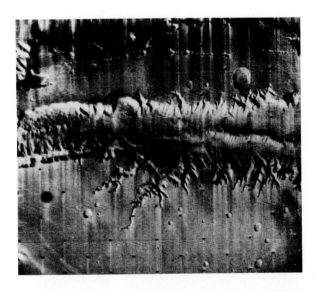

Figure 7-22 A segment of Valles Marineris This photograph, taken from an altitude of 2000 km, shows an area 300 by 400 km, which is nearly the same size as the state of Pennsylvania. Earth's Grand Canyon is only as big as one of the tributary canyons seen in this photograph. (NASA)

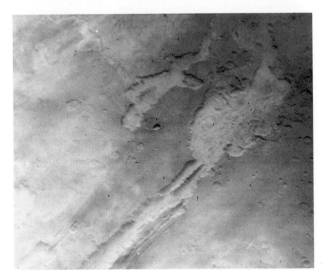

Figure 7-23 Eastern Valles Marineris This mosaic was constructed from fifteen photographs taken by one of the Viking spacecraft and shows an area 1800 by 2000 km. North is toward the upper left. The eastern end of Valles Marineris merges with chaotic, cratered terrain toward the upper right. (NASA)

half that of the Earth, and its mass is one-tenth of the Earth's. Plate tectonics ceased early in Mars's history as its rapidly cooling lithosphere became thick and sluggish.

Although Venus is perpetually shrouded in clouds, radar has enabled scientists to determine the planet's surface features. In 1978, an American spacecraft placed in orbit about Venus mapped 93 percent of the planet's surface using a radar altimeter that bounced microwaves off the ground directly below. By measuring the time delay of the radar echo, scientists determined the heights and depths of hills and valleys. Figure 7-24 displays two views of Venus based on these measurements.

Venus is remarkably flat. About 60 percent of the planet's surface is covered with gently rolling hills varying by less than 1 km from the planet's radius of 6051.4 km. (This value for Venus's radius is commonly used as a reference level, as we use sea level here on Earth.)

Two large "continents" rise well above the generally level surface of Venus. In the northern hemisphere is Ishtar Terra, named after the Babylonian goddess of love. Ishtar is approximately the same size as Australia and consists of a high plateau ringed by high mountains. The highest mountain is Maxwell Montes with a summit 11 km above the reference level. For comparison, Mount Everest on Earth rises 9 km above sea level.

The second major Venusian "continent," Aphrodite Terra (named after the Greek equivalent of Venus), lies just south of the equator. Aphrodite is slightly bigger than Ishtar and has about one-half the area of Africa.

The spacecraft orbiting Venus also discovered some very large volcanoes southwest of Ishtar. These volcanoes, Rhea Mons and Theia Mons, rise to altitudes of 6 km, with gently sloping sides that extend over an area 1000 km in diameter. They are among the largest known volcanoes in the entire solar system and possibly result from hot-spot volcanism.

Figure 7-24 Topographic globes of Venus *These computer-generated globes show the vertical relief on Venus according to a color code. The average elevation (like sea level on Earth) is shaded light blue. Dark blue and violet denote lower elevations. Green and yellow indicate higher elevations, with the highest peaks shown in red. View* **(a)** *has the north pole tilted toward the observer to show Ishtar Terra near the top of the globe. View* **(b)** *has the south pole tilted toward the observer to show Aphrodite Terra. The blank circles cover the polar regions, which were not observed by the spacecraft. (NASA)*

a

b

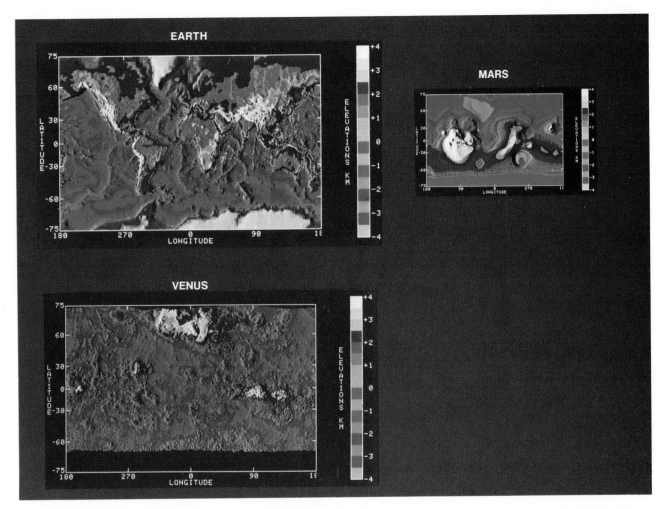

EARTH

MARS

VENUS

Figure 7-25 The surfaces of Earth, Venus, and Mars These maps of Earth, Venus, and Mars are reproduced to the same scale. Each map shows surface elevation (Earth's oceans empty) on a color-coded scale. Elevations up to 4 km above the planet's average radius (sea level for Earth) appear in shades of tan, green, and white. Elevations down to 4 km below the planet's average radius appear in various shades of blue. Venus's two continents are prominent, as are Earth's seven continents. The large protrusion on Mars is a major volcanic region that includes Olympus Mons. (S. P. Meszaros; NASA)

A color-coded map of the Venusian surface is shown in Figure 7-25, with comparable maps of Earth and Mars. One of the surprising results from radar mapping is the discovery that Venus has no long mountain chains resembling the Mid-Atlantic Ridge. Although Venus and Earth have nearly the same mass and size, the Venusian surface shows very few of the sure signs of plate tectonics so prominent on Earth's surface. Nevertheless, the existence of two continents does suggest that some tectonic activity might be beginning there. Venus may just be evolving more slowly than the Earth.

In August 1990, from its orbit about Venus, the American spacecraft *Magellan* will use a sophisticated radar imaging system to map about 90 percent of the Venusian surface, revealing details as small as 250 m across. These high-resolution views of Venus may have a profound effect on our understanding of that planet's geological history.

Interesting theories try to explain the absence of water on both Venus and Mars

Volcanoes provide clues about the gases that probably were spewed into the original atmospheres of Venus, Earth, and Mars. For example, during the Mount St. Helens eruption of 1980 (see Figure 7-26), geologists monitored substantial amounts of sulfuric acid and other sulfur compounds emitted by the volcano. As we have seen, these compounds are common in the Venusian atmosphere.

Figure 7-26 Mount St. Helens eruption of May 1980 A preponderance of data suggests that the terrestrial planets obtained their atmospheres from volcanic outgassing. Geologists study this outgassing process in eruptions of volcanoes on Earth. This spectacular recent eruption in the western United States was particularly well studied. (USGS)

Carbon dioxide and water vapor are among the most abundant gases in volcanic vapors. Venus's atmosphere contains much carbon dioxide but very little water. In contrast, the Earth has abundant water in its oceans and atmosphere, but very little carbon dioxide in its atmosphere. If the atmospheres of both Venus and the Earth developed mainly from the same outgassing process, what happened to all the water on Venus and what became of Earth's carbon dioxide?

In the upper Venusian atmosphere, intense ultraviolet radiation from the Sun breaks water molecules into separate hydrogen and oxygen atoms. The light hydrogen atoms escape into space. Oxygen, which is one of the most chemically active elements, readily combines with other substances in Venus's atmosphere. Thus Venus is left with almost no water and any water outgassed from active volcanoes is soon destroyed.

The carbon dioxide on the Earth is dissolved in the oceans and chemically bound into carbonate rocks such as limestone and marble that formed in those oceans. Were the Earth to become as hot as Venus, so much carbon dioxide would be boiled out of the oceans and baked out of the crust that our planet would soon develop a thick, oppressive carbon dioxide atmosphere much like Venus's.

Some scientists have recently proposed an interesting outline of Venus's early history. Venus and Earth are so similar in size and mass that it is reasonable to suppose that Venusian volcanoes outgassed an amount of water vapor roughly comparable to the total content of Earth's oceans. Although some of this water on Venus might originally have collected in oceans, heat from the Sun soon vaporized the liquid to create a thick cover of water-vapor clouds. Calculations demonstrate that this water vapor would have added 300 atm of pressure to the existing 90 atm of carbon dioxide. Thus the early Venusian atmosphere would have weighed down on the planet's surface with a pressure of three tons per square inch.

This thick, humid atmosphere efficiently trapped heat from the Sun, creating a greenhouse effect far more extreme than the greenhouse effect that operates today on Venus. Calculations demonstrate that the ground temperature would have increased to 1800 K (2700°F), which is hot enough to melt rock. The Venusian surface was probably molten down to a depth of 450 km (280 miles). Soon, however, dissociation of the water molecules and a subsequent loss of hydrogen to space left behind the carbon dioxide atmosphere we find today. Hostile as it may seem, the modern environment on Venus is probably quite mild compared to that of earlier times.

A few photographs of the arid Venusian surface have come from Soviet spacecraft that landed on the planet. A panoramic view taken in 1981 is shown in Figure 7-27. Soviet scientists suggest that this region was covered with a thin layer of lava that fractured upon cooling to create the rounded, interlocking shapes seen in the photograph. This hypothesis agrees with the analysis by the spacecraft's instruments, which indicates that the soil composition is similar to lava rocks called basalt that are common on Earth and the Moon.

Basaltic lava rock is also common on the Martian surface (see Figure 7-28). Chemical analysis of the Martian soil showed a very high iron content. Iron and silicon comprise about two-thirds of a typical Martian rock's content. Surprisingly high concentrations of sulfur were also found—apparently it is 100 times more abundant in Martian rocks than in Earth rocks.

a

b

Figure 7-27 *A Venusian landscape*
(a) *This color photograph from Venera 13 shows that rocks on the Venusian surface appear orange because the thick, cloudy atmosphere absorbs the blue component of sunlight.* **(b)** *When computer processing is used to remove the effects of orange illumination, the true grayish color of the rocks is seen. In this panorama, the rocky plates covering the ground may be fractured segments of a thin layer of lava, or they may be crusty layers of sediment that have been cemented together by chemical and wind erosion. (Courtesy of C. M. Pieters and the U.S.S.R. Academy of Sciences)*

Each Viking lander had a scoop at the end of a mechanical arm for obtaining rock samples for analysis. Bits of rock were observed to cling to a magnet mounted on the scoop, thus confirming a high iron content in the soil. The familiar reddish color of the Martian surface may simply be due to an abundance of rust (iron oxide).

As soon as measurements by spacecraft confirmed a low atmospheric pressure on Mars, scientists realized that the Martian environment must be extremely dry. Water is liquid only over a certain range of temperature and pressure. If the atmospheric pressure above a body of water is very low, molecules escape from the liquid's surface, causing the water to vaporize. Any liquid water of Mars would boil furiously and rapidly evaporate into the thin Martian air. Thus, there are no oceans on Mars.

Figure 7-28 *Digging in the Martian soil*
Viking's mechanical arm with its small scoop protrudes from the right side of this view of the Chryse plains. Several small trenches dug by the scoop in the Martian regolith appear near the left side of the picture. (NASA)

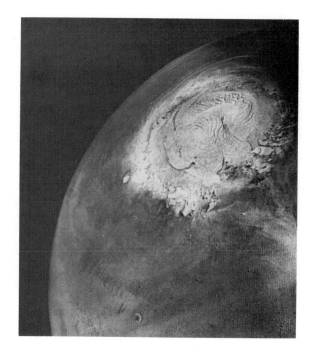

Figure 7-29 [left]   An ancient riverbed
This pair of photographs shows a 700-km-long riverbed in Mars's heavily cratered southern hemisphere. Features such as these were unexpected because liquid water cannot now exist on the Martian surface. This particular riverbed is located about 20° south of the Martian equator. (NASA)

Figure 7-30 [right]   The northern polar cap   *This view of the northern polar cap was taken by a spacecraft in 1972 from a distance of 13,700 km (8500 miles). By observing the rate of shrinkage with the approach of Martian summer, scientists deduced that the residual polar cap contains a substantial amount of frozen water. (NASA)*

Although Valles Marineris was not formed by water erosion, a Mars-orbiting spacecraft did photograph many features that look like dried-up riverbeds (see Figure 7-29). These features were totally unexpected because liquid water cannot exist today on Mars.

In 1976, the Viking spacecraft supplied evidence strongly supporting the idea that liquid water once raged across the Martian surface in great torrents. Numerous Viking photographs show erosional features that look like the results of flash floods in the deserts of Arizona and New Mexico. Scientists were faced with a problem that soon became known as "the mystery of the missing water."

Researchers immediately looked to the Martian polar caps as a probable location of the missing water. It was not clear, though, how much of the polar caps consists of frozen carbon dioxide (dry ice) and how much is water ice.

In 1972, a spacecraft sent back photographs from orbit as summer came to Mars's northern hemisphere. During the spring, the polar cap receded rapidly (see Figure 7-30), suggesting strongly that a thin layer of carbon dioxide frost was evaporating quickly in the sunlight. However, with the arrival of summer, the rate of recession abruptly slowed, suggesting that a thicker layer of water ice had been exposed. Scientists concluded that the **residual polar caps** that survive through the Martian summers contain a large quantity of frozen water. Calculating the volume of water ice in the residual caps is difficult, however, because we do not know how thick the layer of ice is.

Another important clue about water on Mars came from the Viking orbiters in 1976. Many closeup views show flash-flood erosion features where the water appears to have emerged from collapsed, jumbled terrain (see Figure 7-31). These photographs confirmed earlier suspicions that frozen water might form a layer of permafrost under the Martian surface similar to that

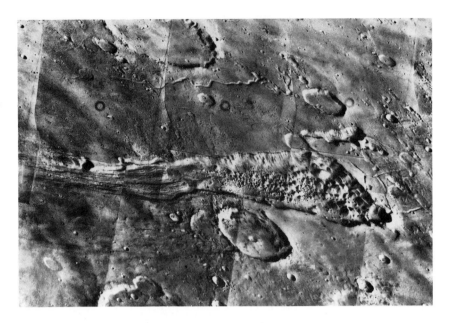

Figure 7-31 Evidence of a flash flood on Mars The melting of subsurface ice and the subsequent collapse, or downfaulting, of the Martian surface easily explain the features seen in the mosaic taken by Viking 1 in 1976. A torrent of water liberated by this process apparently flowed eastward (to the left) across the surrounding plains. The area seen here measures 300 by 300 km. (NASA)

beneath the tundra in far northern regions on Earth. Apparently, heat from volcanic activity occasionally melts this subsurface ice. The ground then collapses as millions of tons of rock push the water to the surface, producing a brief flash flood as the water quickly boils away.

From photographs like that in Figure 7-31, scientists estimated the widths and depths of the Martian flash-flood channels. These dimensions suggest peak flood discharges of 10^7 to 10^9 m^3/sec. For comparison, the average discharge of the Amazon is 10^5 m^3/sec. The largest known flash flood on the earth occurred about 2 million years ago in eastern Washington, when a natural dam gave way. The peak discharge is estimated to have been 10^7 m^3/sec. Most of the Martian flash floods must have been greater than anything known on Earth.

As mentioned earlier, water vapor and carbon dioxide are two of the most common gases in volcanic vapors. Mars's surface gravity, although only one-third as strong as Earth's, is nevertheless capable of keeping these gases from escaping into space. Ultraviolet light from the Sun can dissociate water into hydrogen and oxygen, which are light enough to escape into space. Detailed calculations show, however, that Mars could have lost only a small fraction of its water in this fashion. It is therefore reasonable to suppose that much of the carbon dioxide and water vapor outgassed during Mars's early history still remains on the planet. Some is in the polar caps, some is in subsurface ice and permafrost. Some water and carbon dioxide also may be chemically bound in the rocks and sand that cover the Martian surface.

Earth's interior consists of a crust, a mantle, and an iron-rich core

The densities of rocks you find on the ground are typically 3 to 4 g/cm^3, but the average density of the Earth as a whole is 5.5 g/cm^3. Thus the rocks of the Earth's crust are not representative of our planet's interior, which must be composed of a substance much denser than the crust.

Iron is a good candidate for this substance because it is the most abundant of the heavier elements (recall Table 6-4), and its presence in the

Earth's interior is strongly suggested by the existence of the Earth's magnetic field. Furthermore, iron is common in meteoroids that strike the Earth, suggesting that it was common in the planetesimals from which the Earth formed.

Geologists strongly suspect that the Earth was entirely molten soon after its formation about 4.5 billion years ago. Energy released by the violent impact of numerous meteoroids and asteroids and by the decay of radioactive isotopes melted the solid material collected from the earlier planetesimals. Gravity caused the abundant, dense iron to sink toward the Earth's center, forcing less-dense material to the surface. This process called **chemical differentiation** produced a layered structure within the Earth: a central **core** composed of almost pure iron, surrounded by a **mantle** of dense, iron-rich minerals, which in turn is surrounded by a thin crust of relatively light silicon-rich minerals.

The Earth's interior is as difficult to examine as the most distant galaxies in space. The deepest wells go down only a few kilometers, barely penetrating the surface of our planet. Geologists have, however, deduced basic properties of the Earth's interior by studying earthquakes.

Earthquakes produce several different kinds of **seismic waves** that travel around or through the Earth in different ways and at different speeds. Geologists use sensitive **seismographs** to detect and record these vibratory motions. Seismic waves are bent as the travel through the Earth because of the varying density and composition of the Earth's interior. By studying the deflection of these waves, geologists have discovered properties of the Earth's interior.

By 1906, analysis of earthquake recordings led to the discovery that the Earth's iron core is molten and has a diameter of about 7000 km (4300 miles). For comparison, the overall diameter of our planet is 12,700 km (7900 miles). More careful measurements in the 1930s revealed that inside the molten core is a solid iron core with a diameter of about 2500 km (1550 miles).

The interior of our planet therefore has a curious structure: a liquid core sandwiched between a solid inner core and a solid mantle. To understand why this is so, we must examine the temperature and pressure inside the Earth and their effects on the melting point of rock.

Both temperature and pressure increase with increasing depth below the Earth's surface. The temperature of the Earth's interior rises steadily from about 20°C on the surface to about 4300°C at the center (see Figure 7-32).

The Earth's crust is only about 30 km thick. It is composed of rocks whose melting points are far greater than typical temperatures in the crust. Hence the crust is solid.

The Earth's mantle, which extends to a depth of about 2900 km (1800 miles), is largely composed of minerals rich in iron and magnesium. On the Earth's surface, specimens of these ferromagnesian minerals have melting points slightly over 1000°C. However, the melting point of a substance depends on the pressure to which it is subjected: the higher the pressure, the higher the melting point. As shown in Figure 7-32, the melting point of the mantle's minerals is everywhere higher than the actual temperature, and so the mantle is solid.

At the boundary between the mantle and the outer core, there is a change in chemical composition from ferromagnesian minerals to almost-pure iron with a small admixture of nickel. This iron–nickel material has a lower melt-

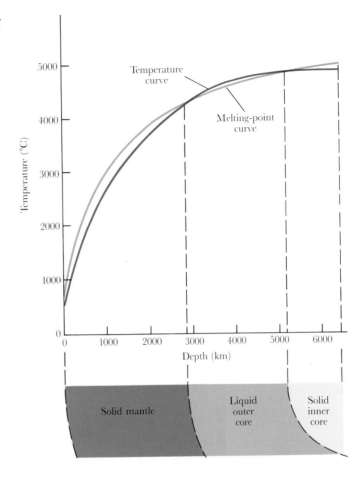

Figure 7-32   The temperature and melting point of rock inside the Earth   The temperature (red curve) rises steadily from the Earth's surface to its center. By plotting the melting point of rock (blue curve) on this graph, we can deduce which portions of the Earth's interior are solid or liquid. (Adapted from F. Press and R. Siever)

ing point than ferromagnesian minerals, and so the melting-point curve on Figure 7-32 dips below the temperature curve as it crosses from the mantle to the outer core. The melting-point curve remains below the temperature curve down to a depth of about 5100 km. Hence, from depths of about 2900 to 5100 km the core is liquid.

At depths greater than about 5100 km, the pressure is more than 3 million atmospheres. This pressure is so great that the melting point of the iron–nickel mixture exceeds the actual temperature (see Figure 7-32). Hence the Earth's inner core is solid.

The Earth's magnetosphere shields us from the solar wind, but neither Venus nor Mars has a magnetic field

Although buried deep within the Earth, the liquid outer core has a great influence on the Earth's outermost environment. Currents in the molten iron in our planet's interior give rise to a planetwide magnetic field through the dynamo effect. As the Earth rotates, these currents produce a magnetic field, just as a loop of wire carrying an electric current generates a magnetic field. The Earth rotates fast enough to produce a magnetic field that dominates space for tens of thousands of kilometers and dramatically affects Earth's interaction with the solar wind.

As mentioned briefly at the end of Chapter 6, the Sun is constantly losing matter as high-speed electrons and protons escape from the Sun's outer layers. This constant streaming of matter away from the Sun is called the

solar wind. The Earth's magnetic field carves out a cavity in the solar wind, deflecting the flow around the Earth as shown in Figure 7-33. This cavity, called the Earth's **magnetosphere,** shields us from the high-speed particles that would otherwise strike the upper atmosphere.

The speed of the particles escaping from the Sun is faster than the speed of sound through the solar wind. Thus we say that the solar wind is **supersonic.** When this supersonic flow encounters a planet, a bow-shaped **shock wave** is formed where particles of the solar wind are abruptly slowed to subsonic speeds. The **magnetopause** is the outer boundary of the Earth's magnetic domain. Most of the particles of the solar wind are deflected around the magnetopause through the turbulent region called the **magnetosheath.** Deep inside Earth's magnetosphere, our planet's magnetic field is strong enough to trap charged particles that manage to leak through the magnetopause. These particles are trapped in two huge, doughnut-shaped rings called the **Van Allen belts.**

These belts were discovered in 1958 during the flight of the United States's first successful Earth-orbiting satellite. They are named after the physicist who insisted that the satellite should carry a Geiger counter to detect charged particles. The inner Van Allen belt extends over altitudes of about 2000 to 5000 km and contains mostly protons. The outer Van Allen belt contains mostly electrons and is about 6000 km thick, centered at an altitude of about 16,000 km above the Earth's surface. It is remarkable that the Van Allen belts—these two vast features that completely encircle the Earth—were totally unknown until a few decades ago.

Occasionally, a violent event on the Sun's surface called a **solar flare** sends a burst of protons and electrons toward the Earth. Many of these

Figure 7-33   Earth's magnetosphere
Earth's magnetic field carves out a cavity in the solar wind. The Earth's magnetosphere consists of a shock wave, a magnetopause, and a magnetosheath. Because of the strength of Earth's magnetic field, our planet is able to trap charged particles in two huge, doughnut-shaped rings called the Van Allen belts.

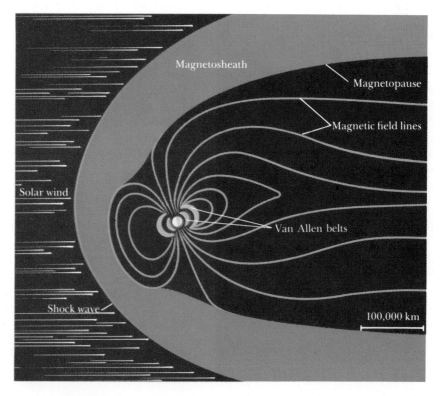

Figure 7-34 The northern lights (aurora borealis) *When a deluge of protons and electrons from a solar flare strikes atoms in Earth's upper atmosphere, the gases glow. Aurorae typically occur at an altitude of about 110 km (70 miles) above the Earth's surface. (Courtesy of S-I. Akasofu, Geophysical Institute, University of Alaska)*

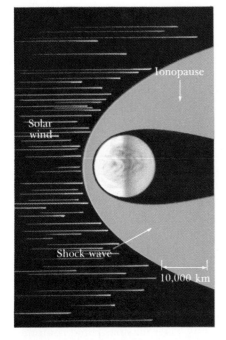

Figure 7-35 Venus's interaction with the solar wind *Venus has no magnetic field, and so the solar wind strikes the uppermost layers of the planet's atmosphere. Interactions with ions in the upper atmosphere produce a shock wave and ionopause, as shown in this scale drawing.*

particles penetrate the magnetopause and overload the Van Allen belts. The excess particles move along the Earth's magnetic field and rain down on the upper atmosphere near the Earth's north and south magnetic poles. As these particles collide with gases in the upper atmosphere, atoms of oxygen and nitrogen fluoresce like the gases in a fluorescent tube. The result is a beautiful, shimmering display called the **northern lights** (*aurora borealis*) or **southern lights** (*aurora australis*), depending on the hemisphere from which the phenomenon is observed (see Figure 7-34).

Numerous American and Soviet space flights have failed to detect any magnetic field around Venus. This absence of a magnetic field might seem surprising since Earth and Venus are so similar in mass, size, and average density. It is in fact reasonable to assume that Venus has an interior structure quite similar to Earth's, with a substantial iron core, some of which is probably molten.

Venus does not have a magnetic field because of the planet's slow rotation. By sending pulses of radar waves toward Venus and analyzing the reflected signals, scientists have determined that Venus rotates backward at a very slow rate. In other words, the Sun rises in the west and sets in the east on Venus. A "day" on Venus (that is, the time from one sunrise to the next) lasts for 116.8 Earth days.

Recall that Earth's magnetic field results from the dynamo effect: currents in the iron core of the rotating Earth produce our planet's magnetism. Although Venus may have a sizable liquid core, the planet's leisurely rotation rate is apparently too slow to induce a planetwide magnetic field.

With virtually no magnetic field, Venus is incapable of producing a magnetosphere to protect itself from the solar wind. The solar wind therefore impinges directly on Venus's upper atmosphere, where many of the atoms become stripped of one or more electrons. The electromagnetic interaction between these ions and the supersonic charged particles of the solar wind produces a shock wave (see Figure 7-35). Along a boundary called the **ionopause**, inside the shock wave, the pressure of the ions just counterbal-

Figure 7-36   The Phobos mission
In 1988, Soviet scientists sent two spacecraft toward Mars's asteroidlike moon, Phobos. The spacecraft were designed to fire a laser beam at Phobos's surface so that on-board instruments could capture and analyze some of the vaporized material. Because of technical difficulties, both spacecraft failed to complete their missions. (Courtesy of M. Carroll)

ances the pressure from the solar wind. The ionopause is analogous to the magnetopause surrounding a planet with a magnetic field.

Although Mars rotates at about the same rate as Earth, Soviet spacecraft to Mars have detected only an extremely weak magnetic field there. The near absence of a Martian magnetic field may mean Mars lacks a molten iron core. Both Venus and Earth have average densities greater than 5 g/cm^3, indicating the presence of substantial iron cores. But Mars's average density (3.9 g/cm^3) is roughly that of ordinary crustal rocks. Thus Mars may not possess an Earth-like iron-rich core. Indeed, Mars may not be as chemically differentiated as Venus or Earth. Mars's iron may be more uniformly distributed throughout the planet as evidenced by the high percentage of iron in rocks on the Martian surface.

Mars's magnetic field is so weak that it is largely ineffectual in warding off the solar wind. In fact, the solar wind may actually impinge directly on the outermost layers of the Martian atmosphere, just as it does on Venus's. Unfortunately, our knowledge of this aspect of the Martian environment is woefully incomplete. An alternative possibility is that Mars's weak field just manages to carve out a magnetosphere just barely enclosing the planet's atmosphere.

It will be up to future missions to search for a Martian magnetosphere and monitor Marsquakes with seismographs. Such missions could continue the search for life on Mars that began in 1976 with the Viking landers. Although the regions examined by the Viking landers seem quite sterile, some scientists point out that the polar regions of Mars may have conditions more suitable for life. These areas might be explored by a rover vehicle that could send back data and pictures from a wide range of sites. We would also learn a lot were a spacecraft to scoop up Martian rocks and return them to

Earth for laboratory analysis. The Soviet Union has been particularly successful with missions of this type to the Moon.

The United States has plans for only one mission to Mars during the rest of the twentieth century: a small satellite scheduled to be launched in the 1990s. In 1988, Soviet scientists sent two spacecraft to Phobos, the larger of Mars's two tiny asteroidlike moons (see Figure 7-36). Unfortunately, both spacecraft experienced malfunctions. Nevertheless, this project is the first phase in a major Soviet effort to explore Mars using robot vehicles as well as manned landings. Soviet scientists ultimately hope to establish a manned base on Mars early in the twenty-first century. A courageous mission of this magnitude would one of the greatest adventures in all human history.

Summary

- The Earth's atmosphere is primarily nitrogen and oxygen whereas the atmospheres of Venus and Mars are almost pure carbon dioxide.

- Earth's atmosphere can be divided into layers. All weather occurs in the lowest layer, the troposphere, which extends up to an altitude of 11 km.

- Venus is similar to the Earth in size, mass, average density, and surface gravity, but it is covered by nearly featureless unbroken clouds.

- The Venusian clouds are confined to a thick layer well above the ground. The clouds consist of droplets of concentrated sulfuric acid.

- On the surface of Venus, the pressure is 90 atm and the temperature is 750 K. The high temperature is caused by the greenhouse effect: carbon dioxide in the atmosphere prevents infrared radiation from escaping into space.

- Mars is smaller then either the Earth or Venus; a day on Mars is slightly longer than a day on the Earth.

- The atmospheric pressure on Mars is about one-hundredth that of the Earth's atmosphere and shows seasonal variations.

- Liquid water would quickly boil away in Mars's thin atmosphere, but the polar caps do contain a considerable amount of frozen water. A layer of permafrost may exist beneath the Martian surface.

- Study of seismic waves shows that the Earth has a small solid inner core surrounded by a liquid outer core; the outer core is surrounded by a dense mantle, which in turn is surrounded by a thin, low-density crust.

    The Earth's inner and outer cores are composed of almost-pure iron; the mantle is composed of iron–magnesium minerals; the crust is largely composed of silicon-rich minerals.

- The Earth's crust and the upper part of its mantle are divided into huge plates. Movements of these plates (called plate tectonics) are driven by convective currents in the mantle. Plate tectonics is responsible for most of the major features of the Earth's surface.

- The surfaces of Venus and Mars show little evidence of the motion of large crustal plates that played a major role in shaping the Earth's surface.

    The surface of Venus is surprisingly flat, mostly covered with gently rolling hills; there are two major continents and some large volcanoes.

    The northern hemisphere of Mars has numerous extinct volcanoes whereas the southern hemisphere has numerous flat-bottomed craters; a huge 4000-km-long canyon stretches along the equator.

- The Earth's magnetic field surrounds our planet with a magnetosphere that shields us from the solar wind; charged particles from the solar wind are trapped in two huge doughnut-shaped rings called the Van Allen belts.

- Venus has no detectable magnetic field or magnetosphere; Mars has only a very weak magnetic field.

Review questions

1 Describe three ways in which the Earth is different from either Venus or Mars.

2 Why is the Earth's surface not riddled with craters like the Moon?

3 Describe the process of plate tectonics. What kinds of features that are a direct result of plate tectonics.

4 Describe the Earth's interior. What causes the Earth's inner core to be solid, whereas its outer core is molten? What gives rise to the Earth's magnetic field?

5 What techniques have astronomers used to examine and map the surface of Venus? What kinds of surface features have they found?

6 Why is it reasonable to suppose that Venus's interior is similar to Earth's? Why doesn't Venus have a planetwide magnetic field as Earth does?

7 Suppose you were in a spacecraft in orbit about Mars. What kinds of surface features would you see? What do these surface features tell you about plate tectonics on Mars?

8 Compare Olympus Mons with the Hawaiian Islands. In what way are they different manifestations of the same physical process?

9 Why does Earth have an abundance of water whereas Venus and Mars are very dry?

10 With carbon dioxide just about as abundant in the Martian atmosphere as in the Venusian atmosphere, why do you suppose there is little or no greenhouse effect on Mars?

11 Is it reasonable to suppose that the polar regions of Mars might harbor life-forms? Explain your answer.

Advanced questions

*12 What fractions of the Earth's total volume are occupied by the core, the mantle, and the crust?

*13 As mentioned in the text, Africa and South America are separating at a rate of about 3 centimeters per year. Assuming that this rate has been constant, calculate when these two continents must have been in contact.

14 Suppose a planet's atmosphere were opaque to visible light but transparent to infrared radiation. How would this affect the planet's surface temperature? Contrast and compare the effect of the atmosphere of this hypothetical planet with the greenhouse effect caused by Venus's atmosphere.

15 Several months after this textbook was published, the Magellan spacecraft was scheduled to arrive at Venus. Read up on this mission in magazines like *Sky & Telescope* and *Science News*. Is the spacecraft functioning properly? Have data been sent back to Earth? What have we learned so far from this mission?

Discussion questions

16   The human population on Earth is currently doubling about every 30 years. Describe the various pressures placed on the Earth by uncontrolled human population growth. Can such growth continue indefinitely? If not, what natural and human controls might arise to curb this growth? It has been suggested that overpopulation problems could be solved by colonizing the Moon or Mars. Do you think this is a reasonable solution? Explain your answer.

17   If you were designing a space vehicle to land on Venus, what special features would be necessary? In what ways would this mission and landing craft differ from a spacecraft designed for a similar mission to Mars?

18   Imagine that you are an astronaut living at a base on Mars. Describe what your day might be like, what you would see, the weather, the space suit you would wear, and so on. Suppose you and your colleagues have a motorized vehicle for exploring the planet. Where would you like to go?

For further reading

Beatty, J. "Report from a Torrid Planet." *Sky & Telescope*, May 1982, p. 452.

Beatty, J. "Radar Views of Venus." *Sky & Telescope*, February 1984, p. 110.

Bazilevskiy, A. "The Planet Next Door." *Sky & Telescope*, April 1989, p. 360.

Calder, N. *The Restless Earth.* Viking, 1972.

Carr, M. "The Surface of Mars: A Post Viking View." *Mercury*, January/February 1983, p. 2.

Carrigan, C., and Gubbins, D. "The Source of the Earth's Magnetic Field." *Scientific American*, Feburary 1979.

Cooper, H. *The Search for Life on Mars.* Holt, Rinehart and Winston, 1980.

Gore, R. "Sifting for Life in the Sands of Mars." *National Geographic*, January 1977.

Haberle, R. M. "The Climate of Mars." *Scientific American*, May 1986.

Hartmann, W. "The Early History of the Planet Earth." *Astronomy*, August 1978, p. 6.

Pettengill, G., et al. "The Surface of Venus." *Scientific American*, August 1980.

Pollack, J. "The Atmospheres of the Terrestrial Planets." In Beatty, J., et al., eds. *The New Solar System*, 2nd ed. Sky Publishing and Cambridge University Press, 1982.

Prinn, R. G. "The Volcanoes and Clouds of Venus." *Scientific American*, March 1985.

Zakharov, A. "Close Encounters with Phobos," *Sky & Telescope*, July 1988.

8

The Jovian planets

SATURN

EARTH

JUPITER

Jupiter, Saturn, and Earth

This montage shows Jupiter, Saturn, and Earth reproduced to the same scale. Jupiter and Saturn are composed primarily of hydrogen and helium, like the Sun. Note that the cloud features on Saturn are much less distinct than those on Jupiter. Saturn's surface gravity is weaker than Jupiter's, and so Saturn's clouds are situated over a greater range of depths than Jupiter's. Sunlight reflected from Saturn's deeper cloud layers undergo a greater amount of absorption, making Saturn look faded compared to Jupiter. (S. P. Meszaros; NASA)

About 70 percent of the mass of the solar system, besides the Sun itself, is concentrated in the single giant planet Jupiter. In this chapter, we learn about the unusual features of this active, vibrant world whose multicolored, turbulent clouds are surrounded by an enormous magnetosphere. We examine Saturn with its spectacular system of thin, flat rings. We find that the rings actually consist of thousands of ringlets composed of ice fragments and ice-covered rock. We then discover that Jupiter and Saturn have similar interior structures and that they have internal sources of heat— but for different reasons. We then turn to Uranus and Neptune. We learn that these outer two Jovian planets differ distinctly from the inner two Jovian planets in many ways. We find that Uranus has a unique orientation and an orbiting system of thin, dark rings. The chapter ends with an overview of Neptune based on the recent flyby of *Voyager 2.*

More than any other single factor, temperatures throughout the young solar nebula dictated the final characteristics of the planets that orbit the Sun. In the warm, inner regions of this ancient nebula, surviving dust grains consisted primarily of metals, silicates, and oxides. The temperature was too high to allow substantial condensation of such volatile substances as water, methane, and ammonia. The four planets that formed close to the Sun were therefore composed almost entirely of rocky material. Their surface gravities were too low and their surface temperatures too high to retain any of the abundant but lightweight hydrogen and helium gases that made up most of the solar nebula.

The four Jovian planets are much farther from the Sun than the terrestrial planets are (recall Figure 6-1). For example, Saturn is roughly 10 times farther from the Sun than Earth is. In the past, as now, it was cold at these vast distances. In the young solar nebula, the dust grains so far from the protosun were coated with a thick layer of frozen water, methane, and ammonia. These volatile substances thus became important constituents of the planets that eventually accreted in the outer reaches of the solar system.

The Jovian planets may have started to form in much the same fashion as the terrestrial ones—by accretion of dust grains (coated with frozen gases, in this case) into a great number of planetesimals, which in turn accreted to form huge protoplanets. Many scientists think, however, that the Jovian planets were created in a two-step process. First, accretion led fairly quickly to the formation of four large protoplanets, each several times more massive than the Earth. Then, the strong gravitational pull of these protoplanets attracted and retained substantial quantities of the hydrogen and helium. Calculations show that this process of gathering lightweight gases would have become very efficient and rapid after the Jovian protoplanets had grown beyond a certain mass. The final result was the largest planets in the solar system.

Jupiter is a huge, colorful world composed largely of hydrogen and helium

Jupiter is huge: its mass is 318 times greater than Earth's. Indeed, the mass of Jupiter is $2\frac{1}{2}$ times the combined masses of not only all the other planets, but also all the satellites, asteroids, meteoroids, and comets in the solar system. Jupiter's equatorial diameter is $11\frac{1}{4}$ times as large as the Earth's and its volume is about 1430 times larger than the Earth's.

Jupiter's average density can be computed from its mass and size: it is only 1.33 g/cm^3. This low average density is entirely consistent with the picture of a huge sphere of hydrogen and helium, compressed by its own gravity. Observations from spacecraft indicate that by weight Jupiter is composed of 82 percent hydrogen, 17 percent helium, and only 1 percent all other elements. This composition closely matches that of the Sun (recall Table 6-4).

Through an Earth-based telescope, Jupiter is a colorful, intricately banded sphere (see Figure 8-1). Figure 8-2 is a closeup view from a spacecraft. The most prominent features are alternating dark and light bands shaded in subtle tones of red, orange, brown, yellow, and blue that are parallel to Jupiter's equator. The dark, reddish bands are called **belts,** the light-colored bands are called **zones.** These are not the only conspicuous markings. A large, reddish oval called the **Great Red Spot** is often visible in Jupiter's southern hemisphere. This remarkable feature, which

Figure 8-1 [left]   Jupiter from Earth
Belts and zones are easily identified in this Earth-based view of the largest planet in the solar system. The Great Red Spot was exceptionally prominent when this photograph was taken. (McDonald Observatory)

Figure 8-2 [right]   Jupiter from a spacecraft   This view was sent back in 1979 from a spacecraft at a distance of only 30 million kilometers from the planet. Features as small as 600 km across can be seen in the turbulent cloudtops of this giant planet. Complex cloud motions surround the Great Red Spot. (NASA)

has been observed since the mid-1600s, appears to be a long-lived storm in the planet's dynamic atmosphere. Many careful observers have reported smaller spots that last for only a few weeks or months in Jupiter's turbulent clouds.

Although it is the largest and most massive planet in the solar system, Jupiter has the fastest rate of rotation. At its equatorial latitudes, Jupiter completes a full rotation in only 9 hours 50 minutes 30 seconds. However, Jupiter does not rotate like a rigid object. The polar regions of the planet rotate a little more slowly than do the equatorial regions, which we can see by following features in the belts and zones. Near the poles, the rotation period is about 9 hours 55 minutes 41 seconds. The first person to notice this **differential rotation** of Jupiter was the Italian astronomer Giovanni Domenico Cassini in 1690. This was the same gifted observer who gave us the first accurate determination of Mars's rotation rate.

Jupiter's colorful cloudtops are the turbulent, uppermost layer of its thick atmosphere. Are the observed rotation rates of these clouds representative of the rotation rates of deeper levels or of a solid central core? Some intriguing clues come from radio waves emitted by Jupiter. In particular, radio emissions with wavelengths in the range of 3 to 75 cm vary slightly in intensity, with a period of 9 hours 55 minutes 30 seconds. This radio emission is believed to be directly associated with Jupiter's magnetic field, which is anchored deep inside the planet. In other words, radio observations reveal Jupiter's **internal rotation period,** which is slightly different from the atmospheric rotation we observe through a telescope.

**The Great Red Spot is a
high-pressure anticyclone**

During the 1970s, four American spacecraft flew past Jupiter and sent back spectacular closeup pictures of its dynamic atmosphere. Short-term changes in Jupiter's cloud cover are most apparent in the vicinity of the Great Red

 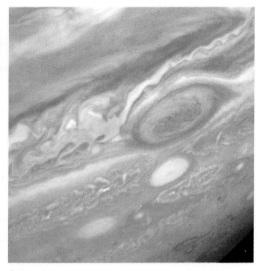

Figure 8-3 Changes in the Great Red Spot These two views, taken $4\frac{1}{2}$ years apart, show major changes in Jupiter's clouds around the Great Red Spot. **(a)** This view was taken in 1974 at a distance of 545,000 km from the cloudtops. **(b)** This view was taken in 1979 from a distance of 6 million kilometers. Notice the dramatic increase in turbulent cloud activity that occurred between the two spaceflights. (NASA)

a

b

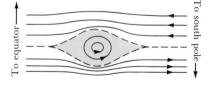

Figure 8-4 Circulation around the Great Red Spot The Great Red Spot spins counterclockwise, completing a full revolution in about six days. Meanwhile, winds to the north and south of the spot blow in opposite directions. Consequently, circulation associated with the spot resembles a wheel spinning between two oppositely moving surfaces. (Adapted from Andrew P. Ingersoll)

Spot. Over the past three centuries, Earth-based observers have also reported many long-term variations in the spot's size and color. At its largest, the Great Red Spot was so huge that three Earths could have fit side by side across it. At other times (as in 1976–1977), the spot faded from view. During two flybys in 1979, the Great Red Spot was only slightly larger than the Earth.

Figure 8-3 shows two contrasting views of the Great Red Spot. In 1974, the Great Red Spot was embedded in a broad white zone that dominated the planet's southern hemisphere. A few years later, however, the cloud structure had changed dramatically. A dark belt had broadened and encroached on the Great Red Spot from the north, and the entire region was apparently embroiled in much greater turbulence.

Careful examination of cloud motions in and around the Great Red Spot reveals that the spot rotates counterclockwise with a period of about six days. Furthermore, winds to the north of the spot blow to the west, whereas winds south of the spot move toward the east. The circulation around the Great Red Spot is thus like a wheel spinning between two oppositely moving surfaces (see Figure 8-4). This surprisingly stable wind pattern has survived for at least three centuries.

The Great Red Spot is a high-pressure system that protrudes above the surrounding cloudtops. Anyone familiar with weather forecasting knows that the Earth's weather is dramatically affected by high-pressure and low-pressure systems. A **high-pressure system** (commonly called a "high") is simply a place where more than the usual amount of air happens to be located. This excess air weighs down on the Earth's surface, resulting in high atmospheric pressure on the ground. A **low-pressure system** (commonly called a "low") is a place having less than the normal amount of air, resulting in low atmospheric pressure on the ground. If we could see air, highs would look like bulges in the atmosphere (where extra amounts of air are piled up) and lows would look like depressions or troughs.

Gravity has a strong influence on the basic dynamics of the Earth's weather. Gravity causes the air to flow "downhill" from the high-pressure bulges into the low-pressure troughs. But because the Earth is rotating, the

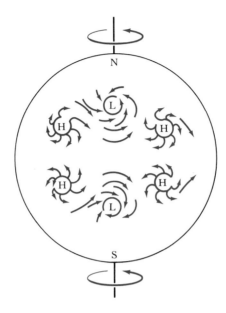

Figure 8-5   Cyclonic and anticyclonic wind flows   Because of a planet's rotation, cyclonic wind flowing into a low-pressure region or anticyclonic winds flowing out of a high-pressure region rotate either clockwise or counterclockwise, depending on the hemisphere in which the weather system is located.

wind flow from the highs toward the lows is not along straight lines. Instead, the Earth's rotation causes a deflection of the winds, producing either clockwise or counterclockwise flow about the high- and low-pressure regions.

Figure 8-5 shows the resulting wind flow about highs and lows. In Earth's northern hemisphere, winds blowing toward a low-pressure region rotate counterclockwise about the low, forming a **cyclone.** Winds blowing away from a high-pressure region rotate clockwise about the high, forming an **anticyclone.** In the southern hemisphere, the directions of rotation are reversed: cyclonic winds rotate clockwise and the anticyclonic winds rotate counterclockwise.

This basic pattern of wind flow applies to any other rotating planet. The six-day rotation of the Great Red Spot is counterclockwise. Thus the Great Red Spot is an anticyclone, a long-lasting high-pressure bulge in Jupiter's southern hemisphere.

Studies using computers help explain some of the phenomena in Jupiter's clouds

The flyby pictures showed other anticyclones in Jupiter's southern hemisphere. These features appear as **white ovals** like those in Figure 8-3b. The wind flow in these ovals clearly is counterclockwise.

Most of the white ovals are observed in Jupiter's southern hemisphere. **Brown ovals** are more common in its northern hemisphere (see Figure 8-6). The white ovals are the high-altitude cloudtops of high-pressure systems, but the brown ovals result from holes in Jupiter's cloud cover that permit us to see down into warmer regions of its atmosphere. Like the Great Red Spot, a white oval is apparently long-lived—Earth-based observers have found them in the same locations since 1938. However, a brown oval lasts for only a year or two.

The distribution of ovals is best seen in computer-generated Figure 8-7, which shows how Jupiter would look if you were located directly over the

Figure 8-6   A brown oval   Large brown ovals in Jupiter's northern hemisphere are caused by openings in the upper clouds that reveal warm, dark-colored gases below. The length of this oval is roughly equal to the Earth's diameter. The spacecraft Voyager 1 was 4 million kilometers from Jupiter when this picture was taken. (NASA)

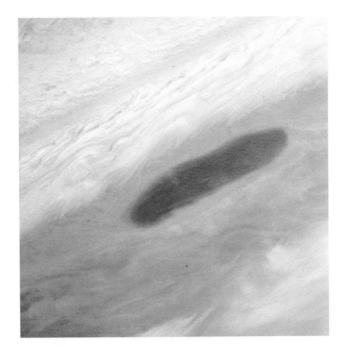

planet's north or south pole. Also note the regular spacing of cloud features such as ripples, plumes, and light-colored wisps.

Computer processing was also used to "unwrap" Jupiter, producing a map like the views of the planet in Figure 8-8. Note the changes that occurred during the four months between the flybys of *Voyager 1* and *Voyager 2*. The regular spacing of light-colored plumes is apparent in the equatorial regions of both pictures. The Great Red Spot moved westward and the white ovals moved eastward during the interval between the two flybys.

These beautiful Voyager photographs might suggest that Jupiter's clouds are the result of incomprehensible turmoil. You may wonder if there is anything constant in the Jovian atmosphere. Surprisingly, there is. Telescopic observations over the past 80 years and the Voyager data demonstrate that

*Figure 8-7   The northern and southern hemispheres   Computers were used to construct these views that look straight down onto Jupiter's north and south poles. **(a)** In the northern view, notice that light-colored plumes are evenly spaced around the equatorial regions. Several brown ovals are visible. **(b)** In the southern view, notice that the three biggest white ovals are separated from each other by almost exactly 90° of longitude. In both views, the banded belt–zone structure is absent near the poles. The ragged black spot is an area not photographed by the spacecraft. (NASA)*

a North pole

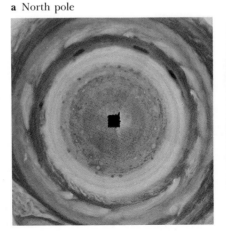

b South pole

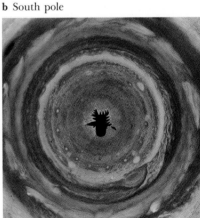

a *Voyager 1* view

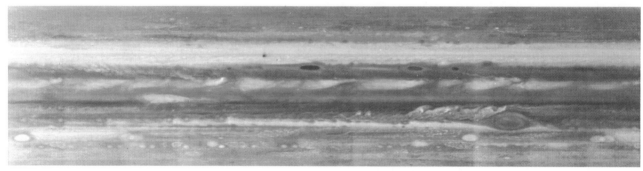

b *Voyager 2* view

Figure 8-8 A comparison of **Voyager 1** *and* **Voyager 2** *views Computer processing produced these two "unwrapped" views of Jupiter from* **(a)** *Voyager 1 and* **(b)** *Voyager 2. Each view was aligned with respect to Jupiter's magnetic axis so that displacements to the right or left represent real cloud motions. Notice that the Great Red Spot moved westward while the white ovals moved eastward during the four months between the two flybys. (NASA)*

the wind speeds in the Jovian atmosphere are remarkably stable. Although Jupiter's colorful bands change quite rapidly, the underlying wind patterns do not.

Jupiter's persistent wind patterns consist of broad streams of counterflowing eastward and westward winds. Computer simulations involving whirlpools and eddies caught between counterflowing streams help us understand both the long-term and short-term features in Jupiter's clouds. Andrew P. Ingersoll and his colleagues at the California Institute of Technology have pioneered the development of these calculations. Figure 8-9, for example, shows the behavior of a small, unstable whirlpool, technically called a **vortex.** This whirlpool is spinning too slowly to remain intact and so is torn apart by the counterflowing winds. Larger, rapidly rotating vortices do survive, however, in these simulations. The white ovals and the Great Red Spot endure by simply rolling with the wind currents (recall Figure 8-4). Figure 8-10 shows a simulation in which two stable vortices merge to form a larger one. The long-lived white ovals apparently maintain themselves in this fashion.

Saturn's spectacular rings are composed of fragments of ice and ice-coated rock

The magnificent rings of Saturn make this planet one of the most spectacular objects in the nighttime sky for an amateur astronomer using a small telescope. Saturn is so far away, however, that our Earth-based telescopes can reveal only the coarsest, large-scale features.

Figure 8-9 [left] The demise of an unstable vortex *In this computer simulation, a small vortex is rotating too slowly to remain intact. After slightly more than one week, the vortex is pulled apart between counterflowing zonal jets. (Adapted from Andrew P. Ingersoll)*

Figure 8-10 [right] The merging and maintenance of stable vortices *This computer simulation shows the collision and merger of two rapidly spinning, stable vortices. The result is a larger vortex—and the ejection of some material. Mergers of this type occur around the Great Red Spot. (Adapted from Andrew P. Ingersoll)*

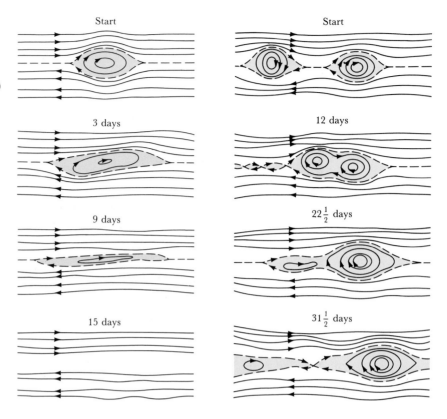

In 1675, G. D. Cassini discovered a dark division in the rings that looks like a gap about 5000 km wide. Astronomers also discovered stripes in Saturn's clouds similar to the belts and zones on Jupiter (see Figure 8-11). The contrast between Saturn's own belts and zones is less dramatic than the colorful patterns in Jupiter's atmosphere. Also, Saturn's stripes are not restricted to the equatorial regions as are Jupiter's; Saturn's alternating dark and light bands extend into the polar regions.

Figure 8-11 Saturn from the Earth *This view is one of the best ever produced of Saturn by an Earth-based observatory. Sixteen original color images taken during the same night in 1974 with the 1.5-m telescope at the Catalina Observatory were combined to make this photograph. Note the prominent Cassini division in the rings and the belts and zones in the Saturnian atmosphere. (NASA)*

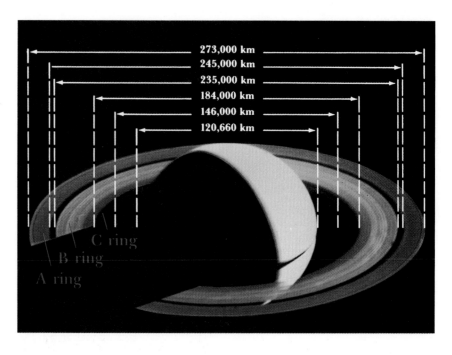

Figure 8-12 Saturn's classic rings *The three broad rings of Saturn are clearly visible in this photograph sent back by Voyager 1. The faint C ring exists in the region of 1.21 to 1.53 Saturn radii from the planet's center. The bright B ring occupies the region from 1.53 to 1.95 Saturn radii. The 5000-km-wide Cassini division lies between the B ring and the A ring, which occupies the region between about 2.03 and 2.26 Saturn radii. (NASA)*

After Cassini's discovery of a gap in Saturn's ring, astronomers began to view the ring as a system of rings. The **Cassini division** separates the outer **A ring** from the brighter **B ring** closer to the planet. By the mid-1800s, astronomers using improved telescopes were able to detect a faint **C ring** (or crepe ring) just inside the B ring (see Figure 8-12).

Earth-based views of the Saturnian ring system change dramatically as Saturn orbits slowly about the Sun (a Saturnian year is equal to $29\frac{1}{2}$ Earth years). This change can be observed because the rings, which lie in the plane of Saturn's equator, are tilted 27° from the plane of Saturn's orbit. Thus, over the course of a Saturnian year, the rings are viewed from various angles by an Earth-based observer (see Figure 8-13). At one time, the observer looks "down" on the rings; half a Saturnian year later, the "underside" of the rings is exposed to view from Earth. At intermediate times, the rings are seen edge on, when they disappear entirely from our view. Thus we conclude that the rings are very thin—less than 2 km in thickness according to recent estimates.

Astronomers have known for more than a century that Saturn's rings cannot possibly be solid, rigid, thin sheets of matter. The Scottish physicist James Clerk Maxwell proved mathematically in 1857 that such a broad, thin, rigid sheet would break apart and he concluded that Saturn's rings are composed of "an indefinite number of unconnected particles." In fact, the rings consist of millions of tiny moonlets, each circling Saturn along its own individual orbit.

Saturn's rings are very bright so the particles that form the rings must be highly reflective. Astronomers had long suspected the rings to be made of ice and ice-coated rocks, but confirming evidence was not obtained until the early 1970s, when astronomers identified the spectral features of frozen water in the near-infrared spectrum of the rings. In the early 1980s, two spacecraft—*Voyager 1* and *Voyager 2*—traveled past Saturn. Measurements

Figure 8-13 *The changing appearance of Saturn's rings* *Saturn's rings are tilted 27° from the plane of Saturn's orbit. Earth-based observers thus see the rings at various angles as Saturn moves around its orbit. Note that the rings seem to disappear entirely when they are viewed edge on. (Lowell Observatory)*

from these spacecraft tell us that the temperature of the rings ranges from only −180°C (−290°F) in the sunshine to less than −200°C (−300°F) in Saturn's shadow. Water ice is in no danger of melting or evaporating at these temperatures.

Voyager observations also showed that the ring particles range in size from snowflakes less than 1 mm in diameter up to icy boulders tens of meters across. It seems reasonable to suppose that all of this material is ancient debris that has failed to accrete into satellites.

Saturn's rings consist of thousands of closely spaced ringlets

During 1980 and 1981, the Voyager cameras sent back pictures showing unexpected details in the ring structures of Saturn. Its broad rings were seen to consist of hundreds upon hundreds of closely spaced thin bands, or

Figure 8-14 Saturn's rings from **Voyager 1** Voyager 1 *took this view of Saturn's rings from a distance of about 1.5 million kilometers. The C ring scatters light differently than the A or B rings do, so that it has a bluer color. The broad Cassini division is clearly visible, as is the narrow Encke division within the A ring. The thin F ring is visible just beyond the outer edge of the A ring. (NASA)*

ringlets, of particles (see Figure 8-14). Although intriguing suggestions have been proposed, scientists still do not understand just why Saturn's A, B, and C rings are divided into these thousands of ringlets.

The Voyager cameras also sent back the first high-quality pictures of the **F ring,** a thin ring visible just beyond the outer edge of the A ring in Figure 8-14. Closeup views revealed a startling and mysterious fact: the F ring is kinky and braided, actually consisting of several intertwined strands (see Figure 8-15). One Voyager image shows a total of five strands each about 10 km across. Voyager scientists were at a loss to explain this complex structure. The braids, kinks, knots, and twists in the F ring pose one of the most challenging puzzles in modern astronomy.

Through Earth-based telescopes, we see only the sunlit side of Saturn's rings. From this perspective, the B ring appears very bright, the A ring moderately bright, the C ring dim, and the Cassini division dark. The proportion of sunlight reflected back toward the Sun is directly related to the density of the fragments or particles in the ring. The B ring is bright because it has a high density of ice and rock fragments, whereas the darker Cassini division has a lower density of fragments.

Figure 8-15 Details of the F ring This Voyager 1 *photograph of the F ring was taken from a distance of 750,000 km. The total width of the F ring is about 100 km. Within this span are several discontinuous strands each roughly 10 km across. (NASA)*

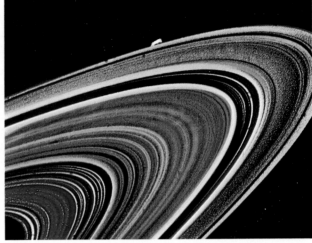

Figure 8-16 [left] Details of the A ring
This view of the underside of Saturn's rings
was taken from a distance of 740,000 km by
Voyager 1. *The thin F ring is clearly visible*
toward the right side of the picture. The
Cassini division and the outer edge of the B
ring are on the left side. The Cassini division
appears bright in this view of the shaded side
of the rings. (NASA)

Figure 8-17 [right] False-color view of
ring details Computer processing greatly
exaggerates subtle color variations in this view
from Voyager 2 *of the sunlight side of the*
rings. Note that the C ring and the Cassini
division appear bluish. Also note distinct color
variations across the A and B rings. (NASA)

Eight hours after it had crossed from the northern to the southern side of the rings, *Voyager 1* took the photograph in Figure 8-16. The Sun was shining down on the northern side of the rings at that time, and so Figure 8-16 shows the sunlight that passes *through* the rings. As expected, the B ring looks darkest here because little sunlight gets through its high concentration of fragments, whereas the Cassini division looks bright because sunlight passes relatively freely through its low concentration of fragments. However, the fact that the Cassini division does appear bright is clear evidence that it does contain some fragments. If it contained no fragments at all, we would see the black of space through it. The brightness in Figure 8-16 must be sunlight scattered by some fragments in the Cassini division.

Subtle color differences from one ring to the next give important clues about the chemical composition of the particles in the rings. These differences are clearly visible in Figure 8-17, in which the colors have been exaggerated by computer processing. The main chemical constituent is frozen water, but trace amounts of other chemicals (perhaps coating the surfaces of the ice particles) are probably the cause of the colors seen in this computer-enhanced view. Although these trace chemicals have not been identified, the existence of color variations that have probably persisted for millions of years suggests that the icy particles do not wander around or migrate substantially from one ringlet to the next.

Saturn's innermost satellites affect the appearance and structure of the rings

Astronomers had realized before the Voyager flybys that one of Saturn's moons, Mimas, has an effect on the ring system. Mimas is a moderate-sized satellite that orbits Saturn every 22.6 hours (see Figure 8-18). According to Kepler's third law, particles in the Cassini division should orbit Saturn every 11.3 hours. Consequently, on every second orbit, particles in the Cassini division line up between Saturn and Mimas. During these repeated alignments, the combined gravitational forces of Saturn and Mimas cause small fragments to deviate from their original orbits. In this way, Mimas depletes the Cassini division of dust that would otherwise scatter sunlight back toward Earth, which is why Earth-based astronomers see the division as a dark band.

Figure 8-18 Mimas Mimas is the smallest and innermost of Saturn's six moderate-sized satellites. This view was taken by Voyager 1 at a range of nearly 500,000 km. The huge impact crater, named Herschel after the satellite's discoverer, is 130 km in diameter. Mimas is only 400 km in diameter. (NASA)

The Voyager cameras also discovered three new ring systems: the D, E, and G rings. The **D ring** is Saturn's innermost ring system. It consists of a series of extremely faint ringlets located between the inner edge of the C ring and the Saturnian cloudtops. The **E ring** and the **G ring** both lie quite far from the planet, well beyond the outer edge of the A ring. Both of these outer ring systems are extremely faint, fuzzy, and tenuous. Each lacks the ringlet structure so prominent in the main ring systems. The E ring lies along the orbit of Enceladus, one of Saturn's icy satellites (see Figure 8-19). Some scientists suspect that geysers on Enceladus are the source of ice particles in the E ring.

The Voyager cameras also discovered two tiny satellites that follow orbits on either side of the F ring (see Figure 8-20). The gravitational forces of these two satellites keep the F ring particles in place. The outer satellite moves around Saturn at a slightly slower speed than that of the ice particles in the ring. As the ring particles pass by this satellite, they experience a tiny gravitational tug that tends to slow them down. These particles thus lose a little energy, which would cause them to fall into orbits a little closer to Saturn—except for the effect of the inner satellite. This satellite orbits the planet a little faster than the F ring particles do. As the satellite moves past the particles, its gravitational pull tends to speed them up, thus trying to nudge them into a slightly higher orbit. The combined effect of these two satellites focuses the icy particles into a well-defined narrow band about 100 km wide. Because of their confining influence, these two moons are called **shepherd satellites.**

A shepherd satellite that circles Saturn just beyond the outer edge of the A ring is responsible for the sharp outer edge of the A ring. As particles near the edge of the A ring pass by the slowly moving shepherd satellite, they feel a gravitational drag that slows them down slightly, preventing them from wandering into orbits farther from Saturn.

Figure 8-19 [left] Enceladus
This high-resolution image of Enceladus was obtained by Voyager 2 from a distance of 191,000 km. Ice flows and cracks strongly suggest that the surface has been subjected to recent geological activity. The youngest crater-free ice flows are estimated to be less than 100,000 years old. (NASA)

Figure 8-20 [right] The F ring and its two shepherds Two tiny satellites, each measuring about 50 km across, orbit Saturn on either side of the F ring. The gravitational effects of these two shepherd satellites focus and confine the particles in the F ring to a band about 100 km wide. This Voyager 2 picture was taken from a range of 10.5 million kilometers. (NASA)

The atmospheres of Jupiter and Saturn each have three main cloud layers

It is warm deep inside the atmospheres of Jupiter and Saturn but cooler at their cloudtops. Infrared measurements of Jupiter confirm that the temperature rises as the descent into the Jovian cloud cover increases. This situation is analogous to the Earth's atmosphere, which is cool at the highest cloudtops but warmer near the ground.

Over the range of temperatures in the Jovian atmosphere, gases emit energy primarily as infrared radiation. Figure 8-21 shows nearly simultaneous photographs of Jupiter at infrared and visible wavelengths. In the infrared picture, the brighter parts of the image correspond to hotter temperatures. There is also a striking correlation between brightness in the infrared image and color in the visible-light image. In other words, the various colors in Jupiter's clouds correspond to differing temperatures and hence to differing depths in the atmosphere. (It is customary to discuss these features in terms of depths measured from the cloudtops rather than altitudes above the surface because the exact location of any solid surface on Jupiter is unknown.) Bluish clouds correspond to the brightest parts of the infrared picture, and so these clouds must be the warmest and hence the deepest layers that we can see in the Jovian atmosphere. Brown clouds form the next highest layer, followed by whitish clouds and red clouds in the highest layer.

The Jovian atmosphere has a minimum temperature of about $-160°C$ ($-260°F$) at an altitude above the cloudtops where the atmospheric pressure is about 0.1 atm. A similar minimum occurs at the 0.1-atm level in Saturn's atmosphere where the temperature is $-180°C$ ($-290°F$). By analogy with the Earth's atmosphere (recall Figure 7-2), we can call this level the boundary between the stratosphere and the troposphere. As on the Earth, all the weather on Jupiter and Saturn takes place below the stratosphere.

Spectroscopic observations reveal that the atmospheres of Jupiter and Saturn contain methane (CH_4), ammonia (NH_3), and water vapor (H_2O). These compounds are the simplest combinations of carbon, nitrogen, and oxygen with hydrogen. From calculations of the behavior of these chemicals under various conditions of temperature and pressure, scientists conclude that Jupiter and Saturn both have three main cloud layers. The uppermost layer, with red and whitish clouds, is composed of crystals of frozen ammonia. Deeper in the troposphere, ammonia (NH_3) and hydrogen sulfide (H_2S) combine to produce a mid-level cloud layer of ammonium hydrosul-

Figure 8-21 Infrared and visible views
(a) This infrared photograph was taken through the 200-in. Palomar telescope. The brightest parts of the image correspond to holes in the clouds where deeper and warmer regions of the Jovian atmosphere are visible. Dark parts of the image correspond to the cool cloudtops. (b) This image in visible light was taken by Voyager 1 at almost the same time the infrared picture was taken. Comparisons between the two photographs show that cloud color is correlated with depth in the Jovian atmosphere. The bluish and brown clouds are roughly 100°C warmer and 100 km lower than the red and whitish clouds. (NASA)

a
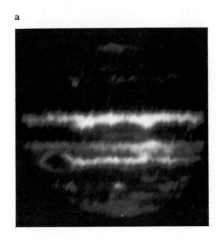

b

Figure 8-22 Temperature profiles of Jupiter and Saturn The structure of the upper atmospheres of Jupiter and Saturn is displayed in these graphs of temperature versus depth. Note that Saturn's atmosphere is more "spread out" than Jupiter's, which is a direct result of Saturn's weaker surface gravity. (Adapted from Andrew P. Ingersoll)

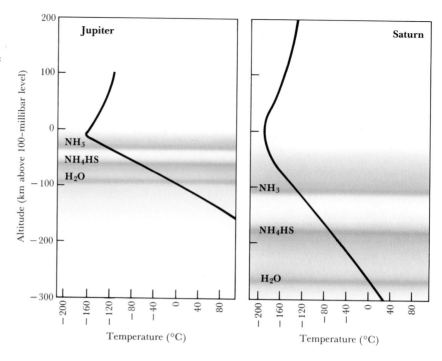

fide (NH_4SH) crystals. The third and deepest cloud layer has a bluish tint and is composed of water droplets and snowflakes of frozen water (see Figure 8-22).

Although their atmospheres are similar in structure and composition, Saturn and Jupiter are not identical in appearance. Saturn's clouds lack the colorful contrast of Jupiter's, although some of the Voyager photographs do show faint belts and zones (see Figure 8-23). After computer processing to enhance the colors in the photographs, details such as storm systems and ovals became visible (see Figure 8-24).

Jupiter and Saturn differ in appearance partly because they differ in mass. Jupiter's strong surface gravity compresses its three cloud layers into a range of 75 km in the upper atmosphere. Saturn's somewhat weaker surface

Figure 8-23 [left] Saturn's clouds from Voyager 1 This view of Saturn's cloudtops was taken by Voyager 1 at a range of 1.8 million kilometers. Note that there is substantially less contrast between belts and zones here than on Jupiter. The shadow of Dione appears at the bottom of the picture. (NASA)

Figure 8-24 [right] Eddy currents in Saturn's atmosphere Computer processing exaggerates the colors in this Voyager 2 picture of Saturn's northern mid-latitudes. The wavy line in the light blue ribbon is a pattern moving eastward at 150 m/sec (300 miles per hour). The dark oval and two puffy, blue-white spots below it are eddies drifting westward at roughly 20 m/sec (40 miles per hour). (NASA)

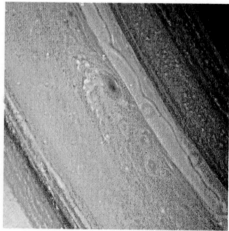

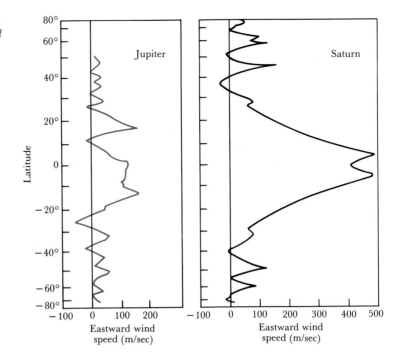

Figure 8-25 Wind speeds on Jupiter and Saturn Average wind speeds on Jupiter and Saturn are here plotted for latitudes ranging from 80° north to 80° south. The positive numbers are eastward velocities; negative are westward. Although both planets exhibit counterflowing currents, Saturn's equatorial zonal jet is much broader and faster than Jupiter's.

gravity subjects its atmosphere to less depression, and so the same three cloud layers are spread out over a range of nearly 300 km (see Figure 8-22). The colors of Saturn's clouds are less dramatic because the deeper layers are partly obscured by the thick atmosphere above them.

By following features in the Jovian and Saturnian clouds, scientists have determined wind speeds in their upper atmospheres. Both planets exhibit counterflowing eastward and westward currents. However, Saturn's equatorial jet is much broader and faster than Jupiter's (see Figure 8-25). In fact, wind speeds near Saturn's equator approach 500 m/sec (1000 miles per hour), approximately two-thirds the speed of sound.

Jupiter and Saturn emit more radiation than they receive from the Sun

Both Jupiter and Saturn have internal sources of energy. Each planet radiates more energy than it receives in the form of sunlight. Many scientists believe that the excess heat escaping from Jupiter is energy left over from the formation of the planet. As gases from the solar nebula fell into the protoplanet, vast amounts of gravitational energy were converted into thermal energy and became trapped far below Jupiter's clouds. For the past 4.5 billion years, Jupiter has been slowly cooling off as this trapped energy escapes in the form of infrared radiation.

Saturn is both smaller and less massive than Jupiter. One would thus expect Saturn to cool more rapidly than Jupiter and hence to emit less energy today. But, in fact, Saturn radiates about $2\frac{1}{2}$ times as much energy as it receives from the Sun, whereas Jupiter emits only about $1\frac{1}{2}$ times what it absorbs from sunlight. What might explain why Saturn emits so much more heat than Jupiter?

For some time before space probes were sent to the Jovian planets, astronomers had suspected that both Jupiter and Saturn have compositions similar to that of the original solar nebula and the Sun's atmosphere today. Each of these giant planets is massive enough and cool enough to have retained all of the gases that originally accreted from the solar nebula. The Voyager flybys did confirm that Jupiter's atmosphere has the same abundance of elements as the Sun: by weight, 82 percent hydrogen, 17 percent helium, and 1 percent all other elements. Surprisingly, however, the Voyager spacecraft reported that Saturn's atmosphere has less helium than expected. The chemical composition of Saturn's atmosphere, by weight, is 88 percent hydrogen, 11 percent helium, and 1 percent all other elements.

A brilliant hypothesis links Saturn's apparent deficiency of helium to the excess heat radiated by the planet. According to this theory, Saturn did indeed cool more rapidly than Jupiter. This cooling triggered a process analogous to the development of a rainstorm here on Earth. When the air is cool enough, humidity in the Earth's atmosphere condenses into raindrops that fall to the ground. On Saturn, according to this explanation, helium droplets rain downward from the planet's atmosphere toward its core. Helium thus appears to be deficient in Saturn's upper atmosphere merely because it has fallen farther down into the planet. Furthermore, as the helium droplets descend through the molecular hydrogen, the two gases rub against each other. The resulting friction produces heat that eventually escapes from Saturn.

The precipitation of helium from Saturn's clouds is calculated to have begun 2 billion years ago. The resulting release of energy adequately accounts for the extra heat radiated by Saturn since that time. Similar calculations for Jupiter indicate that only now it is reaching the stage where a significant amount of helium precipitation can begin in its outer layers. Saturn has therefore given us important clues about the probable course of Jupiter's future evolution.

The internal structures of Jupiter and Saturn can be deduced from their slightly flattened shapes

Jupiter's rapid rotation profoundly affects the overall shape of the planet. Even a casual glance through a small telescope shows that Jupiter is slightly flattened, or oblate. The diameter across Jupiter's equator (143,800 km) is 6.37 percent larger than the diameter from pole to pole (135,200 km). Thus Jupiter is said to have an **oblateness** of 6.37 percent, or 0.0637. Saturn is even more oblate than Jupiter (examine Figure 8-26). Saturn's equatorial diameter is about 10 percent larger than its polar diameter, and so its oblateness is about 0.10.

If Jupiter and Saturn were not rotating, they would be perfect spheres. A massive, nonrotating object naturally settles into a spherical shape in which every atom on its surface experiences the same intensity of gravity aimed directly at the object's center. However, Jupiter and Saturn are rotating rather rapidly. Saturn's equatorial rotation period (10 hours 14 minutes) is only slightly longer than Jupiter's (9 hours 50 minutes). Thus every part of each planet also experiences an outward-directed centrifugal force that is proportional to the distance from the axis of rotation. Equatorial regions are farther from the planet's axis of rotation than are the polar areas, and so the equatorial region experiences a stronger centrifugal force. As a result, the equatorial diameter is slightly larger than the polar diameter. This centrifugal stretching of the equatorial dimensions gives Jupiter and Saturn their characteristic oblate shapes.

Figure 8-26   Saturn from Voyager 2
Saturn is the most oblate planet in the solar system. Saturn's equatorial diameter is 12,000 km larger than its diameter from pole to pole. Voyager 2 sent back this picture when the spacecraft was 34 million kilometers away from the planet. (NASA)

At every point throughout each planet, the inward force of gravity is exactly balanced by the outward pressure of the compressed material plus the outward centrifugal effects of the planet's rotation. This balance is called **hydrostatic equilibrium.** The shape and density of a planet adjust themselves to ensure that this balance is maintained.

The shape of a planet is an excellent indicator of its internal structure. Two planets with the same mass, average density, and rotation will have slightly different oblateness if one planet has a compact core and the other does not. All other things being equal, a planet with a dense core will be more oblate than will a planet without a central core.

Detailed calculations strongly suggest that 4 percent of Jupiter's mass is concentrated in a dense, rocky core. Jupiter's oblateness is in fact consistent with a rocky core nearly 13 times as massive as the entire Earth. Some of this core was probably the original "seed" around which proto-Jupiter accreted.

Jupiter's rocky core is probably not much bigger than the Earth, even though it is 13 times more massive than our planet. The tremendous crushing weight of the remaining 305 Earth masses of Jupiter's bulk compresses the core down to a sphere 20,000 km in diameter (Earth's diameter is 12,800 km). The pressure at Jupiter's center is about 80 million atmospheres, which squeezes the rocky material of Jupiter's core to a density of about 20 g/cm^3. The temperature at the planet's center is probably about 25,000 K. In contrast, the temperature at Jupiter's cloudtops is only 165 K.

Jupiter is largely composed of hydrogen. A hydrogen atom consists of a single proton orbited by a single electron. Deep inside Jupiter, however, pressures are so great that the electrons are stripped from their protons. The electrons, no longer bound to protons, are free to wander around and thereby create electric currents. In other words, the highly compressed hydrogen deep inside Jupiter behaves like a metal. It is therefore called **liquid metallic hydrogen.**

Detailed calculations strongly suggest that molecular hydrogen is transformed into liquid metallic hydrogen when the pressure exceeds 3 million atmospheres. This transition occurs at a depth of approximately 17,000 km below Jupiter's cloudtops. Thus Jupiter's internal structure divides into three distinct regions: (1) a rocky core, surrounded by (2) a 44,000-km-thick layer of liquid metallic hydrogen, which is in turn surrounded by (3) an

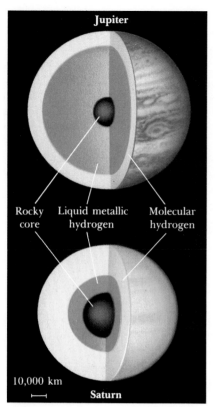

Figure 8-27 *Internal structure of Saturn and Jupiter* *There are three distinct layers in Saturn's interior, as in Jupiter's. The planet's rocky core is surrounded by a layer of liquid metallic hydrogen, which in turn is enveloped in a thick layer of molecular hydrogen. The two diagrams are drawn to the same scale.*

outer 17,000-km-thick layer of ordinary molecular hydrogen. All of the colorful cloud patterns visible through telescopes are located in the outermost 100 km of this outer layer.

Information about the average density, oblateness, and the probable chemical composition of Saturn, leads astronomers to infer that Saturn's interior structure resembles that of Jupiter: a solid, rocky core surrounded by a mantle of liquid metallic hydrogen, surrounded in turn by a layer of molecular hydrogen.

Saturn is less massive than Jupiter and thus the pressures inside it are not as high as those inside Jupiter. Saturn's rocky core is therefore less compressed and larger than Jupiter's. Also, Saturn's weaker gravity is unable to convert as large a quantity of hydrogen into a liquid metallic as occurs in Jupiter. Thus the relative thicknesses of the layers differ in the models of Saturn's and Jupiter's internal structure (see Figure 8-27). Saturn's rocky core is about 32,000 km in diameter but its layer of liquid metallic hydrogen is only 12,000 km thick.

It is clear from Figure 8-27 that a large percentage of Jupiter's enormous bulk is electrically conductive liquid metal. Because of Jupiter's rapid rotation, electric currents in this thick layer of liquid metallic hydrogen generate a powerful magnetic field, in much the same way that liquid portions of the Earth's core produce the Earth's magnetic field. The intrinsic strength of Jupiter's magnetic field is 19,000 times greater than the Earth's. The Jovian magnetic field is so much stronger than ours because Jupiter's liquid metallic region is so much larger than the Earth's and because Jupiter rotates so much faster than the Earth does.

Jupiter's powerful magnetic field surrounds the planet with an enormous magnetosphere, large enough to envelop the orbits of many of its moons. For Earth-based astronomers, the only evidence of this magnetosphere is a faint hiss of radio static. However, four spacecraft that journeyed to Jupiter in the 1970s revealed the awesome dimensions of the Jovian magnetosphere. The volume surrounded by the shock wave is nearly 30 million kilometers across. In other words, if you could see Jupiter's magnetosphere from the Earth, it would cover an area in the sky 16 times larger than the full Moon.

Figure 8-28 *Jupiter's magnetosphere* *Like other planetary magnetospheres, Jupiter's has a bow shock wave, a magnetosheath, and a magnetopause. In Jupiter's case, gas pressure from a hot plasma keeps the magnetosphere inflated, thereby holding off the solar wind. Particles trapped inside the magnetosphere are spewed out into a vast current sheet by the planet's rapid rotation. Jupiter's axis of rotation is inclined to its magnetic axis by about 11°.*

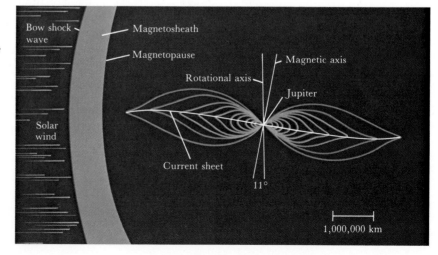

The inner regions of our magnetosphere are dominated by two huge Van Allen belts (recall Figure 7-34) filled with charged particles. The same sort of belts would probably exist around Jupiter were it rotating slowly. But because Jupiter rotates rapidly, centrifugal forces spew out the particles into a huge electrically charged **current sheet** (see Figure 8-28). This current sheet lies in the plane of Jupiter's magnetic equator. Jupiter's magnetic axis is inclined slightly from the planet's axis of rotation and the orientation of its magnetic field is the reverse of Earth's (a compass would point toward the south pole on Jupiter).

Saturn's mantle of liquid metallic hydrogen, like Jupiter's, produces a planetwide magnetic field. However, Saturn's magnetic field is somewhat weaker than Jupiter's because of Saturn's slightly slower rotation and much smaller volume of liquid metallic hydrogen. Data from spacecraft indicate that Saturn's magnetosphere resembles that of Jupiter but that it contains radiation belts similar to those of Earth instead of a huge current sheet like that in Jupiter's magnetosphere.

Earth-based observations provide limited information about Uranus and Neptune

At first glance, Uranus and Neptune appear to be twins. They have nearly the same size, mass, density, and chemical composition. Until only a few years ago, however, the Earth and Venus were thought to be near-twins. Probes of the Venusian environment startled us with evidence of how different Venus and Earth are despite their superficial similarities. So we curbed our desire to speculate about similarities between Uranus and Neptune until the *Voyager 2* flyby of Neptune in 1989.

Through Earth-based telescopes, Uranus and Neptune are dim and uninspiring sights. Each appears as a small, hazy, featureless disk with a faint greenish-blue tinge (see Figures 8-29 and 8-30). Earth-based observations have provided only the most basic information about these two worlds simply because they are so far away.

The outer two Jovian planets are significantly different from the inner two. The large bulks of Jupiter and Saturn are composed primarily of

Figure 8-29 [left] **Uranus** *Nearly 3 billion kilometers from the Sun, Uranus receives only $\frac{1}{400}$ the intensity of sunlight that we experience here on Earth. Uranus is therefore a dim and frigid world. At its brightest, Uranus appears as bright as the faintest stars we can see with our unaided eyes from Earth. Even though its diameter is about four times Earth's, Uranus always shows us a disk less than 4 arc sec across. (New Mexico State University Observatory)*

Figure 8-30 [right] **Neptune** *At 4.5 billion kilometers from the Sun, Neptune's surface receives only $\frac{1}{900}$ the intensity of sunlight we receive on Earth. Neptune is therefore dimmer than Uranus. Although about the same size as Uranus, Neptune looks smaller through Earth-based telescopes because it is farther away. The maximum possible angular diameter of Neptune's disk is 2.2 arc sec. That is roughly the same size as a dime seen from a distance of 1 km. (New Mexico State University Observatory)*

hydrogen and helium, like the Sun. Uranus and Neptune, however, are distinctly smaller and less massive. If Uranus and Neptune also had solar abundances of the elements, their smaller masses would produce less compression and therefore lower average densities than those of Jupiter and Saturn. In fact, however, Uranus and Neptune have average densities comparable to or greater than those of Jupiter or Saturn. We must conclude therefore that Uranus and Neptune contain greater proportions of the heavier elements such as oxygen, nitrogen, carbon, silicon, and iron in addition to abundant hydrogen and helium.

Astronomers who have calculated the internal structures of Uranus and Neptune agree that both planets must have substantial rocky cores like those at the centers of Jupiter and Saturn. Unlike Jupiter and Saturn, however, neither Uranus nor Neptune has the mass needed to achieve the pressure to produce liquid metallic hydrogen. Some scientists suspect that extremely dense atmospheres may extend all the way down to the rocky cores of Uranus and Neptune.

Neptune emits slightly more radiation than it receives from the Sun, whereas Uranus does not. Although farther from the Sun, Neptune's cloudtops register nearly the same temperature as Uranus's (57 K = $-216°C$ = $-357°F$). We saw that both Jupiter and Saturn have internal energy sources. Jupiter is still releasing energy trapped inside it during its formation and Saturn's excess heat probably comes from the precipitation of helium. Neither of these mechanisms would work on Uranus or on Neptune because these outer planets are too small and contain too little hydrogen and helium. The source of Neptune's internal heat therefore remains a mystery.

Voyager 2 revealed many details about Uranus, its rings, and its satellites

After nearly $8\frac{1}{2}$ years of having coasted through interplanetary space, _Voyager 2_ arrived at Uranus in January 1986. As pictures and data poured in, scientists learned more about this remote world than all the knowledge that had accumulated since Uranus was discovered two centuries ago.

It was already known from Earth-based observations that Uranus's axis of rotation lies very nearly in the plane of its orbit (see Figure 8-31). Consequently, as Uranus moves along its 84-year orbit, the planet's north and south poles alternately point toward or away from the Sun, producing exaggerated seasons. During the summertime, near Uranus's north pole the Sun remains high above the horizon for many years while southern latitudes are subjected to a continuous, frigid winter night. Forty-two years later, the situation is reversed.

Uranus's north pole was aimed almost directly at the Sun as _Voyager 2_ approached the planet. Thus Figure 8-32 shows its illuminated northern hemisphere. No clouds or any other atmospheric features are seen, but computer processing did reveal a smoglike haze over the pole.

Voyager's instruments discovered that Uranus has a magnetic field roughly half as strong as Earth's. Furthermore, Uranus's magnetic axis (that is, the line joining the north and south magnetic poles) is tilted away from Uranus's axis of rotation by 55°. This inclination is surprising because most planets have their magnetic and rotational axes nearly aligned. For example, Earth and Jupiter each have their magnetic axes inclined by only 11° from their rotational axes. Uranus is thus unique because of the unusual orientations of both its rotational and magnetic axes.

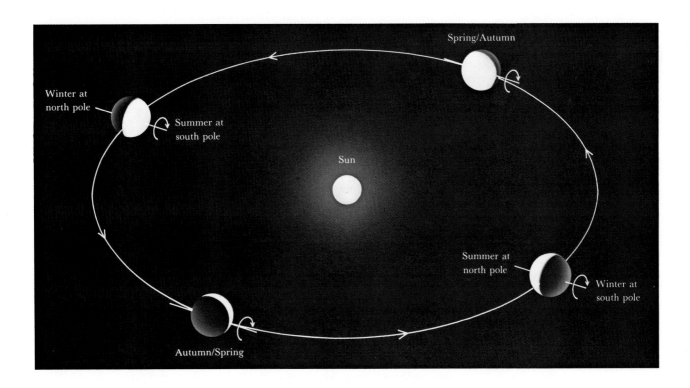

Figure 8-31 Exaggerated seasons on Uranus *Uranus's axis of rotation is tilted so steeply that it lies nearly in the plane of the planet's orbit. Seasonal changes on Uranus are thus severely exaggerated. For example, during midsummer at Uranus's south pole, the Sun appears nearly overhead for many Earth years, while the planet's northern regions are subjected to a long, continuous winter night.*

For many years, Uranus's rotation period had been a topic of controversy among astronomers whose Earth-based observations gave conflicting results. *Voyager 2* was able to detect regular changes in radio emission from Uranus's magnetosphere that repeated every 16.8 hours. These changes are caused by the motion of Uranus's oblique magnetic field as it is carried around by the planet's rotation. The magnetic field is presumably anchored

Figure 8-32 Uranus from Voyager 2 *No distinctive cloud patterns are seen in any of the Voyager views of Uranus. The blue-green appearance of Uranus comes from methane in the planet's atmosphere that absorbs red wavelengths from the incoming sunlight. The spacecraft was 18 million kilometers from the planet when this picture was taken. (NASA)*

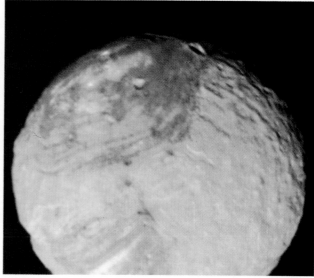

Figure 8-33 [left] The rings of Uranus
Voyager 2 *was in the shadow of Uranus when it took this 1½-minute exposure of Uranus's rings. The rings are much darker than Saturn's, and this long exposure revealed many very thin rings and dust lanes not previously seen. The short streaks are star images blurred because of the spacecraft's motion during the exposure. (NASA)*

Figure 8-34 [right] Miranda
Curious banded features cover much of Miranda, the smallest of Uranus's five major satellites. High-resolution photographs taken by Voyager *show many valleys and ridges parallel to these bands. Miranda is 480 km in diameter. A few impact craters 10 to 30 km in diameter can be seen. (NASA)*

deep inside the planet, and so 16.8 hours must be the rotation period of Uranus's core.

Revolving around Uranus in the plane of the planet's equator are numerous satellites and a system of thin, dark rings. Nine of these rings, ranging in width from 10 to 100 km, were discovered in 1977 when Uranus passed in front of a star. The star's light was momentarily blocked by each of the rings, which proved their existence to astronomers. A 1½-minute exposure taken while *Voyager* was in Uranus's shadow revealed numerous additional very thin rings (see Figure 8-33).

Five of Uranus's satellites, ranging in diameter from 480 to nearly 1600 km, were known prior to the Voyager mission. *Voyager's* cameras discovered 10 additional satellites, each less than 50 km across. Several of these tiny moons are shepherd satellites whose gravity confines particles to the thin rings that circle Uranus.

Of all these moons, Miranda is the most fascinating because it is covered with unusual wrinkled and banded surface features (see Figure 8-34). Miranda is the smallest of Uranus's five main satellites and the impact of an asteroid could have temporarily broken it into several pieces. Perhaps Miranda was shattered by such a collision and the blocklike features on its surface show how various chunks of Miranda came back together.

Neptune is a cold, bluish world with Jupiter-like surface features and a Uranus-like interior

The arrival of *Voyager 2* at Neptune in August 1989, capped one of NASA's most ambitious and most successful missions. Through Earth-based telescopes, Neptune presents a tiny, featureless disk that is barely distinguishable from a star. Scientists were therefore overjoyed at the detailed, close-up pictures and wealth of data sent back to Earth by the spacecraft.

At first glance, Neptune looks like a bluish Jupiter (see Figure 8-35). Methane in Neptune's atmosphere absorbs longer visible wavelengths but not the shorter ones. Sunlight reflected from Neptune's clouds is thus depleted of reds and yellows, making the planet appear quite blue.

The most prominent feature in Neptune's atmosphere, called the **Great Dark Spot,** gives the planet its distinctly Jupiter-like appearance. The Great

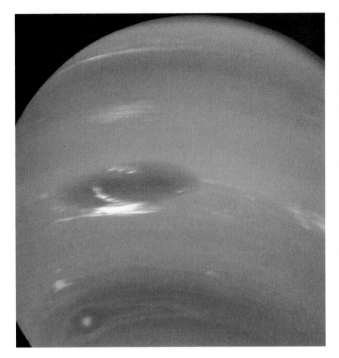

Figure 8-35 [left]   Neptune   *This view from* Voyager 2 *looks down on the southern hemisphere of Neptune. The Great Dark Spot, which is about the same size as the Earth, is near the center of this picture. Note the white, wispy methane clouds. Toward the lower left is a smaller dark spot, located 54° south of Neptune's equator. (NASA)*

Figure 8-36 [right]   Cirrus clouds over Neptune   *This picture shows vertical relief in Neptune's bright, methane cloud streaks.* Voyager 2 *photographed these clouds north of Neptune's equator near the terminator (the border between day and night on the planet). Note the shadows cast by the clouds onto the main cloud deck. (NASA)*

Dark Spot falls at about the same latitude on Neptune as the Great Red Spot falls on Jupiter, and occupies a proportionate amount of the planet's surface area.

The wind flow in the Great Dark Spot is anticyclonic, just as it is in Jupiter's Great Red Spot. Both features are high-pressure systems out of which winds flow counterclockwise. This anticyclonic flow on Neptune is difficult to observe, because high-elevation clouds hover over the Great Dark Spot. These clouds do not move along with the winds in the atmosphere and may be compared with the so-called lenticular clouds on Earth, which seem to hang over mountain tops. On Earth, winds do blow through these clouds at high speeds. The clouds seem not to move, however, because moisture carried by the winds condenses into visible ice crystals only where flow over the mountain pushes the gases to cooler regions of the upper atmosphere. As the winds descend down the leeward side of the mountain, the ice crystals turn back into vapor and the cloud disappears. Neptune's whitish, cirruslike clouds are made of methane ice crystals. As prevailing winds on Neptune carry the methane to elevated heights over the Great Dark Spot, the gas condenses into visible clouds.

The high elevation of Neptune's cirrus clouds was confirmed by photographs showing these clouds casting shadows on the main cloud deck (see Figure 8-36). Knowing the angle of the Sun when the pictures were taken, scientists estimate that these clouds are 50 to 70 km above the cloud deck. (In comparison, Earth's cirrus clouds hover about 7 to 9 km above sea level.) Between the cirrus and the cloud deck, Neptune's atmosphere is quite clear.

Neptune's belts and zones are much fainter than Jupiter's. Most prominent is a broad, darkish band at high southern latitudes. Embedded in this band is a smaller dark spot, about a third the size of the Great Dark Spot. White wispy clouds seem to hover over the smaller dark spot just as they do over the Great Dark Spot.

Figure 8-37 Neptune's rings *Two main rings are easily seen in this view alongside an overexposed image of Neptune. Careful examination of this picture also reveals a faint inner ring. A faint sheet of particles, whose outer edge is located between the two main rings, extends inward toward the planet. (NASA)*

Temperatures in the Neptunian atmosphere range from about 60 K at the equator and the poles to about 50 K at mid-latitudes. This peculiar situation, wherein the poles and the equator are both warmer than are mid-latitudes, is perhaps caused by the movement of gases in Neptune's upper atmosphere. At mid-latitudes, gases cool as they rise upward, but at the poles and equator the gases warm as they descend to lower elevations.

The dominant wind flow in the cloud layer that constitutes Neptune's visible surface is unlike that on any other Jovian planet. *Voyager 2* detected periodic variations in radio emission from Neptune every 16 hours 3 minutes. We speculate that these emissions reveal the underlying rotation period of Neptune assuming they are driven by the planet's magnetic field, which is embedded in its interior. Most of Neptune's cloud cover revolves about the planet more slowly than the planet rotates. Some cloud features take as long as 18 hours to go once around the planet. On Jupiter, Saturn, and Uranus, the clouds revolve around the planets more rapidly than the planets themselves rotate. Thus, Neptune's atmospheric circulation is retrograde.

While its surface resembles that of Jupiter, Neptune's interior resembles that of Uranus. Since Uranus and Neptune have nearly the same size, mass, and density, it is reasonable to suppose that they have roughly the same chemical composition and internal structure. Both planets probably have a rocky core surrounded by a watery mantle and a thick atmosphere primarily composed of hydrogen and helium with an admixture of methane.

While probing Neptune's magnetosphere, *Voyager 2* made the surprising discovery that the planet's magnetic axis is inclined to its axis of rotation by 50°. This is also reminiscent of Uranus, whose magnetic axis is inclined by 59° to its rotational axis. The magnetic fields of all four Jovian planets are oriented opposite to that of Earth; a compass on a Jovian planet points southward.

Like Uranus, Neptune is surrounded by a system of thin, dark rings (see Figure 8-37). It is so cold at Uranus and Neptune that the ring particles can retain methane ice. Scientists speculate that eons of radiation damage have converted this methane ice into darkish carbon compounds, thus accounting for the low reflectivity of the rings.

Having completed its mission to the outer planets, *Voyager 2* is headed out of the solar system. For many years to come, scientists will continue to monitor the spacecraft's instruments as they probe the solar wind at extreme distances from the Sun. At present there are no plans to return to either Uranus or Neptune.

Summary

· Jupiter is by far the largest and most massive planet in the solar system.

· Jupiter and Saturn are primarily composed of hydrogen and helium. Both planets have an overall chemical composition very similar to that of the Sun.

· Because of their rapid rotation, Jupiter and Saturn are noticeably oblate, which provides important clues about their internal structure.

   Jupiter and Saturn both probably have rocky cores surrounded by a thick layer of liquid metallic hydrogen and an outer layer of ordinary hydrogen gas.

- The visible features of Jupiter (belts, zones, the Great Red Spot, ovals, and colored clouds) exist in the outermost 100 km of its atmosphere. Saturn has similar features, but they appear much fainter.

  There are three cloud layers in the upper atmospheres of both Jupiter and Saturn. Saturn's are spread out over a greater altitude range than those of Jupiter, and so the colors of the Saturnian atmosphere are somewhat obscured.

  The colored ovals visible in the Jovian atmosphere represent gigantic storms, some of which (such as the Great Red Spot) are stable and persist for years. The ovals are cyclonic or anticyclonic storms at the boundaries between wind streams moving in opposite directions around the planet.

- Jupiter and Saturn both emit more heat than they receive from the Sun. Presumably Jupiter is still cooling. On Saturn, the precipitation of helium downward into the planet is probably the cause of its excess heat.

- Jupiter has a strong magnetic field created by currents in the metallic-hydrogen layer. Its huge magnetosphere contains a vast current sheet of electrically charged particles. Saturn's magnetosphere is similar to Jupiter's but has Earthlike radiation belts instead of a current sheet.

- Saturn is circled by a system of thin, broad rings lying in the plane of the planet's equator. Each of Saturn's major rings is composed of a great many narrow ringlets consisting of numerous fragments of ice and ice-coated rock.

  Some of the ring boundaries are produced by shepherd satellites, whose gravitational pull restricts the orbits of the ring fragments.

- Uranus and Neptune are quite similar to each other in appearance, mass, size, and chemical composition. They each have a substantial rocky core possibly surrounded by an extremely dense atmosphere.

- Uranus is unique in that its axis of rotation lies nearly in the plane of its orbit, producing greatly exaggerated seasonal changes on the planet.

- Uranus has a system of thin, dark rings and five satellites similar to the moderate-sized moons of Saturn.

- Neptune's surface features resemble those of a bluish Jupiter, having a Great Dark Spot, and faint belts and zones.

- Neptune's magnetic field is like Uranus's in that its magnetic axis is inclined steeply to the planet's axis of rotation.

- Like Uranus, Neptune is surrounded by a system of thin, dark rings. The low reflectivity of the ring particles may be due to radiation-damaged methane ice.

....

Review questions

1 Describe the appearance of Jupiter's atmosphere. Which features are long-lived and which are fleeting?

2 Describe the structure of Saturn's rings. What are they made of?

3 If Jupiter does not have any observable solid surface and its atmosphere rotates at different rates, how are astronomers able to determine the planet's rotation rate?

4 Why do features in Saturn's atmosphere appear to be much fainter and "washed out" compared to features in Jupiter's atmosphere?

5 Why do astronomers believe that Jupiter does not have a large iron-rich core even though the planet possesses a strong magnetic field?

6 Compare and contrast Jupiter's magnetosphere with the magnetosphere of the Earth.

7 Explain how shepherd satellites operate. Is "shepherd satellite" an appropriate term for these objects? Explain.

8 Could astronomers of antiquity see Uranus? If so, why do you suppose it was not recognized as a planet?

9 Compare the ring systems of Saturn and Uranus. Why were Uranus's rings unnoticed until the 1970s?

10 Compare and contrast the internal structures of Jupiter and Saturn with the internal structures of Uranus and Neptune. Can you propose an explanation to account for the differences between the inner and outer Jovian planets?

Advanced questions

11 Explain why Saturn is more oblate than Jupiter even though Saturn rotates more slowly than Jupiter.

12 What sort of experiment would you design in order to establish whether Jupiter has a rocky core?

Discussion questions

13 Describe some of the semipermanent features in Jupiter's atmosphere. What factors influence their longevity? Compare and contrast these long-lived features with some of the transient phenomena seen in Jupiter's clouds.

14 Suppose that you were designing a mission to Jupiter that involved an airplanelike vehicle that would spend many days (months?) flying through the Jovian clouds. What observations, measurements, and analyses should this aircraft make? What dangers might it encounter and what design problems would you have to overcome?

15 NASA and the Jet Propulsion Laboratory have tentative plans to place spacecraft in orbit about Uranus and Neptune early in the twenty-first century. What kinds of data should be collected and what questions would you like to see answered by these missions?

For further reading

Belton, M. "Uranus and Neptune." *Astronomy*, February 1977, p. 6.

Elliot, J., et al. "Discovering the Rings of Uranus." *Sky & Telescope*, June 1977, p. 412.

Esposito, L. "The Changing Shape of Planetary Rings." *Astronomy*, September 1987, p. 6.

Gore, R. "Voyager Views Jupiter." *National Geographic*, January 1980.

Gore, R. "Saturn: Riddle of the Rings." *National Geographic*, July 1981.

Ingersoll, A. "Jupiter and Saturn." *Scientific American*, December 1981.

Johnson, T., and Yeates, C. "Return to Jupiter: Project Galileo." *Sky & Telescope*, August 1983, p. 99.

Morrison, D. *Voyages to Saturn*. NASA SP-451, 1982.

Morrison, D., and Samz, J. *Voyage to Jupiter*. NASA SP-439, 1980.

Pollack, J., and Cuzzi, J. "Rings in the Solar System." *Scientific American*, November 1981.

Washburn, M. *Distant Encounters: The Exploration of Jupiter and Saturn*. Harcourt Brace Jovanovich, 1983.

9 The smaller terrestrial worlds

Moon

Io

Europa

Mercury

Callisto

Titan

Ganymede

The smaller terrestrial planets   Mercury, our Moon, and the largest satellites of Jupiter and Saturn are shown here to the same scale. Each of these worlds has its own unique characteristics and all are large enough to be classified as planets. Only Titan possesses an atmosphere. Not shown is Neptune's largest satellite, Triton, which is slightly smaller than our Moon. (NASA)

In this chapter we examine Mercury along with seven giant satellites large enough to qualify as terrestrial planets. We find that Mercury has a cratered, lunarlike surface but an Earthlike, iron-rich core with a magnetic field. We then follow the astronauts to the Moon and examine lunar rocks containing important clues about the Moon's history. Orbiting Jupiter are four more unique terrestrial worlds: Io with its sulfur volcanoes, Europa covered by a thin layer of shifting ice, and Ganymede and Callisto, each surrounded by a thick mantle of water and ice. At Saturn we find Titan, the only satellite with a substantial atmosphere, thicker even than Earth's. We then discuss Triton, the largest satellite of Neptune, and summarize the discoveries made during the *Voyager 2* flyby in 1989. Finally we look briefly at Pluto, the smallest and most distant planet in the solar system.

In recent years spacecraft have revealed seven worlds that are roughly the same size as Mercury. Each of these seven giant satellites (recall Table 6-3) has its own unique geology. Along with Mercury, these seven worlds give us a new perspective on the variety of which nature is capable.

Mercury has a Moonlike surface and an Earthlike interior

Until 1974, we knew very little about Mercury, the small planet that formed in the warm, inner regions of the solar nebula. Information about Mercury was difficult to obtain for two simple reasons: it is very small, and it is very near the Sun. Indeed, Mercury is so close to the Sun that most people (including many astronomers) have never seen it.

The best opportunities to see Mercury occur when the planet is as far from the Sun as it can be, at greatest eastern or western elongation (recall Figure 3-4). For a few days near the time of greatest eastern elongation, Mercury appears as an "evening star," hovering low over the western horizon for a short time after sunset. Near the time of greatest western elongation, it can be glimpsed as a "morning star," heralding the rising Sun in the brightening eastern sky.

Mercury travels around the Sun faster than any other object in the solar system, taking only 88 days to complete a full orbit. Thus Mercury passes through inferior conjunction at least three times a year, and you might expect occasionally to see Mercury silhouetted against the Sun in what is called a solar **transit.** Transits of Mercury across the Sun are not very common, though, because Mercury's orbit is tilted 7° to the plane of the Earth's orbit. As a result, Mercury usually lies well above or below the Sun at the moment of inferior conjunction. Only 14 solar transits of Mercury are scheduled for the entire twentieth century (see Figure 9-1).

Naked-eye observations of Mercury are best made at dusk or dawn, but the best telescopic views are obtained at midday when it is high above the degrading atmospheric effects near the horizon. The photographs in Figure 9-2, among the finest Earth-based views of Mercury ever recorded, were taken at midday. Because of its small size and nearness to the Sun, you cannot see much surface detail on Mercury through an Earth-based telescope. At best only a few faint, hazy markings can be identified.

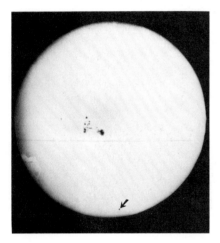

Figure 9-1   A transit of Mercury   Roughly a dozen transits of Mercury occur each century. This photograph shows the tiny planet silhouetted against the Sun during a transit on November 14, 1907. (Yerkes Observatory)

Figure 9-2   Earth-based views of Mercury   These two views are among the finest photographs of Mercury ever produced with an Earth-based telescope. Hazy markings are faintly visible on the tiny planet. (New Mexico State University Observatory)

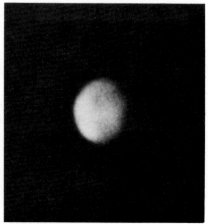

Figure 9-3   Mercury and our Moon
Mercury (left) and our Moon (right) are shown here to the same scale. Mercury's diameter is 4878 km and the Moon's is 3476 km. For comparison, the distance from New York to Los Angeles is 3944 km (2451 miles). Both worlds have heavily cratered surfaces and virtually no atmospheres. Mercury has a substantial iron-rich core and a magnetic field; the Moon does not. Daytime temperatures at the equator on Mercury reach 430°C (800°F), hot enough to melt lead or tin. (NASA; Lick Observatory)

Our first detailed knowledge about Mercury's surface was acquired in 1974, when *Mariner 10* coasted to within 756 km (470 miles) of the planet's surface. As the spacecraft closed in on Mercury, scientists were surprised by the Moonlike pictures appearing on their television monitors. It became obvious that Mercury is a barren, desolate, heavily cratered world. Figure 9-3 shows a typical closeup view of Mercury and a picture of our Moon for comparison.

Although our first impression is of a lunarlike landscape, closer scrutiny of Mercury's surface revealed some significant nonlunar characteristics. Lunar craters are densely packed, one overlapping the next. In sharp contrast, Mercury's surface has extensive **intercrater plains** (see Figure 9-4).

Astronomers believe that most craters on both Mercury and the Moon were produced during the solar system's first 700 million years. The strongest evidence comes from the direct analysis and dating of Moon rocks brought back by the Apollo astronauts. Debris remaining after the planets had formed rained down on these young worlds, gouging out most of the craters we see today.

Astronomers agree that the Moon and the terrestrial planets must have been completely molten spheres of liquid rock at first. After a few hundred million years, their surfaces solidified as the rock cooled. Nevertheless, large meteoroids could still easily puncture the thin cooling crusts, allowing

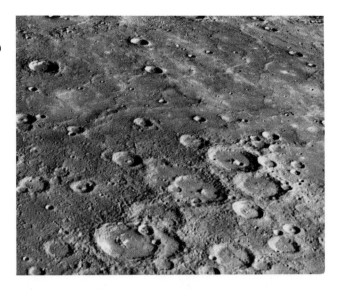

Figure 9-4 *Mercurian craters and intercrater plains* *This view of Mercury's northern hemisphere was taken by Mariner 10 at a range of 55,000 km (34,000 miles) from the planet's surface. Numerous craters and extensive intercrater plains appear in this photograph, which covers an area 480 km (300 miles) wide. (NASA)*

molten lava to well up from their interiors. Older craters were obliterated as seas of molten rock flooded portions of the planets' surfaces. Areas of extensive lava flooding are visible on the Moon today.

The planets did not cool down at the same rate, however. A small planet can more easily radiate its internal heat into space and cool off more rapidly than a big planet. Because Mercury is larger than the Moon, it took longer for a thick protective crust to form on Mercury than on the Moon. Throughout Mercury's early history, molten rock seeped up through cracks in its young, frail crust, and volcanism was probably pervasive. The resulting lava flows inundated many older craters, leaving behind the broad, smooth intercrater plains seen by Mariner 10.

Mariner 10 measured the surface temperature on Mercury and searched for traces of an atmosphere. Temperatures on Mercury vary from 700 K (= 427°C = 800°F) at noon on the equator to 100 K (= −173°C = −280°F) at midnight. This 600 K temperature range is greater than that of any other planet or satellite in the solar system.

Because of its high daytime surface temperature and low surface gravity, Mercury is not able to retain a substantial atmosphere. Nevertheless, *Mariner 10* did detect a thin scattering of particles around the planet. This extremely thin atmosphere consists of particles captured from the solar wind as well as atoms dislodged from surface rocks by the solar wind.

The average density of Mercury (5.4 g/cm^3) is nearly the same as that of the Earth (5.5 g/cm^3). Yet, typical rocks from the surfaces of the terrestrial planets have a density of only 3 g/cm^3 and are composed primarily of silicon and other lightweight elements. The higher average density of the Earth and Mercury is attributable to abundant quantities of iron that sank toward the planets' centers while the planets were still entirely molten. As we saw in Chapter 7, this process is called **chemical differentiation.** As the result of the action of gravity, dense elements sink toward a planet's center and force less dense material toward the surface. Chemical differentiation must have occurred during and immediately after the formation of the solar system, while the terrestrial planets were still entirely molten and internal mass motion could occur on a large scale.

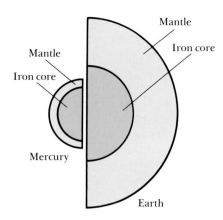

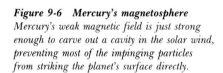

Figure 9-5 The internal structures of Mercury and Earth Mercury is the most iron-rich planet in the solar system. Its iron core occupies an exceptionally large fraction of the planet's interior.

Because Mercury's average density is slightly less than Earth's, you might suspect that Mercury's iron core is proportionally smaller than Earth's. This is not the case. Earth is 18 times more massive than Mercury. This larger mass pushing down on the Earth's interior compresses the Earth's core much more than Mercury's core is compressed. In fact, Mercury is the most iron-rich planet in the solar system, with iron accounting for 65 to 70 percent of its mass. The scale drawing in Figure 9-5 shows the interior structures of Mercury and Earth.

Independent evidence of Mercury's large iron core came from *Mariner 10*'s magnetometers, which discovered that Mercury has a magnetic field. Iron is the only common element whose atoms possess a built-in magnetic field, and so it is the only substance that could account for the magnetic field of a terrestrial planet.

As we saw in Chapter 7, the Earth's magnetic field occurs because of the dynamo effect. Magnetism arises whenever electrically charged particles are in motion. The Earth's rotation generates a planetwide magnetic field similar to the magnetism that surrounds a coil of wire in which electricity is flowing. Many geologists suspect that the Earth's magnetic field originates with electric currents flowing in the liquid portions of our planet's iron core. These currents, carried around by the Earth's rotation, create the planetwide magnetic field.

Because Mercury rotates much more slowly than the Earth (59 days versus 24 hours), most scientists believed that there would be no dynamo effect to induce a magnetic field. They were surprised, therefore, to find that Mercury does have a weak magnetic field (the Earth's field is 100 times stronger).

Figure 9-6 Mercury's magnetosphere Mercury's weak magnetic field is just strong enough to carve out a cavity in the solar wind, preventing most of the impinging particles from striking the planet's surface directly.

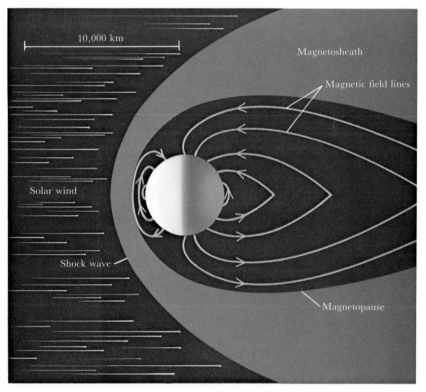

Mariner 10's magnetometers could not prove that Mercury's field is produced by the dynamo effect. Mercury's magnetic field may be frozen in the planet's completely solid iron core, like the magnetism of an iron magnet you can buy in a store.

Charged particle detectors on *Mariner 10* mapped the structure of Mercury's magnetosphere. The results are shown in Figure 9-6. As we saw in Chapter 7 (recall Figure 7-33), when supersonic particles in the solar wind first encounter a magnetic field, they slow abruptly, producing a bow-shaped shock wave at the boundary where this sudden decrease in velocity occurs. Closer to the planet, there is another well-defined boundary, the magnetopause, where the outward magnetic pressure of the planet's field exactly counterbalances the impinging gas pressure of the solar wind. Between the shock wave and the magnetopause is the turbulent region called the magnetosheath in which most of the subsonic particles from the solar wind are deflected around the planet, just as water is deflected by the bow of a ship. The region inside the magnetopause is the true magnetic domain of the planet. Mercury's magnetic field is not strong enough to capture particles permanently, and so it has nothing comparable to Earth's Van Allen belts.

The Moon's early history can be deduced from the craters and plains visible on its surface

Earth's own satellite consistently provides one of the most dramatic sights in the nighttime sky. The Moon is so large and so near that some of its surface features are readily visible to the naked eye. Even casual observation reveals that the Moon perpetually keeps the same side facing the Earth. This is a stable situation resulting from the gravitational interaction between Earth and Moon.

With a small telescope you can see several different types of major lunar terrain (see Figure 9-7). Most prominent are the large, dark, flat areas called **maria** (pronounced MAR-ee-uh). The singular form, **mare** (pronounced MAR-ee), meaning "sea" in Latin, was introduced in the seventeenth century when observers using early telescopes thought they had seen large bodies of water on the Moon. In fact, bodies of liquid water could not possibly exist on our airless satellite. Because there is no atmospheric pressure, a lake or ocean would boil furiously and evaporate rapidly as its atoms rush to escape into the vacuum of space. The maria were actually formed by huge lava flows that inundated low-lying regions of the lunar surface 3.5 billion years ago. They have still, however, retained their fanciful names such as Mare Tranquillitatis (Sea of Tranquillity), Mare Nubium (Sea of Clouds), Mare Nectaris (Sea of Nectar), and Mare Serenitatis (Sea of Serenity).

The largest of the 14 maria is Mare Imbrium (Sea of Showers). Roughly circular, it measures 1100 km (700 miles) in diameter. Although the maria seem quite smooth in telescopic views from the Earth, closeup photographs by the Apollo astronauts reveal small craters and occasional cracks called **rilles** (see Figure 9-8).

Perhaps the most familiar and characteristic features on the Moon are its **craters**. With an Earth-based telescope, some 30,000 of them are visible, from 1 km to more than 100 km across. Following a tradition established in the seventeenth century, the most prominent craters are named after famous philosophers and scientists.

Figure 9-7   The Moon   *Our Moon is one of seven large satellites in the solar system. The Moon's diameter (3476 km = 2160 miles) is slightly less than the distance from New York to San Francisco. This photograph is a composite of first-quarter and third-quarter views, and so elongated shadows enhance all surface features. (Lick Observatory)*

Craters smaller than about 1 km in diameter cannot be seen from Earth, but photographs from lunar orbit reveal millions of craters that escape the scrutiny of Earth-based observers. Virtually all craters, both large and small, are the result of bombardment by meteoritic material.

Many of the youngest craters are surrounded by light-colored streaks called **rays** that were formed by material violently ejected during impact. In

Figure 9-8   Details of Mare Tranquillitatis
Closeup views of the lunar surface reveal numerous tiny craters and cracks on the maria. This photograph was taken from lunar orbit in 1969 by astronauts during a final photographic reconnaissance of potential landing sites. (NASA)

Figure 9-9 *Details of a lunar crater* This photograph, taken from lunar orbit by Apollo 11 *astronauts in 1969, shows a typical view of the Moon's heavily cratered far side. The large crater near the middle of the picture is approximately 80 km (50 miles) in diameter. Note the crater's central peak and the numerous tiny craters that pockmark the lunar surface. (NASA)*

addition, many large craters have a pronounced central peak formed during a high-speed impact by a sizable meteoroid (see Figure 9-9).

The flat, low-lying, dark maria cover only 15 percent of the lunar surface. The remaining 85 percent of the surface is light-colored, heavily cratered terrain at elevations generally higher than those of the maria. This second kind of terrain is called the **terrae,** or **highlands.** (Terra means "land" in Latin, and so in this fanciful terminology, the entire lunar surface is covered by either "land" or "sea.")

One of the surprises arising from lunar exploration is that there are no maria on the Moon's far side, which consists entirely of heavily cratered highlands. Detailed observations by astronauts in lunar orbit demonstrated that the maria on the Moon's Earth-facing side are 2 to 5 km below the average lunar elevation. In contrast, the cratered terrae on the far side are typically at elevations up to 5 km above the average surface elevation. These elevation differences imply that the Moon's crust is thinner on the Earth-facing side than on the Moon's far side, as shown in Figure 9-10.

Large meteoroids easily punctured the thin, cooling crust on the Moon's Earth-facing side shortly after the solar system formed 4.5 billion years ago. Lava welled up, flooding the low-lying areas and producing the maria. Not much has happened since those ancient days; the entire lunar surface has remained almost unchanged for billions of years.

Because of the extraordinary thickness of the Moon's rigid lithosphere, no tectonic plate movement is possible. If the Moon has an iron-rich core, it does not endow our satellite with a magnetic field. Thus the Moon does not have a magnetosphere, and particles of the solar wind strike the lunar surface directly.

Figure 9-10 *The internal structure of the Moon* Like the Earth, the Moon probably has a crust, a mantle, and a core. The lunar crust has an average thickness of about 60 km on the Earth-facing side but about 100 km on the far side. The crust and solid upper mantle form a lithosphere about 800 km thick. The plastic (nonrigid) asthenosphere probably extends all the way to the base of the mantle. If the Moon has an iron-rich core, it is solid and less than 700 km in diameter. Although the main features of the Moon's interior are analogous to those of the Earth's interior, the proportions are quite different. The information here is based on analyses of data from seismographs left on the Moon by astronauts.

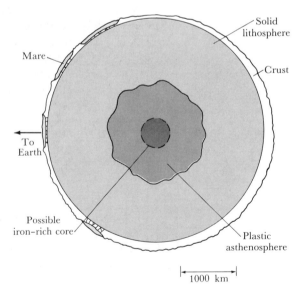

Solid lithosphere

Mare

Crust

To Earth

Possible iron–rich core

Plastic asthenosphere

|⊢ 1000 km ⊣|

Lunar rocks were formed 3 to 4.5 billion years ago

There were six successful manned lunar landings. The first two, *Apollo 11* and *Apollo 12,* set down in maria. The remaining four (*Apollo 14* through *Apollo 17*) were made in progressively more challenging terrain culminating

Figure 9-11 An Apollo astronaut on the Moon *The Moon is a desolate, barren, lifeless world. This typical view of the lunar surface shows an* Apollo 17 *astronaut near a large rock. Since the Moon has no atmosphere, lunar rocks have not been subjected to weathering and thus contain unaltered information about the early history of the solar system. From the six manned lunar landings between 1969 and 1972 astronauts brought back a total of 382 kg (843 lb) of moon rocks. (NASA)*

in rugged mountains just east of Mare Serenitatis (see Figure 9-11). The major factors in the choice of the landing sites were concern for the astronauts' safety and the desire to explore a wide variety of geologically interesting features.

The Apollo astronauts brought back 382 kg (843 pounds) of lunar rocks that proved to be a very important source of information about the early history of the Moon and Earth. All the lunar rock samples appear to have formed through the cooling of molten lava. The samples are almost completely composed of the same minerals found in terrestrial volcanic rocks. In addition, the entire lunar surface is covered with a layer of fine powder and rock fragments produced by 4.5 billion years of relentless meteoritic bombardment. This layer, which ranges in thickness from 1 m to 20 m; is called the **regolith** rather than soil, because the term "soil" as used on Earth normally suggests the presence of decayed biological matter.

The astronauts who visited the maria discovered that these dark regions of the Moon are covered with basaltic rock similar to the dark-colored rocks formed by lava flows from volcanoes in Hawaii and Iceland. The rock of these low-lying lunar plains is called **mare basalt** (see Figure 9-12).

In contrast to the dark maria, the lunar highlands are covered with a light-colored rock called **anorthosite** (see Figure 9-13). On Earth, anorthositic rock is found only in such very old mountain ranges as the Adirondacks in the eastern United States. Anorthosite is rich in calcium and aluminum in comparison to the mare basalts, which have more of the heavier elements such as iron, magnesium, and titanium. Anorthosite therefore has a lower density than basalt. The anorthositic magma apparently floated to the lunar surface when the Moon was molten, solidifying as it cooled to form the lunar crust. The denser mare basalts formed later from lava that oozed out of the interior to fill the mare basins.

By carefully measuring the abundances of trace amounts of radioactive elements in lunar samples, geologists confirmed that anorthosite is more

Figure 9-12 [left] Mare basalt
This 1531-g (3⅜-pound) specimen of mare basalt was brought back by Apollo 15 astronauts. This particular sample is called a vesicular basalt because of the tiny holes, or vesicles, that cover 30 percent of the rock's surface. Gas must have been dissolved under pressure in the lava from which this rock solidified. When the lava reached the airless lunar surface, bubbles formed as the pressure dropped. Some of the bubbles were frozen in place as the rock cooled. (NASA)

Figure 9-13 [right] Anorthosite *The light-colored lunar terrae are covered with an ancient type of rock called anorthosite. Anorthositic rock is believed to be the material of the original lunar crust. This particular sample, called the "Genesis rock" by the Apollo 15 astronauts who picked it up at the base of the Apennine Mountains, has an age of approximately 4.1 billion years. (NASA)*

ancient than the mare basalts. This result had been expected because the lunar highlands are densely cratered, whereas the basaltic surfaces of the mare show relatively few craters. Typical anorthositic specimens from the highlands are between 4.0 and 4.3 billion years old (one rock brought back by *Apollo 17* is nearly 4.6 billion years old). All these ancient specimens represent material from the Moon's original crust. In contrast, all the mare basalts are between 3.1 and 3.8 billion years old. Apparently, the mare basalts solidified from lavas that gushed up from the Moon's mantle and flooded the mare basins between 3.1 and 3.8 billion years ago, just about the time the oldest rocks in the Earth's present surface layers were being formed.

The Apollo astronauts brought back many specimens of **impact breccias,** which are various rock fragments that have been cemented together by meteoritic impact (see Figure 9-14). In addition to making breccias and churning up the regolith, meteoritic impacts also melt rocks to produce glass. Many lunar samples are coated with a thin layer of smooth, dark glass created when the surface of the rock was suddenly melted, then rapidly solidified. Small black glass beads are also common in the lunar regolith. These

Figure 9-14 A lunar breccia *Meteoritic impacts can cement rock fragments together to form breccias. This particular impact breccia was collected by Apollo 16 astronauts from the rim of a crater near their landing site. (NASA)*

glass spheres were presumably formed from droplets of molten rock hurled skyward by the impact of a meteoroid.

Many lunar samples bear the scars of meteoritic dust grains traveling at thousands of kilometers per hour, which produce tiny craters on moon rocks (see Figure 9-15). Because of these tiny, glass-lined craters, Moon rocks often seem to sparkle when held in the sunshine.

Meteoritic bombardment is the only source of "weathering" for lunar rocks. The rate of this weathering is actually quite slow. Geologists estimate that it takes tens of millions of years to wear away a layer of rock only 1 mm thick. Features formed 3 billion years ago are well preserved today, and the astronauts' footprints will remain sharply imprinted on the lunar surface for millions of years to come.

Although lunar rocks bear a strong resemblance to terrestrial rocks, there are some important differences. Every terrestrial rock contains some water, but lunar rocks are totally dry. There is absolutely no evidence that water ever existed on the Moon. In the absence of both an atmosphere and water, it is not surprising that the astronauts found no traces of life.

Volatile elements such as potassium and sodium melt and boil at relatively low temperatures, whereas **refractory elements** like titanium, calcium, and aluminum melt and boil at much higher temperatures. Compared to terrestrial rocks, lunar rocks have slightly greater proportions of refractory elements and slightly lower proportions of volatile elements. The implication is that the Moon formed from Earthlike material that had been baked at temperature high enough to boil away some of the volatile elements, leaving the young Moon relatively enriched in refractory elements.

These characteristics of moon rocks are consistent with the collision-ejection theory of the Moon's formation discussed in Chapter 6. This theory postulates that a huge asteroid struck the Earth about 4.5 billion years ago.

Figure 9-15   A microscopic crater
The upper surfaces of many moon rocks are covered with nearly microscopic craters produced by the impact of high-speed meteoritic dust grains. These glass-lined craters are typically less than 1 mm in diameter. (NASA)

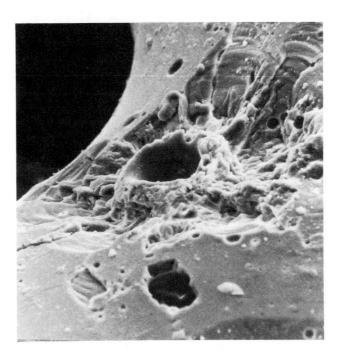

Figure 9-16 Eratosthenes Eratosthenes is a young crater 61 km in diameter on the southern edge of Mare Imbrium. Another young crater, Copernicus, is near the horizon in this photograph taken by the Apollo 17 *astronauts in 1972. (NASA)*

During that impact, a large quantity of vaporized rock was blasted out of the Earth (recall Figure 6-16). The temperature of the ejected rock was high enough to drive off water and other volatile chemicals. As rock fragments cooled and solidified in orbit about the Earth, they gradually accreted to form the Moon.

At first, the young Moon was largely molten because of heat from the impacts of rock fragments still falling onto the lunar surface and from the decay of radioactive isotopes. As the Moon cooled over the next few hundred million years, low-density lava floating on the lunar surface began to solidify into the anorthositic crust that exists today. The heavy barrage of large rock fragments ended about 4 billion years ago, with the final impacts producing the ancient craters that cover the lunar highlands.

At the end of this crater-making era more than a dozen asteroid-sized objects, each possibly as large as 100 km across, rained down on the young Moon, blasting out the vast mare basins. Then, from 3.8 to 3.1 billion years ago, great floods of molten rock gushed up out of the lunar interior, filling the impact basins and creating the maria we see today.

Very little has happened on the Moon since those ancient times. A few fresh craters have been formed (see Figure 9-16), but the world the astronauts visited has remained largely unchanged for over 3 billion years.

The formation of the Galilean satellites probably mimicked the formation of the solar system

Galileo Galilei was the first person to see the four largest satellites of Jupiter (recall Figure 3-10). He called them the "Medicean Stars" to attract the attention of the Medici family, who were wealthy Florentine patrons of the arts and sciences. Since 1610, these four giant moons have become important to our understanding of the solar system. To Galileo, they provided observational evidence supporting the heretical Copernican cosmology. The modern astronomer saw in the Voyager flyby photos of 1979 four extraordinary terrestrial worlds, now called the **Galilean satellites,** different from anything astronomers had ever seen or imagined. They are named after the mythical lovers and companions of Zeus: Io, Europa, Ganymede, and Callisto.

When viewed through an Earth-based telescope, the Galilean satellites look like mere pinpoints of light. With patience, you can follow these four worlds as they orbit Jupiter. Because their orbital periods are fairly short—from 1.8 days for Io to 16.7 days for Callisto—major changes in the positions of the satellites are easily noticeable from one night to the next.

The brightness of each of the moons varies slightly as it moves along its orbit. These variations can logically be attributed to dark and light surface areas that are alternately exposed to or hidden from view as the satellite rotates. Careful measurements show that the brightness of each satellite varies with a period equal to the satellite's orbital period. In other words, each Galilean satellite rotates exactly once on its axis during each trip around its orbit. We can thus conclude that each Galilean satellite keeps the same hemisphere perpetually facing Jupiter just as our Moon keeps the same side facing the Earth.

Accurate measurements of the diameters of the Galilean satellites came from Voyager photographs (see Figure 9-17). The two inner Galilean satellites, Io and Europa, are approximately the same size as our Moon. The two outer satellites, Ganymede and Callisto, are comparable in size to Mercury.

| Io | Europa | Ganymede | Callisto |

Figure 9-17 *The Galilean satellites*
The four Galilean satellites are shown here to the same scale. Io and Europa have diameters and densities comparable to our Moon and are primarily composed of rocky material. Although Ganymede and Callisto are roughly as big as Mercury, their average densities are low. Each of these two outer satellites is covered with a thick layer of water and ice. (NASA)

Deflections in the trajectories of the Voyager spacecraft as they passed the Galilean satellites provided important data from which their masses were computed. Europa, the smallest Galilean satellite, is also the least massive. Ganymede, the largest Galilean satellite, is the most massive. Data about the Galilean satellites are listed in Table 9-1.

As soon as reliable mass and diameter measurements were available, it became apparent that the average densities of the satellites are related to their distances from Jupiter. The innermost satellite, Io, has the highest average density (3.55 g/cm^3, slightly denser than our Moon). The next satellite, Europa, has an average density of 3.04 g/cm^3, which is not quite as dense as our Moon. Recalling that rocks in the Earth's crust typically have densities around 3 g/cm^3, it is reasonable to suppose that both Io and Europa are made primarily of rocky material.

The pattern of decreasing density with increasing distance from Jupiter continues with the outer two satellites. Ganymede and Callisto each have an average density of less than 2 g/cm^3, indicating that these two satellites are composed of roughly equal amounts of rock and ice.

Certain parallels exist between the arrangement of the Galilean satellites about Jupiter and the planets about the the Sun. For example, moving outward from the Sun, average density declines from more than 5 g/cm^3 for Mercury to less than 1 g/cm^3 for Saturn (recall Table 6-2). Scientists therefore began to suspect that the same general processes that had formed the solar system were at work during the formation of the Galilean satellites, though on a much smaller scale.

Table 9-1 The Galilean satellites

Name	Distance from Jupiter (km)	Orbital period (days)	Diameter (km)	Mass (Moon = 1)	Average density (g/cm³)
Io	412,600	1.77	3632	1.21	3.55
Europa	670,900	3.55	3126	0.66	3.04
Ganymede	1,070,800	7.16	5276	2.03	1.94
Callisto	1,880,000	16.69	4820	1.44	1.81
Mercury	——	——	4878	4.49	5.42
Moon	——	——	3476	1.00	3.34

NASA scientists recently computed the conditions necessary for the formation of the Galilean satellites, including in their simulation the fact that Jupiter emits twice as much energy as it receives from the Sun. They calculated that frozen water could be retained and incorporated into satellites at the distances of Ganymede and Callisto but that only rocky material would condense at the orbits of Io and Europa because of Jupiter's warmth. Jupiter's gravity and heat thus produced two distinct classes of Galilean satellites, just as warmth from the protosun caused a dichotomy between the small, dense, rocky inner planets and the huge, gaseous, low-density outer planets.

Io is covered with colorful deposits of sulfur compounds ejected from numerous active volcanoes

Within a few hours after *Voyager 1* passed near Jupiter, Io loomed into view and the probe began sending back a series of strange and unexpected pictures such as the one shown in Figure 9-18. Baffled by what they were seeing, scientists jokingly compared Io to pizzas and rotten oranges.

A major clue to these puzzling vistas was uncovered several days after the Jupiter flyby when a navigation engineer noticed a large umbrella-shaped cloud protruding from Io in one photograph—an erupting volcano! No one had expected to obtain photographs of erupting volcanoes on Io. After all, a probe making a single trip past the Earth would be very unlikely to catch a large volcano in the act of erupting.

Careful re-examination of the closeup photographs revealed eight giant ongoing eruptions. These volcanoes are named after gods and goddesses traditionally associated with fire in Greek, Norse, Hawaiian, and other mythologies. Figure 9-19 shows two views of the symmetric plume of Prometheus.

As it orbits Jupiter, Io is repeatedly caught in a gravitational tug-of-war between the huge planet on one side and the other Galilean satellites on the other. This gravitational battle distorts Io's orbit, varying its distance from Jupiter. As the distance varies, tidal stresses on Io alternately squeeze and flex the satellite. This constant tidal flexing in turn causes frictional heating

Figure 9-18 Io This closeup view of Io was taken by Voyager 1. *Notice the extraordinary range of colors from white, yellow, and orange to black. Scientists believe that these brilliant colors occur because of surface deposits of sulfur ejected from Io's numerous volcanoes. (NASA)*

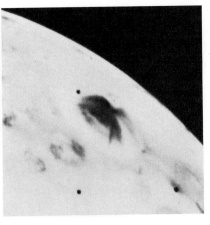 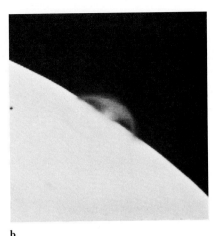

a   b

Figure 9-19 *Prometheus on Io* *These two Voyager views were taken two hours apart and show details of the plume of the volcano called Prometheus.* **(a)** *When viewed against the light background of Io's surface, jets of material give the plume a spiderlike appearance.* **(b)** *The plume's characteristic umbrella shape is seen silhouetted against the blackness of space. Prometheus's plume rises to an altitude of 100 km above Io's surface.* (NASA)

of Io's interior. Calculations show that the heat pumped into Io this way is equivalent to 2000 tons of TNT exploding every second. Eventually this energy makes its way to Io's surface, producing the numerous volcanoes.

The plumes and fountains of material spewing from Io's volcanoes rise to astonishing heights of 70 to 280 km above the surface. To reach these altitudes, the material must emerge from the volcanic vents with speeds between 300 and 1000 m/sec, a speed much greater than is found in the most violent terrestrial volcanoes. For example, Vesuvius, Krakatoa, and Mount St. Helens have eruption velocities of only about 100 m/sec. Scientists therefore began to suspect that Io's volcanoes operate in a fundamentally different way from volcanoes on Earth. The evidence of these differences came from Voyager's pictures and data.

No impact craters like those on our Moon were seen on Io. Material from the volcanoes apparently obliterates impact craters soon after they are created. The absence of craters indicates that Io's surface is extremely young—perhaps less than 100 million years old.

The Voyager cameras revealed numerous black dots on Io, which are apparently the volcanic vents from which the eruptions occur. These black spots are typically 10 to 50 km in diameter and form 5 percent of Io's surface. Lava flows radiate from many of these black dots (see Figure 9-20), some of which have volcanic plumes.

Evidence supporting the volcanic nature of the black spots came from Voyager instruments that measured the intensity of infrared radiation across Io's surface. Some of the black spots have temperatures as high as 20°C, in sharp contrast to the surrounding surface temperature of only −146°C.

After the discovery of widespread volcanic activity on Io, scientists soon concluded that sulfur ejected from the volcanoes is responsible for Io's brilliant colors. Sulfur is normally bright yellow. When it is heated and then suddenly cooled, however, it can assume a range of colors from orange and red to black.

Voyager detected sulfur dioxide (SO_2) in the plumes from Io's volcanoes. Sulfur dioxide is an acrid gas commonly discharged from volcanic vents here on Earth. When this gas is released into the cold vacuum of space from eruptions on Io, it crystallizes into white snowflakes. It is likely that the whitish deposits on Io (see Figure 9-18) are sulfur dioxide frost or snow.

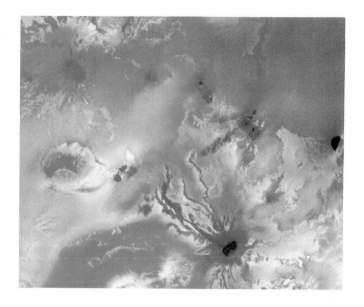

Figure 9-20 A volcanic center on Io
No impact craters are seen in closeup pictures such as this one taken by Voyager 1. *Long, meandering lava flows radiate from many of the black dots that apparently are the sites of intense volcanic activity. This photograph covers an area 1000 by 800 km, approximately twice the size of California. (NASA)*

The abundant sulfur and sulfur dioxide on Io suggest an explanation for the mechanism of its volcanoes. Indeed, the term volcano may be the wrong word altogether. We have seen that material is ejected from volcanic vents on Io at much higher velocities than is observed in even the most explosive volcanic eruptions on Earth. Another major difference is that Io's vents are not located at the tops of tall volcanic mountains. Many volcanoes on Earth and Mars have an easily recognizable conical shape with a caldera at the summit. Few of Io's calderas (the black spots) are associated with major topographical relief.

In all these respects, Io's volcanoes are more similar to terrestrial geysers than volcanoes. In a geyser, such as those in Yellowstone Park in Wyoming, water seeps down to volcanically heated rocks, is suddenly changed to steam, and erupts explosively through a vent. If the Old Faithful geyser were to erupt under the low gravity and vacuum that surround Io, it would send a plume of water and ice to an altitude of 40 km.

Both sulfur and sulfur dioxide are molten at depths of only a few kilometers below Io's surface because of the heat generated by tidal flexing. Geologists have pointed out that sulfur dioxide could be the principal propulsive agent driving Io's eruptions. Just as the explosive conversion of water into steam produces a geyser on Earth, the sudden conversion of liquid sulfur dioxide into a high-pressure gas could produce an eruption on Io. Calculations indicate that this explosive expansion of sulfur dioxide could result in eruption velocities up to 1000 m/sec.

The material erupting from Io's geyserlike volcanoes is composed primarily of sulfur and sulfur dioxide. Voyager's instruments failed to detect any other abundant gases such as the water vapor and carbon dioxide emitted from terrestrial volcanoes. Io has apparently been completely outgassed by volcanic activity over hundreds of millions of years. Io has not been able to retain its volatile gases and has almost no atmosphere because its surface gravity is comparable to that of our Moon.

It is estimated that each volcano on Io ejects roughly 10,000 tons of material per second. Although this material is a hot mixture of molten sulfur and sulfur dioxide gas under high pressure as it gushes from a volcanic vent, the gas–liquid mixture rapidly cools and solidifies in the cold, nearly perfect vacuum around Io. It then takes about half an hour for the fine particles of sulfur dust and sulfur dioxide snow to fall back down onto the surface.

Altogether, Io's volcanoes and vents eject an estimated 100 billion tons of matter each year. This produces a sulfur-rich layer 10 m thick over Io's entire surface each year. Thus the surface is constantly changing, and it is probably safe to say that there are no long-lived or even semipermanent features on Io.

Europa is covered with a smooth layer of ice, crisscrossed with many cracks

Voyager 1 did not pass near Europa, but *Voyager 2* captured the excellent view shown in Figure 9-21. Europa is a very smooth world with no mountains and very few craters, crisscrossed with a spectacular series of streaks and cracks. Most of the cracks appear to be filled with dark-colored material, but some cracks have light-colored substance in them.

We have seen that Europa's average density is about 10 percent less than the average density of our Moon. Spectroscopic observations from Earth indicate that Europa has frozen water on its surface. These two facts together suggest that Europa's surface may be covered with an ice layer 100 km thick. That would be consistent with its remarkable smoothness. An "ocean" of ice 100 km deep would certainly hide mountain ranges and other topographic features. But what causes the network of cracks, and why are impact craters so rare?

Tidal squeezing is responsible for the volcanism on Io, which gives Io many of its extraordinary characteristics. Europa is caught in a similar tidal tug-of-war, with Jupiter to one side and the two largest Galilean moons periodically passing on the opposite side. However, Europa is much farther from Jupiter than Io is, and the tidal effects on Europa are considerably less than those on Io.

The tidal flexing of Europa is responsible for the network of cracks that cover its surface. Some of the darkest streaks in fact follow paths along

Figure 9-21 Europa Europa's ice surface is covered by numerous streaks and cracklike features that give the satellite a fractured appearance. The streaks are typically 20 to 40 km wide. This picture taken by Voyager 2 allows surface features as small as 5 km across to be seen. (NASA)

which the tidal stresses are calculated to be strongest. The tidal flexing of Europa is far too weak to produce volcanoes, but it is thought to supply enough energy to jostle and churn the satellite's icy coating. This activity would explain why only a few small impact craters have survived to the present time, even though Europa's surface is much older than Io's.

The streaks on Europa are presumably caused by cracks in the icy coating through which water gushed up and then froze. These cracks are a few tens of kilometers wide. To accommodate all of them, Europa's surface area would have to have increased by 10 to 15 percent since the cracks first began to appear. It seems unreasonable to suppose that Europa is expanding like an inflating balloon, though. Apparently, more is happening on Europa than meets the eye. Perhaps old surface material is somehow being pulled back down into the mushy layer below the ice coating and then recycled, just as the Earth's crust is pulled back down into the Earth's mantle in subduction zones (recall Figure 7-18). Europa's surface may represent a water-and-ice version of plate tectonics.

Ganymede and Callisto have heavily cratered icy surfaces

Figure 9-22 [left] **Ganymede** *This view from* Voyager 2 *shows the hemisphere that always faces away from Jupiter. The surface is dominated by a huge, dark circular region called Galileo Regio, which is the largest remnant of Ganymede's ancient crust. (NASA)*

Figure 9-23 [right] **Callisto** *This view from* Voyager 2 *shows the second largest Galilean satellite, Callisto. Its diameter is almost exactly the same as Mercury's. (NASA)*

Europa's density can be explained by 100 km of ice and water on top of an otherwise rocky world. A much thicker layer of water and ice must surround Ganymede and Callisto, however. To be consistent with average densities slightly less than 2 g/cm^3, the rocky cores of these two outer satellites must be enveloped in mantles of water and ice nearly 1000 km thick.

Closeup views of Ganymede and Callisto are seen in Figures 9-22 and 9-23. Both satellites have the kind of ancient, cratered surface normally associated with Moonlike landscapes. Of course, the craters on both worlds are of ice rather than rock.

Ganymede is the largest satellite in the solar system. It is 5270 km in diameter, slightly larger than Mercury. The largest single feature on Ganymede is a vast, dark, circular island of ancient ice called Galileo Regio (see

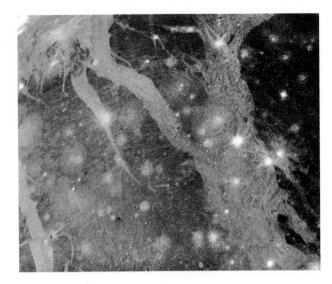

Figure 9-24 Young and old terrain on Ganymede *This closeup view of Ganymede was taken by* Voyager 2. *Features as small as 5 km across can be seen. Dark, angular islands of Ganymede's ancient crust are separated by younger, light-colored, grooved terrain. The southwest edge of Galileo Regio appears at the right side of the picture. (NASA)*

Figure 9-22). It measures 4000 km in diameter and covers nearly one-third of the hemisphere of Ganymede that faces away from Jupiter. This surface feature is the only one on the Galilean satellites that can be detected with Earth-based telescopes.

Ganymede has two very different kinds of terrain, which are distinguished by both appearance and age (see Figure 9-24). Dark, polygon-shaped regions are presumed to be the oldest surface features because they exhibit a high density of craters. Light-colored, heavily grooved terrain found between the dark angular islands is much less cratered and therefore younger.

It is easy to distinguish young craters on Ganymede. The youngest craters are surrounded by bright rays of freshly exposed ice. Older craters clearly have been covered with deposits of dark meteoritic dust. The most ancient craters are barely visible on Galileo Regio and on the other dark, angular island remnants of old crust. The degradation and near obliteration of the oldest craters probably also involved the slow plastic flow of Ganymede's icy surface.

Craters also tell us about the history of the younger, light-colored terrain covered with numerous grooves. In some places the cratering is about as dense as that on the ancient crust, but in other places it is only one-tenth that amount. We thus suspect that this grooved terrain was formed over a long period of time. The process probably began quite early in Ganymede's history and continued through the period of intense meteoritic bombardment. The age of the grooved terrain therefore probably ranges from about 4.5 to 3.5 billion years.

High-resolution photographs such as Figure 9-25 show that this grooved terrain actually consists of parallel mountain ridges up to 1 km high and spaced 10 to 15 km apart. These features suggest that plate tectonics may have dominated Ganymede's early history. Water seeping upward through cracks in the original crust would freeze and force apart fragments of the original crust, producing jagged, dark islands of old crust separated by bands of younger, light-colored, heavily grooved ice. The cracks play a role in Ganymedean plate tectonics analogous to the role of the oceanic rifts on

Figure 9-25 Grooved terrain on Ganymede *This picture, taken by* Voyager *1, shows an area roughly as large as the state of Pennsylvania. The smallest visible features are about 3 km across. Numerous parallel mountain ridges are spaced 10 to 15 km apart and have heights up to 1 km. (NASA)*

the Earth. But unlike thinner-crusted Europa, where tectoniclike activity perhaps occurs even today, tectonics on Ganymede bogged down 3 billion years ago as its cooling crust froze to unprecedented depths.

When Ganymede formed 4.5 billion years ago, it may have been completely covered with an ocean roughly 1000 km deep. During the next 200 million years, the water cooled and a thick coating of ice developed. Today this layer of solid ice is probably about 100 km thick. Beneath it lies a 900-km-thick slushy mantle of water and ice.

Callisto, Jupiter's outermost Galilean satellite, looks very much like Ganymede: numerous impact craters scattered over an ancient, dark, icy crust. There is one obvious difference, however—Callisto has no younger, grooved terrain. We thus infer that tectonic activity never began on Callisto. Perhaps because of its greater distance from Jupiter, the ocean that enveloped young Callisto 4.5 billion years ago froze more rapidly and to a greater depth than on Ganymede, forever preventing tectonic processes. Callisto's icy crust may in fact be several times thicker than Ganymede's, extending to depths of several hundred kilometers. It is bitterly cold on Callisto. *Voyager*'s instruments measured a noontime temperature of −118°C (−180°F), and the nighttime temperature plunges to −193°C (−315°F).

Voyager 1 photographed a huge impact feature on Callisto's Jupiter-facing hemisphere (see Figure 9-26). This feature, called the Valhalla Basin, consists of a large number of concentric rings, separated by 50 to 200 km and having diameters ranging up to 3000 km. Valhalla was produced by an asteroid-sized object very early in the satellite's history.

The great age of Valhalla is inferred both from the presence of overlying impact craters and the absence of substantial vertical relief of the concentric rings. The Valhalla impact probably occurred around 4 billion years ago, when the satellite's young, relatively thin crust was still plastic enough to flow and reduce the height of the upraised rings in the ice.

The Voyager pictures also showed traces of a Valhallalike impact on Ganymede's Galileo Regio. Segments of a system of concentric rings cover a

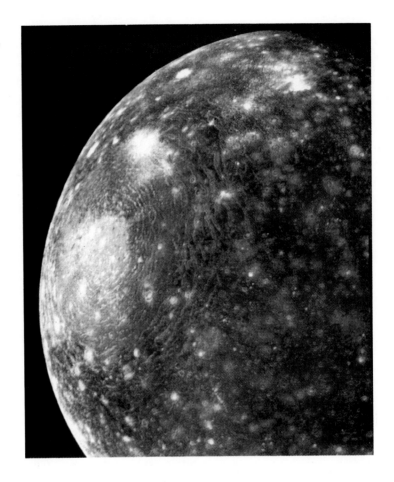

Figure 9-26 Callisto *Numerous craters pockmark Callisto's icy surface, as seen in this mosaic of views from Voyager 1. A huge impact basin called Valhalla dominates the Jupiter-facing hemisphere of this frozen, geologically inactive world. A series of concentric rings up to 3000 km in diameter surrounds the impact site. (NASA)*

large portion of this dark island of ancient crust. However, no obvious impact feature is found at the center of this ring system. Apparently the development of grooved terrain completely obliterated the impact basin.

Titan has a thick, opaque atmosphere rich in methane, nitrogen, and hydrocarbons

Long before the Voyager flybys, astronomers knew Saturn's largest satellite to be an extraordinary world. It was discovered in 1655. By the early 1900s, several scientists suspected that Titan might have an atmosphere because it is cool enough and massive enough to retain heavy gases. Confirming evidence came in 1944 when astronomers discovered spectral lines of methane in the sunlight reflected from Titan. Titan is the only satellite in the solar system known to have an appreciable atmosphere.

Because of this atmosphere, Titan was a primary target for the Voyager missions. To everyone's chagrin, however, the Voyagers spent hour after precious hour sending back featureless images such as Figure 9-27. Titan's cloud cover is so thick that it blocks any view of the surface and allows very little sunlight to penetrate down to the ground; the surface of Titan must be a dark, gloomy place.

In size, mass, and average density, Titan is quite similar to the largest Jovian satellites. We would thus expect its internal structure to resemble those of Ganymede and Callisto—a rocky core surrounded by a mantle of frozen water nearly 1000 km thick.

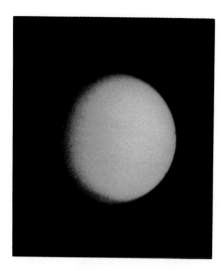

Figure 9-27 Titan *This view of Titan was taken by Voyager 2. Very few features are visible in the thick, unbroken haze that surrounds this large satellite. The main haze layer is located nearly 300 km above Titan's surface. (NASA)*

Titan's thick atmosphere distinguishes it from all other satellites. The atmospheric pressure at Titan's surface is 1.6 atm, or 60 percent greater than the atmospheric pressure at sea level on Earth, even though Titan's surface gravity is lower than the Earth's. Considerably more gas must be weighing down on Titan than on Earth. About 10 times more gas lies above each square centimeter of Titan's surface than above Earth's surface.

What factors leading to the formation of Titan's atmosphere did not exist for Ganymede or Callisto? For one, Titan formed in a much cooler part of the solar nebula. In contrast to the materials available in the warmer conditions near Jupiter, the ices from which Titan accreted probably contained substantial amounts of frozen methane and ammonia. As Titan's interior became warm through the decay of naturally occurring radioactive isotopes, these ices vaporized, producing an atmosphere around the young satellite. Methane (CH_4) is stable in sunlight and remained in the atmosphere, but ammonia (NH_3) is easily broken down into nitrogen and hydrogen by the Sun's ultraviolet radiation. Titan's gravity is too weak to retain hydrogen, and so it escaped into space. Even today, hydrogen is escaping from Titan at a substantial rate.

The breakup of ammonia and the resulting loss of hydrogen leaves Titan with an abundant supply of nitrogen. Voyager data suggest that roughly 90 percent of Titan's atmosphere is nitrogen. The two next most-abundant gases are argon and methane.

The interaction of sunlight with methane induces chemical reactions that produce a variety of carbon–hydrogen compounds called **hydrocarbons.** Voyager's instruments detected small amounts of many hydrocarbons such as ethane (C_2H_6), acetylene (C_2H_2), ethylene (C_2H_4), and propane (C_3H_8) in Titan's atmosphere. Nitrogen combines with these hydrocarbons to produce other compounds such as hydrogen cyanide (HCN), some of which are the building blocks of the organic molecules on which life is based. There is

Figure 9-28 Titan's atmosphere
The Voyager data suggest that Titan's atmosphere has three distinct layers. The uppermost layer absorbs ultraviolet radiation from the Sun. The middle layer is opaque to visible light. The lowest layer is an aerosol of suspended particles. Methane rain clouds may exist near Titan's surface, which is at a temperature of about 95 K. (Adapted from Tobias Owen)

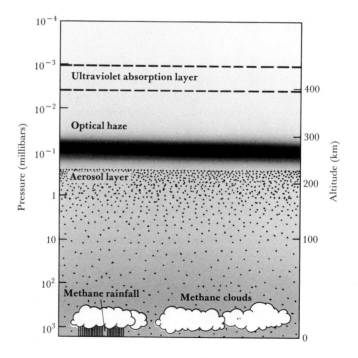

little reason to suspect life on Titan—its surface temperature is 95 K (= −178°C = −288°F)—but a more detailed study of its chemistry may shed light on the origins of life on Earth.

On Earth, the atmospheric pressure and temperature are near the **triple point** of water, meaning that water is found in all three phases: liquid, solid, and gas. The atmospheric pressure and temperature at Titan's surface are near the triple point of methane, however. Thus methane may play a role on Titan similar to that which water plays on Earth. Methane snowflakes may fall onto frozen methane polar caps and methane raindrops may descend into methane rivers, lakes, and seas in warmer areas.

Some molecules are capable of joining together in long, repeating molecular chains to form substances called **polymers.** Many of the hydrocarbons and carbon–nitrogen compounds in Titan's atmosphere form such polymers. Droplets of some polymers remain suspended in the atmosphere to form the kind of mixture called an **aerosol** (see Figure 9-28), but the heavier polymer particles settle down onto Titan's surface, probably covering it with a thick layer of sticky, tarlike goo. Some scientists estimate that the deposits of hydrocarbon sludge on Titan may be $\frac{1}{2}$ km deep.

Triton is a frigid, icy world with a young surface and a tenuous atmosphere

Triton, the outermost giant satellite of the solar system, was discovered in 1846. Because of its great distance from Earth, however, astronomers were unable to learn much about this remote world until the *Voyager 2* flyby of Neptune in 1989. Even through a large telescope Triton is a faint, featureless, starlike pinpoint of light.

An important property of Triton, known since its discovery, is its retrograde orbit. Triton goes around Neptune opposite to the direction in which the planet rotates. No other giant satellite has such an orbit and it is difficult to imagine how any satellite might form near a planet in an orbit opposing the direction of the planet's rotation. Only a few of the outer satellites of Jupiter and Saturn have retrograde orbits and these bodies are probably captured asteroids. Some scientists have therefore hypothesized that Triton may have been captured long ago by Neptune's gravity.

The orbit of Triton is tilted with respect to the plane of Neptune's equator (by about 20°), which in turn is tilted to the plane of Neptune's orbit about the Sun (by about 29°). Because of its 16-hour rotation, Neptune is slightly oblate and so the planet possesses an equatorial bulge in the plane of its equator. The gravitational pull of this equatorial bulge causes the satellite's orbit to precess, resulting in significant changes in Triton's seasons over the years. There are years in which Triton experiences modest seasons like those on Earth, and others (as when *Voyager 2* passed by Triton in 1989) in which it has exaggerated, Uranus-like seasons with the Sun shining down onto the satellite's south pole.

A high resolution image of Triton's south polar region is shown in Figure 9-29. Note that very few craters are seen, indicating that some process must have obliterated the scars of numerous, ancient impacts left over from the first billion years of the solar system. A major modification of Triton's surface is consistent with the idea that the satellite once orbited the Sun on its own and was captured by Neptune, perhaps 3 or 4 billion years ago. Upon being captured, Triton most likely would have started off in a highly elliptical orbit, but today the satellite's orbit is remarkably circular. Triton's original elliptical orbit would have been "circularized" by tidal forces exerted on

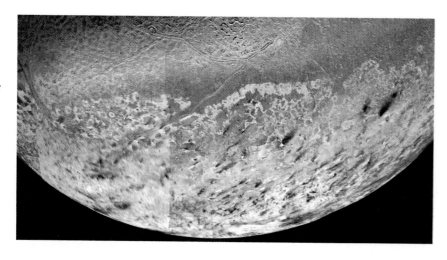

Figure 9-29   Triton's south polar cap
Approximately a dozen high-resolution images were combined to produce this view of Triton's southern hemisphere. The pinkish polar cap is probably made of nitrogen frost. A notable scarcity of craters suggests that Triton's surface was either melted or flooded by icy lava after the era of bombardment that characterized the early history of the solar system. (NASA)

Figure 9-30   A frozen lake on Triton?
Some scientists believe that this lakelike feature is the caldera of an ice volcano. The flooded basin is about 200 km wide and 400 km long, and so covers an area about the size of the state of West Virginia. (NASA)

the satellite by Neptune's gravity. With each revolution about that original elliptical orbit, the changing distance to Neptune would have stretched and flexed Triton, causing its orbit to become more circular while depositing significant energy into the satellite itself. This energy would have melted much of the satellite's interior and the resulting volcanism would have obliterated Triton's original surface features, including craters.

The pinkish ice that constitutes Triton's south polar cap is probably a layer of nitrogen frost. With the arrival of summer at Triton's south pole, this frost layer is slowly evaporating and many of the surface markings seen in Figure 9-29 are possibly the result of the northward flow of the resulting nitrogen gas. These winds are very tenuous, however, because Triton's atmosphere is so very thin. Atmospheric pressure on the satellite's surface is only 10^{-5} of that at sea level on Earth.

Triton exhibits some surface features seen on other icy worlds, such as long cracks resembling those on Europa and Ganymede. Other features are unique to Triton and are quite puzzling. For example, near the terminator in Figure 9-29, you can see a wrinkled terrain that resembles the skin of a cantaloupe. Triton also has a few frozen lakes like the one shown in Figure 9-30. Some scientists speculate that these lakelike features are the calderas of extinct ice volcanoes. A mixture of methane, ammonia, and water can have a melting point far below that of pure water, and could have constituted a kind of "cold lava" on Triton. It is unlikely that any such volcanoes are erupting on Triton today, however, because the satellite is so very cold. *Voyager's* instruments measured a surface temperature of 37 K ($-236°$ C $= -395°$ F), making Triton the coldest world we have ever visited.

Pluto may be an escaped satellite from Neptune

Pluto was discovered in 1930 by Clyde W. Tombaugh, who used a wide-field camera to photograph sections of the sky. He recognized the ninth planet from the Sun as a faint starlike object that slowly shifts it position from night to night. Pluto's only moon, Charon, was discovered in 1978. On modern photographs, the images of Pluto and Charon are blended together (see Figure 9-31).

The average distance between Charon and Pluto is a scant 19,700 km—less than $\frac{1}{20}$ the distance between the Earth and our Moon. Furthermore, Charon's orbital period of 6.3874 days is the same as the rotational period of

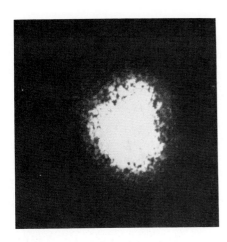

Figure 9-31 Pluto and Charon *Pluto's moon, Charon, appears as a slight elongation or lump on one side of this greatly enlarged image of the planet. Pluto and Charon are separated by only 19,700 km. The Pluto–Charon system deserves to be called a double planet because these two objects resemble each other in mass and size more closely than do any other planet–satellite pair in the solar system. (Courtesy of James Christy and Robert Harrington)*

Pluto. In other words, Pluto always keeps the same side facing Charon. As seen from the satellite-facing side of Pluto, Charon neither rises nor sets but instead seems to hover in the sky, as if perpetually suspended above the horizon.

From 1985 through 1990, the orbit of Charon was positioned so that Earth-based observers could view mutual eclipses of Pluto and its satellite. Astronomers used observations of these eclipses to determine that Pluto's diameter is 2290 km and Charon's is 1280 km. For comparison, our Moon's diameter (3476 km) is about $1\frac{1}{2}$ times as large as Pluto's.

Pluto's orbit about the Sun is more elliptical than the orbit of any other planet in the solar system. In fact, Pluto's orbit is so eccentric that it is sometimes closer to the Sun than Neptune is. In view of Pluto's unusual orbit about the Sun, it is reasonable to wonder whether Pluto might be an escaped satellite that once orbited Neptune. Perhaps some sort of cataclysmic event in the ancient past reversed the direction of Triton's orbit, flung Nereid into its highly elliptical orbit, and catapulted Pluto away from Neptune.

The idea that Pluto is an escaped satellite of Neptune was first proposed in 1936. Problems with this proposal have, however, long been recognized. For one, the present-day orbits of Neptune and Pluto do not intersect—the two planets are never closer than 384 million km. If Pluto was catapulted away from Neptune, why does Pluto's orbit not pass through Neptune's orbit at the point of this event?

With the discovery of Charon, some astronomers proposed that an unknown, massive, planetlike object may have been involved in Pluto's escape from Neptune. Perhaps Triton, Nereid, and Pluto all orbited Neptune along well-behaved regular orbits in the ancient past. Then, in a near-collision, the unknown planet severely perturbed the orbits of all three satellites. Triton and Nereid were swung into their present unusual orbits. Tidal forces from the unknown planet tore Pluto into two pieces and catapulted them into an orbit about the Sun. Astronomers will have to be content using Earth-based observations to probe the mysteries of Pluto because no space missions to this remote world are contemplated in the foreseeable future.

Summary

· Mercury's surface is pocked with craters like those of the Moon, but there are extensive, smooth intercrater plains. These features appear to have formed as the crust of the planet solidified.

· Surface temperatures on Mercury range from 100 to 700 K, the greatest range known on any of the planets.

· Mercury has an iron core much like that of the Earth.

· Mercury's magnetic field produces a magnetosphere that blocks the solar wind from the surface of the planet.

· The Earth-facing side of the Moon displays light-colored, heavily cratered highlands (terrae) and dark-colored, smooth-surfaced maria. The Moon's far side has no maria.

· Lunar rocks contain no water and also differ from terrestrial rocks by being relatively enriched in the refractory elements and depleted in the volatile elements.

· The Moon may have formed from rock torn from the Earth by a collision with a large asteroidlike object shortly after the formation of the solar system.

The Moon was molten in its early stages, and the anorthositic crust solidified from low-density magma that floated to the lunar surface. The mare basins were created later by the impact of planetesimals that released lava from the lunar interior.

The Moon's surface has undergone very little change in the past 3 billion years.

· The inner two Galilean moons, Io and Europa, are roughly the size of our Moon and have densities similar to that of the Moon. The outer two Galilean moons, Ganymede and Callisto, are roughly the size of Mercury and are lower in density than the Moon or Mercury.

· Io is covered with a colorful layer of sulfur compounds deposited by frequent explosive eruptions from volcanic vents.

Io's volcanic eruptions resemble terrestrial geysers. The energy heating Io's interior comes from tidal forces that flex the moon as it passes between the planet and the other large moons.

· Europa is covered with a smooth layer of frozen water that is crisscrossed by an intricate pattern of long cracks, probably produced by tidal flexing of the moon.

· The heavily cratered surface of Ganymede is composed of frozen water. Large polygons of dark, ancient surface are separated by regions of heavily grooved, lighter-colored, younger terrain. Plate tectonics apparently operated during the early history of Ganymede.

· Callisto also has a heavily cratered crust of frozen water, but plate tectonics apparently never operated on this moon, presumably because it quickly developed a thick, solid crust.

· The Galilean satellites probably formed through a process of accretion similar to the process that formed the solar system about the Sun, but on a smaller scale.

· The largest satellite of Saturn, Titan, has a dense nitrogen atmosphere in which methane may play a role similar to that of water on Earth. A variety of hydrocarbons are formed by the interaction of sunlight with methane, creating an aerosol layer in Titan's atmosphere and probably a thick sludge on its surface.

· The largest satellite of Neptune, Triton, moves in a retrograde orbit that is decaying.

· Triton is an icy satellite with a tenuous nitrogen atmosphere.

The scarcity of craters on Triton suggests that its surface was either molten or flooded with icy lava sometime after the era of bombardment that left numerous craters on such worlds as Mercury and our Moon.

· Pluto, the smallest planet in the solar system, may have been a satellite of Neptune.

Review questions

1 Why are naked-eye observations of Mercury best made at dusk or dawn, whereas telescopic observations are best made around noon?

2 Why do astronomers believe Mercury to be "the most iron-rich planet in the solar system"?

3 Why is more lunar detail visible through a telescope when the Moon is near quarter phase than when it is at full phase?

4 How would you prove to someone that the Moon has no atmosphere?

5 Why do you suppose that no Apollo mission landed on the far side of the Moon?

6 How does the Galilean satellite system resemble the solar system? How is it different?

7 With all its volcanic activity, why doesn't Io possess an atmosphere?

8 Compare and contrast the surface features of the four Galilean satellites, discussing the relative geological activity and evolution of these satellites.

Advanced questions

9 Why is it reasonable to presume that none of the large satellites in the solar system, including our Moon, possesses a substantial magnetic field?

10 Long before the Voyager flybys, Earth-based astronomers reported that Io appeared brighter than usual for a few hours after emerging from Jupiter's shadow. Based on what we know about the material ejected from Io's volcanoes, explain this brief anomalous brightening of Io.

11 Using the diameter of Io (3630 km) as a scale, estimate the height to which the plume of Prometheus rises above the surface of Io in Figure 9-19.

12 Why are all the maria on the Earth-facing side of the Moon?

Discussion questions

13 What evidence do we have that the surface features on Mercury were not formed during recent geological history?

14 Imagine that you are planning a lunar landing mission. What type of landing site would you select in order to obtain bedrock? Where might you land to search for evidence of recent volcanic activity?

15 The idea has been advanced that without the presence of the Moon in our sky astronomy would have developed far more slowly. Please comment.

16 Compare the advantages and disadvantages of exploring the Moon with astronauts as opposed to mobile robots.

17 Speculate on the possibility that Europa, Ganymede, or Callisto might harbor some sort of marine life.

For further reading

Cooper, H. *Apollo on the Moon* and *Moon Rocks*. Dial, 1970.

French, B. *The Moon Book*. Penguin, 1977.

Hartmann, W. "The Moon's Early History." *Astronomy*, September 1976, p. 6.

Hartmann, W. "The Significance of the Planet Mercury." *Sky & Telescope*, May 1976, p. 307.

Johnson, T. "The Galilean Satellites." In Beatty, J., et al., eds. *The New Solar System*, 2nd ed. Sky Publishing and Cambridge University Press, 1982.

Johnson, T., and Soderblom, L. "Io." *Scientific American*, December 1983.

Morrison, D. "Four New Worlds: The Voyager Exploration of Jupiter's Satellites." *Mercury*, May/June 1980, p. 53.

Murray, B., and Burgess, E. *Flight to Mercury*. Columbia University Press, 1977.

Murray, B. "Mercury." *Scientific American*, May 1976.

Owen, T. "Titan." *Scientific American*, February 1982.

Soderblom, L. "The Galilean Moons of Jupiter." *Scientific American*, January 1980.

10
Interplanetary vagabonds

The head of Comet Halley

This photograph shows the bluish head of Halley's Comet as it approached the Sun in December 1985. Comet Halley orbits the Sun with an average period of 76 years along a highly elliptical path that stretches from just inside the Earth's orbit to slightly beyond the orbit of Neptune. This color photograph was constructed from three black-and-white photographs taken in rapid succession with red, blue, and green filters on the same telescope. Because the comet moved slightly with respect to the background stars while being photographed, the images of the stars are blurred. (Anglo-Australian Observatory)

The planets are not the only objects that move in orbits about the Sun. In this chapter we discuss the asteroids, meteoroids, and comets that are small but significant members of the solar system. We learn that these objects provide major clues about the history of the solar system. We also discuss speculations that meteoroids and comets may have had a significant effect on our planet, including the extinction of more than one-half the species living on Earth some 63 million years ago.

Many rocks and chunks of ice that condensed out of the primordial solar nebula still continue to orbit the Sun. Just as heat from the protosun produced two classes of planets—terrestrial and Jovian—two main types of interplanetary material were created. Near the Sun, interplanetary debris consists of rock fragments called asteroids or meteoroids. Far from the Sun, thick ice coatings cover the debris which are then called comets.

Bode's law led to the discovery of numerous asteroids between the orbits of Mars and Jupiter

In the late 1700s, a young German astronomer, Johann Elert Bode, popularized a simple rule that describes the distances of the planets from the Sun. This rule is usually known today as **Bode's law**—an unfortunate name because it is not a physical law and was not invented by Bode. It had first been published in 1766 by Johann Titius, a German physicist and mathematician. Most astronomers now regard this "law" as merely a coincidence, but it did lead directly to the discovery of a large number of previously unknown objects that orbit the Sun.

Bode's rule for remembering the distances of the planets from the Sun goes like this:

1 Write down the sequence of numbers 0, 3, 6, 12, 24, 48, 96, ... (Note that each number after the second one is simply twice the preceding number.)

2 Add 4 to each number in the sequence.

3 Divide each of the resulting numbers by 10.

As shown in Table 10-1, the final result is a series of numbers that corresponds remarkably well to the distances (in AU) of the planets from the Sun.

Astronomers regarded Bode's rule as merely a useful trick for remembering the planetary distances until 1781 when William Herschel discovered

Table 10-1 Bode's law

Bode–Titius progression	Planet	Actual distance (AU)
(0 + 4)/10 = 0.4	Mercury	0.39
(3 + 4)/10 = 0.7	Venus	0.72
(6 + 4)/10 = 1.0	Earth	1.00
(12 + 4)/10 = 1.6	Mars	1.52
(24 + 4)/10 = 2.8	?	
(48 + 4)/10 = 5.2	Jupiter	5.20
(96 + 4)/10 = 10.0	Saturn	9.54
(192 + 4)/10 = 19.6	Uranus	19.18
(384 + 4)/10 = 38.8	Neptune	30.06
(768 + 4)/10 = 77.2	Pluto	39.44

Uranus, whose average distance from the Sun is very near that predicted by Bode's scheme. Suddenly it seemed far more likely that Bode's rule might actually represent some physical property of the solar system.

Astronomers now looked with new interest at the "missing planet" in the sequence of Bode's law—the gap between the orbits of Mars and Jupiter. On January 1, 1801, the Sicilian astronomer Giuseppe Piazzi noticed a dim, previously uncharted star that shifted its position slightly over the next several nights. Later that year, the orbit of this object was determined to lie between the orbits of Mars and Jupiter. At Piazzi's request, the object was named Ceres (pronounced SEE-reez) after the patron goddess of Sicily.

Ceres orbits the Sun once every 4.6 years at an average distance of 2.77 AU. This orbit is in remarkable agreement with the distance "predicted" by Bode's law for the missing planet. Ceres is very small, however—its diameter is estimated to be a scant 1000 km. Thus Ceres does not qualify as a full-fledged planet, and astronomers continued the search.

In 1802, the German astronomer Heinrich Olbers discovered another faint, starlike object that moved against the background stars. He called it Pallas. Like Ceres, Pallas orbits the Sun every 4.6 years at an average distance of 2.77 AU. Pallas is even dimmer and smaller than Ceres, with an estimated diameter of only 600 km. Obviously, Pallas is not the missing planet either.

The discovery of these two small objects with similar orbits at the distance expected for the missing planet led astronomers to suspect that Bode's missing planet might have somehow broken apart or exploded. The search for other small objects therefore continued. Only two more were found—Juno and Vesta—until the mid-1800s, when telescopic equipment and techniques had improved. Astronomers then began to stumble across many more such objects circling the Sun between the orbits of Mars and Jupiter. These objects are today called **asteroids** or **minor planets.**

The next major breakthrough came in 1891 when the German astronomer Max Wolf began using photographic techniques to search for asteroids. A total of 300 asteroids had been found up to that time, each painstakingly discovered by scrutinizing the skies for faint, uncharted stars that would shift their positions slowly from one night to the next. With the advent of astrophotography, however, the floodgates of data were opened. Astronomers could simply aim a camera-equipped telescope at the stars and take long exposures. If an asteroid happened to be in the field of view, it left a distinctive, blurred trail on the photographic plate because of its movement along its orbit during the time exposure (see Figure 10-1). Using this technique, Wolf alone discovered 228 asteroids.

Although thousands of asteroids have been sighted, only 3000 have well-determined orbits. An additional 6000 asteroids have "passable" orbits, but the orbits of 20,000 more have never been determined. The orbits of all officially discovered asteroids are published annually in the famous Soviet catalogue *Ephemerides of Minor Planets.*

To become the official discoverer of an asteroid, you must do a lot more than just produce one photograph of a blurred trail whose path does not match any known orbit listed in the *Ephemerides*. You must track the asteroid long enough to compute for it an accurate and reliable orbit (important data may be available from colleagues who may have inadvertently photographed it on earlier occasions). Then you must prove the accuracy of the

Figure 10-1 Two asteroids *Asteroids are detected by their blurred trails on time-exposure photographs of the stars. The images of two asteroids are seen in this picture. Astronomers sometimes find asteroids accidentally while photographing various portions of the sky for other purposes. (Yerkes Observatory)*

orbit by locating the asteroid again on at least one succeeding opposition. At that time, an official number will be assigned to your asteroid (Ceres is 1, Pallas is 2, and so forth). You will also be given the privilege of selecting a name for your asteroid.

Ceres is unquestionably the largest asteroid. With its diameter of nearly 1000 km, Ceres accounts for about 30 percent of the mass of all the asteroids combined. Only two others (Pallas and Vesta) have diameters greater than 300 km. Thirty other asteroids have diameters between 200 and 300 km, and there are 200 more asteroids bigger than 100 km across.

Astronomers estimate that roughly 100,000 asteroids exist that are bright enough to appear on photographs taken from Earth. The vast majority are less than 1 km across. Like Ceres, Pallas, and Juno, most asteroids circle the Sun at distances between 2 and $3\frac{1}{2}$ AU. This region of the solar system between the orbits of Mars and Jupiter is called the **asteroid belt** (see Figure 10-2). Asteroids whose orbits lie entirely within this region are called **belt asteroids.**

The combined matter of all the asteroids (including an estimate for those not yet officially known) would produce an object barely 1500 km in diameter, considerably smaller than our Moon. Bode's missing planet would not have been large enough to rank with the terrestrial planets. It seems more reasonable that the asteroids are debris left over from the formation of the solar system out of the solar nebula.

Constant gravitational perturbations caused by the enormous mass of Jupiter probably kept planetesimals from accreting into larger objects in the region between Mars and Jupiter. The missing planet simply never had a chance to form. What remains today in the gap between the orbits of Jupiter and Mars appears to be merely a remnant of scattered debris from the original solar nebula that elsewhere accreted into planets.

Figure 10-2 The asteroid belt *Most asteroids orbit the Sun in a $1\frac{1}{2}$-AU-wide belt between the orbits of Mars and Jupiter. The orbits of Ceres, Pallas, and Juno are indicated. Apollo and Icarus are asteroids that cross Earth's orbit.*

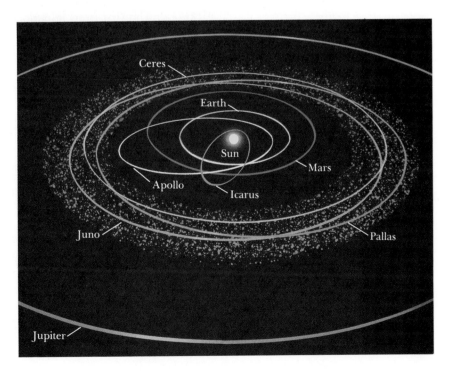

Jupiter's gravity affects the structure of the asteroid belt and captures asteroids along its orbit

The orbits of the asteroids are affected slightly by the gravitational pulls of the various planets. Most notable are the effects of Jupiter because of its large mass and proximity to the asteroid belt. Mars also perturbs asteroid orbits, but to a much lesser extent than Jupiter. The combined effect of such gravitational perturbations over the ages causes the asteroids to prefer certain orbits while avoiding others. It would be a monumental task to compute the resulting distribution of asteroid orbits because the gravitational forces of many bodies would have to be included. Nevertheless, we can appreciate some basic characteristics of the asteroid belt by considering the effects of Jupiter alone.

Imagine a belt asteroid moving along its orbit between the orbits of Jupiter and Mars. Each time the faster-moving asteroid catches up with and passes massive Jupiter, it experiences a slight gravitational tug toward Jupiter. This tug alters the asteroid's orbit slightly. However, over the ages, these close passes occur at different points along the asteroid's orbit, and so their effects tend to cancel each other.

Now imagine an asteroid circling the Sun once every 5.93 years, which is exactly half of Jupiter's orbital period. On every second trip around the Sun, the asteroid finds itself lined up between Jupiter and the Sun again and again, always at the same location and with the same orientation. These gravitational effects add up to deflect the asteroid from its original 5.93-year orbit, leaving a gap in the asteroid belt. According to Kepler's third law, a period of 5.93 years corresponds to a semimajor axis of 3.28 AU. Because of Jupiter, there are no asteroids that orbit the Sun at this average distance.

Similarly, we would expect to find a gap corresponding to an orbital period of one-third Jupiter's period, or 3.95 years. Comparable gaps should exist for other simple relationships between the periods of asteroids and Jupiter. The data graphed in Figure 10-3 show that such gaps do exist. They are called **Kirkwood gaps** in honor of the American astronomer Daniel Kirkwood, who first drew attention to them.

Figure 10-3 [left]   The Kirkwood gaps
This histogram displays the numbers of asteroids at various distances from the Sun. Notice that very few asteroids have orbits whose orbital periods correspond to simple fractions (such as $\frac{1}{2}$, $\frac{2}{5}$, $\frac{2}{3}$, $\frac{1}{3}$) of Jupiter's orbital period. Gravitational perturbations caused by repeated alignments with Jupiter have deflected asteroids away from these orbits.

Figure 10-4 [right]   The Trojan asteroids
Asteroids are trapped at the two Lagrange points along Jupiter's orbit by the combined gravitational forces of Jupiter and the Sun. Asteroids at these locations are named after Homeric heroes of the Trojan War.

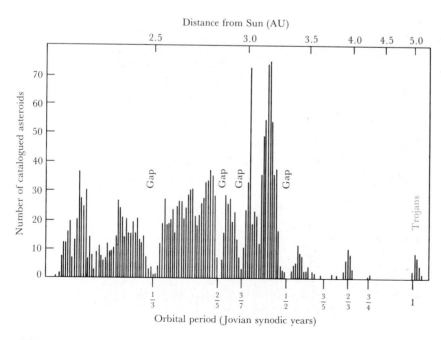

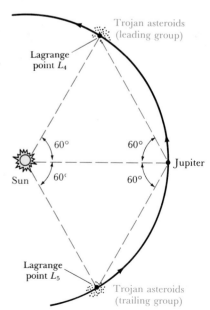

Although Jupiter's gravitational pull depletes certain orbits in the aster-oid belt, it captures asteroids at certain other locations much farther from the Sun. There are two specific points along Jupiter's orbit where the gravi-tational forces of the Sun and Jupiter work together to hold asteroids in orbit. These two locations are called the **Lagrange points** L_4 and L_5 in honor of the French mathematician whose calculations revealed this gravitational effect. Point L_4 is located one-sixth of the way around Jupiter's orbit ahead of the planet, and point L_5 occupies a similar position behind the planet (see Figure 10-4).

The asteroids trapped at Jupiter's Lagrange points are called **Trojan as-teroids,** named individually after heroes of the Trojan War. Nearly two dozen Trojan asteroids have been catalogued, and some astronomers believe that there may be as many as 700 rock fragments orbiting near each La-grange point.

Asteroids occasionally collide with each other and with the inner planets

In addition to the belt asteroids and the Trojan asteroids, there are other asteroids distinguished by highly elliptical orbits that bring them into the inner regions of the solar system. Occasionally one of these asteroids passes quite close to Earth. Figure 10-5 shows Eros as it passed within 23 million kilometers of our planet in 1931. In 1968, Icarus passed Earth at a distance of only 6 million kilometers. One of the closest near-misses in recent history occurred in 1937, when Hermes passed us at a distance of 900,000 km— only a little more than twice the distance to the Moon. A similar close-call occurred on March 23, 1989, when an asteroid called 1989FC passed within 800,000 km of the Earth. If this asteroid had struck the Earth, the impact would have been equivalent to the explosion 20,000 hydrogen bombs.

During these close encounters, astronomers can examine the details of the asteroids. For example, an asteroid's brightness is often observed to vary in a periodic fashion, presumably because different surfaces are turned to-ward us as the asteroid rotates. Periodic brightness variations thus reveal the

Figure 10-5 Eros *Eros is one of the asteroids that occasionally passes near the Earth. This photograph was taken in February 1931, when the distance between Eros and the Earth was only 23 million kilometers. The dimensions of Eros are roughly 10 by 20 by 30 km. This asteroid rotates with a period of 5.27 hours. (Yerkes Observatory)*

asteroid's rate of rotation. Typical asteroid rotation periods are in the range of 5 to 20 hours.

Careful scrutiny of an asteroid's brightness variations can also reveal the asteroid's shape and dimensions. Only the largest asteroids, such as Ceres, Pallas, and Vesta, have enough gravity to pull themselves into a spherical shape. Smaller asteroids permanently retain the odd shapes produced by interasteroid collisions. A small asteroid looks dim when seen end on but appears brighter when seen broadside. Measurements of an asteroid's brightness variations thus tell astronomers a lot about its shape.

There is ample evidence that interasteroid collisions successfully fragment asteroids into small pieces. In 1918, for instance, the Japanese astronomer Kiyotsugu Hirayama drew attention to groups of asteroids that share nearly identical orbits. These groupings presumably resulted from the fragmentation of parent asteroids.

The collision of kilometer-sized asteroids must be an awesome event. Typical collision velocities are estimated at 1 to 5 km/sec (2000 to 11,000 miles per hour), which is more than sufficient to shatter rock. Only in a high-velocity collision is there enough energy to shatter an asteroid permanently. In collisions at low velocities, the resulting fragments may not achieve escape velocity from each other and will reassemble because of their mutual gravitational attraction. Alternatively, several large fragments may end up orbiting each other, which is probably what happened to both Pallas and Victoria. They are **binary asteroids,** each consisting of a main asteroid and a large satellite.

Interasteroid collisions produce numerous chunks of rock, many of which eventually rain down on Venus, Earth, and Mars. Fortunately for us, the vast majority of these asteroid fragments (usually called **meteoroids**) are quite small. On rare occasions, however, a large fragment does collide with our planet. The result is an **impact crater** whose diameter depends on the mass and speed of the impinging object.

One of the most impressive and best-preserved terrestrial impact craters is the famous Barringer Crater near Winslow, Arizona. The crater measures 1.2 km across and is 200 m deep (see Figure 10-6). The crater was formed 25,000 years ago when an iron-rich object measuring roughly 50 m across

Figure 10-6 The Barringer Crater
An iron meteoroid measuring 50 m across struck the ground in Arizona 25,000 years ago. The result was this beautifully symmetrical impact crater measuring 1.2 km in diameter and 200 m deep at its center. (Meteor Crater Enterprises)

struck the ground with a speed estimated at 11 km/sec (25,000 miles per hour). The resulting blast was equal to the detonation of a 20-megaton hydrogen bomb.

Iron is one of the more abundant elements in the universe (recall Table 6-4) as well as one of the most common rock-forming elements (recall Table 6-5), and so it is not surprising that iron is an important constituent of asteroids and their fragments called meteoroids. Another element, iridium—one of several elements that geologists call **siderophiles,** or "iron lovers"—is common in iron-rich minerals but is rare in ordinary rocks. Measurements of iridium in the Earth's crust can therefore tell us about the rate at which meteoritic material has been deposited on the Earth over the ages. A team of geologists including Walter Alvarez and his physicist father Luis Alvarez from the University of California at Berkeley made such measurements in the late 1970s.

Working at a site of exposed marine limestone in the Apennine Mountains in Italy, the Alvarez team discovered an exceptionally high abundance of iridium in a dark-colored layer of clay between the limestone strata (see Figure 10-7). Since this 1979 discovery, a comparable layer of iridium-rich material has been uncovered at a variety of sites around the world. In all cases, geological dating reveals that this apparently worldwide layer of iridium-rich clay is about 63 million years old.

Paleontologists were quick to realize the profound significance of this particular date. Around 63 million years ago, all the dinosaurs became extinct. In fact, at that time a staggering 65 percent of all the species on Earth disappeared within a relatively brief span of time.

The Alvarez discovery suggests a startling explanation for the dramatic extinction of more than half the life-forms that inhabited our planet at the end of the Mesozoic era: perhaps an asteroid hit the Earth. An asteroid

Figure 10-7   The iridium-rich layer of clay   *This photograph of strata in the Apennine Mountains of Italy shows a dark-colored layer of iridium-rich clay sandwiched between white limestone (below) from the late Mesozoic era and grayish limestone (above) from the early Cenozoic era. This iridium-rich layer may be the result of an asteroid impact that caused the extinction of the dinosaurs. The coin is the size of a U.S. quarter. (Courtesy of Walter Alvarez)*

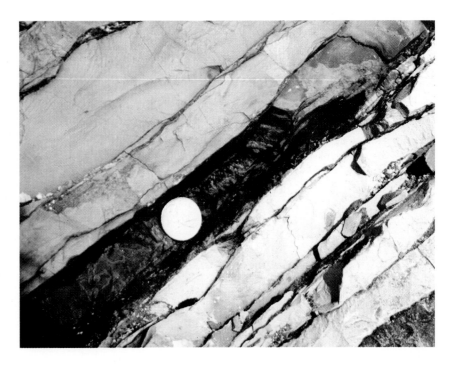

10 km in diameter slamming into the Earth would have thrown enough dust into the atmosphere to block out sunlight for several years. As plants died for lack of sunshine, the dinosaurs would have starved to death along with many other creatures in the vegetation food chain. Eventually the dust settled, depositing an iridium-rich layer around the world. Tiny, rodentlike creatures that could ferret out seeds and nuts were prominent among the animals that managed to survive the holocaust, setting the stage for the rise of mammals in the Cenozoic era.

Some geologists and paleontologists are not yet convinced that such a meteoroid impact did produce "the great dying out" at the end of the Mesozoic era, but many scientists agree that this hypothesis fits the available evidence better than other explanations that have been offered.

Meteorites are classified as stones, stony irons, or irons, depending on their composition

A meteoroid, like an asteroid, is a chunk of rock in space. There is no official dividing line between meteoroids and asteroids, but the term **asteroid** is generally applied only to objects larger than a few hundred meters across.

A **meteor** is the brief flash of light (sometimes called a shooting star) that is visible at night when a meteoroid strikes the Earth's atmosphere (see Figure 10-8).

If a piece of rock survives its fiery descent through the atmosphere, the object that reaches the ground is called a **meteorite.** People have been finding specimens for thousands of years, and descriptions of meteorites appear in ancient Chinese, Greek, and Roman literature. Our ancestors placed special significance on these "rocks from heaven."

Meteorites are classified into three broad categories: stones, irons, and stony-irons. As their name suggests, **stony meteorites,** or **stones,** look like ordinary rocks at first glance, but they are sometimes covered with a **fusion crust** (see Figure 10-9). This crust is produced by the momentary melting of the meteorite's outer layers during its fiery descent through the atmosphere. When a stony meteorite is cut in two and polished, tiny flecks of iron are sometimes found in the rock (see Figure 10-10).

Figure 10-8 A meteor *A meteor is produced when a piece of interplanetary rock or dust strikes the Earth's atmosphere at high speed. Exceptionally bright meteors, such as the one shown in this long exposure (notice the star trails), are usually called fireballs. (Courtesy of Ronald A. Oriti)*

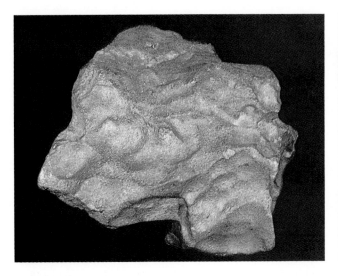

Figure 10-9 [above left] A stony meteorite Ninety-three percent of all meteorites that fall on the Earth are stones. Many freshly discovered specimens, like the one shown here, are coated with dark fusion crusts. This particular stone fell in Texas. (From the collection of Ronald A. Oriti)

Figure 10-10 [above right] A stone (cut and polished) When cut and polished, some stony meteorites are found to contain tiny specks of iron mixed in the rock. This specimen was discovered in California. (From the collection of Ronald A. Oriti)

Figure 10-11 A stony-iron meteorite Stony-irons account for slightly less than 2 percent of all meteorites that fall on the Earth. This particular specimen is a variety of stony-iron called a pallasite. (From the collection of Ronald A. Oriti)

Although stony meteorites account for nearly 93 percent of all meteoritic material that falls on the Earth, stones are the most difficult specimens to find. If they go undiscovered and are exposed to the weather for a few years, they become almost indistinguishable from common terrestrial rocks. Meteorites with a high iron content are much easier to find because they can be located with a metal detector. Consequently, iron and stony-iron meteorites dominate most museum collections.

As their name suggests, **stony-iron meteorites** consist of roughly equal amounts of rock and iron. Olivine commonly is the mineral suspended in the matrix of iron, as in the sample shown in Figure 10-11.

Iron meteorites (see Figure 10-12), or irons, account for nearly 6 percent of the material that falls on the Earth. Iron meteorites may contain from 10 to 20 percent nickel.

In 1808, Count Alois von Widmanstätten discovered a conclusive test for the authenticity of the most common type of iron meteorite. About 75 percent of all iron meteorites are of a type called **octahedrites.** When an octahedrite is cut, polished, and briefly dipped into a dilute acid solution, its unique crystalline structure is revealed. These crystalline designs are appropriately called **Widmanstätten patterns** (see Figure 10-13).

Widmanstätten patterns constitute conclusive proof of a meteorite's authenticity because nickel–iron crystals can grow to lengths of several centimeters only if the molten metal cools slowly over many millions of years. Octahedrites cool at rates of 1 to 10 K per million years during the time the crystals are forming. Widmanstätten patterns can thus never be found in counterfeit meteorites, or "meteorwrongs," as they are humorously called.

The existence of Widmanstätten patterns strongly suggests that some asteroids were partly molten for a substantial period after their formation. Furthermore, the size of an octahedrite's parent asteroid can be estimated by calculating how much rock must have insulated the molten iron–nickel interior to produce its long-term cooling rate. The results of such calculations imply that typical meteorites are fragments of parent asteroids 200 to 400 km in diameter.

The three main types of meteorites may have come from different parts of a parent asteroid in the following manner. As soon as the asteroid had

Figure 10-12 [left]   An iron meteorite
Irons are composed almost entirely of nickel–iron minerals. The surface of a typical iron is covered with thumbprintlike depressions caused by the melting away of the meteorite's outer layers during its high-speed descent through the atmosphere. This specimen was found in Australia. (From the collection of Ronald A. Oriti)

Figure 10-13 [right]   Widmanstätten patterns   When cut, polished, and etched with a weak acid solution, most iron meteorites exhibit interlocking crystals in designs called Widmanstätten patterns. These patterns appear only in the type of iron meteorite called an octahedrite. This particular octahedrite was found in Australia. (From the collection of Ronald A. Oriti)

accreted from planetesimals 4.5 billion years ago, rapid decay of short-lived radioactive isotopes heated the asteroid's interior to temperatures above the melting point of rock. Over the next few million years, chemical differentiation occurred. As iron and other heavy elements sank toward the asteroid's center, they displaced the lighter elements (such as silicon) upward toward the asteroid's surface. After the asteroid cooled and its core solidified, inter-asteroid collisions fragmented the parent body into meteoroids. Iron meteorites are specimens from the asteroid's core, stones being samples of its crust. Stony-irons presumably come from intermediate regions between the asteroid's core and its crust.

It is clear that meteorites derived from the fragmentation of large asteroids were subjected to substantial processing during the first billion years of solar-system formation. These meteoritic specimens are therefore not representative of the primordial material from which the solar system was originally created. In order to find primordial meteorites, we must search for specimens that show no evidence of having been subjected to the metamorphic processes that occurred inside the asteroids.

Such specimens—called **carbonaceous chondrites**—do exist. About 6 percent of all the stones that fall on the Earth are carbonaceous chondrites. Their primordial nature is inferred from their high content of volatile compounds, sometimes including as much as 20 percent water. Furthermore, carbonaceous chondrites are rich in complex organic compounds. The water and volatile chemicals would have been driven out and the large organic molecules broken down if these meteorites had been subjected to any significant heating.

Shortly after midnight on February 8, 1969, the night sky around Chihuahua, Mexico, was illuminated by a brilliant blue-white light moving across the heavens. The dazzling display was witnessed by hundreds of people, many of whom thought that the world was coming to an end. As the light moved across the sky, it exploded in a spectacular, noisy detonation that dropped thousands of rocks and pebbles over the terrified onlookers. Within hours, teams of scientists were on their way to collect specimens, from what was collectively called the Allende meteorite after the locality of the finds.

Perhaps the most significant discovery to come from the Allende meteorite was made by Gerald J. Wasserburg and his colleagues at the California

Institute of Technology. They found unmistakable evidence of the former presence of a radioactive isotope of aluminum, ^{26}Al. This particular isotope has a half-life of only 720,000 years, which means that after 720,000 years half of the ^{26}Al in a sample has changed into a stable isotope of magnesium, ^{26}Mg. On either an astronomical or geological time scale, ^{29}Al is therefore quite short-lived.

Detectable amounts of ^{26}Al could have been included in the meteorite only if the radioactive aluminum was created shortly before the formation of the solar system. If only a few hundred million years had elapsed, virtually all the radioactive aluminum would have changed into magnesium before the meteorite formed. Astronomers were therefore faced with evidence of energetic nuclear processes that occurred in our vicinity roughly 4.5 billion years ago, about the time the Sun was born.

One of nature's most violent and spectacular phenomena, called a **supernova explosion,** occurs during the death of a massive star. As we shall see in greater detail in Chapter 14, the doomed star blows itself apart in a cataclysm that hurls matter outward at tremendous speeds. During this detonation, violent collisions between nuclei produce a host of radioactive isotopes, including ^{26}Al. It thus seems inescapable that a supernova occurred very near the Sun's birthplace 4.5 billion years ago. In addition to contaminating the interstellar medium with radioactive ^{26}Al, the supernova's shock wave would have compressed the interstellar gas and dust, triggering the birth of the solar system.

Besides telling us about the creation of the solar system, the study of meteorites may shed light on the origin of life on Earth. **Amino acids,** the building blocks of proteins on which terrestrial life is based, are among the organic compounds found inside carbonaceous chondrites. Perhaps interstellar organic material falling on Earth played a role in the appearance of simple organisms on our planet nearly 4 billion years ago.

A comet is a dusty chunk of ice that is partly vaporized as it passes near the Sun

Many rocks and chunks of ice that originally condensed out of the primordial solar nebula still continue to orbit the Sun. Just as heat from the protosun produced two classes of planets (the terrestrial and the Jovian), two main types of interplanetary material were created. Near the Sun, interplanetary debris consists of the rocks called asteroids and meteoroids. Far from the Sun, there are numerous dust-covered chunks of ice called **comets.**

As mentioned earlier in this chapter, asteroids travel around the Sun along roughly circular orbits that are largely confined to the asteroid belt and to the plane of the ecliptic. In sharp contrast, comets travel around the Sun along highly elliptical orbits inclined at random angles to the ecliptic plane.

As a comet approaches the Sun, solar heat begins to vaporize the ices. The liberated gases surrounding the icy nucleus soon begin to glow, producing a fuzzy, luminous ball called the **coma** that can eventually expand to a million kilometers in diameter. Continued action by the solar wind and radiation pressure blows these luminous gases outward into a long, flowing **tail.** The result is one of the most awesome sights ever visible in the nighttime sky (see Figure 10-14).

The solid part of a comet, the **nucleus,** is a chunk of ice typically measuring a few kilometers across. Frozen ammonia, methane, and water are the

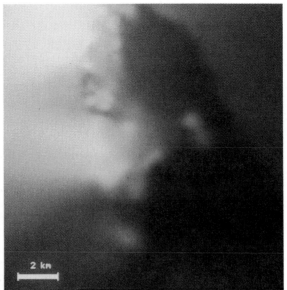

2 km

Figure 10-14 [above left]   Comet West
A comet is always named after the person who first sees it. Astronomer Richard M. West first noticed this comet on a photograph taken with a telescope in 1975. After passing near the Sun, Comet West became one of the brightest comets in recent years. This photograph shows the comet in the predawn sky in March 1976. (Courtesy of H. Vehrenberg)

Figure 10-15 [above right]   The nucleus of Comet Halley   *In March 1986, five spacecraft passed near Comet Halley. This closeup picture was taken by a camera on board the Giotto spacecraft and shows the potato-shaped nucleus of the comet. The nucleus is darker than coal and measures 15 km in the longest dimension and about 8 km in the shortest. The Sun illuminates the comet from the left. Two bright jets of dust extend 15 km from the nucleus toward the Sun, suggesting that major activity emanates from the sunlit side. (Max Planck Institut für Aeronomie)*

primary components of cometary ices, as indicated by their spectral lines. Harvard astronomer Fred L. Whipple, a pioneer in comet research, coined the description "dirty iceberg" to reflect the fact that bits and pieces of dust and rocky material are mixed in with the ices. A closeup view of the nucleus of Halley's Comet is shown in Figure 10-15. The overall structure of a comet is diagrammed in Figure 10-16.

Figure 10-16   The structure of a comet
The solid part of a comet (the nucleus) is roughly 10 kilometers in diameter. The coma can be as large as 100,000 km to 1 million km across. The hydrogen envelope is typically 10 million kilometers in diameter. A comet's tail can be as long as 1 AU, long enough to reach all the way from the Earth to the Sun.

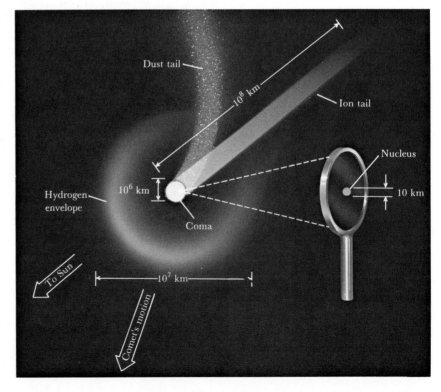

Figure 10-17 Comet Kohoutek and its hydrogen envelope These two photographs of Comet Kohoutek are reproduced to the same scale. (Left) The comet in visible light. (Right) This view of the comet in ultraviolet wavelengths reveals a huge hydrogen cloud surrounding the comet's head. (Johns Hopkins University; Naval Research Laboratory)

Not visible to the human eye is the **hydrogen envelope,** a tenuous sphere of gas surrounding the comet's nucleus and measuring as much as 10 million kilometers in diameter. Figure 10-17 shows two views of Comet Kohoutek: as it appeared to Earth-based observers in 1973, and as photographed by an ultraviolet camera from a rocket. From the ultraviolet view, astronomers first discovered the enormous extent of the hydrogen envelope.

Comets come in a wide range of shapes and sizes. For example, the comet shown in Figure 10-18 had a large, bright coma but a short, stubby tail. In contrast, the comet seen in Figure 10-19 had an inconspicuous coma, but its tail had an astonishing length of 1 AU, long enough to reach all the way from the Earth to the Sun.

It has long been known that comet tails always point away from the Sun (see Figure 10-20), regardless of the direction of the comet's motion. In fact, the Sun usually produces two comet tails: an **ion tail** and a **dust tail.** Ionized

Figure 10-18 [left] The head of Comet Brooks A comet is always named after the person who first sights it. This comet, named Comet Brooks after its discoverer, had an exceptionally large, bright coma. It dominated the night skies during October 1911. (Lick Observatory)

Figure 10-19 [right] Comet Ikeya–Seki This comet, named after its two Japanese codiscoverers, dominated the predawn sky during late October 1965. Although its coma was tiny, its tail was 1 AU long. (Lick Observatory)

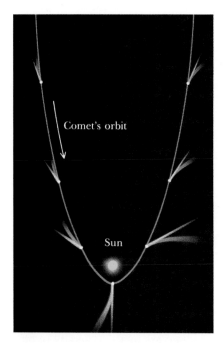

Figure 10-20 The orbit and tail of a comet *The solar wind and radiation pressure from sunlight blow a comet's dust particles and ionized atoms away from the Sun. Consequently, the comet's tail always points away from the Sun.*

atoms (that is, atoms missing one or more electrons) are swept directly away from the Sun by the solar wind. Micrometer-sized dust particles are blown away from the comet's coma by radiation pressure. The relatively straight ion tail can exhibit a dramatic structure that changes from night to night (see Figure 10-21). The more amorphous dust tail is typically arched.

Astronomers discover at least a dozen comets in a typical year. Some are **short-period comets,** which circle the Sun in less than 200 years. Like the famous Halley's Comet, they appear again and again at predictable intervals. The majority of comets discovered each year are **long-period comets,** however, which take 100,000 to 1 million years to complete one orbit of the Sun. These comets travel along extremely elongated orbits and consequently spend most of their time at distances of 40,000 to 50,000 AU from the Sun—about one-fifth of the way to the nearest star.

Because astronomers discover long-period comets at a rate of roughly one per month, it is reasonable to suppose that there is an enormous population of comets out there at 50,000 AU from the Sun. This reservoir of cometary nuclei surrounding the Sun is called the **Oort cloud** after the Dutch astronomer Jan Oort, who first proposed its existence in the 1950s. Estimates of the numbers of "dirty icebergs" in the Oort cloud range from 1 million to more than 100,000 million. Only a large reserve of cometary nuclei can explain why we see so many long-period comets even though each one takes up to 1 million years to travel once around its orbit.

Comets cannot survive very many passages near the Sun. Eventually, a comet's ices are completely vaporized and only a swarm of meteoritic dust and pebbles remains. A comet's nucleus will disintegrate more quickly if it happens to pass near the Sun. Figure 10-22 shows the breakup of the nu-

Figure 10-21 [right] The two tails of Comet Mrkos *Comet Mrkos dominated the evening sky during August 1957. These three views, taken at two-day intervals, show dramatic changes in the comet's ion tail. In contrast, the slightly curved dust tail remained fuzzy and featureless. (Palomar Observatory)*

Figure 10-22 [below] The fragmentation of Comet West *Shortly after passing near the Sun in 1976, the nucleus of Comet West broke into four pieces. This series of five photographs clearly shows the disintegration of the comet's nucleus. Figure 10-14 shows a wide-angle view of this comet. (New Mexico State University Observatory)*

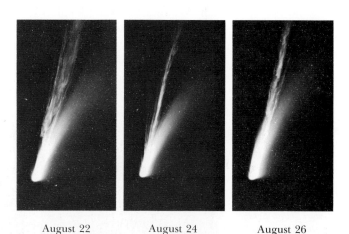

August 22   August 24   August 26

March 8   March 12   March 14   March 18   March 24

Table 10-2 Meteor showers

Shower name	Date of maximum display	Meteors per hour
Quadrantid	January 3	40
Eta Aquirid	May 4	20
Delta Aquarid	July 28	20
Perseid	August 12	50
Orionid	October 21	30
Geminid	December 14	50

cleus of a comet shortly after it passed the Sun at the distance of Mercury's orbit.

After a comet dies, its remaining dust and rock fragments spread out in a loose collection of debris that continues to circle the Sun along the comet's orbit. If the Earth's orbit happens to pass near or through the swarm, a **meteor shower** is seen on the Earth.

Nearly a dozen meteor showers can be seen each year. Like comets, meteor showers are not confined to the plane of the ecliptic. Table 10-2 lists some of the major meteor showers, the date of maximum display, and the average number of meteors per hour that an observer might see under ideal conditions.

As incredible as it may seem, a total of 300 tons of extraterrestrial rock and dust is estimated to fall on the Earth each day. The fluffy, low-density material from comets burns up in the atmosphere, and only denser specimens related to asteroids typically reach the ground. Nevertheless, there is evidence that a comet struck the Earth in the recent past.

On June 30, 1908, a spectacular explosion occurred over the Tunguska region of Siberia. Hundreds of square kilometers of forest were devastated (see Figure 10-23), and the blast was audible 1000 km away. The explosion

Figure 10-23 Aftermath of the Tunguska event In 1908, a piece of a comet's nucleus struck the Earth's atmosphere over the Tunguska region of Siberia. Trees were blown down for many kilometers in all directions around the impact site. (Courtesy of Sovfoto)

was equivalent to the detonation of a tactical nuclear warhead with the destructive power of several hundred kilotons of TNT.

The most likely explanation of this event is that a small comet (perhaps a 100-m fragment of the short-period Comet Encke) collided with the Earth. No impact crater was formed, and the trees at "ground zero" were left standing upright, but they were completely stripped of branches and leaves. This phenomenon is what would be expected from a loosely consolidated ball of cometary ices that vaporized with explosive force before striking the ground. The Tunguska event is a good example of the kind of devastation that can be wreaked by interplanetary debris.

Summary

Bode's law is a numerical sequence that gives the distances (in AU) of the planets Mercury through Uranus from the Sun. This "law" inspired nineteenth century astronomers to search for a planet in the gap between the orbits of Mars and Jupiter.

Thousands of belt asteroids with diameters from a few kilometers to 1000 km circle the Sun between the orbits of Mars and Jupiter.

  Gravitational perturbations by Jupiter deplete certain orbits within the asteroid belt. The resulting gaps, called Kirkwood gaps, occur at simple fractions of Jupiter's orbital period.

  Jupiter's gravity also captures asteroids in two locations, called Lagrange points, along Jupiter's orbit.

Some asteroids move in elliptical orbits that cross the orbits of Mars and Earth. Many of these asteroids will eventually strike one of the inner planets.

  An asteroid may have struck the Earth 63 million years ago causing the extinction of the dinosaurs and many other species.

Small rocks in space are called meteoroids. If a meteoroid enters the Earth's atmosphere, it produces a fiery trail called a meteor. If part of the object survives the fall, the fragment that reaches the Earth's surface is called a meteorite.

  Meteorites are grouped in three major classes according to their composition: iron, stony-iron, or stony meteorites.

  Rare stony meteorites called carbonaceous chondrites may be relatively unmodified material from the solar nebula. These meteorites often contain organic material and may have played a role in the origin of life on Earth.

  An analysis of isotopes in certain meteorites suggests that a nearby supernova explosion triggered the formation of the solar system 4.5 billion years ago.

A comet is a chunk of ices and rock fragments that generally moves in a highly elliptical orbit about the Sun at a great inclination to the plane of the ecliptic.

  As a comet approaches the Sun, its icy nucleus develops a luminous coma surrounded by a vast hydrogen envelope. An ion tail and a dust tail extend from the comet, pushed away from the Sun by the solar wind and radiation pressure.

Fragments of "burned-out" comets produce meteor swarms. Millions of cometary nuclei probably exist in the Oort cloud some 50,000 AU from the Sun.

Review questions

1   Why do you suppose there are many small asteroids, but only a few very large asteroids?

2   Can you think of another place in the solar system in which occurs a phenomenon similar to the Kirkwood gaps in the asteroid belt? Explain.

3   Describe the three main classifications of meteorites. How might these different types of meteorites have originated?

4   What types of surface features would you expect to see on a closeup photograph of an asteroid?

5   Where on Earth might you find large numbers of stony meteorites that are not significantly weathered?

6   Suppose you found a rock that you suspect might be a meteorite. Describe some of the things you could do to see if it were a meteorite or a "meteorwrong."

7   Explain why comets are generally brighter after passing their perihelion.

8   What is the relationship between comets and meteor showers?

Advanced questions

9   Since there are different types of meteorites, would you expect different types of asteroids to exist also? What sorts of observations might an Earth-based astronomer make in order to discover chemical differences between asteroids?

10   Some astronomers have recently argued that passage of the solar system through an interstellar cloud of gas could perturb the Oort cloud, causing many comets to deviate slightly from their original orbits. What might be the consequences for Earth?

Discussion questions

11   Suppose it were discovered that the asteroid Hermes had been perturbed in such a way as to put it on a collision course with Earth. Describe what you would do to counter such a catastrophe within the framework of present technology.

12   From the abundance of craters on the Moon and Mercury, we know that numerous asteroids and meteoroids struck the inner planets early in the history of the solar system. Is it reasonable to suppose that numerous comets also pelted the planets 3 to 4 billion years ago? Speculate about the effects of such a cometary bombardment, especially with regard to the evolution of the primordial atmospheres of the terrestrial planets.

For further reading

Berry, R. "Giotto Encounters Comet Halley." *Astronomy*, June 1986, p. 6.

Chapman, C. "The Nature of Asteroids." *Scientific American*, January 1975.

Dodd, R. *Thunderstones & Shooting Stars: The Meaning of Meteorites.* Harvard University Press, 1986.

Falk, S., and Schramm, D. "Did the Solar System Start with a Bang?" *Sky & Telescope*, July 1979, p. 18.

Hutchinson, R. *The Search for Our Beginnings.* Oxford University Press, 1983.

Morrison, D. "Asteroids." *Astronomy*, June 1976, p. 6.

Seargent, D. *Comets: Vagabonds of Space.* Doubleday, 1982.

Whipple, F. "The Nature of Comets." *Scientific American*, February 1974.

11

Our star

The solar corona *This composite view of the Sun's outer atmosphere combines a white-light view taken during a solar eclipse with an X-ray image captured by a camera on board a rocket launched just before the eclipse. The white-light photograph shows streamers extending quite far above the solar surface. The X-ray view shows near-surface features in shades of yellow, orange, and red. By matching these two views, scientists can study coronal structures over a range of wavelengths and altitudes, as well as construct a three-dimensional picture of solar activity. This is one example of new techniques that astronomers are using to explore the workings of the Sun. (High Altitude Observatory, NCAR, University of Colorado)*

We conclude our studies of the solar system by taking a close look at the Sun. This solar survey also prepares us for understanding stars in general. We first discuss the thermonuclear reactions that occur in the core of the Sun and the ways that this energy moves from the core to the surface, from where it radiates into space. This introduces a theoretical model of the Sun's interior structure. We then turn to the Sun's atmosphere and examine a variety of phenomena observed on the solar surface. We find that the 11-year sunspot cycle is only one aspect of a more basic 22-year solar cycle that affects many properties of the Sun. Finally, we mention the new technique of helioseismology that astronomers are using to study the Sun in detail.

The Sun is an average star. Its mass, size, surface temperature, and chemical composition lie roughly midway between the extremes exhibited by other stars. Unlike other stars, however, the Sun is available for detailed, closeup examination. Studying the Sun therefore offers excellent insights into the nature of stars in general.

Understanding the Sun is important to humanity because the Sun is our source of heat and light. Life would not be possible on Earth without the energy provided by the Sun. Even a small change in the Sun's size or surface temperature could dramatically alter conditions on the Earth, either melting the polar caps or producing another ice age.

Although the Sun is a commonplace star, it is a dramatic arena where we can observe the fascinating and complicated interaction of matter and energy on a colossal scale. Beautiful and bewildering phenomena occur as columns of hot gases gush up to the solar surface, interact with the Sun's magnetic field, and dissipate energy into the Sun's outer atmosphere. The source of all this energy lies buried at the Sun's center.

The Sun's energy is produced by thermonuclear reactions in the core of the Sun

During the nineteenth century, geologists and biologists found convincing evidence that the Earth must have existed in more or less its present form for hundreds of millions of years. This fact posed severe problems for physicists because it seemed impossible to explain how the Sun has been shining for so long, radiating immense amounts of energy into space. If the Sun were made of coal, for example, it could burn for only 3000 years.

In 1905, Albert Einstein published his special theory of relativity, which offered some hope of explaining this mystery. One of the implications of Einstein's theory is that matter and energy are interchangeable according to the simple equation

$$E = mc^2$$

In other words, a mass (m) can be converted into an amount of energy (E) equivalent to mc^2, where c is the speed of light. Because c is a large number and c^2 is huge, a small amount of matter can be converted into an awesome amount of energy.

Inspired by Einstein's work, astronomers began to wonder if the Sun's energy output might come from the conversion of matter into energy. But exactly what kind of mechanism would transform matter into energy?

In the 1920s the British astronomer Arthur Eddington showed that temperatures at the center of the Sun must be much greater than had previously been thought. Under such high-temperature conditions at the Sun's center hydrogen nuclei can fuse together to produce helium nuclei in a reaction that transforms a tiny amount of mass into a very large amount of energy. Because the nuclei fuse together under high temperatures, this process is called **thermonuclear fusion.**

Recall that the nucleus of a hydrogen (H) atom consists of a single proton. The nucleus of a helium atom (He) consists of two protons and two neutrons. In the nuclear process

$$4H \rightarrow He$$

two of the four protons from the hydrogen atoms are changed into neutrons to produce a single helium nucleus.

In the conversion of hydrogen into helium, matter is lost because the ingredients (four hydrogen nuclei) weigh very slightly more than the product (one helium nucleus). Specifically, the mass lost during this reaction may be calculated as follows:

$$
\begin{array}{rl}
\text{4 hydrogen atoms} = & 6.693 \times 10^{-24} \text{ g} \\
-\text{1 helium atom} = & \underline{-6.645 \times 10^{-24} \text{ g}} \\
\text{mass lost} = & 0.048 \times 10^{-24} \text{ g}
\end{array}
$$

This lost mass is converted into energy in the amount predicted by the equation $E = mc^2$.

The Sun's total mass is 2×10^{33} g (333,000 Earth masses) and the Sun's total power output, called its **luminosity,** is 3.9×10^{26} watts. To produce this luminosity, 600 million metric tons of hydrogen must be converted into helium within the Sun each second. This prodigious rate is possible because the Sun contains a vast supply of hydrogen—enough to continue the present rate of energy output for another 5 billion years.

In this process of nuclear fusion, the conversion of hydrogen into helium at the Sun's center is called **hydrogen burning** even though nothing is actually "burned" in the conventional sense of that word. The ordinary burning of wood, coal, or any flammable substance is a chemical process involving only the electrons that orbit the nuclei of atoms. Thermonuclear fusion is a far more energetic process that involves violent collisions between the nuclei of atoms.

Hydrogen burning occurs at the Sun's core, where it is very hot (15 million K). Normally, the positive electric charge on protons is quite effective in keeping the protons far apart, because like charges repel each other. In the extreme heat of the Sun's center, however, the protons are moving so fast that they can penetrate each other's electric fields and stick together. In Chapter 13 we shall learn that most of the stars you can see in the sky have hydrogen burning occurring at their centers.

A theoretical model of the Sun can tell us how energy gets from the Sun's center to its surface

Although the Sun's interior is hidden from our view, we can use the laws of physics to calculate what is going on below the solar surface (see Figure 11-1). The resulting **model** of the Sun tells us about the Sun's internal characteristics such as its pressure, temperature, and density.

In developing a model of the Sun or of any other stable star, we should note that the Sun is not undergoing any dramatic changes. The Sun is not exploding or collapsing, nor is it significantly heating up or cooling off. The Sun is thus in balance in two ways: mechanically and thermally.

Mechanical balance, often called **hydrostatic equilibrium,** means simply that a star is supporting its own weight. Because of gravity, the tremendous weight of the Sun's outer layers pressing inward from all sides tries to make the star contract. As gravity compresses the star, however, gas pressure inside the star increases. The greater the compression, the higher the internal pressures. Hydrostatic equilibrium is achieved when the pressure at every depth within the star is exactly sufficient to support the weight of the overlying layers.

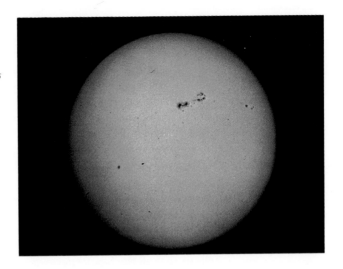

Thermal balance, often called **thermal equilibrium,** means simply that a star keeps shining. Vast amounts of energy escape from the Sun's surface each second. This energy is constantly resupplied by thermonuclear fusion in the Sun's core at a steady, persistent rate that maintains thermal equilibrium. But exactly how is energy transported from the Sun's center to its surface?

Experience teaches us that energy always flows from hot regions to cooler regions. If you heat one end of a metal bar with a blowtorch, the other end of the bar eventually becomes warm. This method of energy transport is called **conduction.** Conduction varies significantly from one substance to another (copper is a good heat conductor, wood is not), depending on the arrangement and interaction of the atoms. Calculations have demonstrated that conditions inside stars like the Sun are not favorable for conduction, and so this means of energy transport is not efficient in ordinary stars. (In Chapter 14 we shall see that conduction is important in very compact stars called white dwarfs.)

Two other means of energy transport—convection and radiative diffusion—do operate inside stars like the Sun to move energy from a star's center to its surface. **Convection** involves the circulation of gases between hot and cold regions. Just as a hot-air balloon drifts skyward, so hot gases rise toward a star's surface while cool gases sink back down toward its center. The net effect of this physical movement of gases is to transfer heat energy from the center toward the surface.

In **radiative diffusion,** photons migrate outward from the star's hot core, where they are constantly created, toward the cooler surface, where they escape into space. The path of an individual photon is quite random as it is knocked about between the atoms and electrons inside a star. In all, it takes nearly a million years for energy created at the Sun's center to reach the solar surface finally and escape as sunlight.

The concepts of hydrostatic equilibrium, thermal equilibrium, and energy transport can be expressed in the form of a set of mathematical equations that are collectively called the **equations of stellar structure.** These equations can be solved to yield the conditions of pressure, temperature, and density that must exist inside a star to maintain equilibrium. Astrophysicists use high-speed computers to solve the equations of stellar structure, thus developing a detailed theoretical model of the structure of a star. The astrophysicist begins with astronomical data about the star's surface such as, for example, that the Sun's surface temperature is 5800 K, its luminosity 3.9×10^{33} erg/sec, and its gas pressure and density almost zero. The equations of stellar structure are used to calculate conditions layer by layer toward the star's center. In this way, the astrophysicist can discover how temperature, pressure, and density increase with increasing depth below the star's surface. This information constitutes a stellar model. We have learned through this technique that the temperature at the Sun's center is 15.5 million K and the pressure is 3.4×10^{11} atmospheres.

Figure 11-2 presents a theoretical model of the Sun. The four graphs show how the Sun's luminosity, mass, temperature, and density vary from the Sun's center to its surface. For instance, the upper graph gives the percentage of the Sun's luminosity. Note that the luminosity rises to 100 percent at about one-quarter of the way from the Sun's center to its surface. This tells us that all the Sun's energy is produced within a volume extending out to $\frac{1}{4}$ solar radius.

Also note that the mass rises to nearly 100 percent at about 0.6 solar radius from the Sun's center. Most of the Sun's mass must therefore be confined to a volume extending only slightly more than half the way from the Sun's center to its surface. The outer layers of the Sun thus contain very little matter.

From the Sun's center out to 0.8 solar radius, energy is transported by radiative diffusion, as we have seen. This inner region is therefore called the

Figure 11-2 A theoretical model of the Sun   *The Sun's internal structure is displayed here with graphs that show how the density, temperature, mass, and luminosity vary with the distance from the Sun's center. A solar radius (the distance from the Sun's center to its surface) equals 696,000 km.*

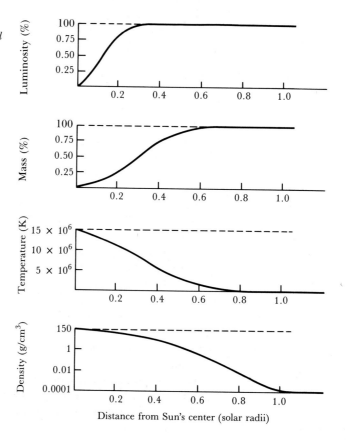

Figure 11-3   The Sun's internal structure
Thermonuclear reactions occur in the Sun's core, which extends to a distance of 0.25 solar radius from the center. Energy from the core is transported outward via radiative diffusion to a distance of 0.83 solar radius. Convection is responsible for energy transport in the Sun's outer layers.

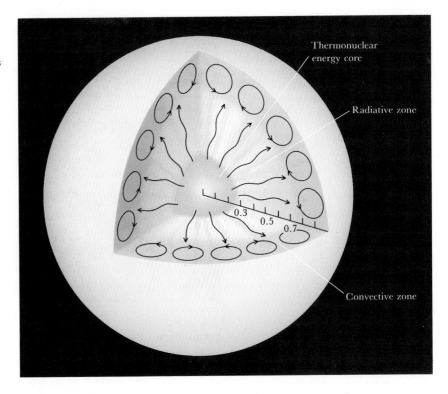

radiative zone. From 0.8 solar radius to the Sun's surface, the density of matter in the Sun is so low (less than 0.01 g/cm^3) that convection dominates the energy flow. We thus say that the Sun has a **convective zone,** or **convective envelope.** These aspects of the Sun's internal structure are sketched in Figure 11-3.

The photosphere is the lowest of three main layers in the Sun's atmosphere

Although astronomers often speak of the solar surface, the Sun really does not have a surface at all. As you move in toward the Sun, you encounter ever more dense gases but no sharp boundary like the surface of the Earth or Moon.

The Sun appears to have a surface (see Figure 11-1) because there is a specific layer in the Sun's atmosphere from which most of the visible light comes. This layer, which is probably not more than 500 km thick, is appropriately called the **photosphere** ("sphere of light"). The photosphere shines with a nearly perfect blackbody spectrum corresponding to an average temperature of 5800 K (recall Figure 5-4).

The photosphere is the lowest of three layers that together constitute the Sun's **atmosphere.** Above the photosphere are two additional layers, the chromosphere and the corona, which are discussed later in this chapter. At visible wavelengths you cannot see through the shimmering gases of the photosphere, and so everything below the photosphere is called the Sun's **interior.**

Convection in the Sun's outer layers affects the appearance of the photosphere. Under good observing conditions with a telescope and using special dark filters to protect your eyes, you can often see a blotchy pattern called

Figure 11-4   Solar granulation

High-resolution photographs of the Sun's surface reveal a blotchy pattern called granulation. Granules, each measuring about 1000 km across, are convection cells in the Sun's outer layers. (San Fernando Observatory)

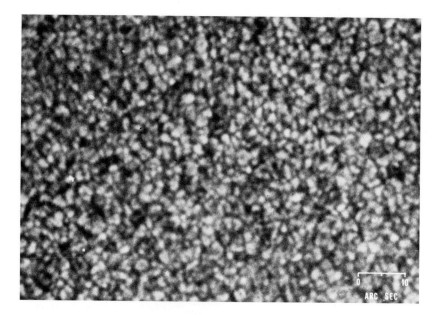

granulation (see Figure 11-4). Each light-colored **granule** measures about 1000 km across and is surrounded by a darkish boundary.

In Chapter 5 we learned that relative motion between a light source and an observer affects the wavelengths of spectral lines (recall Figure 5-18). This phenomenon is called the Doppler effect. A source moving toward you has its spectral lines blueshifted, but a source moving away from you has its spectral lines redshifted. By carefully measuring the wavelengths of spectral lines in various parts of granules, astronomers can determine movements of the solar gases. Astronomers find that hot gases rise upward in the granules, cool off, spill over the edges of the granules, and plunge back down into the Sun along the intergranule boundaries. The difference in brightness between the center and the edge of a granule corresponds to a temperature drop of 300 K.

Granules are individual convection cells in the Sun's outer layers. Time-lapse photography shows that granules form, disappear, and then re-form in cycles lasting several minutes. At any one time, roughly 4 million granules cover the solar surface. Each granule occupies an area roughly equal to Texas and Oklahoma combined (about a million square kilometers).

The photosphere appears darker around the edge, or **limb,** of the Sun than it does toward the center of the solar disk (examine Figure 11-1). This phenomenon, called **limb darkening,** arises because we are looking obliquely at the photosphere near the edge of the disk. We therefore do not see as deeply into the Sun there as we do near the center of the disk. Near the limb, the gas we observe is cooler and thus appears dimmer than the deeper, warmer gas seen near disk center.

The chromosphere is located between the photosphere and the Sun's outermost atmosphere

Immediately above the photosphere is a 10,000-km-thick layer called the **chromosphere** ("sphere of color"), which is the second of the three major levels in the Sun's atmosphere. When the Moon blocks out the photosphere during a total eclipse, the chromosphere is visible as a pinkish strip around the edge of the dark Moon.

The spectrum of the chromosphere is dominated by emission lines. The characteristic pinkish color of the chromosphere is caused by the Balmer line H$_\alpha$, which is the strongest emission line in the red region of the chromosphere's spectrum. The blue end of the chromosphere's spectrum is dominated by two bright emission lines called the H and K lines of ionized calcium.

The spectrum of the photosphere is dominated by numerous absorption lines (recall Figure 5-5), including broad, dark lines at the wavelengths of H$_\alpha$ and the calcium H and K lines. The photosphere emits almost no light at these wavelengths. The chromosphere, however, is especially bright at these wavelengths. Astronomers can thus study details of the chromosphere by viewing the Sun through special filters that are transparent to light only at the wavelengths of H$_\alpha$ or the calcium lines.

Figure 11-5 is a photograph of the chromosphere taken through an H$_\alpha$ filter. Note the many dark, brushlike spikes that protrude upward. These spikes, called **spicules,** are jets of gas surging up out of the Sun. A typical spicule rises at the rate of 20 km/sec, reaches a height of about 7000 km, then collapses and fades away after a few minutes. At any one time, roughly 300,000 spicules exist, covering a few percent of the Sun's surface.

Spicules are generally located on the boundaries between large, organized cells called **supergranules.** Detailed observations of the solar surface in the 1960s revealed the existence of these supergranules, which are about 30,000 km in diameter and contain many hundreds of ordinary granules. Gases rise upward in the middle of a supergranule and move horizontally outward toward its edge, where they descend back into the Sun.

Figure 11-5 Spicules and the chromosphere *Spicules are jets of cool gas that rise up into warmer regions of the Sun's outer atmosphere. Numerous spicules are visible in this H$_\alpha$ photograph, which also shows many details of the chromosphere. Spicules are located along the irregularly shaped boundaries between supergranules. (NOAO)*

The corona is the outermost layer of the Sun's atmosphere

The outermost region of the Sun's atmosphere is called the **corona.** It extends from the top of the chromosphere out to a distance of several million kilometers, where it merges into the solar wind. (As noted at the end of Chapter 6, the solar wind consists of high-speed protons and electrons constantly escaping from the Sun.) The three layers of the Sun's atmosphere are sketched schematically in Figure 11-6.

The total amount of visible light emitted by the solar corona is comparable to the brightness of the Moon at full moon—only about one-millionth as bright as the photosphere. The corona can thus be seen only when the photosphere is blocked out during a total eclipse or in a specially designed telescope called a **coronagraph.** Figure 11-7 is an exceptionally detailed photograph of the corona. Numerous **coronal streamers** are visible, extending outward to a distance of 6 solar radii.

Around 1940, astronomers realized that the spectrum of the Sun's corona contains the emission lines of a number of highly ionized elements. For example, there is a prominent green line caused by the presence of Fe XIV (iron atoms each stripped of 13 electrons). Extremely high temperatures are required to strip that many electrons from atoms, and so it was clear that the corona must be very hot. It is now known that coronal temperatures are in the range of 1 to 2 million K.

Recent observations from space show that the Sun's corona exhibits complicated structure and activity. An example is the huge, bubblelike disturbance seen in Figure 11-8—a **solar transient.** These short-lived protuberances erupt suddenly and expand rapidly outward through the corona. Solar transients probably occur as often as once a day, but they were unknown until the Sun was examined with a coronagraph carried aloft by Skylab.

X-ray photographs of the corona were also obtained during the Skylab missions. Recall from Chapter 5 that Wien's law tells us the temperature of a black body is inversely proportional to the dominate wavelength of the radiation it emits. The corona shines brightly at X-ray wavelengths because its

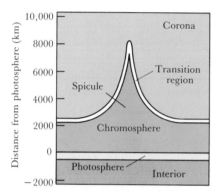

Figure 11-6 The solar atmosphere
The Sun's atmosphere has three distinct layers. The photosphere is roughly 500 km thick. The chromosphere extends to an altitude of about 2000 km above the photosphere, with spicules jutting up to 10,000 km. The corona extends many millions of kilometers out into space and merges with the solar wind. (Adapted from John A. Eddy)

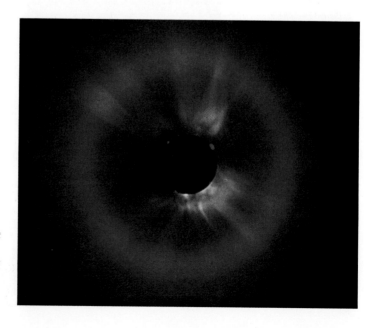

Figure 11-7 The solar corona *This extraordinary photograph was taken from a jet 40,000 feet above Montana during the total solar eclipse of February 26, 1979. Numerous streamers are visible, extending to distances of 4 million kilometers above the solar surface. (Los Alamos Scientific Laboratory)*

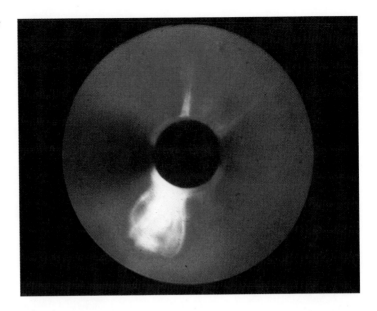

Figure 11-8 A solar transient *During the early 1970s, astronauts on Skylab discovered huge bubbles erupting outward through the corona. This solar transient was photographed in 1973. The leading edge of the bubble moved outward from the Sun at a speed of 500 km/sec. (NASA and the High Altitude Observatory)*

temperature is in the million-K range. The Skylab pictures reveal a very blotchy, irregular inner corona (see Figure 11-9). There are large dark regions, called **coronal holes,** which are nearly devoid of the hot, glowing gases. Many astronomers suspect that coronal holes are the main corridors through which particles of the solar wind escape from the Sun.

X-ray photographs also reveal numerous bright spots that are hotter than the surrounding corona. Temperatures in these bright points occasionally reach 4 million K. Many of the bright coronal hot spots seen in X-ray pictures such as Figure 11-9 are located over sunspots.

The high temperature of the Sun's outer layers has been a major mystery for astronomers. The mechanism for heating the corona is still poorly understood. Convective motions of hot gas just beneath the solar surface involve considerable energy, and many astronomers suspect the corona is somehow heated by this seething and churning. It is not clear, however, how this energy would be transported out into the corona.

Figure 11-9 A coronal hole in X rays *This X-ray picture of the Sun was taken by Skylab astronauts in 1973. A huge, dark, boot-shaped coronal hole dominates this view of the inner corona. Numerous bright points are also visible. (NASA and Harvard College Observatory)*

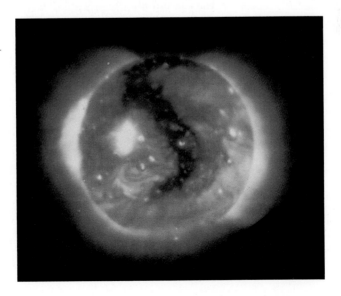

Sunspots are only one of many phenomena associated with the 22-year solar cycle

Superimposed on the basic structure of the solar atmosphere are a host of phenomena that vary with a 22-year period. Irregularly shaped dark regions called **sunspots** (see Figure 11-10) are the most easily recognized of these phenomena because they occur in the photosphere, the most visible

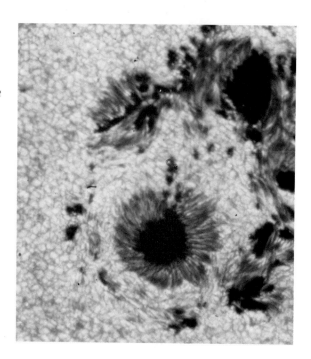

Figure 11-10 [right]   A sunspot group This photograph was taken by a balloon-borne telescope at an altitude of roughly 50,000 ft. Note the featherlike appearance of the penumbrae. Also notice the granulation in the surrounding, undisturbed photosphere. (Project Stratoscope; Princeton University)

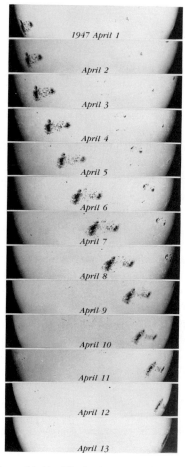

Figure 11-11   The Sun's rotation By observing the same group of sunspots from one day to the next, Galileo found that the Sun rotates once in about four weeks. The equatorial regions of the Sun actually rotate somewhat faster than the polar regions. This series of photographs shows a large sunspot group during half a solar rotation. (Mount Wilson and Las Campanas Observatories)

part of the Sun's atmosphere. The very dark central core of a sunspot is called the **umbra;** it usually is surrounded by a less dark border called the **penumbra.** Although they vary greatly in size, typical sunspots measure a few tens of thousands of kilometers across.

Sunspots look dark against the solar surface because they are cooler than the surrounding photosphere (recall the Stefan–Boltzmann law). Temperatures in a typical sunspot are 4000 to 4500 K, or more than 1000 K cooler than in the undisturbed photosphere.

On rare occasions, a sunspot group will be so large that it can be seen with the naked eye. (Always be sure to use special dark filters or other means to protect your eyes when observing the Sun.) Ancient Chinese astronomers recorded such sightings 2000 years ago. Galileo was the first person to examine sunspots in detail with a telescope, which, of course, gives a much better view. In fact, Galileo discovered that he could determine the Sun's rotation rate by following sunspots as they moved across the solar disk (see Figure 11-11). He found that the Sun rotates once in about four weeks. A typical sunspot group lasts about two months, and so it can be followed for two solar rotations.

Careful observations demonstrate that the Sun does not rotate as a rigid body: the equatorial regions rotate more rapidly than the polar regions. A sunspot near the solar equator takes 25 days to go once around the Sun. However, at 30° north or south of the equator a sunspot takes $27\frac{1}{2}$ days to complete a rotation. The rotation period at 75° north or south of the equa-

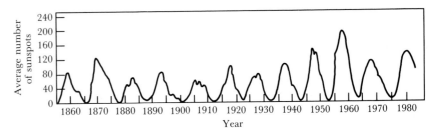

Figure 11-12 The sunspot cycle
The number of sunspots on the Sun varies with a period of about 11 years. Large numbers of sunspots were seen in 1959, 1970, and late 1980. Exceptionally few sunspots were observed in 1954, 1965, and 1976. The most recent sunspot minimum occurred in 1986, and the next sunspot maximum will occur in 1990 or 1991.

tor is about 33 days, and at the poles it may be as long as 35 days. This phenomenon is called **differential rotation,** because different parts of the Sun rotate at slightly different rates.

Observations over many years reveal that the number of sunspots changes in a periodic fashion. In some years, there are many sunspots, in others almost none. This pattern is called the **sunspot cycle.** As shown in Figure 11-12, the average number of sunspots varies with a period of about 11 years. A time of exceptionally many sunspots is called a **sunspot maximum,** such as occurred in 1959, 1970, and 1980. The Sun was almost devoid of sunspots (called times of **sunspot minimum**) in 1965, 1976, and 1986.

Recent observations from the Solar Maximum Mission satellite (see Figure 11-13) demonstrated that the Sun's brightness is linked to the 11-year sunspot cycle. A device called a radiometer on board the spacecraft measured the intensity of light at all wavelengths arriving at the Earth during the 1980s. The data show that the Sun is brightest during sunspot maximum and dimmest during sunspot minimum. The overall change in the solar output is quite small, about 1 part in 2500.

At first this 11-year variation seems to pose a paradox. Why should the Sun be brightest in years when sunspots are most numerous? Apparently, the light blocked by sunspots is more than made up for by an increase in bright patches, called **faculae,** that appear near sunspots and elsewhere on the solar surface.

In 1908, the American astronomer George Ellery Hale made the important discovery that sunspots are associated with intense magnetic fields on the Sun. When Hale focused a spectroscope on sunlight coming from a sunspot, he found that each spectral line in the normal solar spectrum is flanked by additional, closely spaced spectral lines not usually observed (see Figure 11-14). This "splitting" of a single spectral line into two or more lines is called the **Zeeman effect** after the Dutch physicist Pieter Zeeman, who first observed it in 1896 in his laboratory. Zeeman showed that the spectral lines split when the light source is inside an intense magnetic field. The more intense the magnetic field, the wider is the separation of the split lines. The splitting of the Fe I spectral line in Figure 11-14 into three lines corresponds to a magnetic field roughly 5000 times more intense than the Earth's natural magnetic field.

Hale's discovery demonstrates that sunspots are areas where concentrated magnetic field projects through the hot gases of the photosphere. Because of the temperature, many atoms in the photosphere are ionized so that the photosphere is a mixture of electric charges (ions and electrons). This blend, technically called a **plasma,** is an extremely good conductor of electricity that interacts vigorously with magnetic fields. Specifically, a magnetic field restricts and constrains the motions of a plasma. The intense magnetic field in a sunspot greatly inhibits the natural convective motions of

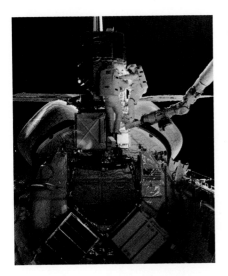

Figure 11-13 Astronauts repairing the SMM satellite *The Solar Maximum Mission satellite was designed to study the Sun from space, above the obscuring effects of the Earth's atmosphere. Shortly after it was launched in 1980, the system that keeps it pointed toward the Sun failed. Two astronauts are seen repairing the satellite above the open cargo-bay doors of the Space Shuttle. (NASA)*

Figure 11-14 Zeeman splitting by a sunspot's magnetic field (a) *The black line drawn across the sunspot indicates the location toward which the slit of the spectroscope was aimed.* (b) *In the resulting spectrogram, one line in the middle of the normal solar spectrum is split into three components. The separation between the three lines corresponds to a magnetic field roughly 5000 times stronger than the Earth's natural magnetic field. (NOAO)*

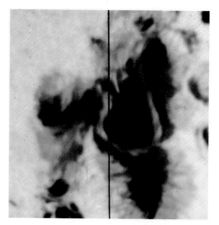

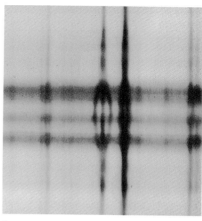

a   b

the gases. Energy cannot flow freely upward from the Sun's convective zone, and the gases in this region of the photosphere cool off. This cool area emits less radiation than the unperturbed photosphere and thus appears dark in contrast to the surrounding solar surface.

A host of exotic phenomena occur around and above sunspots as a direct result of their intense magnetic fields. Huge, arching columns of gas called **prominences** often appear above sunspot regions (see Figure 11-15). Some prominences hang suspended for days above the solar surface while others blast material outward from the Sun at speeds of roughly 1000 km/sec.

The most violent, eruptive events on the Sun, called **solar flares,** occur in complex sunspot groups. During a solar flare, temperatures in a compact region soar to 5 million K. Vast quantities of particles and radiation are blasted out into space. The flare is usually over within 20 minutes. Ultraviolet and X-ray radiation, which take about 8 minutes to reach the Earth, interfere with radio communications. Clouds of high-energy particles arrive a day or two later and often produce beautiful, shimmering lights called

Figure 11-15 A prominence A huge prominence arches above the solar surface in this Skylab photograph taken on December 19, 1973. The radiation that exposed this picture is from singly ionized helium (He II) at a wavelength of 304 Å, corresponding to a temperature of about 50,000 K. (Naval Research Laboratory)

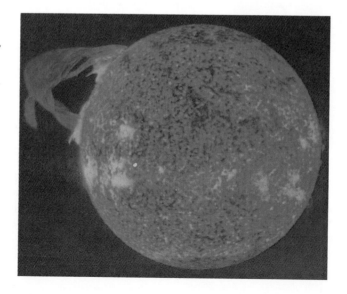

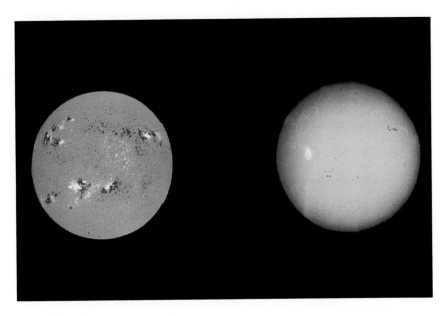

Figure 11-16 Two views of the Sun
The optical view of the Sun (right) was taken at the same time as the magnetogram (left). Notice how regions with strong magnetic fields are related to the locations of sunspots. Careful examination of the magnetogram also reveals that the magnetic polarity of the sunspots in the northern hemisphere is opposite that of those in the southern hemisphere. (NOAO)

aurorae in the night sky. Most astronomers suspect that prominences and flares involve concentrated portions of the Sun's magnetic field.

Astronomers can construct artificial pictures called **magnetograms** that display the magnetic fields in the solar atmosphere by combining two photographs taken at wavelengths on either side of a magnetically split spectral line. The magnetogram in Figure 11-16 shows the entire solar surface, along with an ordinary photograph of the Sun taken at the same time. Dark blue indicates the area of the photosphere covered by north magnetic polarity and yellow designates the area covered by the south polarity.

Many sunspot groups are said to be **bipolar,** meaning that roughly comparable areas are covered by north and south magnetic polarities (see Figure 11-17). The sunspots on the side of the group toward which the Sun is rotating are called the **preceding** members of the group. The remaining spots, which follow behind, are called the **following** members.

Figure 11-17 A magnetogram of a sunspot group This artificially colored picture displays the intensity and polarity of the magnetic field associated with a large sunspot group. One side of a typical sunspot group has one magnetic polarity and the other side has the opposite polarity. (NOAO)

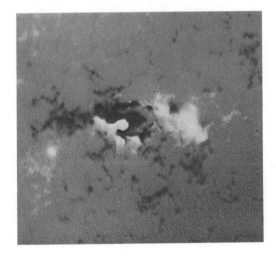

After years of studying the solar magnetic field, Hale was able to piece together a remarkable magnetic description of the solar cycle. First of all, Hale discovered that the preceding spots of all sunspot groups in one solar hemisphere have the same magnetic polarity. We now know that this polarity is the same as that of the hemisphere in which the group is located. In other words, in the hemisphere of the Sun's north magnetic pole, the preceding members of all sunspot groups have north magnetic polarity. In the south magnetic hemisphere, all the preceding members have south magnetic polarity.

In addition, Hale found that the polarity pattern completely reverses itself every 11 years. For instance, the hemisphere that has a north magnetic polarity at one solar maximum has a south magnetic polarity at the next solar maximum. For this reason astronomers prefer to speak of the **solar cycle,** which has a period of 22 years, rather than the 11-year sunspot cycle.

In 1960, the American astronomer Horace Babcock proposed a description that seems to account for many aspects of the 22-year solar cycle. Babcock's scenario, called a **magnetic-dynamo** model, employs two basic properties of the Sun: its differential rotation and its convective envelope. Differential rotation causes the magnetic field in the photosphere to become wrapped around the Sun as shown in Figure 11-18. Convection in the photosphere causes the concentrated magnetic field to become tangled, and kinks erupt through the solar surface. Sunspots appear at locations where the magnetic field projects through the photosphere. Careful examination of how the magnetic field in Figure 11-18 is oriented reveals that the preceding members of a sunspot group should indeed have the same magnetic polarity as that of the hemisphere in which they are located.

The magnetic-dynamo model fails to explain many things. Most embarrassing to astronomers is their ignorance of the basic physics of sunspots. We still do not know what holds a sunspot together week after week. Our best calculations predict that a sunspot should break up and disperse as soon as it forms.

Figure 11-18   Babcock's magnetic dynamo
A possible partial explanation for the sunspot cycle involves the wrapping of a magnetic field around the Sun by differential rotation. Sunspots appear where the concentrated magnetic field has broken through the solar surface.

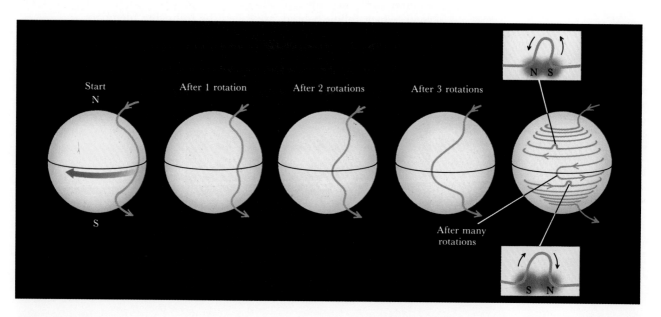

Our understanding of sunspots is further confounded by irregularities in the solar cycle. For example, the overall reversal of the Sun's magnetic field is often piecemeal and haphazard. One pole may reverse polarity long before the other so that for several weeks the Sun may have two north poles but no south pole at all. To make matters worse, there is strong historical evidence that all traces of sunspots and the sunspot cycle have vanished for many years. For example, virtually no sunspots were seen from 1645 through 1715. Similar sunspot-free periods apparently occurred at irregular intervals in earlier times.

Astronomers can study the solar interior by measuring vibrations of the Sun's surface

Puzzles in solar research have prompted astronomers to question many long-standing ideas about the Sun. For instance, the magnetic field that produces sunspots might be bits and pieces of ancient magnetism trapped deep inside the Sun as it formed 4.5 billion years ago. Other speculation includes the possibility that the Sun's core is rotating as much as three times faster than its surface. To search for and study these deeply buried phenomena, astronomers need a way to probe the solar interior.

Geologists are able to study the Earth's interior structure by using a seismograph to record vibrations during earthquakes. Although there are no true sunquakes, the Sun does vibrate at a variety of frequencies, somewhat like a ringing bell. These vibrations, first noticed in 1960, are detected with sensitive Doppler-shift measurements. Portions of the Sun's surface move up and down by about 10 km every 5 minutes (see Figure 11-19). Slower vibrations with periods ranging from 20 minutes to nearly an hour were discovered in the 1970s. More recently, a variety of long-period oscillations has been found.

To conduct long-term studies of these slow pulsations, astronomers have set up telescopes at the South Pole, where the Sun can be observed continuously for several months. This new field of solar research is called **helioseismology.**

The vibrations of the solar surface are like sound waves. If you were within the Sun's atmosphere, you would first notice a deafening roar,

Figure 11-19 Vibrations of the Sun's surface Doppler-shift measurements give the speed at which the Sun's photosphere bobs up and down. These data extend over 2 hours and show numerous oscillations with a period of about 5 minutes. The gradual variation in the size of the oscillations is the result of interference between many vibrations and wavelengths. (Adapted from O. R. White)

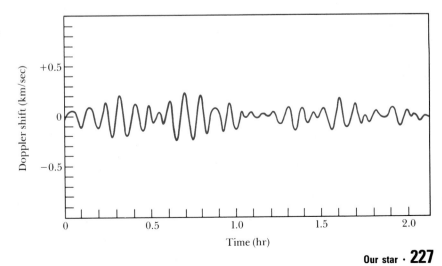

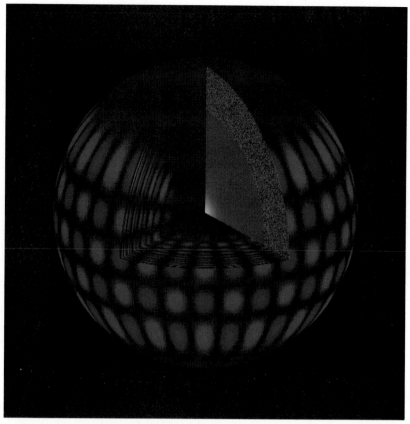

Figure 11-20 Sound waves resonating in the Sun This computer-generated image shows one of the millions of ways in which the Sun vibrates because of sound waves resonating in its interior. The regions that are moving outward are colored blue, those moving inward are red. The cutaway shows how deep these oscillations are believed to extend. (National Solar Observatory)

somewhat like a jet engine, produced by the turbulence associated with convection. Superimposed on this noise is a variety of nearly pure tones. You would be unable to hear these tones, however, for they are 16 octaves below the lowest note on the piano keyboard.

Sound waves moving upward from the solar interior are reflected by the solar surface, which acts somewhat like a mirror to these waves. As a reflected sound wave descends back into the Sun, the increasing density and pressure bend the wave so severely that it is turned around and aimed back out again toward the solar surface. In other words, sound waves bounce back and forth between the solar surface and layers deep within the Sun. These sound waves can reinforce each other and resonate just like sound waves inside an organ pipe.

There are millions of ways in which the Sun can oscillate as a result of sound waves resonating in its interior. Figure 11-20 is a computer-generated illustration that shows one such vibrational mode. Helioseismologists are hopeful that observations of these oscillations will yield detailed information about the solar interior in the coming years.

Summary

· The Sun's energy is produced by the thermonuclear process called hydrogen burning, in which four hydrogen nuclei fuse to produce a single helium nucleus and release energy.

· The energy released in a thermonuclear reaction comes from the conversion of matter into energy according to Einstein's equation $E = mc^2$.

· Thermonuclear reactions occur only at very high temperatures and hydrogen burning occurs only at temperatures of more than about 8 million K.

- A stellar model is a theoretical description of a star's interior derived from calculations based upon the laws of physics. The solar model suggests that hydrogen burning occurs in a core that extends from the Sun's center to about 0.25 solar radius.

- Throughout most of the Sun's interior, energy moves outward from the core by radiative diffusion. In the Sun's outer layers, energy is transported to the Sun's surface by convection.

- The visible surface of the Sun is a layer at the bottom of the solar atmosphere called the photosphere. The gases in this layer shine with almost-perfect blackbody radiation, and convection produces features called granules there.

- Above the photosphere is a layer of cooler and less dense gases called the chromosphere. Jets of gas called spicules rise up into the chromosphere along the boundaries of supergranules.

- The outermost layer of thin gases in the solar atmosphere is called the corona, which blends into the solar wind at great distances from the Sun. The gases of the corona are very hot but at low density. Solar transients, coronal streamers, and coronal holes are other features associated with the corona.

- Surface features on the Sun vary periodically in a 22-year solar cycle.

    Sunspots are cool regions produced by local concentrations of the Sun's magnetic field. The average number of sunspots increases and decreases in a regular 11-year cycle.

    A solar flare is a brief eruption of hot, ionized gases from a sunspot group.

    The magnetic-dynamo model suggests that many features of the solar cycle are caused by the effects of differential rotation and convection on the Sun's magnetic field.

- Helioseismologists probe the interior structure of the Sun by studying vibrations of the solar surface.

Review questions

1 What dangers are inherent in observing the Sun? How have astronomers circumvented these problems?

2 Give an everyday example of hydrostatic equilibrium. Give an everyday example of thermal equilibrium.

3 Give some everyday examples of conduction, convection, and radiative diffusion.

4 Describe the three main layers of the Sun's atmosphere, and how you would best observe them.

5 When do you think the next sunspot maximum and minimum will occur? Explain.

6 What do astronomers mean by "a model of the Sun"?

7 Why is the solar cycle said to have a period of 22 years, even though the sunspot cycle is only 11 years long?

8 What is hydrogen burning? Why does hydrogen burning occur only around the Sun's center?

9 How do astronomers detect the presence of a magnetic field in hot gases, such as those in the solar photosphere?

Advanced questions

10 What would happen if the Sun were not in a state of hydrostatic or thermal equilibrium?

*11 Assuming that the current rate of hydrogen burning in the Sun remains constant, what fraction of the Sun's mass will be converted into helium over the next 5 billion years? How will this affect the chemical composition of the Sun?

12 As this textbook went to press, several solar astronomers were predicting that the next solar maximum would be quite active. By consulting recent issues of astronomy magazines, like *Sky & Telescope* and *Astronomy*, determine whether or not unusually large numbers of sunspots were sighted in the early 1990s. When did the sunspot maximum officially occur? How did this maximum compare with that of previous solar cycles?

Discussion questions

13 Discuss some of the difficulties of correlating solar activity with changes in the terrestrial climate.

14 From 1645 to 1715, virtually no sunspots were seen, and northern Europe experienced a "Little Ice Age" during which record low temperatures were recorded. Could these two phenomena be related? Explain.

15 Describe some advantages and disadvantages of observing the Sun from both space and the South Pole. What kinds of phenomena and issues do solar astronomers want to explore from an Antarctic observatory?

For further reading

Eddy, J. "The Case of the Missing Sunspots." *Scientific American*, May 1977.

Eddy, J. *A New Sun*. NASA SP-402, 1979.

Frazier, K. *Our Turbulent Sun*. Prentice-Hall, 1983.

Leibacher, J., et al. "Helioseismology." *Scientific American*, September 1985.

Levine, R. "The New Sun." In Cornell, J., and Gorenstein, P., eds. *Astronomy from Space*. MIT Press, 1983.

Mitton, S. *Daytime Star*. Scribners, 1981.

Noyes, R. *The Sun, Our Star*. Harvard University Press, 1982.

Robinson, L. "The Disquieting Sun: How Big, How Steady?" *Sky & Telescope*, April 1982, p. 354.

Taylor, M. "Observing from the South Pole." *Sky & Telescope*, October 1988, p. 351.

Wolfson, R. "The Active Solar Corona." *Scientific American*, February 1983.

Zirin, H. "Heading Toward Solar Maximum." *Sky & Telescope*, October 1988, p. 355.

12 The nature of the stars

A cluster of stars By analyzing starlight, an astronomer can determine such details about a star as its surface temperature, chemical composition, and luminosity. This photograph clearly shows color differences in a star cluster called NGC 3293. The reddish stars are comparatively cool, with surface temperatures around 3000 K. These stars are also quite luminous and have very large diameters, typically 100 times as large as the Sun. The bluish and blue-white stars have much higher surface temperatures (15,000 to 30,000 K) and are roughly the same size as the Sun. (Anglo-Australian Observatory)

Although they appear only as brilliant pinpoints of light, stars are huge, massive spheres of glowing gas much like our Sun. In this chapter, we see how astronomers have measured the distances to many of the nearer stars. Once a star's distance is known, its luminosity can be easily deduced. We also learn about ways in which astronomers determine the surface temperatures of stars from their spectra and colors. We make our first acquaintance with the Hertzsprung–Russell diagram, which reveals the various fundamental types of stars. Finally, we turn to the topic of binary stars, those surprisingly common systems in which two stars orbit each other. Binary stars are important because an analysis of their orbital motions provide important information about the masses of stars.

The night sky is spangled with thousands of stars, each appearing as a bright pinpoint of light. Telescopes reveal many thousands of other stars too faint to be seen with the naked eye, but every star appears only as a bright point of light. A star is a huge, massive ball of hot gas like our Sun, held together by its own gravity. We now know that some stars are larger than our Sun, some smaller, some brighter, some dimmer, some hotter, and some cooler.

The quest for information about the mass, luminosity, surface temperature, and chemical composition of different stars has been a major occupation of twentieth century astronomers. A remarkably complete picture has emerged in recent times. By understanding the stars, we gain insight into our relationship to the universe and our place in the cosmic scope of space and time.

Distances to nearby stars are determined by parallax

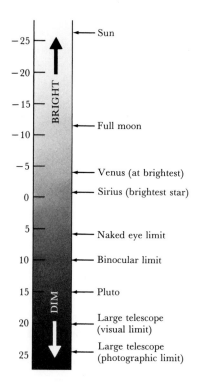

Figure 12-1 The apparent magnitude scale *Astronomers denote the brightness of objects in the sky by the apparent magnitude of the objects. Most stars visible to the naked eye have magnitudes in the range +1 to +6. Photography through large telescopes can reveal stars as faint as magnitude +24.*

Astronomers specify the brightness of a star by its **magnitude,** which is determined by a scheme that dates back to the ancient Greek astronomer Hipparchus. The brightest stars Hipparchus saw in the sky he called first-magnitude stars. The dimmest stars he called sixth-magnitude. To stars of intermediate brightness he assigned numbers on a scale of 1 to 6.

In the nineteenth century, techniques were developed for measuring the amount of light arriving from a star, and then astronomers defined the magnitude scale more precisely. Measurements showed that a first-magnitude star is about 100 times as bright as a sixth-magnitude star. In other words, it would take 100 stars of magnitude +6 to provide as much light energy as we receive from a single star of magnitude +1. The magnitude scale was redefined so that a magnitude difference of 5 corresponds exactly to a factor of 100 in the amount of light received. A magnitude difference of 1 therefore corresponds to a factor of 2.512 in light energy, because

$$2.512 \times 2.512 \times 2.512 \times 2.512 \times 2.512 = 100$$

For example, it takes about $2\frac{1}{2}$ third-magnitude stars to provide as much light as we receive from a single second-magnitude star.

Astronomers also extended the **magnitude scale** to describe the dimmer stars visible through their telescopes. For example, with a good pair of binoculars, you can see stars as faint as tenth magnitude. Through some of the largest telescopes in the world, it is possible to see stars as dim as magnitude 20. Photography with long exposure times reveals even dimmer stars.

Astronomers use negative numbers to extend the magnitude scale so that it includes very bright objects. For example, Sirius—the brightest star in the sky—has a magnitude of $-1\frac{1}{2}$. At its brightest, Venus shines with a magnitude of -4. Of course, the Sun is the brightest object in the sky. Its magnitude is $-26\frac{1}{2}$.

Figure 12-1 illustrates the modern magnitude scale. These magnitudes are properly called **apparent magnitudes** because they describe how bright an object appears to an Earth-based observer. Apparent magnitude is a measure of the energy arriving at the Earth.

Apparent magnitudes do not tell us about the actual brightness of the stars. A star that looks dim in the sky might actually be a brilliant star that

Figure 12-2 Parallax Imagine looking at some nearby object like a tree seen against a distant background such as mountains. If you move from one location to another, the nearby object will appear to shift its location with respect to the distant background scenery. This familiar phenomenon is called parallax.

just happens to be extremely far away. In order to determine the actual brightness of a star, we must first know how far away the star is.

The most straightforward way of measuring stellar distances involves **parallax,** which is the apparent displacement of an observed object caused by a change in the point from which it is viewed. Parallax occurs when nearby objects appear to shift their positions against a distant background as the observer moves from one place to another (see Figure 12-2). Stars also exhibit parallax. As the Earth orbits the Sun, nearby stars appear to move back and forth against the background of more distant stars.

The distance to a star can be determined by measuring the star's parallax. The **parallax** p of a star is defined as one-half of the angle through which a star's position shifts as the Earth moves from one side of its own orbit to the other (see Figure 12-3). The smaller the angle p, the greater the distance d to the star. Astronomers usually denote the distances to stars in light years or parsecs (1 parsec = 3.26 light years).

A simple equation relates the parallax of a star to its distance from Earth. If the angle p is measured in arc sec, then the distance d to the star in parsecs is given by

$$d = \frac{1}{p}$$

For example, a star whose parallax is 0.5 arc sec is 2 parsecs from Earth.

The parallax method of determining stellar distances works only for nearby stars. There are slightly more than 1000 stars within 20 parsecs of the Earth whose parallaxes have been measured with a high degree of precision. Most of these nearby stars are so dim that they are invisible to the unaided eye. The majority of the familiar, bright stars in the nighttime sky are too far away to exhibit measurable parallax as the Earth orbits the Sun.

Figure 12-3 Stellar parallax As the Earth orbits the Sun, a nearby star appears to shift its position against the background of distant stars. The angle p (the parallax of the star) is equal to the angular size of the radius of the Earth's orbit as seen from the star. The smaller the parallax (p), the larger the distance (d) to the star.

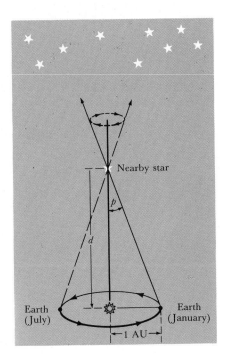

A star's luminosity can be determined from its apparent magnitude and its distance from Earth

The **absolute magnitude** of a star is the apparent magnitude it would have if it were located at a distance of exactly 10 parsecs from the Earth. For example, if the Sun were moved to a distance of 10 parsecs from the Earth, it would have an apparent magnitude of +4.8. Therefore the absolute magnitude of the Sun is +4.8. The absolute magnitudes of other stars range from roughly −10 for the brightest to +15 for the dimmest. The Sun's absolute magnitude is about in the middle of this range, suggesting that the Sun is an average star.

Absolute magnitude is a very informative quantity, because it indicates the actual brightness of a star. In contrast, apparent magnitude tells only how bright the star appears in the sky. To appreciate how astronomers determine the absolute magnitude of a star, you must first understand how the brightness of a light depends on its distance from an observer.

Imagine a source of light like a light bulb or a star. As light moves away from the source, it spreads out over increasingly larger regions of space as shown in Figure 12-4. As the light spreads out, its brightness decreases. That is why the farther away a source of light is, the dimmer it appears. Specifically, the **inverse-square law** tells us that the apparent brightness of a light source is inversely proportional to the square of the distance between the source and the observer. For example, double the distance to a light source and its apparent brightness decreases by 2^2, or a factor of four. Similarly, at triple the distance, the brightness decreases by 3^2, or a factor of nine.

Using the inverse-square law, astronomers have derived a mathematical equation relating three quantities: a star's apparent magnitude m, its absolute magnitude M, and its distance d from the Earth. If you know any two of these quantities (such as apparent magnitude and distance), you can calculate the third one (such as absolute magnitude). Thus astronomers measure

Figure 12-4   The inverse-square law
This drawing shows how the same amount of radiation from a light source must illuminate an ever-increasing area as distance from the light source increases. Because the light becomes spread out as it moves away from the source, the apparent brightness of the source decreases.

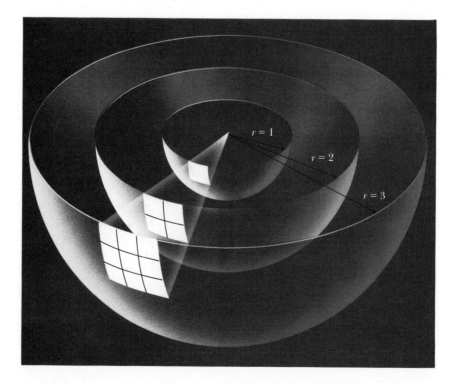

the apparent magnitude of a nearby star, find its distance by measuring its parallax, then calculate its absolute magnitude.

Absolute magnitude is directly related to **luminosity,** the amount of energy escaping from a star's surface each second (usually given in ergs per second). Many scientists prefer to speak of a star's luminosity rather than its absolute magnitude because luminosity is a measure of the star's energy output. A simple equation relates absolute magnitude to luminosity, and astronomers can convert from one to the other as they see fit.

For convenience, stellar luminosities are expressed in multiples of the Sun's luminosity ($L_\odot$), which is 3.90×10^{33} erg/sec. The brightest stars (absolute magnitude = -10) have luminosities of 10^6 $L_\odot$. In other words, each of these stars has the energy output of a million Suns. The dimmest stars (absolute magnitude = $+15$) have luminosities of 10^{-4} $L_\odot$.

A star's color reveals its surface temperature

One of the first things you notice when comparing stars in the nighttime sky is their differences in apparent magnitude. More careful examination, even with the naked eye, reveals that stars also have different colors. For example, in the constellation of Orion, you can see the difference between reddish Betelgeuse and bluish Rigel (see Figure 4-23).

A star's color is directly associated with its surface temperature through relationships such as Wien's law, which is discussed in Chapter 5 (review the blackbody curves in Figure 5-3). The intensity of light from a cool star peaks at long wavelengths, and so the star looks red (see Figure 12-5a). A hot star's intensity curve is skewed toward short wavelengths, making the star look blue (see Figure 12-5c). The maximum intensity of a star of intermediate temperature (such as the Sun) occurs near the middle of the visible spectrum, giving the star a yellow-white color (see Figure 12-5b).

To measure the colors of the stars accurately, astronomers have developed a technique called **photoelectric photometry.** This technique uses a light-measuring device called a **photometer** at the focus of a telescope, with a standardized set of colored filters. By far, the most commonly used filters are the three **UBV filters.** Each filter is transparent in one of three broad wavelength bands: the ultraviolet (U), the blue (B), and the central region (V) of the visible spectrum (see Figure 12-6). The transparency of the V filter mimics the sensitivity of the human eye.

To do photometry, the astronomer aims a telescope at a star and measures the intensity of starlight three times, each time using a different filter.

Figure 12-5 Temperature and color
This schematic diagram shows the relationship between the color of a star and its surface temperature. The intensity of light emitted by three hypothetical stars is plotted against wavelengths (compare Figure 5-3). The range of visible wavelengths is indicated. The way in which a star's intensity curve is skewed determines the dominant color of its visible light.

a This star looks red

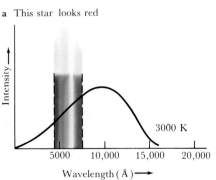

b This star looks yellow–white

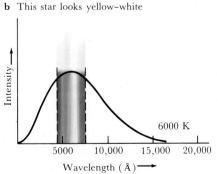

c This star looks blue

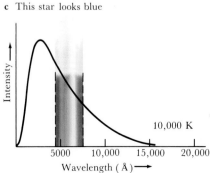

Table 12-1   The UBV magnitudes and color indices of selected stars

Star name	V	B	U	(B − V)	(U − B)	Apparent color
Bellatrix (γ Ori)	1.64	1.42	0.55	−0.22	−0.87	Blue
Regulus (α Leo)	1.35	1.24	0.88	−0.11	−0.36	Blue-white
Sirius (α CMa)	−1.46	−1.46	−1.52	0.00	−0.06	Blue-white
Megrez (δ UMa)	3.31	3.39	3.46	+0.08	+0.07	White
Altair (α Aql)	0.77	0.99	1.07	+0.22	+0.08	Yellow-white
Sun	−26.78	−26.16	−26.06	+0.62	+0.10	Yellow-white
Aldebaran (α Tau)	0.85	2.39	4.29	+1.54	+1.90	Orange
Betelgeuse (α Ori)	0.50	2.35	4.41	+1.85	+2.06	Red

This procedure gives three apparent magnitudes for the star, usually designated by the capital letters U, B, and V. The astronomer then compares the intensity of starlight in each wavelength band by subtracting one magnitude from another, forming the combinations (U − B) and (B − V). These combinations are called the star's **color indices.** The UBV magnitudes and color indices for several representative stars are given in Table 12-1.

A color index tells you how much brighter or dimmer a star is in one wavelength band than in another. For example, the (B − V) color index tells you how much brighter or dimmer a star appears through the B filter than it does through the V filter.

A color index is important because it tells you about the star's surface temperature. If a star is very hot, its radiation is skewed toward the short-wavelength ultraviolet, which makes the star bright through the U filter, dimmer through the B filter, and dimmest through the V filter. The star Bellatrix (see Table 12-1) is an example of this case. If the star is cool, its radiation peaks at long wavelengths, making the star brightest through the V filter, dimmer through the B filter, and dimmest through the U filter. The star Betelgeuse is an example.

The graph in Figure 12-7 gives the relationship between the (B − V) color index and temperature. If you know a star's (B − V) color index, you can use this graph to find the star's surface temperature. For example, the Sun's (B − V) index is +0.62, which corresponds to a surface temperature of 5800 K.

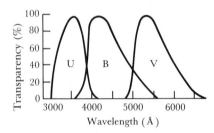

Figure 12-6   Light transmission through the UBV filters   *This graph shows the wavelength ranges over which standardized U, B, and V filters are transparent to light. The U filter is transparent to light from 3000 to 4000 Å, a range called the near-ultraviolet because it lies just beyond the violet end of the visible spectrum. The B filter is transparent from about 3800 to 5500 Å, and the V filter from about 5000 to 6500 Å.*

A star's spectrum is a clue to the star's surface temperature

The field of stellar spectroscopy was born in the 1860s when the Italian astronomer Angelo Secchi attached a spectroscope to his telescope and pointed it toward the stars. Secchi observed stellar spectral lines and made the important discovery that stars can be classified into various **spectral types** according to the appearances of their spectra. In those days, the nature and cause of spectral lines were not well understood. Astronomers classified each star by assigning it a letter from A through P, depending on the strength of the hydrogen Balmer lines in the star's spectrum.

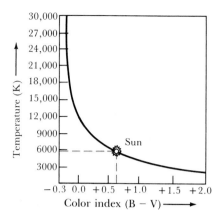

Figure 12-7 *Blackbody temperature versus color index* *The (B − V) color index is the difference between the B and V magnitudes of a star. If the star is hotter than about 10,000 K, it is a very bluish star and its (B − V) index is less than zero. If a star is cooler than about 10,000 K, its (B − V) index is greater than zero. The Sun's (B − V) index is about 0.62, which corresponds to a temperature of 5800 K. After measuring a star's B and V magnitudes, an astronomer can estimate the star's surface temperature from a graph such as this one.*

After Niels Bohr explained the structure of the hydrogen atom in the early 1900s (recall Figure 5-11), astronomers realized that the strength of the lines in a star's spectrum is directly related to the temperature of the gases in the star's outer layers. Hydrogen is by far the most abundant element in the universe, accounting for about three-quarters of the mass of a typical star. Hydrogen lines do not necessarily show up in a star's spectrum, however. If the star is much hotter than 10,000 K, high-energy photons pouring out of the star's interior easily knock electrons out of the hydrogen atoms in the star's outer layers, ionizing the gas. Hydrogen ions have no electrons to produce absorption lines. Conversely, if the star is much cooler than 10,000 K, the majority of photons escaping from the star do not possess enough energy to boost many electrons up from the ground state of the hydrogen atoms. These unexcited atoms also fail to produce Balmer lines. In short, in order to produce Balmer lines, a star must be hot enough to excite electrons out of the ground state—but not hot enough to ionize the atoms. A stellar surface temperature of 10,000 K results in the strongest Balmer lines.

A prominent set of Balmer lines is thus a clear indication that a star's surface temperature is about 10,000 K. At other temperatures, the spectral lines of other elements dominate a star's spectrum. For example, at around 25,000 K the spectral lines of helium are strong because, at this temperature, photons have enough energy to excite helium atoms without tearing away the electrons altogether.

When a hydrogen atom is ionized, its only electron is torn away and no absorption lines can be produced. An atom of a heavier element, however, has two or more electrons. When one electron is knocked away, the remaining electrons produce a new and distinctive set of spectral lines. For example, in stars hotter than about 30,000 K one of the two electrons in a helium atom is torn away. The remaining electron produces a set of spectral lines that is recognizably different from the lines produced by un-ionized helium. When the spectral lines of singly ionized helium appear in a star's spectrum, we know that the star has a surface temperature greater than 30,000 K.

Astronomers designate an un-ionized atom with a Roman numeral I; thus H I is neutral hydrogen. A Roman numeral II is used to identify an atom with one electron missing, and so He II is singly ionized helium (He^+). Similarly, Si III is doubly ionized silicon (Si^{++}), whose atoms are each missing two electrons.

In the early 1900s, Annie Cannon and her colleagues at Harvard Observatory set up the spectral classification scheme we use today. Many of Secchi's A-through-P categories were dropped, and the remaining spectral classes were reordered into the temperature sequence **OBAFGKM.** This sequence has traditionally been memorized as "**O**h, **B**e **A** **F**ine **G**irl, **K**iss **M**e!" The O stars are the hottest; their surface temperatures are in excess of 35,000 K, and their spectra show He II and Si IV. The M stars are the coolest; their surface temperatures of around 3000 K are so cool that atoms can stick together in molecules such as titanium oxide, whose spectral lines are prominent. Figure 12-8 shows representative examples of each spectral type.

Astronomers have found it useful to subdivide the original OBAFGKM sequence further. These finer steps are indicated by the addition of an integer from 0 through 9. Thus, for example, we have F8, F9, G0, G1, G2, . . ., G9, K0, K1, K2, ... The Sun, whose spectrum is dominated by singly ionized metals (especially Fe II and Ca II) is a G2 star.

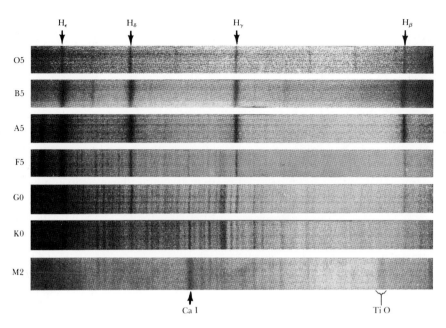

Figure 12-8 Principal types of stellar spectra These seven spectra show the main features of the seven spectral classes. Spectral lines of hydrogen, calcium, and titanium oxide are indicated. The hydrogen lines are strongest in A stars, which have surface temperatures of about 10,000 K. The spectra of G and K stars exhibit numerous lines caused by metals, indicating temperatures from 4000 to 6000 K. The broad, dark bands in the spectrum of an M star are caused by titanium oxide, which can exist only if the temperature is cooler than about 3000 K. (Courtesy of N. Houk, N. Irvine, and D. Rosenbush)

The strength of a particular spectral line depends on both the ionization and the excitation of the atom responsible for that line. Ionization and excitation of atoms in a gas depend on the temperature of the gas as shown in Figure 12-9, allowing us to deduce a star's surface temperature from its spectral type. For example, a star exhibiting strong Ca II and Fe I lines in its spectrum is a K5 star with a surface temperature around 4500 K.

Figure 12-9 Spectral type and temperature The strengths of the absorption lines of various elements are directly related to the temperature of the star's outer layers. For example, the Sun's spectrum has strong lines of singly ionized iron and calcium (Fe II and Ca II), corresponding to a spectral class of G2 and a surface temperature of about 5800 K. Note that hydrogen lines are strongest in A stars, whereas stars cooler than about 3500 K show absorption caused by titanium oxide.

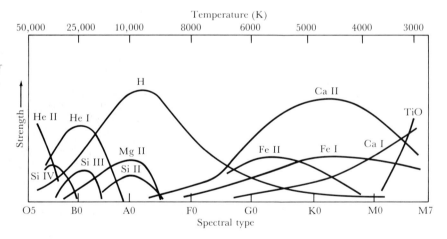

The Hertzsprung–Russell diagram demonstrates that there are different kinds of stars

The first accurate measurements of stellar parallax were made in the mid-nineteenth century, about the same time that astronomers began observing stellar spectra. Observing techniques improved during the next half-century, and the spectral types and absolute magnitudes of many stars became known.

Around 1905, the Danish astronomer Ejnar Hertzsprung pointed out that a regular pattern appears when the absolute magnitudes of stars are

Figure 12-10   A Hertzsprung–Russell diagram   *An H–R diagram is a graph on which the absolute magnitudes of stars are plotted against their spectral types. Each dot on this diagram represents a star in the sky whose absolute magnitude and spectral class have been determined. Note that the data points are grouped in specific regions on the graph. This pattern reveals the existence of different types of stars in the sky: main-sequence stars, giants, supergiants, and white dwarfs.*

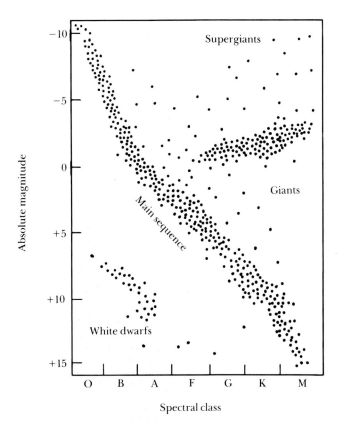

plotted against their color indices on a graph. Almost a decade later, the American astronomer Henry Norris Russell independently discovered this regularity in a graph using spectral types instead of color indices. Plots of this kind are now known as **Hertzsprung–Russell diagrams** or **H–R diagrams.**

Figure 12-10 is a typical Hertzsprung–Russell diagram. Each dot represents a star whose absolute magnitude and spectral type have been determined. Bright stars are near the top of the diagram, dim stars near the bottom. Hot stars (O and B) are toward the left side of the graph, cool stars (M) toward the right.

The most striking feature of an H–R diagram is that the data points are not scattered randomly all over the graph but are grouped in several distinct regions. The band stretching diagonally across the H–R diagram represents the majority of stars we see in the nighttime sky. This band, called the **main sequence,** extends from the hot, bright, bluish stars in the upper-left corner of the diagram down to the cool, dim, reddish stars in the lower-right corner. A star whose properties place it in this region of the H–R diagram is called a **main-sequence star.** For example, the Sun (spectral type G2, absolute magnitude +5) is such a star.

Toward the upper-right side of the H–R diagram is a second major grouping of data points. Stars represented by these points are both bright and cool. From Stefan's law, we know that a cool object radiates much less light per unit of surface area than a hot object does. In order to be so bright, these stars must be huge, and so they are called **giants.** They are typically 10 to 100 times as large as the Sun and have surface temperatures around 3000

Figure 12-11 *An H–R diagram* *On this H–R diagram, stellar luminosities are graphed against the surface temperatures of stars. Note that the various types of stars (e.g., main-sequence, giants, supergiants, etc.) fall in the same regions of the graph as in Figure 12-10, where absolute magnitude is plotted against spectral class. The dashed diagonal lines indicate stellar radii. The Sun's size is midway between the largest and smallest stars we see in the sky.*

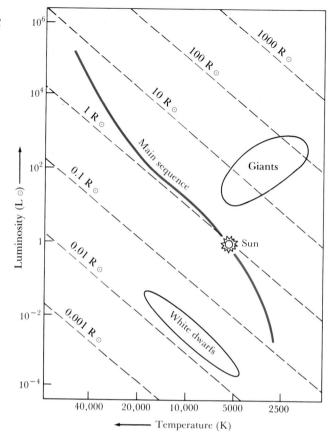

to 6000 K. Cooler members of this class of stars (those with surface temperatures from about 3000 to 4000 K) are often called **red giants** because they appear reddish in the nighttime sky. Aldebaran in the constellation Taurus and Arcturus in Boötes are examples of red giants that you can easily see with the naked eye.

A few rare stars are considerably bigger and brighter than typical red giants. These superluminous stars are appropriately called **supergiants.** Betelgeuse in Orion and Antares in Scorpius are examples of supergiants visible in the nighttime sky.

Finally, there is a third distinct grouping of data points toward the lower-left corner of the Hertzsprung–Russell diagram. These stars are hot, dim, and small. They are appropriately called **white dwarfs.** These stars, which are roughly the same size as the Earth, can be seen only with the aid of a telescope.

There is another useful way to construct an H–R diagram. Instead of absolute magnitude, luminosity is plotted on the vertical axis of the graph. And instead of spectral class, the surface temperature is plotted on the horizontal axis. The resulting graph is still an H–R diagram, but the observational quantities (such as spectral type) have been replaced by calculated quantities (such as temperature in kelvins).

Figure 12-11 shows this type of H–R diagram. Note that the temperature scale on the horizontal axis of the graph increases toward the left, because Hertzsprung and Russell drew their original diagrams with O stars on the

Table 12-2   Stellar luminosity classes

Luminosity class	Types of stars
I	Supergiants
II	Bright giants
III	Giants
IV	Subgiants
V	Main sequence

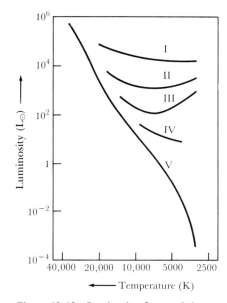

Figure 12-12 Luminosity classes *It is convenient to divide the H–R diagram into regions corresponding to luminosity classes. This breakdown permits finer distinctions between giants and supergiants. Luminosity class V encompasses the main-sequence stars, including the dim red stars called red dwarfs toward the lower-right side of the H–R diagram.*

left and M stars on the right. (They made this choice because the O stars have the simplest spectra.) Having hot stars toward the left and cool ones toward the right is a convention that no one has seriously tried to change.

Also shown in Figure 12-11 are dashed lines that display the radii of stars. The symbol $R_\odot$ stands for 1 solar radius (that is, half the Sun's diameter) and is equal to 6.96×10^5 km, or roughly half a million miles. Notice that most red giants are 10 to 100 times as large as the Sun, whereas white dwarfs are only about $\frac{1}{100}$ the size of the Sun. Also notice that most main-sequence stars are roughly the same size as the Sun.

The stars are classified into spectral types on the basis of the most prominent lines in their spectra. There are minor differences, however, among the spectral patterns of stars having the same spectral type. Based upon these minor differences, a system of **luminosity classes** was developed in the 1930s. Luminosity class I includes all the supergiants, and luminosity class V includes the main-sequence stars. The intermediate classes distinguish giants of various luminosities as indicated in Table 12-2. When the luminosity classes are plotted on the H–R diagram (see Figure 12-12), they provide a useful subdivision of the star types in the upper-right half of the diagram.

Astronomers commonly describe a star by combining its spectral type and its luminosity class into a sort of shorthand description, for example, calling the Sun a G2V star. This notation supplies a great deal of information about the star. Its spectral type is correlated with its surface temperature, and its luminosity class is correlated with its luminosity. Thus an astronomer knows immediately that a G2V star is a main-sequence star with a luminosity of about 1 $L_\odot$ and a surface temperature of nearly 6000 K. Similarly, a description of Aldebaran as a K5III star tells an astronomer that it is a red giant with a luminosity of around 500 $L_\odot$ and a surface temperature of about 4000 K.

An awareness of fundamentally different types of stars is the first important lesson to come from the H–R diagram. As we shall see in the following chapters, the different kinds of stars represent different stages of stellar evolution. We shall learn that a true appreciation of the H–R diagram requires an understanding of the life cycles of stars: how they are born, what happens as they mature, and what becomes of them when they die.

Binary stars provide information about stellar masses

We now know something about the sizes, temperatures, and luminosities of stars. To complete our picture of the physical properties of stars, we need only to know their masses. There is, however, no practical and direct way to measure the mass of an isolated star observed in the sky.

Fortunately for astronomers, about half of the visible stars in the nighttime sky are not isolated individual ones. Instead, they are members of multiple-star systems in which two or more stars orbit about each other. By observing how these stars orbit each other, astronomers can glean important information about the stellar masses.

When two stars can actually be seen orbiting each other, they are called a **visual binary** (see Figure 12-13). After many years of patient observation, astronomers can plot the orbit of one star about another in a binary pair (see Figure 12-14).

The gravitational force between the two stars in a binary star system keeps them in orbit about each other. This means that details of their orbital

1908   1915   1920

Figure 12-13 The binary star Kruger 60
About one-half of the visible stars are double stars. This series of photographs shows the binary star Kruger 60 in the constellation Cepheus. The orbital motion of the two stars about each other is apparent. This binary system has a period of $44\frac{1}{2}$ years. The maximum angular separation of the stars is about $2\frac{1}{2}$ arc sec; their apparent magnitudes are +9.8 and +11.4. (Yerkes Observatory)

motions can be described using the laws of Newtonian mechanics. Specifically, Kepler's third law relates the masses of the stars in a binary to their orbital period and the size of the orbit. Newton proved that a useful form of this law is

$$M_1 + M_2 = \frac{a^3}{P^2}$$

where M_1 and M_2 are the masses of the two stars expressed in solar masses, P is the orbital period in years, and a is the semimajor axis (measured in AUs) of the elliptical orbit of one star about the other. Thus, an astronomer can combine data concerning the orbit of a binary with Kepler's third law to calculate the sum of the masses of the two stars.

Each of the stars in the binary system moves in an elliptical orbit about the **center of mass** of the system (see Figure 12-15). This concept is analogous to two children on a seesaw. In order for the seesaw to balance properly, the center of mass of the two-child system must be located just above the support point or fulcrum. As you no doubt know from experience, this center of mass is offset from the midpoint between the two children toward the heavier child.

Similarly, the center of mass of a binary system could ostensibly be determined by placing the two stars at either end of a huge seesaw and determin-

Figure 12-14 The orbit of 70 Ophiuchi
After plotting the observations of a binary star over the years, astronomers can draw the orbit of one star with respect to the other (either star may be regarded as stationary—the shape and size of the orbit will be the same in either case). Once the orbit is known, Kepler's third law can be used to deduce information about the masses of the stars. This illustration shows the orbit of a faint double star in the constellation Ophiuchus.

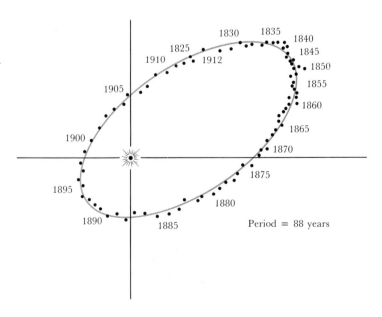

Figure 12-15 Star orbits in a binary
Each star in a binary system follows an elliptical orbit about their common center of mass. The center of mass is always nearer the more massive of the two stars.

ing where the fulcrum must be placed to balance the seesaw. The center of mass is always offset toward the more massive star.

In practice, the center of mass of a visual binary is determined by using the background stars as reference points. The separate orbits of the two stars can then be plotted as in Figure 12-15. The center of mass is located by finding the common focus of the two elliptical orbits. This information is needed to calculate the individual masses of the two stars.

Years of careful, patient observation of binaries have yielded the masses of many stars. As the data accumulated, an important trend began to emerge. For main-sequence stars, there is a direct correlation between mass and luminosity: the more massive the star, the more luminous it is. This **mass–luminosity relation** can be conveniently displayed as a graph (see Figure 12-16). Note that the range of stellar masses extends from one-tenth

Figure 12-16 The mass–luminosity relation For main-sequence stars, there is a direct correlation between mass and luminosity: the more massive a star, the more luminous it is.

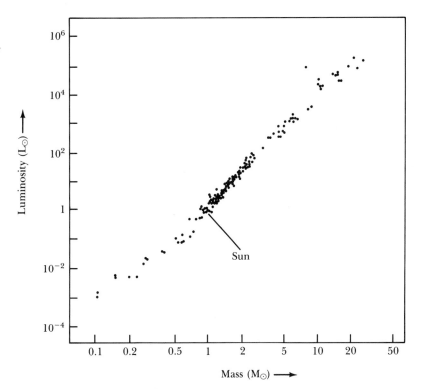

of a solar mass to about 30 solar masses. The Sun's mass lies in the middle of this range, and so we see again that our star is ordinary and typical.

The mass–luminosity relation demonstrates that the main sequence on the H–R diagram is a progression in mass as well as in luminosity and surface temperature. The hot, bright, bluish stars in the upper-left corner of the H–R diagram (see Figure 12-10) are the most massive main-sequence stars in the sky. The dim, cool, reddish stars in the lower-right corner of the H–R diagram are the least massive. The main-sequence stars of intermediate temperature and luminosity have intermediate mass. This relationship of mass to the main sequence will play an important role in our later discussion of stellar evolution.

....

The motion of a star affects the location of its spectral lines

Many binary stars are scattered throughout our Galaxy, but only those that are nearby or that have a large separation between the two stars can be distinguished as visual binaries. Star images in a remote binary are often blended together to produce a visual image that looks like a single star. Spectroscopy provides evidence that some single-appearing stars are in fact binaries.

The spectral analysis of some stars may yield incongruous spectral lines. For example, the spectrum of what appears at first to be a single star may include both strong hydrogen lines (characteristic of a type-A star) and strong absorption bands of titanium oxide (indicating a type-M star). A single star could not have the differing physical properties of these two spectral types, and so we conclude that this star is actually a binary system.

Spectroscopy can also be used to detect the movements of stars. As we saw in Chapter 5, the Doppler effect describes how the wavelength of light is affected by the relative motion between the source and the observer (recall Figure 5-18). If a source of light is coming toward you, you see a shorter wavelength than you would if the source were stationary. All the spectral lines in the spectrum of an approaching source are shifted toward the short-wavelength (blue) end of the spectrum. Conversely, all the spectral lines in the spectrum of a receding source are shifted toward the longer-wavelength (red) end of the spectrum. The size of the shift of a spectral line is proportional to the speed with which a source of light is moving toward or away from you. The greater the speed, the greater the shift.

If the orbital speeds of the two stars in a binary are more than a few kilometers per second, the Doppler effect can be used to calculate important information about the binary, even though two separate stars may not actually be observed. Such binaries yield a spectrum in which two complete sets of spectral lines shift back and forth. They are called **spectroscopic binary stars.** The regular, periodic shifting of the spectral lines is caused by the orbital motions of the stars as they revolve about their center of mass.

Figure 12-17 shows two spectra of a spectroscopic binary taken a few days apart. In Figure 12-17a, two sets of spectral lines are visible, slightly offset in opposite directions from the normal positions of these lines. The star moving toward the Earth has its lines blueshifted; the other star (moving away from the Earth) has its lines redshifted. A few days later, the stars have progressed along their orbits so that one star is moving toward the left and the other toward the right. At this point, neither star has any motion toward

Figure 12-17 A spectroscopic binary
*A spectroscopic binary exhibits spectral lines that shift back and forth as the two stars revolve about each other. These two spectra show the behavior of the spectroscopic binary κ Arietis. **(a)** The stars are moving parallel to the line of sight (one star approaching Earth, the other star receding), producing two sets of shifted spectral lines. **(b)** Both stars are moving perpendicular to our line of sight. (Lick Observatory)*

or away from the Earth, so there is no Doppler shifting and both stars yield spectral lines at the same positions. Thus only one set of spectral lines appears in Figure 12-17*b*.

Significant information about the orbital velocities of the stars in a spectroscopic binary can be deduced from measuring shifts in spectral lines. This information is best displayed as a **radial velocity curve** graphing radial velocity versus time for the binary system (see Figure 12-18). Radial velocity is the portion of a star's motion that is directed parallel to the line of sight between the Earth and the star.

The two curves on the graph in Figure 12-18 establish a pattern that repeats with a period of about 15 days, which is the orbital period of the binary. Note that this pattern is displaced upward from the zero-velocity line

Figure 12-18 A radial velocity curve
The graph displays the radial velocity curve of the binary HD 171978. The drawings indicate the positions of the stars and their spectra at four selected moments during an orbital period.

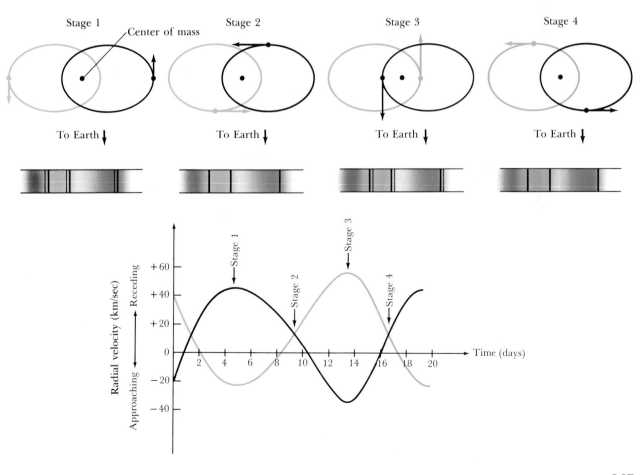

by about 12 km/sec, which is the overall motion of the binary system away from the Earth. Superimposed on this overall recessional motion (an overall redshift of the spectral lines) are the periodic approaches and recessions of the two stars as they orbit about the center of mass.

In many spectroscopic binaries, one of the stars is so dim that its spectral lines cannot be detected. The fact that the star is a binary is obvious, however, because its spectrum shows a single set of spectral lines that shift regularly back and forth. Such a **single-line spectroscopic binary** yields less information about its two stars than does a **double-line spectroscopic binary** like that shown in Figure 12-17.

The orbital speeds of the two stars in a binary are related to the masses of the stars by Kepler's laws and Newtonian mechanics. However, the individual masses of the two stars can be determined only if the tilt of their orbits is known. The angle of the orbits determines how much of the true orbital speeds of the stars appears as radial velocity measured from the Earth.

If the two stars are observed to eclipse each other, their orbits must be nearly edge-on as viewed from the Earth. As we shall see next, individual stellar masses can be determined if a spectroscopic binary also happens to be an **eclipsing binary star.**

Light curves of eclipsing binaries provide detailed information about the stars

A small fraction of all binary systems are oriented so that the two stars periodically eclipse each other as seen from Earth. Such eclipsing binaries can be detected even when the two stars cannot be resolved visually to be two distinct images in a telescope. The apparent brightness of the image of the binary dims momentarily each time one star blocks out the other.

Using a light-sensitive detector at the focus of a telescope, an astronomer can measure light intensity from binaries very accurately. The data for an eclipsing binary are most usefully displayed in the form of **light curves** such as those shown in Figure 12-19. The overall shape of the light curve for an eclipsing binary reveals at a glance such facts as whether the eclipse is total or partial (compare Figures 12-19a and 12-19b).

The light curve of an eclipsing binary provides detailed information about the motion of the two stars. For example, the time between successive eclipses of the same star is the orbital period of the binary, which is related by Kepler's third law to the separation of the two stars. Astronomers can use Newtonian mechanics to analyze such data from the light curve and calculate the combined masses of the two stars. If spectroscopic data are also available (so that the individual orbital speeds of the stars can be calculated from the Doppler effect), we would then have enough information to calculate the individual masses of the two stars.

The duration of an eclipse depends upon the size of the eclipsing star and the speed with which it moves. If the orbital speed is known, the diameter of the star can be computed from its light curve. Even such details as tidal distortion of each star by the other can be detected in the shape of the light curve (see Figure 12-19c).

If one of the stars in an eclipsing binary is very hot, its radiation typically creates a hot spot on its cooler companion star. Every time this hot spot is exposed to our Earth-based view, we receive a little extra light energy, which

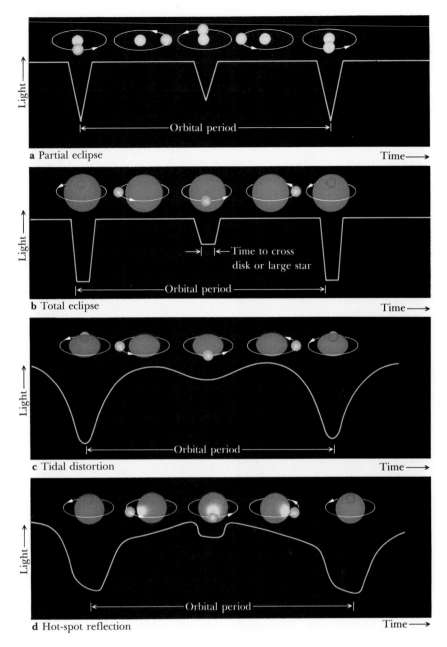

Figure 12-19 Representative light curves of eclipsing binaries *The shape of its light curve usually reveals many details about an eclipsing binary. Examples of* **(a)** *partial eclipse,* **(b)** *total eclipse,* **(c)** *tidal distortion, and* **(d)** *hot-spot reflection are illustrated here.*

a Partial eclipse

Orbital period

Light

Time ⟶

b Total eclipse

Time to cross disk or large star

Orbital period

Light

Time ⟶

c Tidal distortion

Orbital period

Light

Time ⟶

d Hot-spot reflection

Orbital period

Light

Time ⟶

produces a characteristic "bump" on the binary's light curve (see Figure 12-19*d*).

Light curves can also be used to study stellar atmospheres. Suppose that one star of a binary is a white dwarf and the other is a bloated red giant. By observing how the light from the bright white dwarf gradually is cut off as it moves behind the edge of the red giant during the beginning of an eclipse, we can deduce conditions in the upper atmosphere of the red giant, like its pressure and density.

Mass transfer in close binary systems can produce unusual double stars

Binary stars are fascinating objects. A single star leads a straightforward, birth-to-death existence, but exotic things can happen in a **close binary** system, where the stars are separated by only a few stellar diameters. In such binaries, the stars are so close together that the gravity of one can dramatically affect the appearance and evolution of the other.

As we shall see in the next chapter, a star expands dramatically, becoming a red giant, as it grows old. When one member of a close binary becomes a giant, it can dump gas onto its companion. This process, called **mass transfer,** occurs only if the giant becomes sufficiently bloated and its companion star is near enough that the giant's outer layers can be gravitationally captured by the companion star.

In the mid-1800s, the French mathematician Edward Roche pointed out that a figure-eight curve drawn around two stars in a binary can portray the gravitational domain of each star. This figure-eight-shaped boundary is often called the **critical surface.** The two lobes of the critical surface are known as a **Roche lobes.** The more massive star is always located inside the larger Roche lobe. Gases escaping a Roche lobe are no longer bound to that star and are free to leave the binary or to fall onto the companion star. When mass transfer occurs, gases flow across the point where the two Roche lobes touch.

In many binaries, the stars are so far apart that even during their red giant stage the stars' surfaces remain well inside their Roche lobes so that little mass transfer can occur. Each star thus lives out its life independently as if it were single and isolated.

A binary system in which both stars are within their Roche lobes is referred to as a **detached binary** (see Figure 12-20). If the two stars are relatively close together, when one star expands to become a giant it may fill or overflow its Roche lobe, in which case the system is called a **semidetached binary.** If both stars happen to fill their Roche lobes, the system is called a **contact binary,** because the two stars actually touch and may share a common envelope of gas. Semidetached and contact binaries are most easily detected if they happen also to be eclipsing binaries, because their light curves then have a distinctly rounded appearance caused by these tidally distorted egg-shaped stars (recall Figure 12-19c).

The eclipsing binary called Algol (from an Arabic term for demon) is a semidetached binary that can be seen easily with the naked eye in the constellation of Perseus. From Algol's light curve (see Figure 12-21a) astronomers have determined that the binary contains a giant that fills its Roche lobe. Sometime in the past, as this star expanded and became a giant, it dumped a significant amount of gas onto its companion. Astronomers theorize that the giant was originally the more massive star, but as a result of mass transfer, the detached companion is now the more massive star.

Mass transfer is still occurring in a semidetached eclipsing binary called β Lyrae in the constellation of Lyra, the harp. Like Algol, β Lyrae contains a giant that fills its Roche lobe (see Figure 12-21b). For many years astronomers were puzzled by the fact that the detached companion star in β Lyrae is severely underluminous, contributing virtually no light at all to the visible radiation coming from the system. Furthermore, the spectra of β Lyrae contain unusual features, some of which are caused by gas flowing between the stars and around the system as a whole.

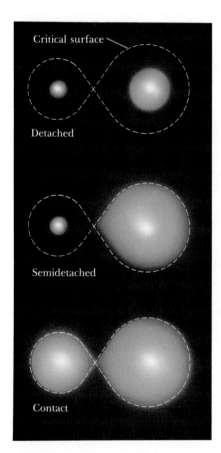

Figure 12-20 Detached, semidetached, and contact binaries *A double star is said to be detached, semidetached, or contact depending on whether either or both stars fills its Roche lobe. Mass transfer is often observed in semidetached binaries. The two stars in a contact binary share the same outer atmosphere.*

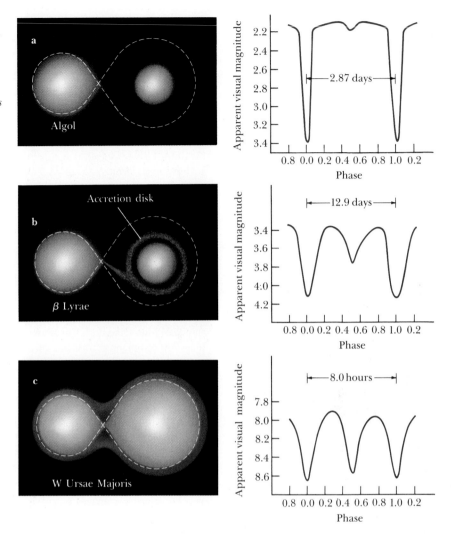

Figure 12-21   Three close binaries
Light curves for and sketches of three eclipsing binaries are shown. The "phase" denotes the fraction of the orbital period from one primary minimum to the next. (a) Algol is a semidetached binary. (b) β Lyrae is a semidetached binary in which mass transfer has produced an accretion disk surrounding the detached star. (c) W Ursae Majoris is a contact binary.

The β Lyrae system was explained in 1963 when Su-Shu Huang of Northwestern University published his interpretation that the underluminous star in β Lyrae is enveloped in a huge rotating disk of gas captured from its bloated companion. The disk is so large and thick that it completely shrouds the secondary star, making it impossible to observe at visible wavelengths.

Observations made since the 1960s have confirmed the existence of a disk of gas, called an **accretion disk,** encircling the underluminous star in β Lyrae. The primary star is overflowing its Roche lobe, with gases streaming onto the disk at the rate of 10^{-5} $M_\odot$ per year. Ultraviolet spectra taken from spacecraft in the 1970s revealed that some gas is constantly escaping altogether from the system.

The fate of a semidetached system like Algol or β Lyrae depends primarily on how fast their stars evolve. If the detached star expands to fill its Roche lobe while the companion star is still filling its own Roche lobe, then the result is a contact binary. An example is W Ursae Majoris in which two stars share the same photosphere (see Figure 12-21*c*). In Chapters 14 and 15 we shall see that mass transfer onto dead stars produces some of the most extraordinary objects in the sky.

Summary

· Determining stellar distances is the first step to understanding the nature of the stars.

Distances to the nearer stars can be determined by parallax—the apparent shift of a star's location against the background stars while the Earth moves around its orbit.

· The apparent magnitude of a star depends on the star's distance and energy output, or luminosity.

The luminosity of a star is the amount of energy escaping from it each second.

The absolute magnitude of a star is the apparent magnitude it would have if viewed from a distance of 10 parsecs. Absolute magnitudes are calculated from the star's apparent magnitude and distance.

· Astronomers use photometers to measure a star's brightness through a set of standard filters like the UBV filters. The color index of a star is the difference in magnitudes obtained with two different filters.

· The surface temperature of a star can be determined from its color indices or from its spectral type (O, B, A, F, G, K, or M). Spectral type is determined by patterns of spectral lines in a star's spectrum.

· The Hertzsprung–Russell (H–R) diagram is a graph plotting absolute magnitudes of stars against their spectral types (or, equivalently, luminosities against surface temperatures). It reveals the existence of various types of stars such as main-sequence stars, giants, supergiants, and white dwarfs.

· Binary stars are surprisingly common in the universe. Those that can be resolved as two distinct star images by an Earth-based telescope are called visual binaries.

The masses of the two stars in a binary system can be computed from measurements of the orbital period and orbital dimensions of the system.

· The mass–luminosity relation expresses a direct correlation between mass and luminosity for main-sequence stars.

· A spectroscopic binary exhibits periodic shifting of its spectral lines due to the Doppler effect as the orbits of the stars carry them alternatively toward and away from the Earth.

· An eclipsing binary is one whose orbits are viewed nearly edge on from the Earth, so that one star periodically eclipses the other.

Detailed information about the stars in an eclipsing binary can be obtained by studying its light curve.

· Mass transfer in a close binary system can affect the appearance and evolution of both stars.

Review questions

1 Explain the difference between apparent magnitude and absolute magnitude.

2 Describe one way in which Wien's law and the Stefan–Boltzmann law can be used to deduce information about stars.

3 How and why is the spectrum of a star related to its surface temperature?

4 Describe the UBV filters and how an astronomer uses them to measure a star's surface temperature.

5 What is the chemical composition of most stars?

6 Which is the hottest star listed in Table 12-1? Which is the coolest?

7 Draw an H–R diagram and sketch the regions occupied by main-sequence stars, red giants, and white dwarfs. Briefly discuss the different ways in which you could have labeled the axes of your graph.

8 What is the mass–luminosity relation, and to what kind of stars does it apply?

9 Sketch the radial velocity curve of a binary whose stars are moving in a circular orbit that is **(a)** perpendicular and **(b)** parallel to our line of sight.

Advanced questions

10 Sketch the light curve of an eclipsing binary having high orbital eccentricity in which **(a)** the major axes are pointed toward the Earth and **(b)** the major axes are perpendicular to our line of sight.

*11 Estimate the mass of a main-sequence star that is 10,000 times as luminous as the Sun. What is the luminosity of a main-sequence star whose mass is one-tenth that of the Sun?

Discussion questions

12 Why do you suppose that stars of the same spectral type but different luminosity class exhibit slight differences in their spectra?

13 How might a star's rotation affect the appearance of its spectral lines?

For further reading

Ashbrook, J. "Visual Double Stars for the Amateur." *Sky & Telescope*, November 1980, p. 379.

Evans, D., et al. "Measuring Diameters of Stars." *Sky & Telescope*, August 1979, p. 130.

Gingerich, O. "A Search for Russell's Original Diagram." *Sky & Telescope*, July 1982, p. 36.

Kaler, J. "Origins of the Spectral Sequence." *Sky & Telescope*, February 1986, p. 129.

Nielsen, A. "E. Hertzsprung—Measurer of Stars." *Sky & Telescope*, January 1968, p. 4.

Page, T., and Page, L. *Starlight—What It Tells About the Stars*. Macmillan, 1967.

Phillip, A., and Green, L. "Henry N. Russell and the H–R Diagram." *Sky & Telescope*, April 1978, p. 306; May 1978, p. 395.

Tomkin, J., and Lambert, D. "The Strange Case of Beta Lyrae." *Sky & Telescope*, October 1987, p. 354.

Upgren, A. "New Parallaxes for Old: Coming Improvements in the Distance Scale of the Universe." *Mercury*, November/December 1980, p. 143.

13

The lives of stars

Reflection and emission nebulae
The two bluish objects (toward the upper right side of this photograph) are reflection nebulae surrounding two young, hot main-sequence stars. Interstellar dust around these two stars efficiently reflects their bluish light. Several smaller reflection nebulae are scattered around the large reddish patch of ionized hydrogen gas. Dust mixed with the gas dilutes the intense red emission of the hydrogen atoms with a soft bluish haze. (Anglo-Australian Observatory)

Astronomers' observations of stars and calculations of stellar models have led to a theory of stellar evolution. In this chapter we learn how protostars form in cold clouds of interstellar dust and gas, then evolve into main-sequence stars. We see that the spiral arms of a galaxy and explosions of supernovae can trigger the birth of stars in these huge interstellar clouds. Next we examine how stars evolve after they leave the main sequence. We find that stars become red giants after their core hydrogen is exhausted, and we discuss the thermonuclear process of helium burning at the centers of aging stars. Throughout, H–R diagrams prove useful in understanding many details of stellar evolution.

The legend of the Ephemera tells of a race of remarkable insects that once inhabited a great forest. These noble creatures were blessed with great intelligence, yet cursed with tragically short life spans of less than a day.

To the Ephemera, the forest seemed eternal and unchanging. Many generations lived out their brief lives without ever noticing any alteration in the surrounding foliage. Nevertheless, careful observation and reasoning led some Ephemera to postulate that the forest is not static. They began to suspect that small green shoots grow to become huge trees and that mature trees eventually die, topple over, and litter the forest with rotting logs, enriching the soil for future trees. Although they were unable to witness this transformation personally, the Ephemera became aware of life cycles stretching over awesome periods of many years.

An analogous situation faces astronomers. Gazing across the galaxy, we can see a vast array of stars and nebulae. The heavens seem eternal and unchanging at first; what we see is virtually indistinguishable from what our ancestors saw. This seeming permanence is an illusion, however. Stars emit vast amounts of radiation, which must produce changes that cause the stars to evolve. From careful observation and calculation, astronomers have assembled a theory of **stellar evolution,** which describes how stars are born in great interstellar clouds of gas and dust. They mature, grow old, and eventually blow themselves apart in death throes that enrich the interstellar gas for future stellar generations. The stars seem unchanging only because major stages in their lives last for millions or billions of years.

Protostars form in cold, dark nebulae

For many years, astronomers have suspected that stars are born in cold, dark clouds of interstellar gas because the temperature of a gas is directly related to the average speed of its atoms and molecules. If an interstellar cloud is warm, its atoms are moving about so rapidly that there is no chance for a protostar to condense from the agitated gases. If the cloud's temperature is low, however, its atoms are moving slowly enough to allow denser portions of the cloud to contract gravitationally into clumps that collapse to form new stars.

Many of these cold clouds are scattered across the Milky Way. In some cases, they appear as dark regions silhouetted against a glowing background nebulosity, such as the famous Horsehead Nebula in Figure 13-1. In other

Figure 13-1 The Horsehead Nebula
Dust grains block the light from the background nebulosity whose glowing gases are excited by ultraviolet radiation from young, massive stars. The nebula is located in Orion at a distance of roughly 1600 light years from Earth. The bright star to the left of center is Alnitak, the easternmost star in the "belt" of Orion. (Royal Observatory, Edinburgh)

Figure 13-2   A dark nebula   *This dark nebula, called Barnard 86, is located in Sagittarius. It is visible in this photograph simply because it blocks out light from the stars beyond it. The cluster of bluish stars to the left of the dark nebula is NGC 6520. (Anglo-Australian Observatory)*

cases, they appear as dark blobs that obscure the background stars (see Figure 13-2). These **dark nebulae** are sometimes called **Barnard objects,** after the American astronomer E. E. Barnard, who discovered many of them around 1900.

The chemical composition (by mass) of these dark clouds is the standard "cosmic abundance" of about 75 percent hydrogen, 23 percent helium, and 2 percent heavier elements (recall Table 6-4). Infrared observations indicate that the internal temperatures of globules are from 5 to 15 K. This range is so cold that dense regions inside the globule contract gravitationally, and the globule gradually coalesces into lumps called **protostars.**

Protostars evolve into young main-sequence stars

As early as the 1950s, astrophysicists began calculating the structure and evolutionary history of protostars. At first, a protostar is a cool blob of gas several times larger than the solar system. Low pressures inside the protostar are incapable of supporting all this cool gas, and so the protostar begins to contract. As it does so, gravitational energy is converted into thermal energy, which causes the gases to heat up and start glowing. After only a few thousand years of gravitational contraction, the surface temperature has reached 2000 to 3000 K. At this point, the protostar is still quite large, and so its glowing gases produce substantial luminosity. After only 1000 years of contraction, a protostar of 1 solar mass would be 20 times larger in diameter and 100 times brighter than the Sun.

Astrophysicists use high-speed computers and the equations of stellar structure described in Chapter 11 to calculate the conditions inside a contracting protostar. The results indicate how the protostar's luminosity and surface temperature change at various stages during its contraction. With this information, we can graph the **evolutionary track** of the protostar on a Hertzsprung–Russell diagram.

When a protostar begins to shine at visible wavelengths, it is both luminous and cool. Evolutionary tracks of protostars thus begin near the upper-right corner of the H–R diagram (see Figure 13-3). Continued gravitational contraction causes the protostar to shift rapidly away from this region of the diagram, however. A protostar more massive than three Suns becomes hotter without much change in overall luminosity. The evolutionary tracks of

**Figure 13-3 Pre–main-sequence
evolutionary tracks** *The evolutionary tracks
of seven stars having different masses are
shown in this H–R diagram. The dashed lines
indicate the stage reached after the indicated
number of years of evolution. Note that all
tracks terminate on the main sequence at
points agreeing with the mass–luminosity
relation. (Based on stellar model calculations
by I. Iben)*

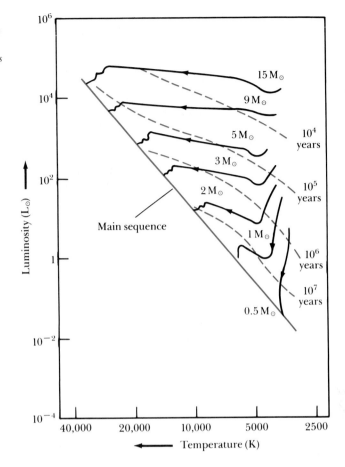

massive protostars thus traverse the H–R diagram horizontally, from right
to left. Less massive protostars dim slightly as they contract, and so their
luminosities decrease somewhat as their surface temperatures rise.

A protostar continues to shrink until the temperature at its center reaches
a few million degrees, when hydrogen burning begins. As we saw in Chapter
11, this thermonuclear process releases enormous amounts of energy. The
outpouring of energy from hydrogen burning creates conditions inside the
protostar that finally halt its gravitational contraction. Hydrostatic and thermal equilibrium are eventually established—and a stable star is born. It is at
this stage that the wandering evolutionary track ends on the main sequence,
as seen in Figure 13-3.

We now know that the main sequence represents relatively young stars
inside of which hydrogen burning has only "recently" begun. This state is
a very stable one for most stars. For example, the Sun will remain on or
near the main sequence, quietly burning hydrogen at its core, for a total of
10 billion years.

Note that the evolutionary tracks in Figure 13-3 end at locations along the
main sequence that agree with the mass–luminosity relation (recall Figure
12-16). The most massive stars are the most luminous, and the least massive
stars are the least luminous. Protostars less massive than about 0.08 solar
masses never manage to develop the necessary pressures and temperatures
to start hydrogen burning at their cores. These small protostars instead

contract to become planetlike objects. On the other hand, protostars with masses greater than about 100 solar masses (100 M$_\odot$) rapidly develop such extremely high temperatures that radiation pressure becomes the dominant force supporting the star against gravitational collapse. Such stars are very unstable and very uncommon.

The evolutionary tracks of protostars begin in the red-giant region of the H–R diagram, but they are not red giants. An H–R diagram such as Figure 12-10 shows where stars spend *most* of their lives, protostars spend only a tiny fraction of their existence in the red-giant region. A 15 M$_\odot$ protostar takes only 10,000 years to become a main-sequence star; a 1 M$_\odot$ protostar takes a few tens of millions of years to ignite hydrogen burning at its core. By astronomical standards, these are such brief intervals that it is unusual to discover a protostar that is in its earliest stages of formation.

We also are unlikely to observe the birth of a star at visible wavelengths because its surrounding globule or interstellar cloud shields the protostar from our view. The vast amount of visible light emitted by the protostar is absorbed by interstellar dust in the surrounding **cocoon nebula,** which becomes heated to a few hundred K. The warmed dust reradiates the energy at infrared wavelengths. Infrared observations can thus reveal what is going on inside a "stellar nursery."

Figure 13-4 shows two views of a stellar nursery taken at infrared and visible wavelengths. The dark cloud near the center of Figure 13-4b contains roughly 1000 solar masses of hydrogen and is a site of active star formation. This same region appears very bright in the infrared view, Figure 13-4a. Interstellar dust in the cloud absorbs nearly all the visible light emitted by newborn stars, but infrared radiation passes easily through the obscuring material. Recent infrared observations reveal a cluster of 20 new stars embedded deep within the dust cloud. Astronomers are understandably excited by the ease with which infrared observations probe these clouds.

Figure 13-4 Formation of new stars near ρ Ophiuchi (a) *A wide-angle infrared view of the region near the star ρ Ophiuchi, covering nearly 13° × 13° at a wavelength of 100 µm. The white rectangle indicates the smaller region covered by photograph* **b** *at visible wavelengths.* **(b)** *The white circle indicates the location of at least 20 newborn stars that are probably less than 1 million years old. The bluish star near the top of this view at visible wavelengths is ρ Ophiuchi. Antares, brightest star in Scorpius, appears at the lower left. Many stars in this region have a fuzzy appearance because of interstellar dust. (National Aerospace Laboratory of the Netherlands; Royal Observatory, Edinburgh)*

a

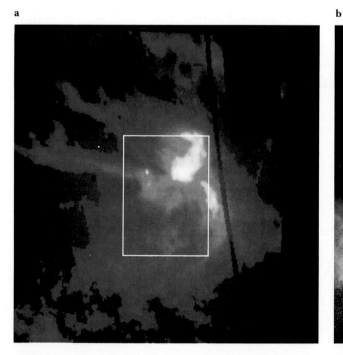

b

Young star clusters are found in H II regions

We can see from the evolutionary tracks of protostars in Figure 13-3 that the most massive stars are the first to form. The more massive the protostar, the sooner it develops the necessary central pressures and temperatures to ignite hydrogen burning. These massive main-sequence stars (spectral types O and B) are also the hottest and most luminous stars in the sky. Their surface temperatures are typically 15,000 to 35,000 K, meaning that they emit vast quantities of ultraviolet radiation.

The energetic ultraviolet photons from newborn massive stars easily ionize the surrounding hydrogen gas. For example, radiation from an O5 star can knock electrons from hydrogen atoms in a volume roughly 1000 light years across. This ionization has a dramatic effect on the globule or dark nebula in which a cluster of stars is forming. While some hydrogen atoms are being knocked apart by ultraviolet photons, other hydrogen atoms are being reassembled as some of the free protons and electrons manage to get back together. During this **recombination** of hydrogen atoms, the captured electrons cascade downward through the atom's energy levels toward the ground state. These downward quantum jumps release numerous photons at many visible wavelengths. The nebula begins to glow. Particularly prominent is the transition from $n = 3$ to $n = 2$, which produces H_α photons at 6563 Å in the red portion of the visible spectrum (recall Figure 5-17). The nebulosity around a newborn star cluster thus shines with a distinctive reddish hue.

Figure 13-5 shows one of these **emission nebulae.** Because these nebulae are predominantly ionized hydrogen, they are also called **H II regions.** The collection of a few hot, bright O and B stars near the core of the nebula that produces the ionizing ultraviolet radiation is called an **OB association.**

Observations of individual stars in a young cluster yield further information about stars in their infancy. Figure 13-6 shows a beautiful emission nebula surrounding the cluster called NGC 2264. By observing each star in

Figure 13-5 An H II region Because of its shape, this emission nebula is called the Eagle nebula. It surrounds the star cluster called M16 in the constellation of Serpens at a distance of 6500 light years from Earth. Several bright, hot O and B stars are responsible for the ionizing radiation that causes the gases to glow. (Anglo-Australian Observatory)

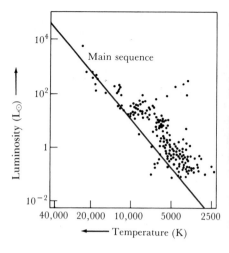

Figure 13-6 A young star cluster and its H–R diagram The photograph shows an H II region and the young star cluster NGC 2264 in the constellation of Monoceros. The nebulosity is located about 2600 light years from Earth and contains numerous stars that are about to begin hydrogen burning in their cores. Each dot plotted on the H–R diagram represents a star in this cluster whose luminosity and surface temperature have been measured. Note that most of the cool, low-mass stars have not yet arrived at the main sequence. This star cluster probably started forming only 2 million years ago. (Based on observations by M. Walker; Anglo-Australian Observatory)

the cluster and measuring its magnitude and color, astronomers can deduce the star's luminosity and surface temperature. The data for all the stars in the cluster can then be plotted on an H–R diagram as shown in Figure 13-6. Note that the hottest stars, with surface temperatures around 20,000 K, are on the main sequence. These stars are the rapidly evolving, massive, hot stars whose radiation is causing the surrounding gases to glow. The stars cooler than about 10,000 K have not yet quite arrived at the main sequence, however. These less massive stars are in the final stages of pre–main-sequence contraction and thermonuclear reactions are just now beginning at their centers. The locations of data points on Figure 13-6 suggest that the cluster is only about 2 million years old.

Spectroscopic observations of the cooler stars in NGC 2264 show that many are vigorously ejecting gas, a very common phenomenon in most stars just before they reach the main sequence. Such gas-ejecting stars are called **T Tauri stars** because the first example was discovered in the constellation of Taurus. Some astronomers suggest that the onset of hydrogen burning is preceded by vigorous chromospheric activity, with enormous spicules and flares that propel the star's outermost layers back out into space. In fact, an infant star going through its T Tauri stage can lose as much as 0.4 solar masses before it settles down on the main sequence.

Sporadic activity and mass loss can continue after a star has arrived on the main sequence. For example, Figure 13-7 shows a young star cluster called the Pleiades in the constellation of Taurus that is easily visible to the unaided eye. All the stars in this picture are on the main sequence. The cluster is presumably about 100 million years old because that is how long it takes for the least massive stars finally to begin hydrogen burning in their cores. One of the brighter stars in the cluster is still experiencing minor outbursts.

Note the distinctly bluish color of the nebulosity around the Pleiades. This glow is called a **reflection nebula** because it is caused by starlight reflected from dust grains in the interstellar material. The nebulosity is blue for the same reason Earth's daytime sky is blue: particles—whether in a nebula or in the air—scatter short-wavelength light much more efficiently

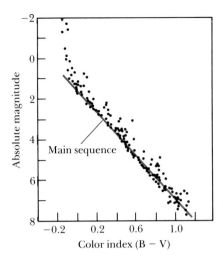

Figure 13-7 The Pleiades and its H–R diagram *This open cluster called the Pleiades, which can easily been seen with the naked eye in the constellation of Taurus, is about 400 light years from Earth. Each dot plotted on the H–R diagram represents a star in the Pleiades whose absolute magnitude and color index have been measured. Note that most of the cool, low-mass stars have arrived at the main sequence, indicating that hydrogen burning has begun in their cores. The cluster has a diameter of about 5 light years and is about 100 million years old. (U.S. Naval Observatory)*

than longer wavelength radiation. Blue light is bounced around and reflected back toward us much more readily than is light of other colors.

In many respects, the Pleiades cluster is a mature version of the H II regions seen in Figures 13-5 and 13-6. Many years of T Tauri activity and vigorous stellar winds have widely dispersed the surrounding gases so that they no longer glow as an emission nebula. These gases will eventually become so widely scattered that the reflection nebulosity will also fade from view.

A loose collection of stars such as the Pleiades or NGC 2264 is called an **open cluster** or a **galactic cluster.** Such clusters possess barely enough mass to hold themselves together by gravity. A star moving faster than the average speed will occasionally escape, or "evaporate," from a cluster. By the time the stars are a few billion years old, they may be so widely separated that a cluster no longer truly exists.

Star birth begins in giant molecular clouds

Astronomers agree that H II regions are stellar nurseries, but where do the H II regions come from? This question was finally answered in the 1970s when radio astronomers began discovering enormous clouds of gas scattered about our Galaxy.

As we saw in Chapter 6, hydrogen is by far the most abundant element in the universe. In the cold depths of interstellar space, hydrogen atoms combine to form hydrogen molecules (H_2). A molecule vibrates and rotates at specific frequencies dictated by the laws of quantum mechanics. As a molecule goes from one vibrational or rotational state to another, it emits or absorbs a photon, just as an atom emits or absorbs a photon when an electron jumps from one energy level to another. Many interstellar molecules emit photons with wavelengths of a few millimeters. In recent years, observations with radio telescopes tuned to wavelengths in this range have greatly increased our knowledge of the **interstellar medium,** or interstellar matter.

Although hydrogen molecules are scattered abundantly across space, they are difficult to detect. The hydrogen molecule consists of two atoms of

Figure 13-8 A spiral galaxy This beautiful spiral galaxy, called NGC 2779, has two arching spiral arms outlined by numerous H II regions and clusters of young, hot stars. Each pinkish speck along the spiral arms of this galaxy is an H II region. The spiral arms of a galaxy are active sites of vigorous star formation. (Anglo-Australian Observatory)

equal mass joined together, and such molecules do not efficiently emit photons at radio frequencies. Radio astronomers can more easily detect asymmetric molecules such as carbon monoxide (CO), which consists of two atoms of unequal mass joined together. Carbon monoxide emits photons at a wavelength of 2.6 mm, corresponding to a transition between two rates of rotation of the molecule.

The existence of carbon monoxide is especially useful in probing the interstellar medium. Astronomers have reason to believe that the ratio of hydrogen to carbon monoxide is reasonably constant in space: about 10,000 H_2 molecules for every CO molecule. Consequently, wherever astronomers detect strong emission from CO, they know that hydrogen gas must be abundant.

In mapping the locations of CO emission, astronomers soon realized that vast amounts of hydrogen are concentrated in **giant molecular clouds.** These clouds have masses in the range of 10^5 to 2×10^6 solar masses and diameters that range from 20 to 80 parsecs. The density inside one of these clouds is about 200 hydrogen molecules per cubic centimeter. Astronomers estimate that our Galaxy contains about 5000 of these enormous clouds.

Many galaxies, including our own and the one shown in Figure 13-8, have **spiral arms,** which are huge, arching lanes of glowing gas and stars. In Chapter 16, when we discuss our Galaxy, we shall learn that these spiral arms are caused by compression waves that squeeze the interstellar gases through which they pass. When one of these waves passes through a giant molecular cloud, it compresses the cloud and vigorous star formation begins in the densest regions. Massive stars, which are the first to form, emit ultraviolet light that soon ionizes the surrounding hydrogen and an H II region is born. An H II region is thus a small, bright "hot spot" in a giant molecular cloud. The famous Orion nebula (see Figure 13-9) is an example. Four hot, massive O and B stars at the heart of the Orion nebula are responsible for the ionizing radiation that causes the surrounding gases to glow. The Orion nebula is embedded in a giant molecular cloud whose mass is estimated at 500,000 solar masses.

The OB association at the core of the H II region affects the rest of the giant molecular cloud. Vigorous stellar winds, along with ionizing ultraviolet radiation from the O and B stars, carve out a cavity in the cloud. Much of this outflow is supersonic, creating a **shock wave** where the outer edge of

Figure 13-9 The Orion Nebula
This famous H II region can be seen with the naked eye. It is located 1600 light years from Earth and has a diameter of roughly 16 light years. The mass of this nebulosity is estimated at 300 solar masses. Four bright, massive stars at the center of the nebula produce the ultraviolet light that causes the gases to glow. These four stars, called the Trapezium, are separated from each other by only 0.13 light year. (Anglo-Australian Observatory)

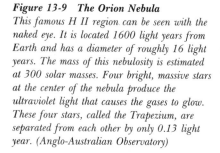

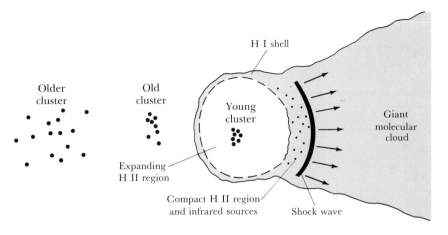

Figure 13-10 The evolution of an OB association *Ultraviolet radiation from young O and B stars produces a shock wave that compresses gas further into the molecular cloud and stimulates new star formation deeper into the cloud. Meanwhile, older stars are left behind. (Adapted from C. Lada, L. Blitz, and B. Elmegreen)*

Figure 13-11 The core of the Orion Nebula *(a) This view at visible wavelengths shows the inner regions of the Orion Nebula (compare with Figure 13-9). At the center are four massive stars, called the Trapezium, which cause the nebula to glow. (b) This infrared view is also centered on the Trapezium, but covers an area slightly smaller than that in a. Infrared radiation can penetrate interstellar dust more easily than can visible photons. Numerous infrared objects, many of which are probably new stars in the early stages of formation, are seen in the infrared view. (Anglo-Australian Observatory)*

the expanding H II region impinges on the rest of the giant molecular cloud. The shock wave compresses the hydrogen gas through which it passes, stimulating a new round of star birth. Compact H II regions and infrared sources are often found immediately behind the shock wave. These newborn H II regions soon form their own O and B stars, which in turn carve out a new cavity deeper into the giant molecular cloud. Meanwhile, the older O and B stars left behind begin to disperse, because their mutual gravitational attraction is not sufficient to keep them together (see Figure 13-10). In this way, an OB association "eats into" a giant molecular cloud, "spitting out" stars in its wake.

Infrared observations reveal many features that resemble protostars in the swept-up layer immediately behind the shock wave from an H II region. For instance, Figure 13-11 shows both optical and infrared views of the core of the Orion nebula. Four O and B stars and glowing gas and dust dominate the center of the view at visible wavelengths. Infrared observations penetrate this material to reveal dozens of infrared objects that may be cocoons of warm dust enveloping newly formed stars.

a

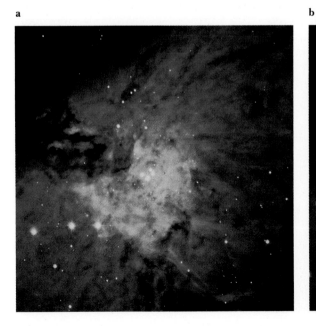

b

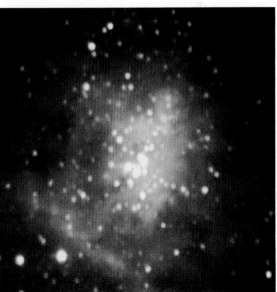

Star birth is also triggered by supernova explosions that compress the interstellar medium

Presumably, any mechanism that compresses interstellar clouds can trigger the birth of stars. As we shall see in detail in Chapter 14, a massive star can end its life with violent detonation called a **supernova explosion.** In a matter of seconds, the doomed star is blown apart and its outer layers are blasted outward into space at speeds of several thousand kilometers per second.

Astronomers find many nebulosities across the sky that are the shredded funeral shrouds of these dead stars. Such nebulae (like the Cygnus Loop shown in Figure 13-12) are called **supernova remnants.** Many supernova remnants have a distinctly arched appearance, as would be expected for an expanding shell of gas. This wall of gas typically is still moving away from the dead star at supersonic speeds. Its passage through the surrounding interstellar medium excites the atoms, causing the gases to glow.

Supersonic motion is always accompanied by a shock wave that abruptly compresses the gas through which it passes. If the expanding shell of a supernova remnant encounters an interstellar cloud, it can squeeze the cloud, stimulating star birth. As we learned in Chapter 10, there is evidence that the Sun was created in this fashion.

Our understanding of star birth has improved dramatically in recent years, primarily because of infrared and millimeter-wavelength observations. Nevertheless, many mysteries remain. For example, astronomers had generally assumed that there must be a lot of interstellar dust shielding a stellar nursery from the disruptive effects of external sources of ultraviolet light, which is what we seem to find in our own Galaxy. However, in a neighboring galaxy called the Large Magellanic Cloud there are young OB associations with virtually no dust. Does the process of star birth differ slightly from one galaxy to another?

We also cannot explain why different methods of star birth tend to produce different percentages of different kinds of stars. Why, for instance, does the passage of a spiral arm through a giant molecular cloud tend to

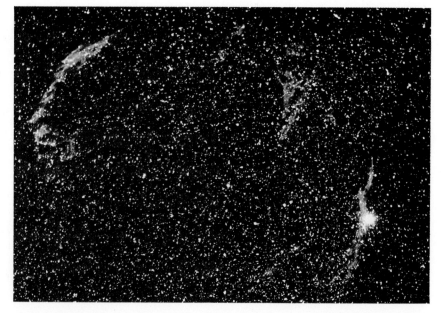

Figure 13-12 A supernova remnant
This remarkable nebula, called the Cygnus Loop, is the remnant of a supernova explosion that occurred about 20,000 years ago. The expanding spherical shell of gas now has a diameter of about 120 light years. (Courtesy of H. Vehrenberg)

Figure 13-13 The core of the Rosette Nebula *The Rosette Nebula is a large, circular emission nebula near one end of a sprawling giant molecular cloud in the constellation of Monoceros. Radiation from young, hot stars has blown gas away from the center of this nebula. Some of this gas has become clumped in dark globules that appear silhouetted against the glowing background gases. (Anglo-Australian Observatory)*

produce an abundance of massive O and B stars, whereas a shock wave from a supernova seems to produce fewer O and B stars and many more of the less massive A, F, G, and K stars?

Additional mechanisms of star birth may await discovery. For example, a simple collision between two interstellar clouds should create new stars. When two such clouds collide, compression must occur at the interface, with vigorous star formation to follow. Another possibility is that light from a star or group of stars may exert strong enough radiation pressure on the surrounding interstellar medium to cause compression, followed by star formation. The Rosette Nebula shown in Figure 13-13 may be an example of this process.

In spite of unanswered questions, it is now clear that star birth involves mechanisms on a colossal scale that we have just begun to appreciate—from the deaths of massive stars to the rotation of the entire galaxy. The study of cold, dark stellar nurseries will certainly be an active and exciting area of astronomical research for many years to come.

When core hydrogen burning ceases, a main-sequence star becomes a red giant

A main-sequence star is a young star whose radiated energy comes from the thermonuclear process of hydrogen burning in its core. A main-sequence star is in thermal equilibrium, with energy liberated in its core balanced by energy radiated from its surface. Eventually, however, all the hydrogen in the core of the star will be used up. **Core hydrogen burning** then must cease, with dramatic effects upon the star's equilibrium, structure, and evolution.

We have seen that a massive protostar quickly builds up the necessary core temperature and pressure for hydrogen burning to commence, so that the most massive protostars are the first to become main-sequence (O and B) stars. Furthermore, the more massive main-sequence stars are the most luminous stars; this rapid emission of energy must correspond to a rapid depletion of hydrogen in the core of the massive stars. Thus, even though a massive O or B star contains much more hydrogen fuel than a less massive main-sequence star, it consumes its hydrogen far more rapidly. The main-sequence lifetime of a massive star thus is considerably shorter than that of a less massive star.

Hydrogen burning has continued in the Sun's core for the past 5 billion years. Initially, the Sun's chemical composition by mass was roughly 75 percent hydrogen and 25 percent helium (plus a smattering of heavy elements). The continued fusion of hydrogen into helium in the Sun's core, however, has dramatically altered the core composition. Indeed, there is now more helium than hydrogen at the Sun's center.

Enough hydrogen remains in the Sun's core for another 5 billion years of core hydrogen burning. That brings the Sun's total lifetime on the main sequence to 10 billion years. Table 13-1 shows how long other stars take to exhaust the supplies of hydrogen in their cores. Note that high-mass stars gobble up their hydrogen fuel in only a few million years, whereas low-mass stars take hundreds of billions of years to accomplish the same thing.

As the supply of hydrogen at a star's center dwindles, the star begins to have difficulty supporting the weight of its outer layers. This enormous weight pressing inward from all sides compresses the star's core slightly. The compressed gases become warmer, allowing hydrogen burning to move outward from the core. In other words, during its final years on the main sequence a star makes a final attempt to maintain thermal equilibrium by enlarging its hydrogen-burning region. There is still plenty of fresh hydrogen surrounding the star's center. By tapping this supply, the star manages to eke out a few million more years on the main sequence.

Finally, all the hydrogen in the core of an aging main-sequence star is used up, and hydrogen burning ceases in the core. Hydrogen burning does continue, however, in a thin spherical shell surrounding the core. This **shell hydrogen burning** initially occurs only in the hot region just outside the core, where the hydrogen fuel has not yet been exhausted.

Because no thermonuclear reactions are now producing energy in the core, the heat flowing out of it is no longer replaced. The core gradually contracts, converting gravitational energy into thermal energy to maintain thermal equilibrium. As the hydrogen-burning shell slowly works its way

Table 13-1 Main-sequence lifetimes

Mass (M$_\odot$)	Surface temperature (K)	Luminosity (L$_\odot$)	Time on main sequence (10^6 years)
25	35,000	80,000	3
15	30,000	10,000	15
3	11,000	60	500
1.5	7,000	5	3,000
1.0	6,000	1	10,000
0.75	5,000	0.5	15,000
0.50	4,000	0.03	200,000

Figure 13-14   The Sun today and as a red giant   In about 5 billion years, when the Sun expands to becomes a red giant, its diameter will increase while its core becomes more compact. Today, the Sun's energy is produced in a hydrogen-burning core whose diameter is about 300,000 km. When the Sun becomes a red giant, it will draw its energy from a hydrogen-burning shell surrounding a compact helium-rich core. This helium core will have a diameter of only 10,000 km.

The Sun as a main-sequence star
(diameter = 1.4×10^6 km $\approx \frac{1}{100}$ AU)

The Sun as a red giant
(diameter $\approx$ 1 AU)

outward from the original core, more helium is added to the core, causing further core contraction and heating.

As the core shrinks and becomes hotter, its energy output rises and the star's outer layers expand, eventually producing a hundredfold increase in the star's diameter. As the star's outer atmosphere expands into space, its gases cool. Soon the temperature of the star's bloated surface has fallen to about 3500 K and its gases glow with a reddish hue. Such stars are appropriately called **red giants.**

In about 5 billion years, the Sun will have finished converting hydrogen into helium in its core. As the Sun's core contracts, its atmosphere will expand to encompass Mercury, then Venus, and possibly reach our own planet. The red-giant Sun will swell to a diameter of about 1 AU, and its surface temperature will decline. As a full-fledged red giant (see Figure 13-14), our star will shine with the brightness of a hundred Suns. While the inner planets are vaporized, the thick atmospheres of the outer planets will boil away to reveal tiny, rocky cores. Thus, in its later years the aging Sun will destroy the planets that have accompanied it since its birth.

Helium burning begins at the center of a red giant

Helium is the "ash" of hydrogen burning. When a star first becomes a red giant, its hydrogen-burning shell surrounds a small, compact core of almost pure helium. In a moderately low-mass red giant (like the Sun 5 billion years from now), the dense helium core is about the same size as the Earth and the star's bloated surface has roughly the same diameter as the Earth's orbit.

At first, no thermonuclear reactions occur in the helium core of a red giant. The hydrogen-burning shell continues to move outward in the star, adding mass to the helium core, which slowly contracts, thereby forcing the star's central temperature to climb. Finally, when the central temperature reaches 100 million K, **core helium burning** is ignited at the star's center. This new thermonuclear reaction occurs in two steps. First, two helium nuclei combine. Then a third helium nucleus is added to this combination, resulting in a carbon nucleus:

$$3\,{}^4\mathrm{He} \rightarrow {}^{12}\mathrm{C} + \gamma$$

with the release of a gamma-ray photon (γ). Some of the carbon created in this process can fuse with an additional helium nucleus to produce oxygen:

$$^{12}\mathrm{C} + {}^4\mathrm{He} \rightarrow {}^{16}\mathrm{O} + \gamma$$

Thus, carbon and oxygen make up the "ash" of helium burning.

The creation of carbon and oxygen by helium burning releases gamma rays. For the first time since leaving the main sequence, the aging star again has a central energy source. Core helium burning establishes thermal equilibrium, thereby preventing any further gravitational contraction of the star's core. A mature red giant burns helium in its core for about 5 to 20 percent of the time it spent burning hydrogen as a main-sequence star. For example, in the distant future the Sun will consume helium in its core for about 1 billion years.

The way in which helium burning begins at a red giant's center depends on the mass of the star. For high-mass stars (those with masses greater than 2 $M_\odot$), helium burning begins gradually as temperatures in the star's core finally reach 100 million degrees. In low-mass (less than 2 $M_\odot$) stars, however, helium burning begins explosively and suddenly, an event called the **helium flash.** The helium flash occurs because of unusual conditions that develop in the core of a low-mass star on its way to becoming a red giant.

In most circumstances, the gas inside a star acts the way most gases do. Usually if gas is compressed, it heats up; if gas expands, it cools down. In a star, this behavior serves as a "safety valve," to ensure that the star does not explode. For example, if energy production overheats the star's core, the core expands, cooling the gases and slowing the rate of thermonuclear reactions. If too little energy is created to support the star's overlying layers, the core becomes compressed and the resulting increase in temperature speeds up the thermonuclear reactions, and hence the energy output.

In a low-mass red giant, the core must undergo considerable gravitational compression to drive temperatures high enough to ignite helium burning. At the extreme pressures and temperatures deep inside the star, atoms are completely torn apart into nuclei and electrons. Under these conditions, a law of quantum mechanics called the **Pauli exclusion principle** becomes very important. This principle, formulated in 1925 by the Austrian physicist Wolfgang Pauli, is analogous to the idea that you can't have two things in the same place at the same time. Further compression of the star's core would be like trying to squeeze one electron inside another.

Just before the onset of helium burning, the electrons in the core of a low-mass star are so closely crowded together that any further compression would violate the Pauli exclusion principle. Because the electrons cannot be squeezed any closer together, they produce a powerful pressure that resists further core contraction. This pressure is largely independent of the temperature and so the usual safety valve does not operate.

This phenomenon, whereby closely packed particles resist compression in accord with the Pauli exclusion principle, is called **degeneracy.** Astronomers say that the helium-rich core of a low-mass red giant is **degenerate** and is supported by **degenerate-electron pressure.**

When the temperature in the core of a low-mass red giant reaches the high level required for helium burning, gamma-ray photons begin to be released by this reaction. This release of energy heats the core, but since degenerate-electron pressure is not affected by temperature, the usual safety valve does not operate and the core does not expand. Instead, the increased temperature causes more helium to burn which heats the core still further. The result is a runaway process called the **helium flash.** Soon, however, the temperature becomes so high that degeneracy is overcome and the star's core expands, thereby terminating the helium flash. These events occur extremely rapidly and the helium flash is over in only a few seconds.

Evolutionary tracks on the H–R diagram reveal the ages of star clusters

The post–main-sequence evolutionary tracks of six mature stars are plotted on the Hertzsprung–Russell diagram shown in Figure 13-15. The **zero-age main sequence** (or ZAMS) is the location on the H–R diagram where stars first begin core hydrogen burning. In subsequent years, the evolutionary tracks slowly inch away from the ZAMS as the hydrogen-burning core grows in search of fresh fuel. The dashed line on Figure 13-15 shows the locations

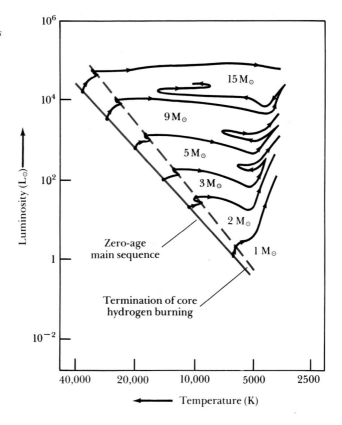

Figure 13-15   Post–main-sequence evolution   The evolutionary tracks of six stars are shown on this H–R diagram. In the high-mass stars, core helium burning ignites where the evolutionary tracks make a sharp turn in the red-giant region of the diagram. The evolutionary tracks for low-mass stars (1 $M_\odot$ and 2 $M_\odot$) are shown only up to the points where the helium flash occurs at their centers. (Adapted from I. Iben)

Figure 13-16 A globular cluster

A globular cluster is a spherical cluster that typically contains a few hundred thousand stars. This particular cluster, called M13, is located in the constellation of Hercules roughly 25,000 light years from Earth. (U.S. Naval Observatory)

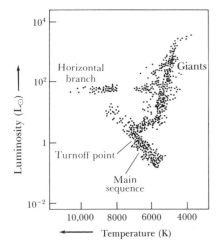

Figure 13-17 An H–R diagram of a globular cluster *Each dot on this graph represents a star in the globular cluster, called M3, whose luminosity and surface temperature have been measured. Note that the upper half of the main sequence is missing. The horizontal branch stars are believed to be low-mass stars that recently experienced the helium flash in their cores. (Adapted from H. L. Johnson and A. R. Sandage)*

of the stellar models when all the core hydrogen has been consumed. After core hydrogen burning ceases, the tracks move rapidly toward the red giant region of the H–R diagram.

Soon after a star becomes a red giant, helium burning begins in its core. The evolutionary track of a high-mass star then wanders back and forth in the red-giant region as the star adjusts to its new energy source. The evolutionary tracks of the two low-mass stars graphed on Figure 13-15 are shown only to the point where the helium flash occurs. Although the helium flash releases a sudden flood of energy, the effects of the outburst are not seen for thousands of years, because it takes that long for photons from the flash to reach the star's surface.

Examples of red giants and post–helium-flash stars are found in old star clusters, called **globular clusters** because of their spherical shape. A typical globular cluster, like the one shown in Figure 13-16, contains up to 1 million stars in a volume less than 100 parsecs across. Astronomers know that such clusters are old because they contain no high-mass main-sequence stars. If you measure the luminosity and surface temperature of many stars in a globular cluster and plot the data on an H–R diagram as shown in Figure 13-17, you find that the upper half of the main sequence is missing. All the high-mass main-sequence stars have evolved long ago into red giants, leaving behind only low-mass, slowly evolving stars still undergoing core hydrogen burning.

The H–R diagram of a globular cluster typically shows a horizontal grouping of stars in the upper-left to upper-center portion of the diagram. As shown in Figure 13-17, these stars form a horizontal row at luminosities in the range of 50 $L_\odot$ to 100 $L_\odot$. Called **horizontal-branch stars,** they are believed to be post–helium-flash low-mass stars. In years to come, these stars will move back toward the red-giant region as both core helium burning and shell hydrogen burning continue to devour fuel.

An H–R diagram of a cluster can be used to determine the age of the cluster. In the diagram for a very young cluster (recall Figures 13-6 and

Figure 13-18 A composite H–R diagram
The blue bands indicate where data from various star clusters fall on the H–R diagram. The age of a cluster can be estimated from the location of the turnoff point where the cluster's most massive stars are just leaving the main sequence for the red-giant region.

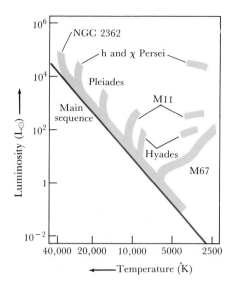

13-7), the entire main sequence is intact. As a cluster gets older, however, stars begin to leave the main sequence. The high-mass, high-luminosity stars are the first to become red giants, and thus the main sequence starts to burn down like a candle. As the years pass, the main sequence gets shorter and shorter. The top of the surviving portion of the main sequence is called the **turnoff point.** Stars at the turnoff point are just now exhausting the hydrogen in their cores, and their main-sequence lifetime (recall Table 13-1) is equal to the age of the cluster. For example, in the case of the cluster M3 (see Figure 13-17), 0.8 $M_\odot$ stars have just left the main sequence, and so the cluster's age is roughly 15 billion years.

Data for several star clusters are plotted on Figure 13-18. All the young clusters (those with their main sequences still intact) are open clusters in the disk of our Galaxy, where star formation is an ongoing process. Stars in these young clusters are said to be **metal-rich** because their spectra contain many prominent spectral lines of heavy elements. This material came from dead stars that long ago exploded, enriching the interstellar gases with the heavy elements formed in their cores. The Sun is a young, metal-rich star.

Most of the oldest clusters are globular clusters. Globular clusters are generally located outside the disk of our galaxy. Their spectra show only weak lines of heavy elements. These ancient stars are therefore said to be **metal-poor.** They were created long ago from interstellar gases that had not yet been substantially enriched with heavy elements. Spectra of a metal-poor star and of the Sun are compared in Figure 13-19.

Figure 13-19 Spectra of a metal-poor and a metal-rich star *These spectra compare (a) a metal-poor star and (b) a metal-rich star (the Sun). Numerous spectral lines prominent in the solar spectrum are caused by elements heavier than hydrogen and helium. Note that corresponding lines in the metal-poor star's spectrum are weak or absent. Both spectra cover a wavelength range that includes H_γ and H_δ. (Lick Observatory)*

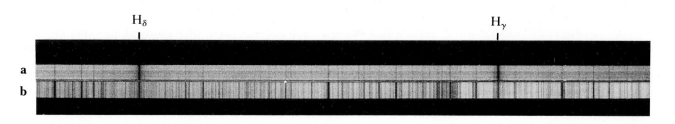

Red giants typically show mass loss

Red-giant stars are so enormous that their bloated outer layers constantly leak gases into space. At times, this **mass loss** is quite significant (see Figure 13-20).

Mass loss can be detected spectroscopically. Escaping gases coming toward us exhibit narrow absorption lines that are slightly blueshifted. According to the Doppler effect, this small shift toward shorter wavelengths corresponds to a speed of 10 km/sec. This value is typical of the expansion velocities with which gases leave the tenuous outer layers of red giants. A typical mass-loss rate for a red giant is roughly 10^{-7} solar masses per year.

Figure 13-20 [right] A mass-loss star Old stars become red giants whose bloated outer atmospheres shed matter into space. This star is losing matter at a high rate and is surrounded by a reflection nebula caused by starlight reflected from dust grains. These dust grains may have condensed from material shed by the star. A typical red giant can lose 10^{-7} solar masses per year, and many are surrounded by circumstellar shells of matter they have shed. (Anglo-Australian Observatory)

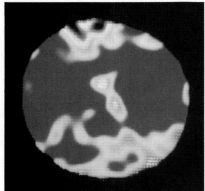

Figure 13-21 Betelgeuse Betelgeuse is a bright red giant in the constellation of Orion (see Figure 4-23). It is one of the largest stars, with a diameter about 800 times the Sun's. This image, obtained by a technique called speckle interferometry, shows temperature differences on the surface of Betelgeuse. The bluish regions are cool, the red areas are warmest. The two blue spots near the center of the image are each about the same size as the Earth's orbit. (NOAO)

Betelgeuse in the constellation of Orion is a good example of a red supergiant experiencing mass loss. Betelgeuse is 470 light years away and has a diameter roughly equal to the diameter of Mars's orbit (see Figure 13-21). Recent spectroscopic observations show that this star is losing mass at the rate of 1.7×10^{-7} solar masses per year and is surrounded by a huge **circumstellar shell** that is expanding at 10 km/sec. These escaping gases have been detected out to distances of 10,000 AU from the star. Consequently, the expanding circumstellar shell has an overall diameter of $\frac{1}{3}$ light year.

Supergiant stars (that is, stars brighter than 10^5 Suns) sustain mass loss throughout most of their existence and have mass-loss rates comparable to those of red giants. In the next chapter, we will see that dying stars also eject vast quantities of material into space. Nevertheless, mass loss from supergiants and red giants accounts for roughly one-fifth of all the matter returned by stars to the interstellar medium.

Many mature stars pulsate

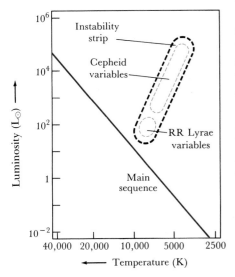

Figure 13-22 The instability strip
The instability strip occupies a region between the main sequence and the red-giant branch on the H–R diagram. A star passing through this region along its evolutionary track becomes unstable and pulsates.

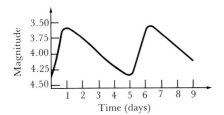

Figure 13-23 The light curve of a Cepheid variable *This graph shows the light curve of δ Cephei, which is the prototype of an important class of variable stars. All Cepheid variables exhibit the same type of periodic changes in luminosity as shown here: rapid brightening followed by gradual dimming.*

After core helium burning begins, the evolutionary tracks of mature stars crisscross the middle of the H–R diagram, as shown in Figure 13-15. Post–helium-flash, low-mass stars on the horizontal branch also cross the middle of the H–R diagram as they return to the red-giant region.

During these excursions across the H–R diagram, a star can become unstable and pulsate. In fact, there is a region on the H–R diagram between the main sequence and the red-giant branch that is called the **instability strip** (see Figure 13-22). When a star passes through this region on its evolutionary track, the star pulsates. As it pulsates, its brightness varies periodically.

Low-mass, post–helium-flash stars pass through the lower end of the instability strip as they move in the horizontal branch along their evolutionary tracks. These stars become **RR Lyrae variables,** named after the prototype in the constellation of Lyra. RR Lyrae variables all have periods shorter than one day and all have roughly the same average brightness as stars on the horizontal branch. High-mass stars pass back and forth through the upper end of the instability strip on the H–R diagram. These stars become **Cepheid variables.**

A Cepheid variable is recognizable by the characteristic way in which its light output varies: rapid brightening followed by gradual dimming. This behavior is most easily displayed in a light curve, which is a graph of a star's brightness plotted against time. The light curve of δ Cephei is shown in Figure 13-23. This star was the first pulsating variable to be discovered and serves as the prototype for this important class of stars, often simply called Cepheids.

A Cepheid variable brightens and fades because of a cyclic expansion and contraction of the whole star. This behavior is deducible from spectroscopic observations. Spectral lines in the spectrum of δ Cephei shift back and forth with the same 5.4-day period as that of the magnitude variations. According to the Doppler effect, these shifts mean that the star's surface is alternately approaching and receding from us.

When a Cepheid variable pulsates, the star's surface oscillates up and down like a spring. During these cyclical expansions and contractions, the star's gases alternately heat up and cool down. Thus the characteristic light curve of a Cepheid variable, like that shown in Figure 13-23, results from both changing size and changing surface temperature.

Just as a bouncing ball eventually comes to rest, a pulsating star would soon stop pulsating without some sort of mechanism to keep its oscillations going. In 1941, the British astronomer Arthur Eddington explained that a Cepheid variable feeds energy into its pulsations by a valvelike action involving the periodic ionization and deionization of gas in the star's outer layers. When the star is at its greatest diameter, its surface temperature is at a minimum, and so ions and electrons can combine to form neutral atoms of

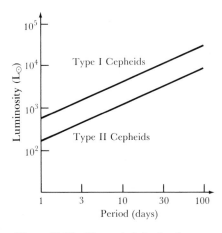

Figure 13-24 *The period–luminosity relation* *The period of a Cepheid variable is directly related to its average luminosity. Metal-rich (Type I) Cepheids are brighter than the metal-poor (Type II) Cepheids.*

hydrogen and helium. These electrically neutral gases allow the star's energy to radiate freely into space, causing the gas pressure to decline and so the star's outer layers start to fall inward. As the gases in the shrinking star become compressed, they absorb energy from the star's interior and become reionized. This absorbtion of energy heats the gases and raises their pressure, pushing the star's surface back out to restart the cycle.

There is a direct relationship between a Cepheid's period and its average luminosity. Dim Cepheid variables pulsate rapidly, have periods of 1 to 2 days, and have an average brightness of a few hundred Suns. The most luminous Cepheids are the slowest variables, with periods of 100 days and average brightnesses equal to 10,000 Suns. This connection between period and brightness is called the **period–luminosity relation.** As we shall see in Chapter 17, this relationship played an important role in determining the overall size and structure of the universe.

The abundance of heavy elements in a Cepheid's atmosphere affects the star's luminosity. The average luminosity of metal-rich Cepheids is roughly four times greater than the average luminosity of metal-poor Cepheids having the same period. Thus there are two classes: **Type I Cepheids,** which are the brighter, metal-rich stars, and **Type II Cepheids,** which are the dimmer, metal-poor stars. The period–luminosity relation for both types of variables is shown in Figure 13-24.

The size of stellar pulsations can in rare cases be quite substantial. Sometimes the expansion velocity is so high that the star's outer layers are ejected completely. As we shall see in the next chapter, significant mass ejection accompanies the death of stars in a sometimes violent process that renews and enriches the interstellar medium for future generations of stars.

Summary

- Enormous cold clouds of gas, called giant molecular clouds, are scattered about our Galaxy. Star formation begins when gravitational attraction causes a protostar to coalesce within a giant molecular cloud.

- As a protostar contracts, its gases begin to glow. When its core temperature becomes high enough to begin hydrogen burning, the protostar becomes a main-sequence star.

  The most massive protostars are the first to become main-sequence stars (O and B stars). They emit strong ultraviolet radiation that ionizes hydrogen in the surrounding cloud, creating reddish emission nebulae called H II regions.

  Ultraviolet radiation and stellar winds from the OB association at the core of an H II region create shock waves that compress the gas cloud, triggering formation of more protostars. Supernova explosions also compress gas clouds and trigger star formation.

  In the final stages of pre–main-sequence contraction, when hydrogen burning is about to begin in the core of a protostar, the star may undergo vigorous chromospheric activity that ejects large amounts of matter into space. Such gas-ejecting stars are called T Tauri stars.

- A collection of newborn stars is called an open cluster, or galactic cluster. Occasionally, a rapidly moving star will escape, or "evaporate," from such a cluster.

- The more massive a star is, the shorter is its main-sequence lifetime. The Sun has been a main-sequence star for about 5 billion years old should remain so for about another 5 billion years. Less massive stars evolve more slowly and have longer lifetimes.

- Core hydrogen burning ceases when hydrogen is exhausted in the core of a main-sequence star, leaving a core of nearly pure helium surrounded by a shell where hydrogen burning continues.

- Shell hydrogen burning adds more helium to the star's core, which contracts and becomes hotter. The outer atmosphere expands considerably and the star becomes a red giant.

- When the central temperature of a red giant reaches about 100 million K, the thermonuclear process of helium burning begins there. This process converts helium to carbon and oxygen.

    In a massive red giant, helium burning begins gradually. In a less massive red giant, it begins suddenly in a process called a helium flash.

- The age of a stellar cluster can be estimated by plotting its stars on an H–R diagram. The upper portion of the main sequence will be missing because more massive main-sequence stars have become red giants.

    Relatively young stars are metal-rich; ancient stars are metal-poor.

- Red giants undergo extensive mass loss, sometimes producing circumstellar shells of ejected material around the stars.

- When a star's evolutionary track carries it through a region called the instability strip in the H–R diagram, the star becomes unstable and begins to pulsate.

    Cepheid variables are high-mass pulsating variables exhibiting a regular relationship between period of pulsation and luminosity.

    RR Lyrae variables are low-mass pulsating variables with short periods.

Review questions

1   Why are low temperatures necessary for protostars to form inside dark nebulae?

2   Explain why thermonuclear reactions occur only at the center of a main sequence star and never on its surface.

3   What is an "evolutionary track" and in what way can it help us interpret the H–R diagram?

4   Why do you suppose the vast majority of the stars we see in the sky are main-sequence stars?

5   Draw the pre–main-sequence evolutionary track of the Sun on an H–R diagram. Briefly describe what was probably occurring throughout the solar system at various stages along this track.

6   On what grounds are astronomers able to say that the Sun has about 5 billion years remaining in its main-sequence stage?

7   Draw the post–main-sequence evolutionary track of the Sun on an H–R diagram, up to the point when the Sun becomes a red giant. Briefly describe what might occur throughout the solar system as the Sun undergoes this transition.

8 What does it mean when an astronomer says that a star "moves" from one place to another on an H–R diagram?

9 What are Cepheid variables and how are they related to the instability strip?

Advanced questions

10 If you took a spectrum of a reflection nebula, would you see absorption lines, emission lines, or no lines? Explain your answer and explain how the spectrum demonstrates that the light is reflected from nearby stars.

11 Speculate on why a shock wave from a supernova seems to produce relatively few high-mass O and B stars compared to the lower-mass A, F, G, and K stars.

12 How would you distinguish a newly formed protostar from a red giant, in view of their identical location on the H–R diagram?

Discussion questions

13 What do you think would happen if the solar system passed through a giant molecular cloud? Do you think that the Earth has ever passed through such clouds?

14 Speculate about the possibility of life forms and biological processes occurring in giant molecular clouds. In what ways might conditions in giant molecular clouds favor or hinder biological evolution?

For further reading

Blitz, L. "Giant Molecular Cloud Complexes in the Galaxy." *Scientific American*, April 1982.

Bok, B. "Early Phases of Star Formation." *Sky & Telescope*, April 1981, p. 284.

Cohen, M. "Stellar Formation." *Astronomy*, September 1979, p. 66.

Johnson, B. "Red Giant Stars." *Astronomy*, December 1976, p. 26.

Kaler, J. "Journeys on the H–R Diagram." *Sky & Telescope*, May 1988, p. 482.

Kaufmann, W. *Stars and Nebulas*. W. H. Freeman and Company, 1979.

Lada, C. "Energetic Outflows from Young Stars." *Scientific American*, July 1982.

Robinson, L. "Orion's Stellar Nursery." *Sky & Telescope*, November 1982, p. 430.

Wyckoff, S. "Red Giants: The Inside Scoop." *Mercury*, January/February 1979, p. 7.

Zeilik, M. "The Birth of Massive Stars." *Scientific American*, April 1978.

The deaths
of stars

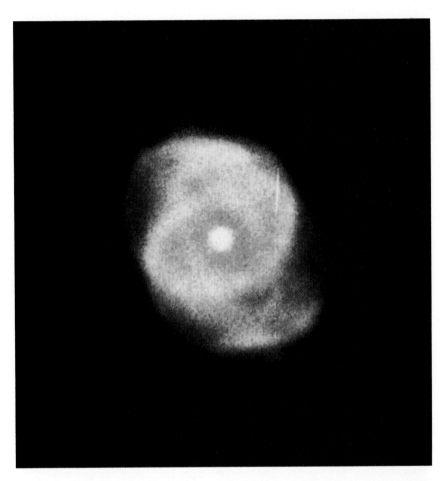

*A **planetary nebula*** *Dying stars often eject their outer layers. A low-mass star can lose half its mass in a comparatively gentle process that produces a planetary nebula. The exposed stellar core typically has a surface temperature of about 100,000 K and is roughly one-tenth the size of the Sun. Ultraviolet radiation from the hot stellar core causes the surrounding gases to glow. The greenish color of this planetary nebula comes from oxygen ions. The central star in this nebula has exhausted all its nuclear fuel and is contracting to become a white dwarf. (Lick Observatory)*

Having experienced old age as a red giant, a star approaches the end of its life. In this chapter, we learn that stars eject significant amounts of matter into space as they die. We find that low-mass stars eject their outer layers relatively gently, producing planetary nebulae, whereas high-mass stars explode violently as supernovae. In 1987, an historic supernova visible to the naked eye gave astronomers an unprecedented opportunity to study the explosive death of a massive star. The core of a dead low-mass star contracts to become a white dwarf, whereas the corpse of a high-mass star can become a neutron star. In examining white dwarfs and neutron stars, we discover that there are upper limits to the masses that these stellar corpses can have. Finally, we discuss the various fascinating forms in which neutron stars have been observed, such as pulsars, pulsating X-ray sources, and bursters.

From infancy through adulthood, a star leads a fairly placid life burning hydrogen in its core. Upon approaching old age, however, the star takes on a contradictory character: the star's core contracts and becomes highly compressed while its atmosphere expands and becomes very extended. As a red giant or supergiant, the aging star exhibits erratic and fitful behavior as it devours the remaining nuclear fuels.

Low-mass stars die by gently ejecting their outer layers, creating planetary nebulae

As we saw in the last chapter, carbon and oxygen are the "ashes" of helium burning. After the helium flash in a low-mass red giant, substantial amounts of these two elements begin to accumulate at the star's center as a result of core helium burning. Eventually, all the helium at the center of a low-mass star is used up and core helium burning ceases. By this time, thermonuclear reactions have also ceased in the hydrogen-burning shell. As a result, the star's core begins to contract. This compression heats the helium-rich gases surrounding the carbon–oxygen core, igniting **shell helium burning** around the core. The star's internal structure now consists of a thin helium-burning shell inside the former hydrogen-burning shell, all within a volume roughly the size of the Earth (see Figure 14-1).

When shell hydrogen burning first began inside this star, the outpouring of energy caused it to expand and become a red giant. Astronomers say that the star ascended the red-giant branch on the H–R diagram. Then came the helium flash, and the star shifted over to the horizontal branch. With shell helium burning, a renewed outpouring of energy causes the star to expand again. It ascends the red-giant branch for a second and final time to become a **red supergiant.** Such stars typically have diameters as big as the orbit of Mars and shine with the brightness of 10,000 Suns. A star with this furious rate of energy loss cannot live much longer.

The star's impending death is signaled by instabilities that develop in its helium-burning shell. As with the helium flash discussed in the previous chapter, there is again a sudden outpouring of energy, but the details are quite different. The helium flash occurred because the star's core was degenerate. However, the helium-burning shell is not compressed to a density high enough to be degenerate. Instead, a **helium-shell flash** occurs because

Figure 14-1 The structure of an old low-mass star Near the end of its life, a low-mass star becomes a red supergiant, with a diameter almost as large as the diameter of the orbit of Mars. The star's dormant hydrogen-burning shell and active helium-burning shell are contained within a volume roughly the size of the Earth.

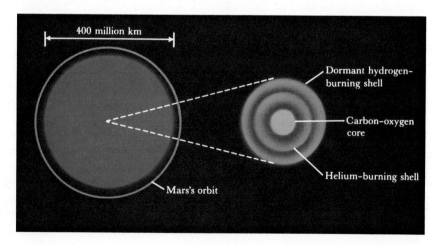

400 million km

Dormant hydrogen-burning shell

Carbon-oxygen core

Helium-burning shell

Mars's orbit

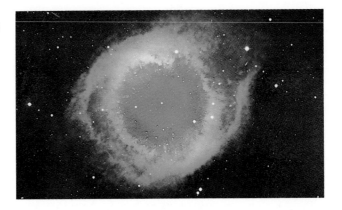

Figure 14-2   The planetary nebula NGC 7293   *This beautiful object, often called the Helix Nebula, covers an area of the sky equal to half the full moon. The star that ejected these gases is seen at the center of the glowing shell. The greenish color comes from oxygen ions, while the pink and red comes from nitrogen and hydrogen ions. This nebula is in the constellation of Aquarius, about 400 light years from Earth. (Anglo-Australian Observatory)*

the shell is thin. The flash ends only after vigorous helium burning causes the shell to become thick enough to relieve the pressure of the star's outer layers.

During the flash, the helium-shell's energy output jumps from 100 Suns to roughly 100,000 Suns in a rapid series of brief bursts called **thermal pulses.** These bursts are separated by relatively quiet intervals lasting about 300,000 years.

During one of these periods of thermal oscillation, the dying star's outer layers can separate completely from the carbon–oxygen core. As the ejected material expands into space, dust grains condense out of the cooling gases. Radiation pressure from the star's hot, burned-out core acts on the specks of dust to continue propelling them outward, and the star sheds its outer layers altogether. A star can lose more than one-half of its mass in this fashion.

As a dying star ejects its outer layers, its hot core is exposed, emitting ultraviolet radiation intense enough to ionize the expanding shell of ejected gases. This ionization causes the gases to glow, producing a so-called **planetary nebula.** Planetary nebulae have nothing to do with planets. This unfortunate term was introduced in the eighteenth century when these glowing objects were thought to look like distant planets when viewed through small telescopes.

Many planetary nebulae, such as the one shown in Figure 14-2, have a distinctly spherical appearance arising from the symmetrical way in which the gases were ejected. In other cases, as in Figure 14-3, the rate of expan-

Figure 14-3   The planetary nebula NGC 6302   *At the end of their lives, low-mass stars shed much of their mass. Although most cases involve a fairly symmetrical ejection of gases, there are many examples in which the gas has expanded unevenly. An irregular planetary nebula such as NGC 6302 in Scorpius (shown here) is the result. Spectroscopic observations of NGC 6302 indicate that the gases are moving toward us at 400 km/sec, indicating a particularly violent initial ejection. (Anglo-Australian Observatory)*

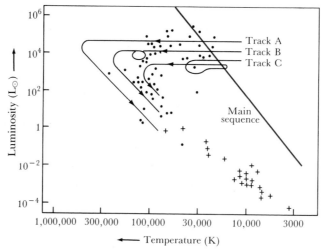

Figure 14-4 *Evolution from red supergiants to white dwarfs*
The evolutionary tracks of three low-mass red supergiants are shown as they eject planetary nebulae. The table gives the extent of mass loss in each of the three cases. The dots on this graph represent the central stars of planetary nebulae whose surface temperatures and luminosities have been determined. The crosses are white dwarfs for which similar data exist. (Adapted from B. Paczynski)

Evolutionary track on Figure 14-4	Supergiant mass ($M_\odot$)	Mass of ejected nebula ($M_\odot$)	White dwarf mass ($M_\odot$)
Track A	3.0	1.8	1.2
Track B	1.5	0.7	0.8
Track C	0.8	0.2	0.6

sion is not the same in all directions, and the resulting nebula takes on an hourglass or dumbbell appearance.

Planetary nebulae are quite common: astronomers estimate there to be 20,000 to 50,000 in our Galaxy alone. Spectroscopic observations of these nebulae show bright emission lines of ionized hydrogen, oxygen, and nitrogen. From the Doppler shifts of these lines, astronomers conclude that the expanding shell of gas is moving outward from the dying star with speeds of 10 to 30 km/sec. A typical planetary nebula has a diameter of roughly 1 light year, which means that it must have begun expanding about 10,000 years ago.

By astronomical standards, a planetary nebula is a very short-lived entity. After about 50,000 years, the nebula has spread over distances so far from the cooling central star that its nebulosity simply fades from view. The gases then mingle and mix with the surrounding interstellar medium. Astronomers estimate that all the planetary nebulae in the Galaxy return a total of 5 $M_\odot$ to the interstellar medium each year. This amounts to about 15 percent of all matter expelled by all sorts of stars each year. This contribution is so significant that planetary nebulae are thought to play an important role in the evolution of the Galaxy as a whole.

The burned-out core of a low-mass star cools and contracts to become a white dwarf

Stars less massive than about 3 $M_\odot$ never develop the central pressures or temperatures necessary for thermonuclear reactions using carbon or oxygen as fuel. Instead, mass ejection exposes the carbon–oxygen core, which simply cools off.

The evolutionary tracks of three burned-out stellar cores are shown in Figure 14-4. The initial red supergiants had masses between 0.8 $M_\odot$ and 3.0 $M_\odot$. During their final spasms, these dying stars eject between 25 and 60 percent of their matter. During the ejection phase, the outward appearance of these stars changes rapidly and they race along their evolutionary tracks across the H–R diagram, sometimes executing loops corresponding to thermal pulses. Finally, as the ejected nebulae fade and the stellar cores cool, their evolutionary tracks take a sharp turn downward toward the white-dwarf region of the diagram.

There is no possibility of igniting additional nuclear fuels inside one of these dead stars, and so the crushing weight of gases pressing inward from all sides severely compresses the stellar corpse. The density inside the dead star skyrockets until the electrons are so closely packed that they become degenerate. The degenerate-electron pressure is strong enough to support the star, and so the gravitational contraction halts and the star is roughly the same size as the Earth. Such a star is called a **white dwarf.**

The density of matter in one of these Earth-sized stellar corpses is typically 10^6 g/cm^3. In other words, a teaspoonful of white-dwarf matter brought to Earth would weigh as much as a truck.

Calculations by Subrahmanya Chandrasekhar at the University of Chicago prove that there is an upper limit to the mass that a white dwarf can have. This maximum mass, called the **Chandrasekhar limit,** is equal to 1.4 $M_\odot$. Above this limit, degenerate-electron pressure cannot support the weight of the star's matter pressing inward from all sides. Thus all white dwarfs must have masses less than 1.4 $M_\odot$.

Several hundred white dwarfs are scattered across the sky. All are too faint to be seen with the naked eye. One of the first white dwarfs to be discovered is a companion to the bright star Sirius. The binary nature of Sirius was first deduced in 1844 by the German astronomer Friedrich Bessel, who noticed that the star was moving back and forth slightly, as if orbited by an unseen object. This companion, called Sirius B (see Figure 14-5), was first glimpsed in 1862. Recent satellite observations at ultraviolet wavelengths—where white dwarfs emit most of their light—demonstrate that the surface temperature of Sirius B is about 30,000 K.

As a white dwarf cools, both its luminosity and its surface temperature decline. As billions of years pass, white dwarfs get dimmer and dimmer as their surface temperatures drop toward absolute zero. This condition will be the final fate of our Sun: a cold, dark, dense sphere of degenerate gases rich in oxygen and carbon, about the size of the Earth.

Figure 14-5 Sirius and its white-dwarf companion *Sirius, the brightest-appearing star in the sky, is actually a double star. The secondary star is a white dwarf, seen here at the "five-o'clock" position, in the glare of Sirius. The spikes and rays around Sirius are created by optical effects within the telescope. (Courtesy of R. B. Minton)*

High-mass stars die violently by blowing themselves apart in supernova explosions

High-mass stars end their lives very differently than do low-mass stars. A high-mass star is capable of igniting a host of additional thermonuclear reactions in its core. The more massive a star is, the greater the temperatures that can be achieved in its core by gravitational compression. High temperatures are required to initiate nuclear reactions involving heavy nuclei because the large electric charges of these nuclei exert strong forces tending to keep the nuclei apart. Only at the great speeds associated with high temperatures are the nuclei traveling fast enough to penetrate each other's repulsive electric fields and fuse together.

Recall that carbon and oxygen are the "ashes" of helium burning. When helium burning is finished at the core of a star, gravitational compression drives the star's central temperature up to 600 million K, and **carbon burning** begins. This thermonuclear process produces elements such as neon and magnesium.

If a star is massive enough to drive its central temperature to 1 billion K, **neon burning** begins. This process uses up the neon accumulated from carbon burning, further increasing the concentrations of oxygen and magnesium in the star's core.

If the central temperature of the star reaches about 1.5 billion K, **oxygen burning** begins. The principal product of oxygen burning is sulfur. As the star consumes increasingly heavier nuclei, thermonuclear reactions produce many different elements. For instance, oxygen burning also produces isotopes of silicon, phosphorus, and more magnesium.

When a given nuclear fuel is exhausted in the core of a massive star, gravitational contraction to ever higher densities drives up the star's central temperature, thereby igniting the "ash" of the previous burning stage. Each successive thermonuclear reaction occurs with increasing rapidity. For example, detailed calculations for a 25 $M_\odot$ star demonstrate that carbon burning occurs for 600 years, neon burning for 1 year, but oxygen burning for only 6 months.

After half a year of core oxygen burning in a 25 $M_\odot$ star, gravitational compression forces the central temperature up toward 3 billion K, when **silicon burning** begins. This thermonuclear process proceeds so furiously that the entire core supply of silicon in a 25 $M_\odot$ star is used up in one day.

Silicon burning involves many hundreds of nuclear reactions. The major final product of this process is iron. Iron cannot fuel any further thermonuclear reactions, and so the sequence of burning stages ends with iron. In order for an element to be a thermonuclear fuel, energy must be released when its nuclei collide and fuse. This energy comes from packing the neutrons and protons together more tightly in the ash nuclei than in the fuel nuclei. The protons and neutrons inside the iron nuclei are already so tightly bound together that no further energy can be extracted by fusing still more nuclei with iron.

The buildup of an inert, iron-rich core signals the impending violent death of a massive star. Surrounding this iron core, successive layers of shell burning consume the star's remaining reserves of fuel (see Figure 14-6). The entire energy-producing region of the star is contained in a volume as big as the Earth, whereas the star's enormously bloated atmosphere is nearly as big as the orbit of Jupiter.

Because iron does not "burn," the electrons in the core must now support the star's outer layers by the brute strength of degeneracy pressure alone. Soon, however, the continued deposition of fresh iron from the silicon-burning shell causes the core's mass to exceed the Chandrasekhar limit. Electron degeneracy suddenly becomes unable to support the star's enormous weight, and the core begins to collapse.

Any star with a mass greater than 10 $M_\odot$ is capable of developing an iron core that at some stage will exceed the Chandrasekhar limit, triggering a rapid series of cataclysms that will tear the star apart in a few seconds. Let us see how this happens in the death of a 25 $M_\odot$ star.

First, degenerate-electron pressure fails when the density inside the iron core reaches 1 billion g/cm^3. The core immediately begins to collapse. Cen-

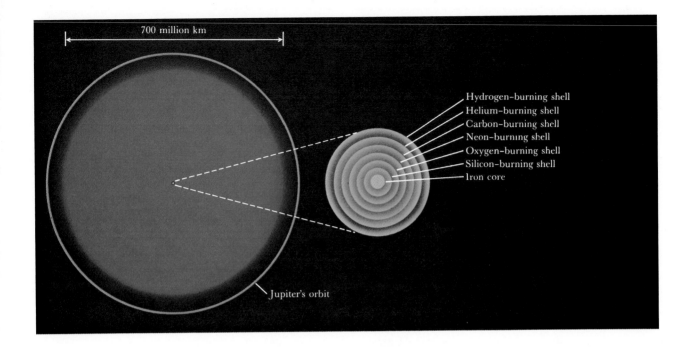

Near the end of its life, a high-mass star becomes a red supergiant almost as big as the orbit of Jupiter. The star's energy comes from six concentric burning shells, all contained within a volume roughly the same size as the Earth.

Labels on figure:
700 million km
Hydrogen-burning shell
Helium-burning shell
Carbon-burning shell
Neon-burning shell
Oxygen-burning shell
Silicon-burning shell
Iron core
Jupiter's orbit

Figure 14-6 **The structure of an old high-mass star** *Near the end of its life, a high-mass star becomes a red supergiant almost as big as the orbit of Jupiter. The star's energy comes from six concentric burning shells, all contained within a volume roughly the same size as the Earth.*

tral temperatures promptly soar to almost inconceivable heights. In roughly a tenth of a second, the temperature exceeds 5 billion K. Photons associated with this intense heat have so much energy that they begin to break up the iron nuclei in a process called **photodisintegration.**

Within another tenth of a second, as densities continue to climb, the electrons are forced to combine with protons to produce neutrons in a process called **neutronization.** About $\frac{1}{4}$ second after the collapse begins, the density in the core reaches 4×10^{14} g/cm^2, equaling the density with which neutrons and protons are packed together inside nuclei.

Matter at nuclear density is virtually incompressible. Thus, when the neutron-rich material of the core reaches this density, it suddenly becomes quite stiff, and the core collapse comes to an abrupt halt.

At this critical stage, the star's unsupported inner regions are plunging inward at speeds of 10 to 15 percent of the speed of light. As this material crashes onto the now-rigid core, enormous temperatures and pressures develop, causing the falling material to bounce. In a fraction of a second, a wave of matter begins to move back out toward the star's surface. This wave accelerates rapidly as it encounters less and less resistance, and soon it becomes a shock wave. After a few days, this shock wave reaches the star's surface, by which time the star's outer layers have begun to lift away from the core. A cataclysm of this magnitude is called a **supernova explosion.**

This particular description of the death of a 25 $M_\odot$ star is based on detailed computer calculations of an evolving stellar model. The key stages in such a star's evolution are summarized in Table 14-1.

The final stages in the evolution of other massive stars probably follow similar scenarios, although the details may be different. This particular 25 $M_\odot$ star ejects 24 $M_\odot$, leaving behind a 1 $M_\odot$ corpse called a **neutron star.** Under slightly different conditions, a massive star might blow itself completely apart, leaving no corpse at all. For example, the bounce and subsequent shock wave might develop at the center of the star rather than at the

Table 14-1 Evolutionary stages of a 25 M$_\odot$ star

Stage	Temperature (K)	Density (g/cm^3)	Duration of stage
Hydrogen burning	4×10^7	5	7×10^6 years
Helium burning	2×10^8	700	5×10^5 years
Carbon burning	6×10^8	2×10^5	600 years
Neon burning	1.2×10^9	4×10^6	1 year
Oxygen burning	1.5×10^9	10^7	6 months
Silicon burning	2.7×10^9	3×10^7	1 day
Core collapse	5.4×10^9	3×10^9	$\frac{1}{4}$ second
Core bounce	2.3×10^{10}	4×10^{14}	Milliseconds
Explosive	About 10^9	Varies	10 seconds

surface of the neutronized core. Alternatively, with a different set of initial conditions inside the star, a massive burned-out core might gravitationally collapse to form a **black hole.** This exotic object is discussed in the next chapter.

A nearby supernova in 1987 gave us a closeup look at the death of a massive star

As the outer layers of a massive dying star are blasted into space, the star's luminosity suddenly increases by a factor of 10^8, equivalent to a jump of 20 magnitudes in brightness. An outburst of this enormity is called a **supernova.** For a few days following the explosion, a supernova can shine as brightly as an entire galaxy.

On February 23, 1987, a supernova was discovered in the Large Magellanic Cloud, which is a companion galaxy to our own Milky Way. The supernova (designated SN 1987A because it was the first to be observed that year) occurred near an enormous H II region called 30 Doradus or the Tarantula Nebula because of its spiderlike appearance (see Figure 14-7). The supernova was so bright that it could be seen with the naked eye.

A bright, nearby supernova is a rare event. In 1885, a supernova in the Andromeda Galaxy was just barely visible to the unaided eye. We have to go back to 1604 to find another supernova bright enough to have been seen without a telescope. SN 1987A gave astronomers the unique opportunity to study the death of a massive, nearby star using modern equipment.

SN 1987A was unusual because it did not promptly rise to its expected maximum brightness. Instead, it stopped at only a tenth of the luminosity typical of an exploding massive star like the one described in Table 14-1. During the next 85 days following the outburst, SN 1987A brightened gradually and then settled into a slow decline characteristic of an ordinary supernova.

Fortunately, the doomed star had been observed prior to becoming a supernova, and these observations helped explain why SN 1987A was not altogether typical. The Large Magallenic Cloud is near enough to us that many of its stars have been individually observed and catalogued. The

Figure 14-7   The supernova SN 1987A
In 1987, a supernova exploded in a nearby galaxy called the Large Magellanic Cloud (LMC), about 160,000 light years from Earth. This photograph shows a portion of the LMC that includes the supernova and a huge H II region called the Tarantula Nebula. The supernova reached third magnitude at maximum brightness and could easily be seen without a telescope by observers at southern latitudes. (European Southern Observatory)

Figure 14-8   The doomed star and SN 1987A *The upper photograph shows a small section of the Large Magellanic Cloud before the outburst, with the doomed star—a blue supergiant—identified by an arrow. The lower view shows the supernova a few days after the explosion. (Anglo-Australian Observatory)*

doomed star was identified as a B3I supergiant (see Figure 14-8). When this star was on the main sequence, its mass was about 20 $M_\odot$, although by the time it exploded it probably had shed a few solar masses.

SN 1987A was initially less luminous than expected because the doomed star was a blue supergiant when it exploded rather than a red supergiant. The evolutionary track for an aging 20 $M_\odot$ star wanders back and forth across the top of the H–R diagram, and so the star alternates between being a hot (blue) supergiant and cool (red) supergiant . The star's size changes significantly as it undergoes these changes in its surface temperature. A blue supergiant is only 10 times larger in diameter than the Sun, but a red supergiant of the same luminosity would be 1000 times larger in diameter than the Sun (recall Figure 12-11). Because more energy was used to expand the doomed blue supergiant than a more usual red supergiant, SN 1987A reached only a tenth of its expected brightness.

Stellar model calculations suggest that the interior of the doomed star contained 6 $M_\odot$ of helium and 2 $M_\odot$ of heavier elements. At the moment of detonation, the star's iron core had a mass of 1.5 $M_\odot$ and a temperature of 10 billion K. About 0.15 $M_\odot$ of radioactive isotopes were also created during the explosion. Energy released by the decay of these isotopes contributed significantly to the supernova's brightness as it began to fade.

Most of the energy of a supernova explosion is carried away by subatomic particles called **neutrinos,** which are produced during the collapse of the star's core. Neutrinos have no electric charge and are generally believed to have no mass either, which means they resemble photons and travel at the speed of light. A neutrino is created every time an electron combines with a proton to produce a neutron in the collapsing core of a supernova (see Figure 14-9).

Neutrinos are very difficult to detect because they seldom interact with ordinary matter. In fact, neutrinos easily pass through the Earth as if it were not there. Scientists have built neutrino detectors consisting of large tanks of water. Water (H_2O) contains many protons, the nuclei of hydrogen atoms. When a high energy neutrino strikes a proton, it produces an electron that emits a flash of light called **Cerenkov radiation.** This type of radiation, first

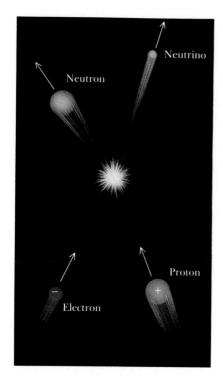

observed by the Russian physicist Pavel A. Cerenkov, occurs whenever a particle moves through water faster than light can. (Note that light is slowed considerably as passes through water, and high-energy particles can exceed this reduced speed without violating the principle that nothing can go faster than the speed of light in a vacuum.) Scientists try to detect neutrinos by observing Cerenkov flashes with light-sensitive photomultipliers mounted in the water (see Figure 14-10).

The day before SN 1987A was observed in the sky, teams of scientists at neutrino detectors in Japan and the United States excitedly reported Cerenkov flashes from a burst of neutrinos. The Kamiokande II detector in Japan detected 12 neutrinos at about the same time that 8 were found by the IMB (Irvine-Michigan-Brookhaven) detector in a salt mine under Lake Erie. Neutrinos preceded the visible outburst because they escaped from the dying star before the shock wave from the collapsing core reached the star's surface. They were detected in the northern hemisphere, where the supernova is always below the horizon, after having passed through the Earth.

The ability to detect neutrinos from a supernova offers a new and exciting opportunity for astronomers. Ordinary telescopic observations of a supernova explosion can show us only the expanding outer layers of the dying star. Even X-ray or radio observations fail to see through the hot gases being blasted into space. Thus, ordinary techniques do not allow us to observe the extraordinary events occurring in and around the doomed star's core. However, neutrinos easily penetrate a star's outer layers. These particles carry

Figure 14-9 [above]   The creation of a neutrino   A collision between an electron and a proton can produce a neutron and a neutrino. The electron and the proton have equal and opposite electric charges; both the neutron and neutrino are electrically neutral. The neutrino may have a very tiny mass, much smaller than that of an electron, or it may be massless.

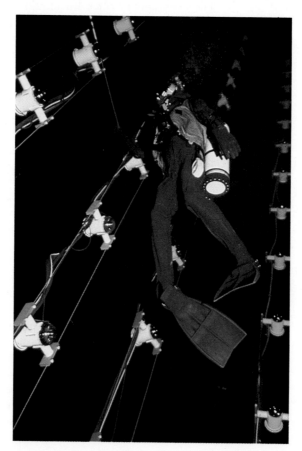

Figure 14-10 [right]   Inside a neutrino detector   A diver is shown servicing one of the photomultiplier tubes in the IMB neutrino detector. Neutrinos from SN 1987A were detected by this apparatus when they struck protons in the water, producing brief flashes of light that were picked up by the photomultipliers. (Courtesy of K. S. Luttrell)

information about the conditions under which they were created. By measuring the energy and momentum carried by the escaping neutrinos, astronomers can learn many details about the star's collapsing core.

Astronomers have been rather lucky with SN 1987A. The doomed star had been studied and its distance from Earth (160,000 ly) was known. The supernova was located in an unobscured part of the sky and neutrino detectors happened to be operating at the time of the outburst. And finally, the supernova was unusual, which advanced our understanding of supernovae in general. Because SN 1987A is such an important object, astronomers will be monitoring its progress for years to come.

Accreting white dwarfs in close binary systems can also become supernovae

Most of our understanding of supernovae comes from observing outbursts in distant galaxies (see Figure 14-11). These observations reveal that supernovae fall into two categories, designated Type I and Type II, which involve very different phenomena. A supernova's spectrum is the best clue to its type: hydrogen lines are prominent in Type II, but are weak or absent in Type I.

Both types of supernovae begin with a sudden rise in brightness (see Figure 14-12). A Type I supernova typically reaches an absolute magnitude of −19 at peak brightness, but a Type II supernova is usually about 2 magnitudes fainter. Type I supernovae then settle into a gradual decline that lasts for over a year, whereas the Type II light curve has a steplike appearance caused by alternating periods of steep and gradual declines in brightness.

A Type II supernova is caused by the death of a massive star. Gravitational energy powers a Type II supernova, because detonation is triggered by the collapse of the star's iron-rich core. Hydrogen lines dominate a Type

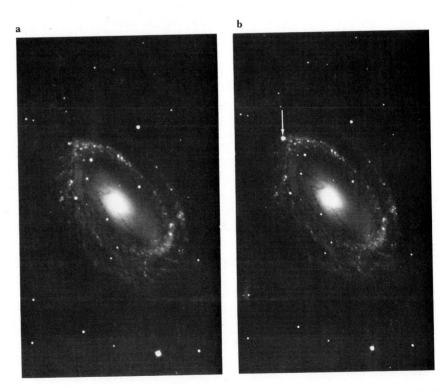

Figure 14-11 A supernova Sometime during 1940, a supernova exploded in this galaxy in the constellation of Coma Berenices. **(a)** *The galaxy before the outburst.* **(b)** *By the time this photograph was taken in 1941, the supernova (indicated by arrow) had faded from its maximum brightness. (Mount Wilson and Las Campanas Observatories)*

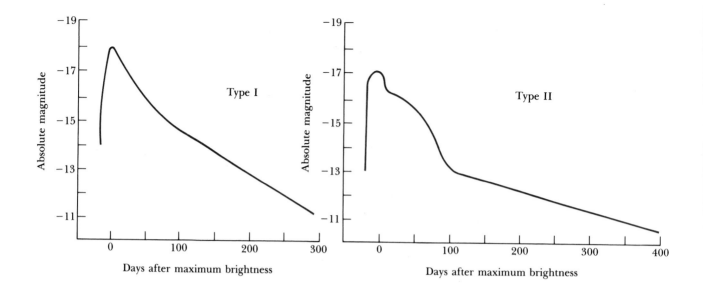

Figure 14-12   Supernova light curves
A Type I supernova, which exhibits a gradual decline in brightness, is caused by an exploding white dwarf in a close binary system. A Type II supernova usually has alternating intervals of steep and gradual decline in brightness. Type II supernovae are caused by the explosive death of massive stars.

II supernova's spectrum simply because that gas is abundant in the star's outer layers being blasted into space. SN 1987A was a Type II supernova.

A Type I supernova begins with a carbon–oxygen-rich white dwarf in a close binary system. As we saw in Chapter 12, mass transfer can occur in a close binary if one star overflows its Roche lobe (recall Figures 12-20 and 12-21). To trigger a Type I supernova, the companion star expands and dumps gas onto the white dwarf. When the white dwarf's mass gets close to the Chandrasekhar limit, carbon burning begins. Because the white dwarf is composed of degenerate matter, the usual safety valve involving pressure and temperature does not operate. In a catastrophic runaway process reminiscent of the helium flash, the rate of carbon burning skyrockets and the star blows up.

A Type I supernova is powered by nuclear energy and the resulting spectacle in the sky is simply the fallout from a thermonuclear explosion. A wide array of radioactive isotopes are produced during the outburst. An unstable isotope of nickel whose nuclei happen to contain equal numbers of protons and neutrons is especially abundant. The entire electromagnetic display of a Type I supernova, including the smooth decline of its light curve, directly results from the radioactive decay of nickel.

Astronomers estimate that in a typical galaxy like the Milky Way, Type I supernovae occur roughly once every 36 years while Type II supernovae occur about once every 44 years. The fact that Type I and Type II supernovae reach nearly the same luminosity at maximum brightness is pure coincidence.

A supernova remnant is detectable at many wavelengths for many years after a supernova explosion

Astronomers find many **supernova remnants** scattered across the skies. A beautiful example is the Veil Nebula seen in Figure 14-13. The doomed star's outer layers were blasted into space with such violence that they are still traveling at supersonic speeds through the interstellar medium. As this expanding shell of gas plows through the interstellar medium, it collides with interstellar atoms, which excites the gas and causes it to glow.

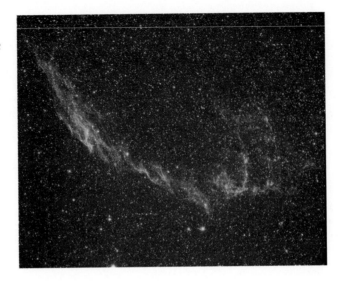

Figure 14-13 The Veil Nebula
This nebulosity is a portion of the Cygnus Loop (see Figure 13-12), which is the remnant of a supernova that exploded about 20,000 years ago. The distance to the nebula is about 1600 light years, and the overall diameter of the loop is about 70 light years. (Palomar Observatory)

Many supernova remnants are quite large and cover sizable fractions of the sky. The largest is the Gum Nebula, with a diameter of 60° (see Figure 14-14). This nebula looks so big because it is so close. Its near side is only about 300 light years from Earth. Studies of the nebula's expansion rate suggest that the supernova exploded around 9000 BC. It could have been witnessed by people then living in such places as Egypt and India. At maximum brilliancy, the exploding star reached a brightness equal to that of the Moon at first quarter.

Many supernova remnants are virtually invisible at optical wavelengths, but as the expanding gases collide with the interstellar medium they do radiate energy at a wide range of other wavelengths, from X rays through radio waves. For example, Figure 14-15 shows both X-ray and radio images of the supernova remnant Cassiopeia A. Optical photographs of this part of the sky reveal only a few small, faint wisps. Thus radio searches for supernova remnants are more fruitful than optical searches. Only two dozen supernova remnants have been found on photographic plates, but more than 100 remnants have been discovered by radio astronomers.

Figure 14-14 The Gum Nebula *The Gum Nebula is the largest known supernova remnant, spanning 60° of the sky, roughly centered on the southern constellation of Vela. The nearest portions of this expanding nebula are only 300 light years from the Earth. The supernova explosion occurred about 11,000 years ago. The supernova remnant now has a diameter of about 2300 light years. Only the central regions of the nebula are shown here. (Royal Observatory, Edinburgh)*

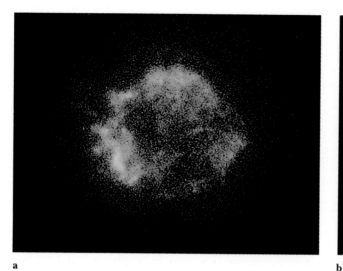

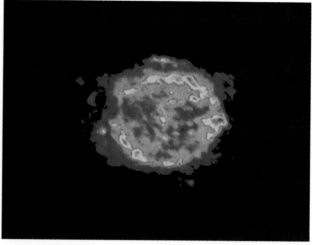

a

b

Figure 14-15 Cassiopeia A Supernova remnants such as Cassiopeia A are typically strong sources of X rays and radio waves. (a) An X-ray picture of "Cas A" taken by the Einstein Observatory. (b) A corresponding radio image produced by the VLA. The supernova explosion that produced this nebula occurred 300 years ago, about 10,000 light years from Earth. (Smithsonian Institution and the Very Large Array)

From the expansion rate of the nebulosity in Cassiopeia A, astronomers conclude that the supernova explosion occurred about 300 years ago. Although telescopes were in wide use by the late 1600s, no one saw the outburst. In fact, the last supernova other than SN 1987A seen in our Galaxy was observed by Johannes Kepler in 1604. In 1572, Tycho Brahe also recorded the sudden appearance of an exceptionally bright star in the sky. To find any other accounts of supernova explosions, we must delve into ancient astronomical records that are almost a thousand years old.

Why have so few nearby supernovae been observed? Astronomers have seen more than 600 supernovae in distant galaxies. A comparable frequency for our own Galaxy should give us about five supernovae per century. Where are they?

The answer is that supernovae probably do erupt every few years in remote parts of our Galaxy, but their detonations are hidden from our view by intervening interstellar debris. Vigorous stellar evolution is occurring primarily in the disk of our Galaxy where, for example, the spiral arms sweep through giant molecular clouds. Our Galaxy's disk is therefore the place where massive stars are born and where supernovae explode. This region of our galaxy is so filled with interstellar gas and dust, however, that we simply cannot see very far into space in directions occupied by the Milky Way.

Pulsars are rapidly rotating neutron stars with intense magnetic fields

The neutron was discovered during laboratory experiments in 1932. Within a year, two astronomers had predicted the existence of neutron stars. Inspired by the realization that white dwarfs are supported by degenerate electron pressure, Fritz Zwicky at the California Institute of Technology and his colleague Walter Baade at Mount Wilson Observatory proposed that a highly compact ball of neutrons could similarly produce a powerful pressure. This **degenerate-neutron pressure,** like the degenerate-electron pressure, could also support a stellar corpse, perhaps even more massive than the Chandrasekhar limit allows. "With all reserve," Zwicky and Baade theorized, "we advance the view that supernovae represent the transition from ordinary stars into **neutron stars,** which in their final stages consist of extremely closely packed neutrons." In other words, there could be at least two types of stellar corpses: white dwarfs and neutron stars.

This proposal was politely ignored by most scientists for years. After all, a neutron star must be a rather weird object. In order to transform protons and electrons into neutrons, the density in the star would have to be equal to nuclear density, 10^{14} g/cm^3. A thimbleful of neutron-star matter brought back to Earth would weigh 100 million tons. Furthermore, an object compacted to nuclear density would be very small. A 1-$M_\odot$ neutron star would have a diameter of only 30 km, about the same size as San Francisco or Manhattan. The surface gravity on one of these neutron stars would be so strong that the escape velocity would equal one-half the speed of light. All these conditions seemed so outrageous that few astronomers paid any serious attention to the subject of neutron stars—until 1968.

As a young graduate student at Cambridge University, Jocelyn Bell had spent many months assisting in the construction of an array of radio antennas covering $4\frac{1}{2}$ acres in the English countryside. By the fall of 1967 the instrument was completed, and Bell and her colleagues began detecting radio emissions from various celestial sources. In November, while scrutinizing data from the new telescope, Bell noticed that the antenna had detected regular "beeps" from one particular location in the sky. Careful repetition of the observations demonstrated that the radio pulses were arriving with a regular period of 1.3373011 sec (see Figure 14-16).

The regularity of this pulsating radio source was so striking that the Cambridge team suspected that they might be detecting signals from an advanced alien civilization. This possibility was soon discarded as several more of these pulsating radio sources, or **pulsars,** were discovered across the sky. In all cases, the periods were extremely regular, from about $\frac{1}{4}$ sec for the fastest to about $1\frac{1}{2}$ sec for the slowest.

When the discovery of pulsars was officially announced in early 1968, astronomers around the world began proposing all sorts of possible explanations for these regular radio pulsations. Many of these theories were bizarre, and arguments raged for months. However, by late 1968, all controversy was laid to rest with the discovery of a pulsar in the middle of the Crab Nebula.

In AD 1054, Chinese astronomers recorded the appearance of a supernova (they called it a "guest star") in the constellation of Taurus. When we turn a telescope toward this location, we find the Crab Nebula shown in Figure 14-17. This object looks like an exploded star and is a supernova remnant. The pulsar at the center of the Crab Nebula is called the Crab pulsar.

The Crab pulsar is one of the fastest pulsars ever discovered. Its period is 0.033 sec, which means that it beeps 30 times each second. The fact that a pulsar is located in a supernova remnant tells us that pulsars probably are associated with dead stars.

Before the discovery of pulsars, most astronomers believed all stellar corpses to be white dwarfs. There seemed to be a sufficient number of white

Figure 14-16 A recording of the first pulsar *This chart recording shows the intensity of radio emission from the first pulsar, discovered by British astronomers in 1967. Note that some pulses are weak and others are strong. Nevertheless, the spacing between the pulses is exactly 1.3373011 sec. (Adapted from Antony Hewish)*

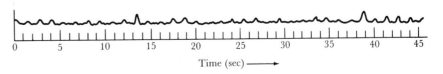

Time (sec) ⟶

Figure 14-17   The Crab Nebula
This beautiful nebula, named for the armlike appearance of its filamentary structure, is the remnant of a supernova seen in AD 1054. The distance to the nebula is about 6000 light years, and its present angular size (4 by 6 arc min) corresponds to linear dimensions of about 7 by 10 light years. (Palomar Observatory)

dwarfs in the sky to account for all the stars that have died since our Galaxy was formed. It was thus generally assumed that all dying stars somehow manage to eject enough matter so that their corpses do not exceed the Chandrasekhar limit.

The discovery of the Crab pulsar proved that not all dead stars become white dwarfs. Calculations involving stellar models of white dwarfs demonstrated that these dead stars are too big and bulky to produce 30 signals per second. For instance, a white dwarf can neither rotate that fast nor vibrate that fast. The Crab pulsar clearly indicates that the stellar corpse at the center of the Crab Nebula is much smaller and more compact than a white dwarf. Astronomers realized that they would have to face the prospect of neutron stars seriously.

As we have seen, a neutron star would be small and dense. It should also be rotating rapidly. All stars rotate, but most of them do so leisurely. For example, our Sun takes nearly one month to rotate once about its axis. A collapsing star speeds up as its size shrinks just as an ice skater doing a pirouette speeds up when she pulls in her arms. This phenomenon is a direct consequence of a law of physics called the **conservation of angular momentum,** which holds that the total amount of angular momentum in a system remains constant. An ordinary star rotating once a month would be spinning faster than once a second if compressed to the size of a neutron star.

In addition to having rapid rotation, we expect a neutron star to have an intense magnetic field. It is probably safe to say that every star has a magnetic field of some strength. In an average star, like our Sun, the strength is typically quite low because the magnetic field is spread out over millions upon millions of square kilometers of the star's surface. However, if a star of solar dimensions collapses down to a neutron star, its magnetic field becomes very compact and its strength increases by a factor of a billion.

Finally, we would expect the axis of rotation of a typical neutron star to be inclined at some angle to the magnetic axis connecting the north and south magnetic poles (see Figure 14-18), much in the same way that the Earth's

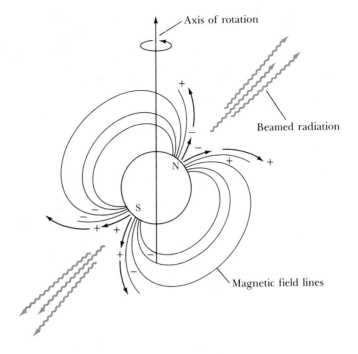

Figure 14-18 A rotating, magnetized neutron star *It is reasonable to suppose that a neutron star is rotating rapidly and possesses a powerful magnetic field. Charged particles are accelerated near the star's magnetic poles and produce two oppositely directed beams of radiation. As the star rotates, the beams sweep around the sky. If the Earth happens to lie in the path of the beams, we see a pulsar.*

Axis of rotation

Beamed radiation

Magnetic field lines

magnetic and rotation axes are inclined to each other. The combination of a powerful magnetic field and rapid rotation operate like a giant electric generator to create intense electric fields near the star's surface. At its surface, protons and electrons are plentiful because the pressures there are too low to combine them into neutrons. The powerful electric fields acting on these charged particles cause them to flow out from the neutron star's polar regions along the curved magnetic field, as sketched in Figure 14-18. As the particles stream along the curved field, they are accelerated and emit energy. The end result is two very thin beams of radiation pouring out of the neutron star's north and south magnetic polar regions.

A rotating, magnetized neutron star is somewhat like a lighthouse beacon. As the star rotates, its beams of radiation sweep around the sky. If the Earth happens to be located in the right direction, a brief flash can be observed each time the beam sweeps past our line of sight. This explanation for pulsars is often called the lighthouse model. Indeed, one of the stars at the center of the nebula is actually flashing on and off 30 times each second (see Figure 14-19).

During the 1970s, radio astronomers discovered about 300 more pulsars scattered across the sky. Presumably, each one is the neutron-star corpse of an extinct massive star. In the 1980s, astronomers found several pulsars with periods of only a few milliseconds. One of them is blinking on and off at visible wavelengths, like the Crab pulsar. In 1989, Carl Pennypacker and his colleagues using the four-meter telescope at Cerro Tololo claimed to detect 2000 flashes per second from a pulsar at the center of SN 1987A, but this discovery remains to be confirmed. Also visibly flashing is the Vela pulsar at the core of the Gum Nebula (see Figure 14-14). The Vela pulsar, with a period of 0.089 sec, is the slowest pulsar ever detected at visible wavelengths.

The Crab pulsar is one of the youngest pulsars, its creation having been observed some 900 years ago. Of course, a few well-known supernovae have been seen since then (such as those noted by Tycho Brahe and Johannes

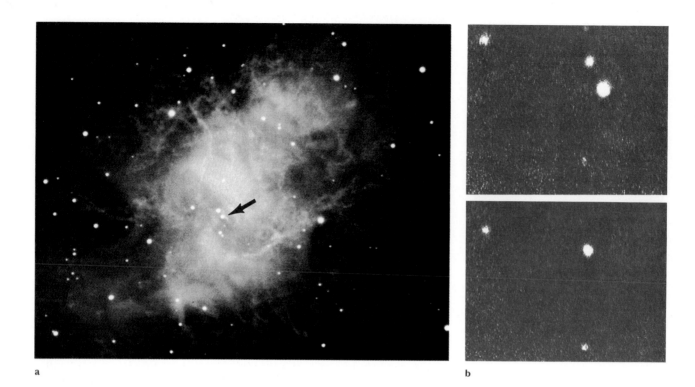

a

b

Kepler), but no pulsars are found at these locations. Perhaps the stellar corpses are not neutron stars, or perhaps their beams simply do not sweep past the Earth. There must be many pulsars that we will never discover because they are oriented at unfavorable angles.

The Vela pulsar, like the Crab pulsar, is quite young, having been created roughly 11,000 years ago. We may conclude that pulsars slow down as they get older and that only the very youngest pulsars are energetic enough to emit optical flashes along with their radio pulses.

Pulsating X-ray sources are neutron stars in close binary systems

During the 1960s, astronomers obtained tantalizing X-ray views of the sky during brief rocket and balloon flights that momentarily lifted X-ray detectors above the Earth's atmosphere. A number of strong X-ray sources were discovered and each was then named after the constellation in which it is located. For example, Scorpius X-1 is the first X-ray source found in the constellation of Scorpius.

Astronomers were so intrigued by these preliminary discoveries that they built and launched *Explorer 42*, an X-ray-detecting satellite that could make observations 24 hours a day (see Figure 14-20). The satellite was launched in 1970 from Kenya to place it in an equatorial orbit. In recognition of the hospitality of the Kenyan people, *Explorer 42* was renamed *Uhuru*, which means "freedom" in Swahili.

Uhuru gave us our first comprehensive look at the X-ray sky. As the satellite slowly rotated, its X-ray detectors swept across the heavens. Each time an X-ray source came into view, signals were transmitted to receiving stations on the ground. Before its battery and transmitter failed in early 1973, *Uhuru* had succeeded in locating 339 X-ray sources.

Figure 14-20 Uhuru (Explorer 42)
Uhuru *was a small satellite designed to detect astronomical sources of X rays. During three years of flawless operation, it observed more than 300 different X-ray–emitting objects across the sky. (NASA)*

The discovery of pulsars was still fresh in everyone's mind when the *Uhuru* team discovered X-ray pulses coming from Centaurus X-3 in early 1971. Figure 14-21 shows data from one sweep of *Uhuru*'s detectors across Centaurus X-3. The pulses have a regular period of 4.84 sec. A few months later, similar pulses were discovered coming from a source called Hercules X-1 that has a period of 1.24 sec. Because the periods of these two X-ray sources are so short, astronomers began to suspect that they had found rapidly rotating neutron stars.

It soon became clear, however, that systems such as Centaurus X-3 and Hercules X-1 are not ordinary pulsars like the Crab or Vela pulsars. Periodically, every 2.087 days, Centaurus X-3 turns off for almost 12 hours. This fact suggests that Centaurus X-3 is an eclipsing binary and that it takes nearly 12 hours for the X-ray source to pass behind its companion star.

The case for the binary nature of Hercules X-1 is even more compelling. It has an "off" state corresponding to a 6-hour eclipse every 1.7 days, and careful timing of the X-ray pulses shows a periodic Doppler shifting every 1.7 days. This information is direct evidence of orbital motion about a companion star. When the X-ray source is approaching us, its pulses are separated by slightly less than 1.24 sec. When the source is receding from us, slightly more than 1.24 sec elapses between the pulses.

Careful optical searches around the location of Hercules X-1 soon revealed a dim star named HZ Herculis. The magnitude of this star varies between 13 and 15 and it has a period of 1.7 days. Because this period is

Figure 14-21 X-ray pulses from Centaurus X-3 *This graph shows the intensity of X rays detected by* Uhuru *as Centaurus X-3 moved across the satellite's field of view. The successive pulses are separated by 4.84 sec. The gradual variation in the height of the pulses from left to right is a result of the changing orientation of* Uhuru's *X-ray detectors toward the source as the satellite rotates. (Adapted from R. Giacconi and colleagues)*

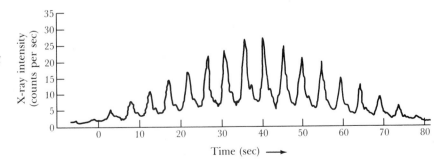

exactly the same as the orbital period of the X-ray source, astronomers conclude that HZ Herculis is the companion star around which Hercules X-1 is orbiting.

Putting all the pieces together, astronomers have now realized that systems such as Centaurus X-3 and Hercules X-1 are examples of double stars in which one of them is a neutron star. All these binaries have very short orbital periods. The distance between the ordinary star and its neutron star must therefore be very small. This proximity puts the neutron star in a position to be able to capture gases escaping from the ordinary companion star.

To explain pulsating X-ray sources such as Centaurus X-3 or Hercules X-1, astronomers assume that the ordinary star either fills or nearly fills its Roche lobe. Either way, matter escapes from the star. This mass loss occurs by direct "Roche-lobe overflow" if the star fills its lobe, as in the case of Hercules X-1, or it occurs by a stellar wind if the star's surface lies just inside its lobe, as with Centaurus X-3 (see Figure 14-22). A typical rate of mass loss from the ordinary star is roughly 10^{-9} solar masses per year.

As with an ordinary pulsar, the neutron star in a pulsating X-ray source rotates rapidly and has a powerful magnetic field inclined to the axis of rotation (recall Figure 14-18). Because of its strong gravity, the neutron star easily captures much of the gas escaping from the companion star. As the gas falls toward the neutron star, the magnetic field funnels the incoming matter down onto its north and south magnetic polar regions. The star's gravity is so strong that the gas is traveling at nearly half the speed of light by the time it crashes onto the star's surface. This violent impact creates hot spots at both poles with temperatures of about 10^8 K, so that these hot spots emit abundant X rays with a luminosity roughly 100,000 times brighter than

Figure 14-22 A model of a pulsating X-ray source Gas escaping from an ordinary star is captured by the neutron star. The infalling gas is funneled down onto the neutron star's magnetic poles, where it strikes the star with enough energy to create two X-ray–emitting hot spots. As the neutron star spins, beams of X rays from the hot spots sweep around the sky.

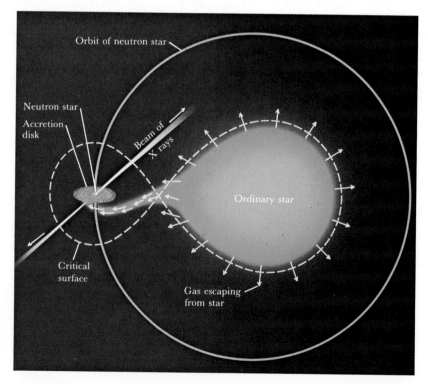

the Sun. As the neutron star rotates, the beams of X rays from the polar caps sweep around the sky. If the Earth happens to be in the path of one of the two beams, we can observe a pulsating X-ray source. The pulse period is thus equal to the neutron star's rotation period. For example, the neutron star in Hercules X-1 is spinning at the rate of once every 1.24 sec, equal to its pulse period.

There is an interesting side effect in this basic scenario. Gases captured by the neutron star's gravity may go into orbit about the neutron star. The result is a rotating disk of material called an **accretion disk,** as shown in Figure 14-22.

Accretion disks have been detected in many close binary systems where mass transfer is occurring between the two stars (recall Figure 12-21*b*). Under certain circumstances, some bizarre things can happen. With ordinary pulsating X-ray sources such as Hercules X-1, the rate at which gas falls onto the neutron star is fairly low. This material falls onto the neutron star from the inner edge of the accretion disk at a rate low enough to allow the resulting X rays to escape.

If the companion star is dumping vast amounts of material onto the neutron star, however, the resulting energy cannot easily escape. Tremendous pressures build up in the gases crowding down onto the neutron star. These pressures meet strong resistance in the plane of the accretion disk, from which newly arrived gases are constantly spiraling in toward the neutron star. The "path of least resistance" is perpendicular to the plane of the accretion disk. The gas pressure around the neutron star is thus relieved by material squirting out along this perpendicular direction. The result can be two powerful beams of high-velocity hot gas, as shown in Figure 14-23. This explanation apparently accounts for the weird star called SS433.

Figure 14-23   A model of SS433
Gas from a normal star is captured into an accretion disk about a neutron star. Two high-speed, oppositely directed jets of gas are ejected from the faces of the disk. Because the disk is tilted, the gravitational pull of the normal star causes the jets to precess with a period of 164 days.

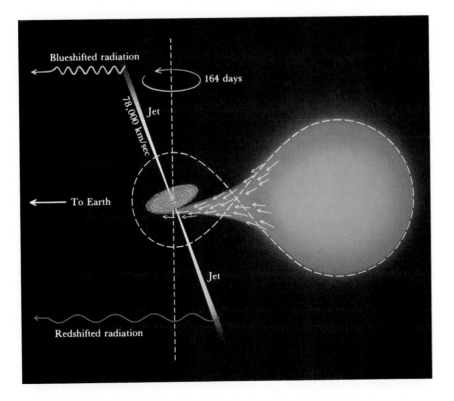

Figure 14-24 Four views of SS433
These four radio views, taken in early 1981, show jets of gas extending out to one-sixth of a light year on either side of SS433. Three-quarters of the radio emission comes from SS433 itself (the red central blob), which is located at the center of a supernova remnant 13,000 light years from Earth. (Very Large Array)

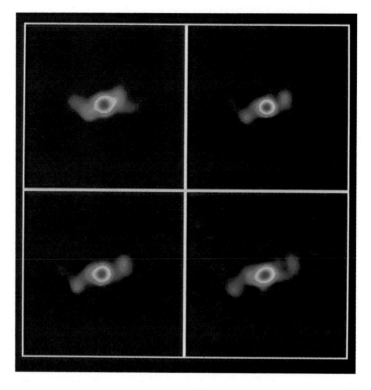

In the autumn of 1978, Bruce Margon and his colleagues at UCLA were observing SS433, which had been noted for strong emission lines in its spectrum. To everyone's surprise, the spectrum of SS433 contained several complete sets of spectral lines. One set was very redshifted from its usual wavelengths, another set was comparably blueshifted. Somehow, SS433 is "coming and going at the same time." To make matters even more puzzling, the wavelengths of these redshifted and blueshifted lines change dramatically from one night to the next.

Astronomers had never seen anything like this, and soon many were observing SS433. By mid-1979, it was clear that the system's redshifted and blueshifted lines are actually moving back and forth across the spectrum of SS433 with a period of 164 days. Astrophysicists were quick to point out that the two sets of spectral lines could be caused by two oppositely directed jets of gas, one tilted toward us and the other away from us. Furthermore, the 164-day variation could be explained by a precession of the two jets. As the two jets circle about the sky every 164 days, we see a periodic variation in the Doppler shift.

All these features come together in the model sketched in Figure 14-23. To explain the large redshifts and blueshifts discovered by Margon, gas in the two oppositely directed jets must have a speed of 78,000 km/sec, roughly one-quarter the speed of light. In addition, the accretion disk must be tilted with respect to the orbital plane of the two stars of the binary system. Just as the tilt of the Earth's axis with respect to the plane of the ecliptic causes the Earth to precess, the tilt of the accretion disk results in the 164-day precession of the two jets.

Figure 14-24 shows four high-resolution radio views of SS433. Note the two oppositely directed appendages emerging from the central source. As we shall see in Chapter 18, many quasars and peculiar galaxies have a similar radio structure, though on a much larger scale. Quasars are incredibly far away and they are thus difficult to study. The real significance of SS433 may be that it gives us a miniature quasarlike object right in our own celestial backyard.

Explosive thermonuclear processes on white dwarfs and neutron stars produce novae and bursters

Low-mass stars are far more common than high-mass stars, and so white dwarfs are far more common than neutron stars. With all the bizarre and fascinating phenomena associated with neutron stars, you might be wondering if white dwarfs do anything more dramatic than simply cool off.

The answer definitely is yes. Occasionally, some star in the sky suddenly brightens by a factor of 10^6. This phenomenon is called a **nova** (not to be confused with a supernova, which involves a much greater increase in brightness). Novae are fairly common. Their abrupt rise in brightness is followed by a gradual decline that may stretch for several months or more (see Figures 14-25 and 14-26).

Painstaking observations of many novae strongly suggest that all novae are members of close binary systems containing a white dwarf. Gradual mass transfer from the ordinary companion star (which presumably fills its Roche lobe) deposits fresh hydrogen onto the white dwarf. Because of the strong gravity, this hydrogen becomes compacted into a dense layer covering the hot surface of the white dwarf. As more gas is deposited, the temperature in the hydrogen layer increases. Finally, when the temperature reaches about 10^7 K, hydrogen burning ignites throughout the layer, embroiling the white dwarf's surface in a thermonuclear holocaust that we see as a nova.

A similar phenomenon occurs with neutron stars. Beginning in late 1975, astronomers analyzing data from X-ray satellites realized that their instruments had detected sudden, powerful bursts of X rays from certain objects

Figure 14-25 Nova Herculis 1934
These two pictures show a nova (a) shortly after peak brightness as a magnitude +3 star and (b) two months later, when it had faded to magnitude +12. Novae are named after the constellation and year in which they appeared. (Lick Observatory)

a

b

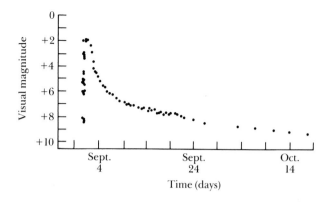

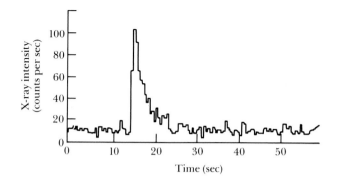

Figure 14-26 [left] The light curve of Nova Cygni 1975 This graph shows the history of a nova that blazed forth in the constellation of Cygnus in September 1975. The rapid rise in magnitude followed by a gradual decline is characteristic of all novae.

Figure 14-27 [right] X rays from a burster A burster emits a constant low intensity of X rays interspersed with occasional powerful bursts of X rays. This particular burst is typical. It was recorded on September 28, 1975, by an X-ray telescope on board the Astronomical Netherlands Satellite while the telescope was pointed toward the globular cluster NGC 6624. About one-third of all known bursters are located in globular clusters. (Adapted from Walter Lewin)

in the sky. The record of a typical burst is shown in Figure 14-27. The source emits X rays at a constant low level, then suddenly, without warning, there is an abrupt increase, followed by a more gradual decline. Typically, an entire burst lasts for only 20 sec. Sources that behave in this fashion are called **bursters.** Several dozen of them have been discovered, most located toward the center of our Galaxy.

Bursters, like novae, are believed to involve close binaries experiencing mass transfer. With a burster, however, the stellar corpse is a neutron star rather than a white dwarf. Gases escaping from the ordinary companion star fall onto the neutron star. The energy released as this gas crashes down onto the neutron star's surface produces the low-level X rays that are continuously emitted by the burster.

Most of the gas falling onto the neutron star is hydrogen, which becomes compressed against the hot surface of the star by the star's powerful surface gravity. In fact, temperatures and pressures in this accreting layer are so high that the arriving hydrogen is promptly converted into helium by the hydrogen-burning process. Constant hydrogen burning soon produces a layer of helium that covers the entire neutron star.

Finally, when the helium layer is about 1 m thick, helium burning ignites explosively, and we observe a sudden burst of X rays. In other words, explosive hydrogen burning on a white dwarf produces a nova, but explosive helium burning on a neutron star produces a burster. In both cases, the burning is explosive because the fuel is so strongly compressed against the star's surface that it is degenerate, like the star itself. As we saw with the helium flash inside red giants, ignition of a degenerate thermonuclear fuel always involves a sudden thermal runaway because the usual "safety valve" between temperature and pressure does not operate.

Just as there is an upper limit to the mass of a white dwarf, there is also an upper limit to the mass of a neutron star. Above this limit, degenerate-neutron pressure cannot support the overpowering weight of the star's matter pressing inward from all sides. The Chandrasekhar limit for a white dwarf is 1.4 $M_\odot$ and the corresponding upper limit for a neutron star is roughly 3 $M_\odot$.

Before the discovery of pulsars, dying stars were thought to eject enough material somehow so that their corpses would be below the Chandrasekhar limit. Obviously, this idea proved incorrect. Inspired by this lesson, astronomers soon began wondering what might happen if a dying massive star failed to eject enough matter to get below the upper limit for a neutron star. What for example, might a 5 $M_\odot$ stellar corpse be like?

The gravity associated with a neutron star is so strong that the escape velocity is roughly one-half the speed of light. With a stellar corpse greater than 3 $M_\odot$, there is so much matter crushed into such a small volume that the escape velocity exceeds the speed of light. Because nothing can travel faster than light, nothing—not even light—can leave the dead star. The star therefore disappears from the universe, its powerful gravity leaving a hole in the fabric of space and time. The discovery of neutron stars thus inspired astrophysicists to examine seriously one of the most bizarre and fantastic objects ever predicted by modern science—the black hole.

Summary

- A low-mass star becomes a red giant when shell hydrogen burning begins. It becomes a horizontal-branch star when core helium burning begins. And it becomes a red supergiant when the helium in its core is exhausted and shell helium burning begins.

- Thermal pulses in the helium-burning shell can eject the star's outer layers. Ultraviolet radiation from the hot carbon–oxygen core ionizes and excites the ejected gases, producing a planetary nebula.

- The burned-out core of a low-mass star becomes a dense sphere about the size of the Earth called a white dwarf.

  The maximum mass of a white dwarf (the Chandrasekhar limit) is 1.4 solar masses.

- After exhausting its central supply of hydrogen and helium, a high-mass star undergoes a sequence of thermonuclear reactions in its core. These are carbon burning, neon burning, oxygen burning, and silicon burning. The star eventually develops an iron-rich core.

- A high-mass star can die in a supernova explosion that ejects most of the star's matter into space at very high speeds (a Type II supernova). If the core of the star survives the explosion, it may become a neutron star or even a black hole.

  The historic naked-eye supernova of 1987 was a Type II supernova. Neutrinos were detected from the explosion of SN 1987A.

- A white dwarf in a close binary can become a Type I supernova when it is engulfed in a brief but explosive episode of carbon burning.

- A neutron star is a very dense stellar corpse consisting of closely packed neutrons in a sphere roughly 30 km in diameter.

- A pulsar is a source of periodic radio pulses. It is caused by a rapidly rotating neutron star with a powerful magnetic field. Energy pours out of the north and south polar regions of the neutron star in intense beams that sweep around the sky.

- Some X-ray sources exhibit regular pulses. These objects are thought to be neutron stars in close binary systems with ordinary stars.

  Material from the ordinary star in a binary pair can fall onto the surface of its companion white dwarf or neutron star to produce a surface layer in which thermonuclear reactions can occur.

  Explosive hydrogen burning may occur in the surface layer of a companion white dwarf, producing the sudden increase in luminosity that we call a nova.

Explosive helium burning may occur in the surface layer of a companion neutron star, producing the sudden increase in X-ray radiation called a burster.

Review questions

1 Why is the temperature in a star's core so important in determining which nuclear reactions can occur?

2 On an H–R diagram, sketch the evolutionary track that the Sun will follow as it leaves the main sequence and becomes a white dwarf. Approximately how much mass will the Sun have when it becomes a white dwarf? Where will the rest of the mass have gone?

3 Why do you suppose that all the white dwarfs known to astronomers are relatively close to the Sun?

4 Why do astronomers suggest that a supernova in our Galaxy visible to the naked eye is long overdue?

5 Why have radio searches for supernovae remnants been more fruitful than optical searches?

6 What is the difference between a nova and a supernova?

7 During the weeks immediately following the discovery of the first pulsar, one explanation for them was that the pulses are signals from an extraterrestrial civilization. Why do you suppose astronomers discarded this idea?

8 Describe what radio pulsars, X-ray pulsars, and bursters have in common. How are they different manifestations of the same type of astronomical object?

Advanced questions

9 Suppose you wanted to determine the age of a planetary nebula. What observations would you make, and how would you use the resulting data?

10 What might explain why the rate of expansion of the gas shells in some planetary nebulae is nonuniform?

11 What kinds of stars would you monitor if you wished to observe a supernova explosion from its very beginning? Look up tabulated lists of the brightest and nearest stars. Which, if any, of these stars are possible supernova candidates? Explain.

12 To determine accurately the period of a pulsar, astronomers must take the Earth's orbital motion about the Sun into account. Explain why.

13 In 1989, as this book was going to press, Carl Pennypacker and his colleagues reported the tentative discovery of a pulsar at the center of SN 1987A. Read up on this topic in popular magazines like *Sky & Telescope* and *Science News*. Has the discovery been confirmed? If so, how fast is the pulsar flashing?

14 In 1989, as this book was going to press, several astronomers reported observing a bright, fast-moving spot near the center of SN 1987A. This may be a reappearance of the "mystery spot" sighted shortly after the supernova exploded. Read up on this topic in popular magazines like *Sky & Telescope* and *Science News*. Has the discovery been confirmed? Are there any theories about what the spot might be?

Discussion questions

15 Suppose that you discover a small glowing disk of light while searching the sky with a telescope. How would you observationally decide if this object is a planetary nebula? Could your object be something else? Explain.

16 Compare novae and bursters. What do they have in common? In what ways are they different?

For further reading

Clark, D. *Superstars*. McGraw-Hill, 1984.

Greenstein, G. *Frozen Stars: Of Pulsars, Black Holes, and the Fate of Stars*. Freundlich Books, 1984.

Grindlay, J. "New Bursts in Astronomy." *Mercury*, September/October 1977, p. 6.

Helfand, D. "Pulsars." *Mercury*, May/June 1977, p. 2.

Kaler, J. "Planetary Nebulae and Stellar Evolution." *Mercury*, July/August 1981, p. 114.

Lattimer, J., and Burrows, A. "Neutrinos from Supernova 1987A." *Sky & Telescope*, October 1988, p. 348.

Margon, B. "The Bizarre Spectrum of SS 433." *Scientific American*, October 1980.

Reddy, F. "Supernovae: Still a Challenge." *Sky & Telescope*, December 1983, p. 485.

Weiler, K. "A New Look at Supernova Remnants." *Sky & Telescope*, November 1979, p. 414.

White, N. "New Wave Pulsars." *Sky & Telescope*, January 1987, p. 22.

Woosley, S., and Weaver, T. "The Great Supernova of 1987." *Scientific American*, August 1989.

15

Black holes

A black hole in a double star
system   This artist's rendition shows the
close binary system that contains Cygnus
X-1. Cygnus X-1 is a strong source of X
rays and is widely believed by astronomers
to be a black hole. Gas from the
companion star is captured into orbit
about the black hole, forming an
accretion disk. As gases spiral in toward
the black hole, they are heated to high
temperatures. At the inner edge of the
accretion disk, the gases are so hot that
they emit X rays. (Courtesy of D. Norton,
Science Graphics)

We have seen that the mass of a neutron star cannot exceed three solar
masses. In this chapter we learn that stellar corpses more massive than 3
$M_\odot$ become compacted to a single point of infinite density, surrounded by
gravity so intense that nothing—not even light—can escape. We discuss
the theory of relativity that predicts the existence of these objects, called
black holes. We find that a black hole has only three physical properties,
and that the stellar corpse in the center of a black hole has disappeared
forever from our direct observation. We see how a black hole or a massive
galaxy can act as a "gravitational lens" that distorts the appearance of
background objects. Finally, we examine evidence that astronomers may
have discovered a few black holes in double star systems.

Suppose that the mass of a dying star's burned-out core exceeds three solar masses (3 $M_\odot$). This mass is well above the Chandrasekhar limit, so degenerate-electron pressure could not possibly support the stellar corpse. Similarly, because 3 $M_\odot$ is also above the mass limit for neutron stars, degenerate-neutron pressure is also incapable of supporting the crushing weight of the burned-out matter pressing inexorably toward the dead star's center. If it can become neither a white dwarf nor a neutron star, what might this massive stellar corpse become?

As we saw in the previous chapter, a typical neutron star consists of roughly 1 $M_\odot$ of matter compressed by its own gravity to nuclear density in a sphere roughly 30 km in diameter. The star's gravity is so strong that the escape velocity from its surface is one-half the speed of light.

A massive stellar corpse, whose weight overpowers degenerate-neutron pressure, easily compresses its matter to densities greater than nuclear density. It does not take much further compression to cause the escape velocity to exceed the speed of light. For example, if 3 $M_\odot$ of matter is squeezed inside a sphere 18 km in diameter, the escape velocity from the object becomes greater than the speed of light. Nothing can travel faster than the speed of light, and thus nothing—not even light—can manage to escape from the dead star. The star has disappeared from the observable universe, although some of its effects can be detected.

The general theory of relativity describes gravity in terms of the geometry of space and time

To appreciate fully the nature of massive stellar corpses, we must use the best theory of gravity at our disposal. The gravitational field around one of these massive dead stars is so strong that Isaac Newton's theories fail to describe it accurately. We must turn instead to Albert Einstein's general theory of relativity.

According to the classical physics of Newton, space is perfectly uniform and fills the universe like a rigid framework. Time passes at a monotonous, unchangeable rate. It is always possible to measure how fast you are moving through this rigid fabric of space and time, and to calculate how your observations depend on your state of motion. Significant differences between observations made by a stationary person and by a moving person arise at speeds near the speed of light.

In 1905, Albert Einstein began a revolution in physics with his **special theory of relativity.** His goal was to reformulate electromagnetic theory so that it did not depend on the motion of an observer. In other words, using the special theory of relativity, both you on Earth and a friend in a rocketship traveling near the speed of light would have the same logical, complete description of electricity and magnetism, unencumbered by pitfalls or paradoxes caused by your relative motions.

Einstein held forth that neither our location in space and time nor our motion through space and time should prejudice our description of physical reality. This meant that Einstein had to abandon traditional, rigid notions of space and time. For example, imagine your friend whizzing across the solar system in a rocketship while you remain here at rest on the Earth. Einstein proved that, in order for both of you to agree on the same coherent description of reality, you must say that your friend's clocks have slowed down and her rulers have shrunk. Specifically, an observer always finds that moving

Figure 15-1 The equivalence principle
The equivalence principle asserts that you cannot distinguish between being at rest in a gravitational field and being accelerated upward in a gravity-free environment. This idea was an important step in Einstein's development of the general theory of relativity.

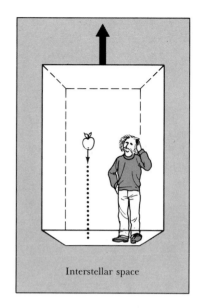

The Earth

Interstellar space

clocks seem to be slowed and moving rulers seem to be shortened in the direction of motion.

After developing the special theory of relativity, Einstein turned his attention to gravity. He began by demonstrating that it is not necessary to think of gravity as a force. According to Newton's theory, an apple falls to the ground because the force of gravity pulls the apple down. Einstein pointed out that the apple would behave exactly the same in free space far from any gravity if the floor were accelerating upward to meet the apple, as sketched in Figure 15-1.

This example of Einstein's **principle of equivalence** explains that, in a small volume of space, the downward pull of gravity can be accurately and completely duplicated by an upward acceleration of the observer. The two people in their closed compartments in Figure 15-1 in fact have no way of telling who is at rest on the Earth and who is in the elevator moving upward at a constantly increasing speed.

This approach allowed Einstein to focus entirely on motion (rather than force) in discussing gravity. From his special theory of relativity, he knew exactly how rulers and clocks are affected by motion, and he could thus describe gravity entirely in terms of its effects on space and time. Far from a source of gravity, the acceleration is small and the effect on clocks and rulers is therefore small; nearer a source of gravity, the acceleration is greater and the distortion of clocks and rulers is greater. In this way, Einstein "generalized" his special theory to arrive at his **general theory of relativity.**

The general theory of relativity describes gravity entirely in terms of the geometry of space and time. Far from a source of gravity, space is flat and clocks tick at their normal rate. Near a source of gravity, space is curved and clocks slow down. The stronger the gravity, the greater are these distortions of the shape of space and the flow of time.

In a weak gravitational field, Einstein's general relativity theory gives the same results as Newton's classical theory. But in stronger gravity, such as that near the Sun's surface, the two theories predict different results. For examples, recall the precession of Mercury's perihelion and the deflection of a light ray grazing the Sun, discussed in Chapter 3 (recall Figures 3-16 and 3-18). In these and other situations, general relativity has withstood numerous tests. It is by far the most elegant and accurate description of gravity ever devised.

A black hole is a very simple object that has only a "center" and a "surface"

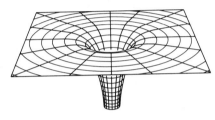

Figure 15-2 *The geometry of a black hole This diagram shows how the shape of space is distorted by the gravitational field of a black hole. Far from the hole, gravity is weak and space is flat. Near the hole, gravity is strong, and space is highly curved. This diagram uses a two-dimensional surface as an analogy for the three-dimensional fabric of space.*

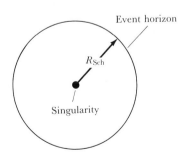

Figure 15-3 *The structure of a black hole A black hole has only two parts: a singularity surrounded by an event horizon. Inside the event horizon, the escape velocity exceeds the speed of light, and so the event horizon is a one-way surface: things can fall in, but nothing can get out. The distance between the singularity and the event horizon is called the Schwarzschild radius,* R_{Sch}.

Imagine a dying star too massive to become either a white dwarf or a neutron star. The overpowering weight of the star's burned-out matter pressing inward from all sides causes the star to contract rapidly. As the star's matter is compressed to enormous densities, the strength of gravity around the collapsing star increases dramatically. According to the general theory of relativity, distortions of space and time become increasingly pronounced around the dying star. Finally, the escape velocity from the star's surface equals the speed of light, and the star seems to disappear from the universe. At this stage, space has become so severely curved that a hole is punched in the fabric of the universe. The dying star disappears into this hole in space, leaving behind only a **black hole.**

The geometry of space around a black hole is sketched in Figure 15-2. Note that space is flat far from the hole because gravity is weak there. Near the hole, however, gravity is strong, and the curvature of space is severe.

The location in space where the escape velocity from the black hole equals the speed of light is called the **event horizon.** This sphere is also sometimes thought of as the surface of the black hole. Once a massive dying star collapses inside its event horizon, it permanently disappears from the universe. This surface is literally a horizon in the geometry of space beyond which we cannot see any events.

In addition to having the effect of curving space, gravity causes time to slow down. If you stood at a safe distance and watched your friend fall toward a black hole, you would note that the friend's clocks tick more and more slowly. In fact, you would conclude that at the event horizon the clocks stop entirely.

Once a dying star has contracted inside its event horizon, no forces in the universe can prevent the complete collapse of the star down to a single point at the center of the black hole. The star's entire mass is crushed to infinite density at this point, called the **singularity,** often considered the center of the black hole.

The structure of a black hole is therefore very simple. As sketched in Figure 15-3, it has only two parts: a singularity (the center) surrounded by an event horizon (the surface). The distance between the singularity and the event horizon is the **Schwarzschild radius,** named after the German astronomer Karl Schwarzschild, who in 1916 became the first to solve Einstein's equations of general relativity.

To understand why the complete collapse of a doomed star is inevitable, think about the nature of space and time here on Earth, far from any black holes. On Earth we have the freedom to move as we wish through the three dimensions of space—up and down, left and right, or forward and back. But we do not have freedom to move at will through the dimension of time. Whether we like it or not, we are all dragged inexorably from the cradle to the grave.

Inside a black hole, powerful gravity distorts the shape of space and time so severely that the orientations of space and time become interchanged. In a limited sense, inside a black hole one could move freely through time. This freedom is compromised, however, by the loss of a corresponding amount of freedom to move through space. Whether we like it or not, we are dragged inexorably from the event horizon to the singularity. Just as no force in the universe can prevent the forward march of time (past to future) outside a black hole, no force in the universe can prevent the inward march of space (event horizon to singularity) inside a black hole.

At the singularity, the strength of gravity is infinite, and so the curvature of space and time is infinite. In other words, space and time at the singularity are all jumbled up. Space and time do not exist there as separate, identifiable entities.

This confusion of space and time has profound implications for what goes on inside a black hole. All the laws of physics require a clear, distinct background of space and time. Without this identifiable background, we could not speak rationally about the arrangement of objects in space or the ordering of events in time. Because space and time are all jumbled up at the center of a black hole, the singularity does not obey the laws of physics. The singularity behaves in a random and capricious fashion, without rhyme or reason.

Fortunately, we are shielded from the singularity by the event horizon. Although irrational things happen at the singularity, none of the effects manage to escape to the outside universe. Consequently, the outside universe remains understandable and predictable.

The irrational, random behavior of the singularity is so disturbing to physicists that, in 1969, the British mathematician Roger Penrose and his colleagues proposed the so-called **law of cosmic censorship:** "Thou shalt not have naked singularities." In other words, every singularity must be completely surrounded by an event horizon. If a naked singularity could exist, it would affect the universe in unpredictable and random ways.

....

The structure of a black hole can be completely described with only three numbers

In addition to shielding us from singularities, the event horizon prevents us from knowing about the material that has fallen into a black hole. For instance, there is no way we can ever discover the chemical composition of the massive star whose collapse produced a black hole. Even if someone went into a black hole to make measurements or chemical tests, it would be impossible to get any of this information back to the outside world. Indeed, a black hole is an "information sink," because in-falling matter carries many properties about it (chemical composition, texture, color, shape, size) that are forever removed from the universe.

A black hole is unaffected by the information it destroys. For example, consider two black holes, one made from the gravitational collapse of 10 $M_\odot$ of iron, the other made from the gravitational collapse of 10 $M_\odot$ of peanut butter. Very different substances went into the creation of these two black holes, but once their event horizons formed, both the iron and the peanut butter permanently disappeared from the universe. As seen from the outside, the two holes are absolutely identical. In this way, information about material inside a black hole has no effect on the structure or properties of the black hole.

Is there anything we can know or measure about a black hole? Specifically, what properties characterize a black hole?

First, we can measure the *total mass* of a black hole, by placing a satellite in orbit about the hole, for example. Recall that Kepler's third law relates a satellite's orbital period and semimajor axis to the mass around which the satellite moves. Thus, after measuring the size and period of the satellite's orbit, we can use this law to determine the mass of the hole. This mass is equal to the total mass of all the material that has gone into the hole.

Science fiction abounds with nasty rumors that black holes are evil things that go around gobbling everything in the universe. Not so! The bizarre effects of highly warped space and time are limited to a volume extending only a few million kilometers from the hole. Farther away, gravity is sufficiently weak that Newtonian physics adequately describes everything. For example, at a distance of only a few astronomical units from a 10 $M_\odot$ black hole, the behavior of gravity is identical to that around any ordinary 10 $M_\odot$ star.

In addition to the total mass, we can also measure the *total electric charge* possessed by a black hole. Like gravitational force, electric force is a long-range interaction, making its effects felt in the space around the hole. Appropriate equipment on a space probe passing near the hole could measure the intensity of the electric field around the hole, from which the electric charge of the hole can be determined.

We do not, however, expect a black hole to possess any appreciable electric charge. If a hole did start off with a sizable positive charge, for example, it would vigorously attract vast numbers of negatively charged electrons from the interstellar medium, which would soon neutralize the hole's charge. For this reason, astronomers generally do not consider electric charge when discussing real black holes.

Although a black hole might have a tiny electric charge, it cannot have any magnetic field whatsoever. It can be mathematically proved from Einstein's equations that a black hole cannot possess the geometry that a magnetic field would require. During the creation of a black hole, the collapsing star would probably possess an appreciable magnetic field, but it would be radiated away in the form of electromagnetic and **gravitational waves** before the dead star could settle down inside its event horizon. Since gravitational waves are ripples in the overall geometry of space, some physicists are exploring the possibility of observing the creation of black holes by detecting the bursts of **gravitational radiation** emitted by massive, dying stars as they collapse.

In addition to its mass and electric charge, we can also measure a black hole's *total angular momentum*. Because of the principle of the conservation of angular momentum, we expect a black hole to be spinning very rapidly. Einstein's theory makes the startling prediction that this rotation causes space and time to be dragged around the hole. A spinning black hole is therefore surrounded by space that rotates with the hole. Around the event horizon of every rotating black hole is a region where this dragging of space and time is so severe that it is impossible to stay fixed in the same place. No matter what you do, you get pulled around the hole along with the rotating geometry of space and time. This region where it is impossible to be at rest is called the **ergosphere** (see Figure 15-4).

To measure a black hole's angular momentum, we could place two satellites in orbit about the hole. Suppose that one satellite circles the hole in the same direction the hole rotates and the other satellite circles in the opposite direction. One satellite is therefore carried along with the rotating geometry

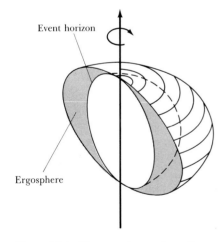

Figure 15-4 The ergosphere *A rotating black hole is surrounded by a region where the dragging of space and time around the hole is so severe that it is impossible for anything to remain at a fixed location. Because the ergosphere (the shaded region in this cross-section diagram) is outside the event horizon, this bizarre region is accessible to us and can be traversed by astronauts or asteroids without having them disappear into the black hole.*

Event horizon

Ergosphere

of space and time, but the other is constantly fighting its way "upstream." The two satellites thus will have different orbital periods. From a comparison of these periods, the total angular momentum of the hole can be deduced.

And that is all there is to black holes. As the famous **no-hair theorem,** first formulated in the early 1970s, put it: "Black holes have no hair." More formally stated, a black hole's structure is completely specified by only three numbers: its mass, its charge, and its angular momentum. Any and all additional properties carried by the matter that fell into the hole have disappeared from the universe and thus have no effect on the structure of the hole.

A black hole distorts the images of background stars and galaxies

Finding black holes in the sky is a difficult business. Obviously, you cannot observe a black hole in the same way you can see a star or a planet. The best you can hope for is to detect the effects of a black hole's powerful gravity.

It might be possible to find a black hole because of the way that it affects the appearance of background objects. For example, suppose that the Earth, a black hole, and a background star are in nearly perfect alignment, as sketched in Figure 15-5. Because of the warped space around the black hole, there are two paths along which light rays can travel from the background star to us here on Earth. Thus, we should see two images of the star.

The distortion of background images by a powerful source of gravity is called a **gravitational lens.** It is virtually the only way we can hope to find an isolated black hole in our Galaxy. Unfortunately, in order for this effect to be noticeable, the alignment between the Earth, the black hole, and a remote star must be almost perfectly straight. Without nearly perfect alignment, the secondary image of the background star is too faint to be noticed.

No one has ever found a gravitational lens within our Galaxy. However, several gravitational lenses have been discovered far from our Galaxy that involve the remote, luminous objects called quasars. As we shall see in Chapter 18, a quasar is a very bright starlike object. Typical quasars shine as brightly as a hundred galaxies and are located a few billion light years from Earth. The 3000 known quasars are uniformly scattered throughout the sky. On the average, one quasar is found in every 15 square degrees.

In 1979, astronomers were surprised to find two quasars separated by only 6 arc sec. They took spectra of both quasars and discovered that they were nearly identical. They thus concluded that they were looking at two images of the same quasar. Further observations revealed that there was a

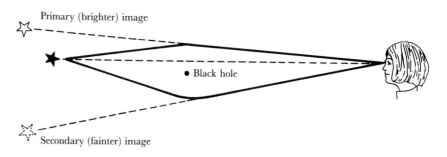

Figure 15-5 A gravitational lens A black hole deflects light rays from a distant star so that an observer sees two images of the star. Several so-called gravitational lenses have been discovered in which light from a remote quasar is deflected by an intervening galaxy. No gravitational lenses caused by black holes have yet been discovered, however.

Primary (brighter) image

● Black hole

Secondary (fainter) image

galaxy between the quasar images (see Figure 15-6). The gravitational field of this galaxy bends the light from the remote quasar and thus acts like a gravitational lens.

More than half a dozen gravitational lenses have been reported. All involve a remote quasar whose starlike image is split by a galaxy located between us and the quasar. One of the most curious cases consists of two bright sources connected by a nearly circular ring (see Figure 15-7). This may be an example of an **Einstein ring** predicted by Einstein in 1936. A gravitational lens can produce a ring image if there is perfect alignment between the observer, the background object, and the intervening source of gravity. Further observations are needed to confirm that the ring in Figure 15-7 is in fact the result of gravitational lensing.

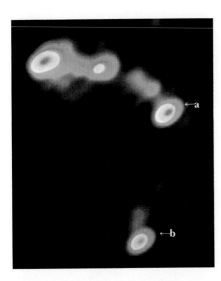

Figure 15-6 [above]   A "double" quasar
*Two images (**a** and **b**) of the same quasar are seen in this radio view. Light from the distant quasar is deflected to either side of a massive galaxy located between us and the quasar. A faint image of the deflecting galaxy is seen directly above image **b**. The jetlike feature protruding from the upper image (**a**) does not appear alongside the lower image because the jet is too far away from the required quasar-galaxy-Earth alignment. (VLA; NRAO)*

Figure 15-7 [right]   The Einstein ring?
A gravitational lens should produce a ringlike image if there is perfect alignment between the background light source, the observer, and the deflecting galaxy or black hole. This radio object in the constellation of Leo may be the first known example of this so-called Einstein ring. (VLA; NRAO)

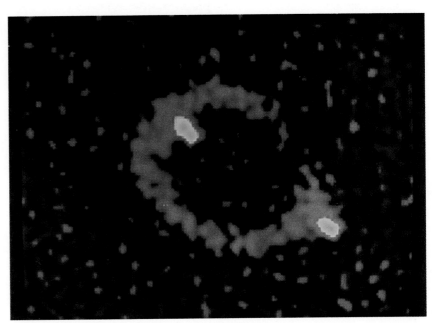

Black holes have been detected in close binary-star systems

Double stars offer the best chance of finding black holes in our Galaxy. The existence of a black hole might be revealed, for instance, if a black hole were to capture gas from a companion star.

Shortly after the launch of the *Uhuru* satellite in the early 1970s, astronomers became intrigued with an X-ray source called Cygnus X-1. The source is highly variable and irregular. Its X-ray emission flickers on time scales as short as a hundredth of a second. One of the fundamental concepts in physics is that nothing can travel faster than the speed of light. As a consequence, an object cannot vary its brightness or flicker faster than the travel time of light across the object. Because light travels 3000 km in 0.01 sec, Cygnus X-1 must be smaller than the Earth.

Cygnus X-1 occasionally emits radio radiation, and in 1971 radio astronomers succeeded in identifying it with the star HDE 226868 (see Figure 15-8). Spectroscopic observations promptly revealed that HDE 226868 is a B0 supergiant. Such stars do not emit significant amounts of X rays, and thus

Figure 15-8   HDE 226868   This star is the optical companion of the X-ray source Cygnus X-1. The star is a B0 supergiant located about 8000 light years from Earth. Many astronomers agree that Cygnus X-1 is probably a black hole. This photograph was taken with the 200-in. telescope at Palomar. (Courtesy of J. Kristian, Mt. Wilson and Las Campanas Observatories)

HDE 226868

HDE 226868 alone cannot be Cygnus X-1. Double stars are very common, and so astronomers began to suspect that the visible star and the X-ray source are in orbit about each other.

Further spectroscopic observations soon showed that the spectral lines in the spectrum of HDE 226868 shift back and forth with a period of 5.6 days. This behavior is characteristic of a single-line spectroscopic binary, wherein the spectral features of only one of the two stars are detected. The companion of HDE 226868 is too dim to produce its own set of spectral lines. The clear implication is that HDE 226868 and Cygnus X-1 are the two components of a double-star system.

Using the mass–luminosity relation for a B0 star, HDE 226868 is estimated to have a mass of roughly 30 $M_\odot$. This information implies that Cygnus X-1 must have a mass greater than 6 $M_\odot$, otherwise it would not exert enough gravitational pull to make the B0 star wobble by the amount deduced from the periodic Doppler shifting of its spectral lines. Six solar masses is too large for Cygnus X-1 to be either a white dwarf or a neutron star, and so the only remaining possibility is a black hole.

Of course, the X rays from Cygnus X-1 do not come from the black hole itself. Gas captured from HDE 226868 goes into orbit about the hole, forming an accretion disk about 4 million kilometers in diameter (see Figure 15-9). As material in the disk spirals in toward the hole, friction heats the gas to temperatures approaching 2 million K. In the final 200 km above the hole, these extremely hot gases emit the X rays that we detect with our satellites. Presumably, the X-ray flickering is caused by small hot spots on the rapidly rotating inner edge of the accretion disk. In this way, the black

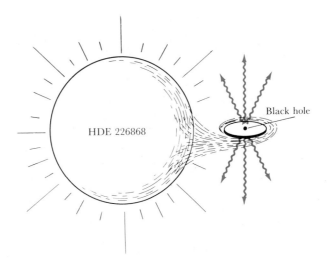

Figure 15-9 *The Cygnus X-1 system*
A stellar wind from HDE 226868 pours matter onto an accretion disk surrounding a black hole. The in-falling gases are heated to high temperatures as they spiral in toward the hole. At the inner edge of the disk, just above the black hole, the gases are so hot that they emit vast quantities of X rays. An artist's rendition of this system is seen on the opening page of this chapter.

HDE 226868

Black hole

hole's existence is announced by doomed gases just before they plunge to oblivion.

In the early 1980s, a binary system similar to Cygnus X-1 was identified in the nearby galaxy called the Large Magellanic Cloud. The X-ray source, called LMC X-3, exhibits rapid fluctuations just like those of Cygnus X-1. LMC X-3 circles a B3 main-sequence star every 1.7 days. From its orbital data, astronomers conclude that the mass of the compact X-ray source is probably about 9 $M_\odot$. Once again, a black hole is the only reasonable candidate for such a massive, compact object.

There are several other rapidly flickering X-ray sources in binary systems. All are excellent black-hole candidates. They include Circinus X-1 and GX 339-4 in Scorpius. With these and similar objects, a great deal of observational effort is necessary to rule out all non–black-hole explanations of the data. Only then can we feel confident that additional black holes have been discovered.

The black hole is one of the most bizarre and fantastic objects ever to emerge from modern physical science. Although the idea of black holes was initially met with skepticism, it is now clear that many of the stars we see in the sky are doomed someday to disappear from the universe, leaving behind only black holes. Even more astounding is the idea that enormous black holes—containing millions or even billions of solar masses—are located at the centers of many galaxies and quasars. Indeed, as we shall see in the next chapter, one of these monstrosities may be lurking at the center of the Milky Way, only 10 kiloparsecs from the Earth.

Summary

· The general theory of relativity asserts that gravity causes space to become curved and time to slow down. These effects are significant only in the vicinity of large masses or very compact objects.

· If a stellar corpse has a mass greater than about 3 $M_\odot$, gravitational compression will make the object so dense that the escape velocity exceeds the speed of light. The corpse then contracts rapidly to a single point called a singularity.

- The singularity is surrounded by a surface called the event horizon (where the escape velocity equals the speed of light). Nothing—not even light—can escape from the region inside the event horizon.

- A black hole (a singularity surrounded by an event horizon) has only three physical properties: mass, electric charge, and angular momentum.

- In the case of a rotating black hole, a region called the ergosphere surrounds the event horizon. In the ergosphere, space and time themselves are dragged along with the rotation of the black hole.

- The general theory of relativity predicts the existence of gravitational radiation. Gravitational waves are ripples in the overall geometry of space and time that are produced by moving masses.

- The gravitational field of a black hole or a galaxy distorts the images of background objects. Several examples of such gravitational lenses have been discovered.

- A few binary-star systems contain a black hole. In such a system, gases captured from the companion star by the black hole emit detectable X rays.

Review questions

1 Under what circumstances are degenerate-electron pressure and degenerate-neutron pressure incapable of preventing the complete gravitational collapse of a dead star?

2 In what way is a black hole blacker than black ink or a black piece of paper?

3 If the Sun suddenly became a black hole, how would the Earth's orbit be affected?

4 According to general relativity, why can't some sort of yet undiscovered degenerate pressure prevent the matter inside a black hole from collapsing all the way down to the singularity?

5 Why do you suppose that all the black-hole candidates mentioned in the text are members of very short period binary systems?

6 If light cannot escape from a black hole, how can we detect X rays from such objects?

Advanced questions

7 As a binary system loses energy by emitting gravitational waves, why do its members *speed up*, and why does the period become *shorter*?

8 If more massive stars evolve and die before less massive stars, why do all black hole candidates have lower masses than their normal stellar companions?

9 Under what circumstances might a white dwarf or neutron star in a double-star system become a black hole?

Discussion questions

10 Describe the kinds of observations you might make in order to locate and identify black holes.

11 Speculate on the effects you might encounter on a trip to the center of a black hole.

For further reading

Goldsmith, D. "When Time Slows Down." *Mercury,* May/June 1975, p. 2.

Kaufmann, W. *Black Holes and Warped Spacetime.* W. H. Freeman and Company, 1979.

Kaufmann, W. *Cosmic Frontiers of General Relativity.* Little, Brown, 1977.

McClintock, J. "Do Black Holes Exist?" *Sky & Telescope,* January 1988, p. 28.

Parker, B. "In and Around Black Holes." *Astronomy,* October 1986, p. 6.

Stokes, G., and Michalsky, J. "Cygnus X-1." *Mercury,* May/June 1979, p. 60.

Thorne, K. "The Search for Black Holes." *Scientific American,* December 1974.

Turner, E. "Gravitational Lenses." *Scientific American,* July 1988.

16

The Milky Way Galaxy

Our Galaxy   *This wide-angle photograph, taken from Australia, spans 180° of the Milky Way, from the Southern Cross at the lower left to Cygnus at the upper right. The center of the Galaxy is in the constellation of Sagittarius, in the middle of this photograph. Figure 16-1 is a comparable photograph showing the northern Milky Way. (Courtesy of D. di Cicco)*

The hazy band across the night sky called the Milky Way is our edge-on view of the disk of our own Galaxy. In this chapter we see how astronomers have learned about the size, shape, and rotation of the Galaxy. We learn about the processes that produce the spiral arms in galaxies, and we probe the mysterious source of energy at the galactic center. These studies of our own Galaxy prepare us to explore other galaxies in the next chapter.

On a clear, moonless night, far from city lights, you can often see a hazy, luminous band stretching across the sky. Ancient peoples devised fanciful myths to account for this "Milky Way" among the constellations. Today we realize that this hazy band is actually our view from inside a vast disk-shaped assemblage of several hundred billion stars that includes the Sun.

In studying the Milky Way, we explore the universe on a large scale and ask comprehensive questions. Instead of focusing on the location and life of an isolated star, we look at the overall arrangement and history of a huge stellar community of which the Sun is a member.

The Sun is located in the disk of the Galaxy, about 25,000 light years from the galactic center

Galileo was the first person to look at the Milky Way through a telescope. He immediately discovered that it is composed of countless dim stars. The Milky Way stretches all the way around the sky in a continuous band that is almost perpendicular to the plane of the ecliptic. Figure 16-1 is a wide-angle photograph showing roughly half of the Milky Way.

Because the Milky Way completely encircles us, astronomers in the eighteenth century began to suspect that the Sun and all the stars in the sky are part of an enormous disk-shaped assemblage called **the Milky Way Galaxy.** In the 1780s, William Herschel attempted to deduce the Sun's location in the Galaxy by counting the number of stars in 683 regions of the sky. He reasoned that the greatest density of stars should be seen toward the Galaxy's center and a lesser density seen toward the edge of the Galaxy. Herschel found roughly the same density of stars all along the Milky Way and concluded that we are at the center of the Galaxy. We now know that he was wrong.

The reason for Herschel's mistake was revealed in the 1930s by R. J. Trumpler. While studying star clusters, Trumpler discovered that remote clusters appear unusually dim, more so than would be expected from their distance alone. Trumpler therefore concluded that interstellar space must not be a perfect vacuum: it contains dust that absorbs light from distant stars. Like the stars, this obscuring material is concentrated in the plane of the Galaxy.

Figure 16-1   The Milky Way   This wide-angle photograph shows the Milky Way extending from Sagittarius on the left to Cassiopeia on the right. Note the dark lanes and blotches. This mottling is caused by interstellar gas and dust that obscure the light from background stars. (Steward Observatory)

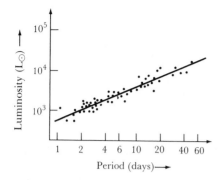

Figure 16-2 The period–luminosity relation *This graph shows the relationship between the periods and luminosities of classical (Type I) Cepheid variables. Each dot represents a Cepheid whose brightness and period have been measured. The line is the "best fit" to the data. (Adapted from H. C. Arp)*

Great patches of this interstellar dust are clearly visible in wide-angle photographs such as the one shown in Figure 16-1. At optical wavelengths, the center of the Galaxy is totally obscured from our view. The absorption of starlight by interstellar dust misled Herschel. Because he was actually seeing only the nearest stars in the Galaxy, he had no true idea either of the enormous size of the Galaxy or of the vast number of stars concentrated around the galactic center.

Because interstellar dust is concentrated in the plane of the Galaxy, the absorption of starlight is strongest in those parts of the sky covered by the Milky Way. However, our view is relatively unobscured to either side of the Milky Way. Knowledge of our position in the Galaxy came from observations of globular clusters in these unobscured portions of the sky. Before we turn to those observations, we must review the nature of variable stars.

In 1912, the American astronomer Henrietta Leavitt published her important discovery of the period–luminosity relation for classical (Type I) Cepheid variables. As we saw in Chapter 13 (recall Figure 13-22), Cepheid variables are pulsating stars whose brightness varies in a characteristic way. Leavitt studied numerous Cepheids in the Small Magellanic Cloud (a small galaxy very near the Milky Way) and found that their periods of light variation are directly related to their average luminosities (see Figure 16-2). Today, astronomers realize that there are two kinds of Cepheid variables: the metal-rich Type I Cepheids that Leavitt studied, and the metal-poor Type II Cepheids. As shown in Figure 13-23, the Type II Cepheids are slightly dimmer than the Type I.

The period–luminosity law is a very important tool in astronomy because it can be used to determine distances. For example, suppose you find a Cepheid variable in the sky. By measuring its period and using a graph such as Figure 16-2, you promptly discover the star's average luminosity. This measure of the star's true brightness can easily be expressed as an absolute magnitude. Meanwhile, you observe the star's apparent magnitude. Since you now know both the apparent and the absolute magnitudes, you can calculate the star's distance.

Shortly after Leavitt's discovery, Harlow Shapley, a young astronomer at the Mount Wilson Observatory in Southern California, became very interested in the family of pulsating stars known as RR Lyrae variables that are quite similar to Cepheid variables. RR Lyrae variables are commonly found in globular clusters (see Figure 16-3). Shapley guessed that these variables are simply short-period classical Cepheids like those Leavitt had studied.

By 1915, Shapley had noticed a peculiar property of globular clusters. Ordinary stars and open star clusters are rather uniformly spread along the Milky Way. The majority of the globular clusters that Shapley studied, however, were preferentially located in one half of the sky widely scattered around the portion of the Milky Way in Sagittarius. Figure 16-4 shows two globular clusters in this part of the sky.

Shapley used the period–luminosity relation to determine the distances to the then-known 93 globular clusters in the sky. From their directions and distances, he mapped out the three-dimensional distribution of these clusters in space. By 1917, Shapley had discovered that the globular clusters form a huge spherical system that is not centered on the Earth. The clusters are instead centered about a point in the Milky Way toward the constellation of Sagittarius. Shapley then made the bold conjecture (subsequently con-

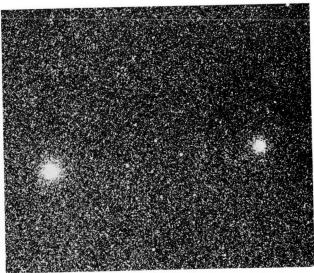

firmed) that the globular clusters outline the true size and extent of the Galaxy.

Astronomers today estimate the **disk** of the Galaxy to be about 80,000 light years in diameter and about 2000 light years thick (see Figure 16-5). These dimensions are somewhat uncertain (perhaps by 10 to 15 percent) and are slightly smaller than sizes believe to be correct only a few years ago.

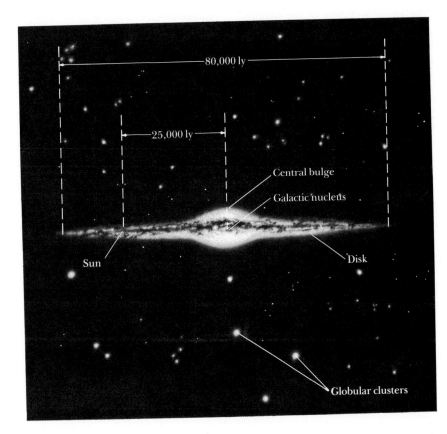

Figure 16-5   Our Galaxy (edge-on view)
There are three major components to our Galaxy: a thin disk, a central bulge, and a halo. The disk contains gas and dust along with young, metal-rich stars. The halo is composed almost exclusively of old, metal-poor stars. The central bulge around the galactic nucleus is a mixture of metal-rich and metal-poor stars.

Figure 16-6 *Edge-on view of a spiral galaxy* *If we could view our Galaxy edge on from a great distance, it would probably look like this galaxy in the constellation of Coma Berenices. A thin layer of dust and gas is clearly visible in the plane of the galaxy. Also note the reddish color of the bulge that surrounds the galaxy's nucleus. (U.S. Naval Observatory)*

The **galactic nucleus** is about 25,000 light years from Earth and is surrounded by a spherical distribution of stars called the **central bulge.** The spherical distribution of globular clusters outlines the **halo** of the Galaxy. If we could view the Galaxy edge on from a very great distance, it would probably look much like the galaxy shown in Figure 16-6.

The spiral structure of the Galaxy has been determined from radio observations

Because interstellar dust effectively obscures our optical views in the plane of the Galaxy, a detailed understanding of the structure of the galactic disk had to await the development of radio astronomy. Because of their long wavelengths, radio waves easily penetrate the interstellar medium without being scattered or absorbed. As we shall see in this section, radio observations reveal that the Galaxy has spiral-shaped concentrations of gas and dust called **spiral arms** unwinding from the center in the shape of a pinwheel.

We have seen that hydrogen is by far the most abundant element in the universe. By looking for concentrations of hydrogen gas, we should therefore detect important clues about the structure of the disk of the Galaxy. Unfortunately, the major electron transitions in the hydrogen atom (recall Figure 5-12) produce photons at ultraviolet and visible wavelengths that do not penetrate the interstellar medium. What hope do we have of detecting all this hydrogen?

In addition to mass and charge, particles such as protons and electrons possess a tiny amount of angular momentum commonly called **spin.** An electron or a proton can be crudely visualized as a tiny spinning sphere.

Figure 16-7 *Electron spin and the hydrogen atom* *In the lowest orbit of the hydrogen atom, the electron and the proton can be spinning either in the same or opposite directions. When the electron flips over, the atom either gains or loses a tiny amount of energy. This energy is either absorbed or emitted as photons with wavelengths of 21 cm.*

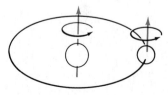

Parallel spins

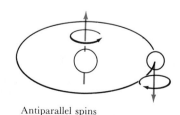

Antiparallel spins

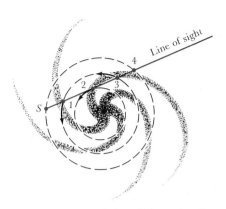

Figure 16-8 A technique for mapping the Galaxy *Hydrogen clouds at different locations along our line of sight are moving at different speeds. Radio waves from the various gas clouds are therefore subjected to slightly different Doppler shifts, permitting radio astronomers to sort out the gas clouds and map the Galaxy.*

According to the laws of quantum mechanics, the electron and proton in a hydrogen atom can be spinning either in the same or opposite directions (see Figure 16-7), but they can have no other spin orientations. If the electron flips over from one configuration to the other, the hydrogen atom must gain or lose a tiny amount of energy. For example, in going from parallel to antiparallel spins, the atom emits a low-energy photon whose wavelength is 21 cm. In 1951, a team of astronomers succeeded in detecting the faint hiss of 21-cm radio static from interstellar hydrogen.

The detection of 21-cm radio radiation was a major breakthrough that permitted astronomers to map the structure of the galactic disk. To see how this is done, suppose that you aim your radio telescope along a particular line of sight across the Galaxy as sketched in Figure 16-8. Your radio receiver picks up 21-cm emission from hydrogen clouds at points 1, 2, 3, and 4 (point *S* is the location of the Sun). However, the radio waves from these various clouds are Doppler shifted by slightly different amounts because they are moving at different speeds as they travel around the Galaxy.

The various Doppler shifts that occur cause the 21-cm radiation to be smeared out over a range of wavelengths. Because radio waves from gases in different parts of the Galaxy arrive at our radio telescopes with slightly different wavelengths, it is possible to sort out the various gas clouds and thus produce a map of the Galaxy, such as that shown in Figure 16-9.

A 21-cm map of our Galaxy reveals numerous arched lanes of neutral hydrogen gas but gives only a vague hint of spiral structure. Photographs of other galaxies (see Figure 16-10) show spiral arms outlined by bright stars and emission nebulae indicative of active star formation. We thus can see

Figure 16-9 A map of the Galaxy *This map, based on radio-telescope surveys of 21-cm radiation, shows the distribution of hydrogen gas in a face-on view of the Galaxy. Many hints of spiral structure are seen. The Sun's location is indicated by a white arrow near the top of the map. Details in the large, blank, wedge-shaped region toward the bottom of the map are unknown because gas in this part of the sky is moving perpendicular to our line of sight and thus does not exhibit a detectable Doppler shift. (Courtesy of G. Westerhout)*

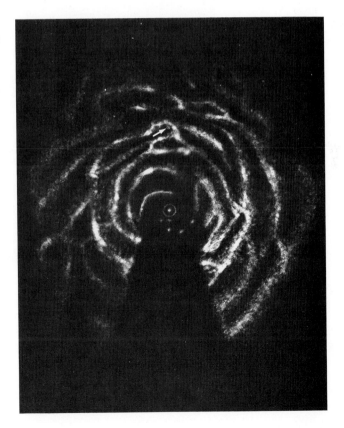

Figure 16-10 A spiral galaxy *This galaxy, called M83, is in the southern constellation of Centaurus about 12 million light years from Earth. (a) This photograph at visible wavelengths clearly shows the spiral arms illuminated by young stars and glowing H II regions. (b) This radio view at a wavelength of 21 cm shows the emission from neutral hydrogen gas. Note that the spiral arms are better distinguished by emission from ionized hydrogen than from neutral hydrogen. (Anglo-Australian Observatory; VLA, NRAO)*

a

b

that the best way to surmise the spiral structure of the Galaxy is to map the locations of star-forming complexes marked by OB associations, H II regions, and molecular clouds.

Interstellar absorption limits the range of visual observations in the plane of the Galaxy to less than 3000 pc from the Earth. Nevertheless, there are enough OB associations and H II regions visible in the sky to plot the spiral arms in the vicinity of the Sun. To locate star-forming regions far from the

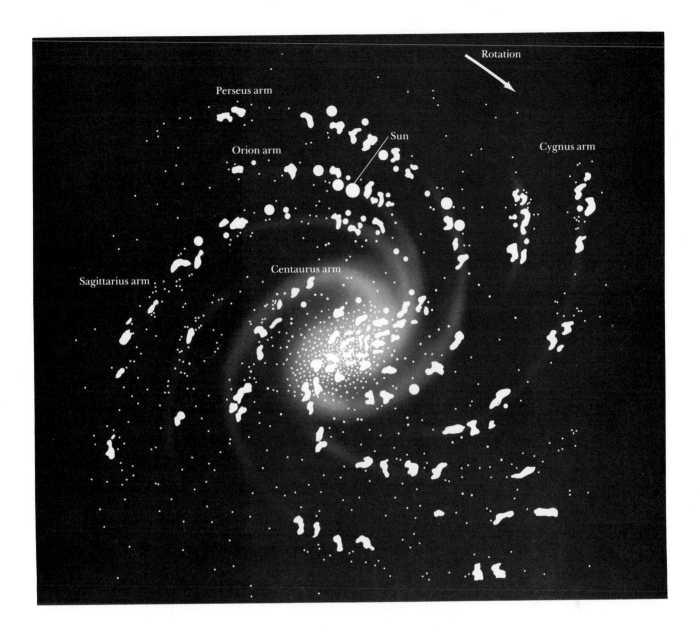

Figure labels: Rotation, Perseus arm, Sun, Cygnus arm, Orion arm, Sagittarius arm, Centaurus arm

Figure 16-11 Our Galaxy (face-on view)
Our Galaxy has four major spiral arms and several shorter segments of arms. The Sun is located on the Orion arm, between two major spiral arms. The Galaxy's diameter is about 80,000 light years, and the Sun is about 25,000 light years from the galactic center.

Sun, radio astronomers search for radio emission from carbon monoxide molecules, which signal the presence of giant molecular clouds. Taken together, these observations indicate that our Galaxy has four major spiral arms and several short arm segments (see Figure 16-11).

The Sun is located on a relatively short arm segment called the Orion arm, which includes the Orion Nebula (see Figure 13-9) and neighboring sites of vigorous star formation in that constellation (see Figure 4-23). Two major spiral arms border either side of the Sun's position. The Sagittarius arm is on the side toward the galactic center. This is the arm you see during the summer months when you look at the portion of the Milky Way stretching across Scorpius and Sagittarius (see photograph on the first page of this chapter). During winter, when our nighttime view is directed away from the galactic center, we see the Perseus arm. The remaining two major spiral arms are usually referred to as the Centaurus arm and the Cygnus arm.

Moving at half a million miles per hour, the Sun takes 200 million years to complete one orbit of our Galaxy

The existence of spiral arms suggests that galaxies rotate. However, detecting and measuring the rotation of the Galaxy has been a difficult business.

Radio observations of 21-cm radiation from hydrogen gas are helpful in studying our Galaxy's rotation. By measuring Doppler shifts, astronomers can determine the speed of objects parallel to our line of sight across the Galaxy. These observations clearly indicate that our Galaxy does not rotate like a rigid body but rather exhibits **differential rotation:** stars at different distances from the galactic center travel at different orbital speeds about the Galaxy.

The motions of stars in the sky reveal some aspects of galactic rotation. Because of differential rotation, the Sun is like a car on a circular freeway with the fast lane on one side and the slow lane on the other. As sketched in Figure 16-12, stars in the fast lane are passing the Sun and thus appear to be moving in one direction, while stars in the slow lane are being overtaken by the Sun and therefore appear to be moving in the opposite direction.

Unfortunately, like the 21-cm observations, this study reveals only how fast things are moving relative to the Sun. Of course, the Sun itself is moving. To get a complete picture of the Galaxy's rotation, we must find out how fast the Sun is traveling.

A method of computing this speed was proposed by the Swedish astronomer Bertil Lindblad. Not all the stars in the sky move in the orderly pattern in Figure 16-12. Distant galaxies and some components of our own, such as globular clusters, do not participate in the general rotation of the Galaxy; instead they seem to have more or less random motions. Using the average of these random motions as a background, astronomers estimate that the Sun moves along its orbit about the galactic center at a speed of 250 km/sec, or about $\frac{1}{2}$ million miles per hour.

Given the Sun's speed and its distance from the galactic center, astronomers can calculate the Sun's orbital period. Traveling at $\frac{1}{2}$ million miles per hour, our Sun takes 200 million years to complete one trip around the Galaxy. This result demonstrates how vast our Galaxy is.

Knowing the Sun's orbit around the Galaxy, we can use Kepler's third law to estimate the mass of the Galaxy. Putting in the numbers, we obtain a mass of 1.1×10^{11} M$_\odot$.

This calculation gives us only the mass inside the Sun's orbit. The matter exterior to the Sun's orbit does not affect the Sun's motion and thus cannot be calculated from Kepler's law. We know there is matter out there. In recent years, astronomers have been astonished to discover just how much matter actually lies beyond the orbit of the Sun.

Knowing the speed of the Sun, we can convert the Doppler shifts measured by radio astronomers into actual speeds of the spiral arms. This computation gives us a **rotation curve,** a graph of the velocity of galactic rotation measured outward from the galactic center (see Figure 16-13).

This graph is surprising because the rotation curve keeps rising, even out to a distance of 60,000 light years from the galactic nucleus. Thus we still have not detected the actual edge of our Galaxy. According to Kepler's third law, the orbital speeds of stars or gas clouds beyond the confines of most of the Galaxy's mass should decrease with their distance away from the Galaxy's center, just as the orbital speeds of the planets decrease with the planets' distance away from the Sun. Instead, galactic orbital speeds continue to climb well beyond the visible edge of the galactic disk. A surprising

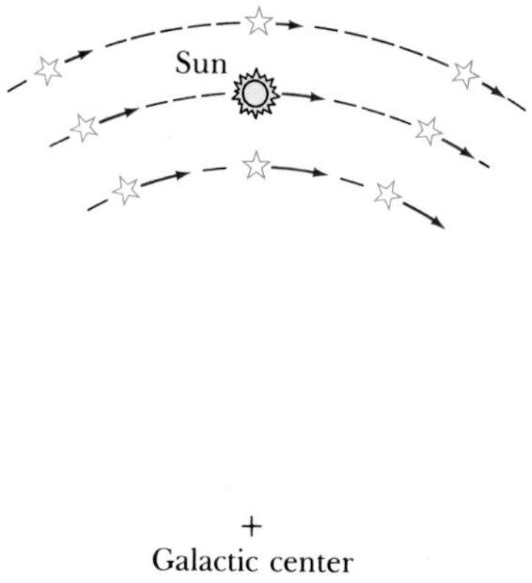

Figure 16-12 *Differential rotation of the Galaxy* *Near the Sun the orbital speeds of stars decrease outward in the Galaxy. Stars inside the solar orbit are overtaking the Sun, while stars farther from the galactic center than us are lagging behind the Sun.*

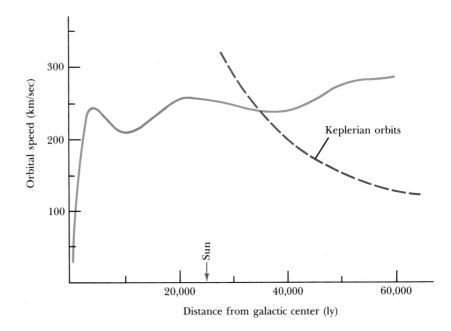

Figure 16-13 The Galaxy's rotation curve
This graph plots the orbital speeds of stars and gas in the Galaxy out to a distance of 60,000 light years from the galactic center. The dashed red line labeled "Keplerian orbits" indicates how the rotation curve should decline beyond the edge of the Galaxy. The fact that the rotation curve continues to rise proves that vast quantities of subluminous matter surrounds the Galaxy.

amount of matter must therefore be scattered around the edges of the Galaxy. In fact, our Galaxy's mass could easily be at least 6×10^{11} M$_\odot$. To make matters even more mysterious, this outlying matter is dark and does not show up on photographs. Astronomers suspect that this dark matter is spherically distributed all around the Galaxy along with the globular clusters. Thus the Galaxy's halo is more massive than previously expected. The nature of this dark mass—whether black holes, gas, or dim stars—is, so far, a complete mystery.

Self-sustaining star formation can produce spiral arms

That spiral arms should exist at all confounded astronomers for many years. Many galaxies exhibit beautiful arching arms outlined by brilliant H II regions and OB associations (recall Figure 16-10). As we think through the effects of a galaxy's rotation, a dilemma arises. All spiral galaxies have rotation curves similar to our own. As Figure 16-13 demonstrated, the velocity of stars and gases is fairly constant over a large portion of a galaxy's disk. However, the farther away that stars are from a galaxy's center, the farther they must travel to complete one orbit of the galaxy. Thus, stars and gases in the outskirts of a galaxy take much longer to complete an orbit than does material near the galaxy's center. Consequently, the spiral arms should eventually "wind up." After a few galactic rotations, the spiral structure should disappear altogether.

Spiral arms vary considerably from galaxy to galaxy. In some galaxies, called **flocculent spirals** from the word meaning "fleecy," the spiral arms are broad, fuzzy, chaotic, and poorly defined (see Figure 16-14*a*). In other galaxies, called **grand-design spirals,** the spiral arms are thin, delicate, graceful, and well defined (see Figure 16-14*b*). This range of shapes suggests that more than one mechanism can give rise to the spiral structure of a galaxy.

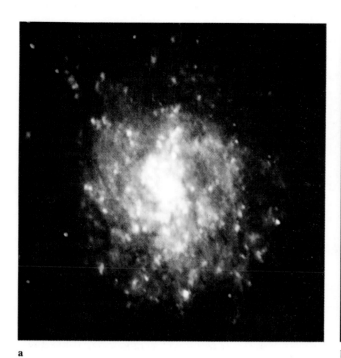

a

b

Figure 16-14 *Variety in spiral arms* The differences from one spiral galaxy to another suggest that more than one mechanism can create spiral arms. **(a)** This galaxy has fuzzy, poorly defined spiral arms. **(b)** This galaxy has thin, well-defined spiral arms. (Courtesy of P. Seiden, D. Elmegreen, B. Elmegreen, and A. Mobarak; IBM)

The theory of **self-propagating star formation** provides a straightforward explanation of spiral arms that takes into account the galaxy's differential rotation. Imagine that star formation begins in a dense interstellar cloud somewhere in the disk of a galaxy that does not yet have spiral arms. As soon as hot, massive stars form, their radiation compresses nearby nebulosities, triggering the formation of additional stars in that gas. The massive stars also become supernovae that produce shock waves, which further compress the surrounding interstellar medium, thus encouraging still more star formation. As the star-forming region grows, the galaxy's differential rotation drags the inner edges ahead of the outer edges. This conglomeration of newly formed stars, which includes bright O and B stars and glowing nebulae that make these areas quite conspicuous, soon becomes stretched out in the form of a spiral arm.

Spiral arms produced by bursts of star formation come and go more or less at random across a galaxy. Bits and pieces of spiral arms appear where star formation has recently begun but fade and disappear at other locations where all the massive stars have died off. Such galaxies thus have a chaotic appearance with poorly defined spiral arms, as was the case in Figure 16-14a. We must turn to an alternative explanation for spiral structure to account for the orderly appearance of other galaxies.

Spiral arms can also be produced by density waves that sweep across a galaxy

In the 1920s, Bertil Lindblad proposed that the spiral arms of a galaxy are a persistent pattern that moves among the stars. A similar situation is the movement of waves on the ocean. As the waves move across the surface of the water, the individual water molecules simply bob up and down in little circles. The waves are simply a pattern that moves across the water; no water actually travels with the wave pattern. Lindblad, in fact, used the term **density waves** in describing a possible cause of spiral structure.

This density-wave theory was greatly elaborated upon and mathematically embellished by the American astronomers C. C. Lin and Frank Shu in the mid-1960s. Lin and Shu argued that density waves passing through the disk of a galaxy cause material to "pile up" temporarily. A spiral arm therefore is simply a temporary enhancement, or compression, of the material in a galaxy.

This situation is analogous to a traffic jam. Imagine workers painting a line down a busy freeway. The cars normally cruise at 55 mph, but the crew of painters causes a temporary bottleneck. The cars must slow down temporarily to avoid hitting anyone. As seen from the air, there is a noticeable congestion of cars around the painters. An individual car spends only a few moments in the traffic jam before resuming its usual 55-mph speed. The traffic jam in the vicinity of the painters lasts all day long, however, inching its way along the road. The traffic jam, seen from an airplane, is simply a temporary enhancement of the number of cars in a particular location.

To better understand how a density wave operates in a galaxy think once again about the ocean. If the water molecules were left completely undisturbed, the surface of the ocean would be perfectly smooth. In reality, however, these molecules are constantly buffeted by small disturbances, or perturbations, such as the wind. Because of a perturbation, one molecule pushes the next, which pushes the next, and so on. The result is a water wave. Individual molecules on the surface of the ocean move in tiny elliptical paths as the wave pattern passes across the water (see Figure 16-15a).

In a galaxy, the stars are separated by vast distances. Collisions between stars virtually never happen. Nevertheless, stars do interact because they are affected by each other's gravity. In water waves or sound waves, molecular forces are responsible for orchestrating the motions of molecules. In a galaxy, the force of gravity controls the interactions between stars.

Seen from above, the undisturbed orbit of a star about the center of a galaxy would be a nearly perfect circle. However, the motions of other

Figure 16-15 Water waves and stellar orbits (a) In a water wave, each molecule rotates about a point on the undisturbed water level in a tiny ellipse. (b) Similarly, a small disturbance in the orbit of a star can cause the star to rotate in tiny ellipses about its original orbit. The actual path of the star is a precessing ellipse. (Adapted from A. Toomre)

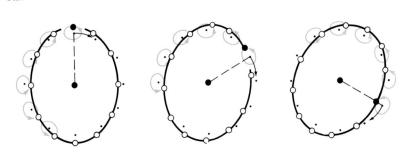

a Water wave

b Star orbit

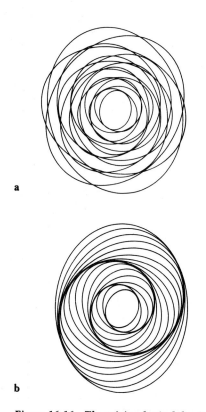

a

b

Figure 16-16 The origin of spiral density waves *Both drawings have exactly the same number of ellipses, each representing the orbit of a star.* **(a)** *Randomly oriented ellipses.* **(b)** *Adjacent ellipses are inclined to each other at a constant angle.*

matter in the galaxy produce small gravitational perturbations that cause the star to deviate from its undisturbed orbit. Just as a water molecule bobs up and down on the surface of the ocean, a star oscillates back and forth about its undisturbed orbit. Lindblad demonstrated that these oscillations can be described by thinking of the star as attached to a tiny epicycle. As sketched in Figure 16-15b, the star rotates counterclockwise around the epicycle while the epicycle itself is moving clockwise along the undisturbed path. The final path of the star is a **precessing ellipse,** an ellipse that rotates. Of course, the gravity of this star affects the motions of its neighbors, and so a wave disturbance propagates from one stellar orbit to the next.

The precessing elliptical orbits of stars in a galaxy are not randomly oriented, as sketched in Figure 16-16a. Instead, there is a correlation between orbits because of their gravitational interaction: each precessing elliptical orbit is tilted with respect to its neighbor through a specific angle. The result is a beautiful spiral pattern shown in Figure 16-16b. This pattern arises where the ellipses are bunched closest together. Although stars are randomly distributed along their respective orbits, a concentration of stars occurs along the spiral patterns because that is where their orbits are close together for long stretches.

The enhancement of stars along these spiral patterns has a profound effect on the interstellar gas and dust. Because of the concentration of stars, there is an increased gravitational attraction all along the spiral. This added force has almost no effect on the massive stars, which simply continue to lumber along their orbits. The lightweight atoms and molecules in the interstellar medium are, however, readily sucked into the gravitational well along the spiral forming the crest of the density wave.

As the spiral patterns precess, the density waves move through the material of a galaxy at a speed of roughly 30 km/sec, which is slower than the stars are moving. On its own, however, the interstellar gas can transport a disturbance, such as a slight compression, at a speed of only 10 km/sec, which is the speed of sound in the interstellar medium. The density wave is thus supersonic, because its speed through the interstellar gas is greater than the speed of sound in that gas. As happens with a supersonic airplane traveling through the air, a shock wave builds up along the leading edge of the density wave. Shock waves are characterized by a sudden, abrupt compression of the medium through which they move, which is why you hear a "sonic boom" from a supersonic airplane.

The density-wave theory seems to explain many of the properties of spiral structure. As spiral density waves sweep through the plane of a galaxy, they recycle the interstellar medium. Old gas and dust left behind from dead stars are compressed into new nebulae to form new stars. The sprawling dust lanes alongside the string of emission nebulae outlining a spiral arm attest to the recent passage of a compressional shock wave (examine Figure 16-14b). Because the material left over from the deaths of ancient stars is enriched in heavy elements, new generations of stars are more metal-rich than were their ancestors.

Many gaps remain in our understanding of spiral structure. For instance, density waves expend an enormous amount of energy to compress the interstellar gas and dust. Where do they get this energy from? One possibility is gravitational interaction with a companion galaxy. When two galaxies pass near each other, one galaxy can perturb the other so as to induce spiral structure. We shall examine how this happens in the next chapter when we discuss collisions between galaxies. Closer to home, a second possible source of energy is the galactic nucleus.

Infrared and radio observations are used to probe the galactic nucleus whose nature is poorly understood

The nucleus of our Galaxy is an active, crowded place. The number of stars in Figure 16-4 gives some hint of the stellar congestion there. If you lived on a planet near the galactic center, you could see a million stars as bright as Sirius, the brightest single star in our own nighttime sky. The total intensity of starlight from all those nearby stars would be equivalent to 200 of our full moons. In effect, night would never really fall on a planet near the center of the Galaxy.

Because of the severe interstellar absorption at visual wavelengths, some of our most important information about the galactic center comes from infrared and radio observations. Figure 16-17 shows three views looking toward the center of the Galaxy. Figure 16-17a is a wide-angle photograph at visible wavelengths covering a 50° segment of the Milky Way through Sagittarius and Scorpius. The same field of view at infrared wavelengths, from the IRAS satellite, is seen in Figure 16-17b. The prominent band across this infrared image is a thin layer of dust in the plane of the Galaxy. The numerous knots and blobs along the dust layer are interstellar clouds heated by young O and B stars. Figure 16-17c is an IRAS view of the galactic center. Numerous streamers of dust (in blue) surround the galactic nucleus. The strongest infrared emission (in white) comes from Sagittarius A, which is a grouping of several powerful sources of radio waves. One of these sources, called Sagittarius A*, is believed to be the galactic nucleus.

Figure 16-17 The galactic center
*(a) This wide-angle view at visible wavelengths shows a 50° segment of the Milky Way centered on the nucleus of the Galaxy. (b) This wide-angle infrared view from IRAS covers the same area as **a**. Black represents the dimmest regions of infrared emission, with blue the next dimmest, followed by yellow and red, and white for the strongest emission. The prominent band across this photograph is a layer of dust in the plane of the Galaxy. Numerous knots and blobs along the plane of the Galaxy are interstellar clouds of gas and dust heated by nearby stars. (c) This closeup infrared view of the galactic center covers the area outlined by the white rectangle in **b**. (Courtesy of D. di Cicco; NASA)*

a

b

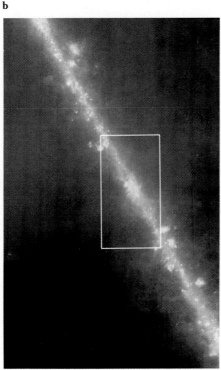

c

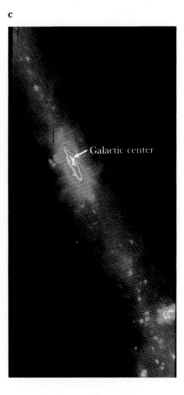

Galactic center

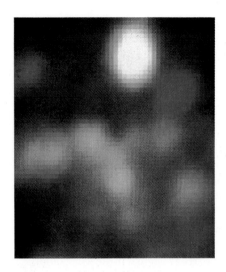

Figure 16-18 The galactic center *This composite infrared photograph shows details of the galactic center. The galactic nucleus is located at the center of the bluish Y-shaped feature just below the middle of the photograph. The area covered is only about 1 light year across. (Anglo-Australian Observatory)*

Details of Sagittarius A can be seen in the high-resolution infrared photograph shown in Figure 16-18, which covers a region at the galactic center that is only 1 light year across. The various colors represent the emission at different infrared wavelengths. The prominent white object near the top of the view is a red supergiant, which shines as brightly as 100,000 Suns. Most of the other bright objects are red giants. The bluish Y-shaped feature is the galactic nucleus, centered on Sagittarius A*.

Radio observations give a different picture of the center of our Galaxy. In 1960, Doppler shifts measurements of 21-cm radiation revealed two enormous arms of hydrogen. One arm, which is located between us and the galactic center, is approaching us at a speed of 53 km/sec. The other arm is on the other side of the galactic nucleus and is receding from us at a rate of 135 km/sec. The total amount of hydrogen in these expanding arms is at least several million solar masses. Something quite extraordinary must have happened about 10 million years ago to expel such an enormous amount of gas from the center of the Galaxy.

In addition to analyzing 21-cm radiation from neutral hydrogen gas, astronomers also detect radio noise coming from the galactic center. This radio emission, which is produced by high-speed electrons spiraling around a magnetic field, is called **synchrotron radiation.** In spite of its small size, Sagittarius A is one of the brightest sources of synchrotron radiation in the entire sky.

Some of the most detailed radio images of the galactic center come from the VLA. Figure 16-19*a* is a wide-angle view of Sagittarius A covering an area about 250 light years across. Huge filaments perpendicular to the plane of the Galaxy stretch 200 ly northward of the galactic center, then abruptly arch southward toward Sagittarius A. The orderly arrangement of these filaments, reminiscent of quiescent prominences on the Sun, suggests that a magnetic field may be controlling the distribution and flow of ionized gas.

The inner core of Sagittarius A, shown in Figure 16-19*b*, covers an area about 30 ly across. A pinwheellike feature surrounds Sagittarius A* at the center of this view. One of the arms of this pinwheel is part of a ring of gas and dust orbiting the galactic center.

The motion of the gas clouds around the galactic center can be deduced from Doppler shift measurements of infrared spectral lines, such as emission from singly ionized neon. In the late 1970s, measurements by Charles H. Townes and his colleagues at the University of California revealed that the Ne II line is severely broadened, perhaps from the orbital speed of the gas around the galactic nucleus. The radiation from gas coming toward us is blueshifted, whereas the radiation from receding gas is redshifted. The final result is a smeared-out spectral line covering a range of wavelengths corresponding to a range of line-of-sight velocities. On one side of the galactic nucleus, gas is coming toward us at speeds up to 200 km/sec, but on the other side, it is rushing away from us at the same speeds.

Something must be holding this high-speed gas in orbit about the galactic center. Using Kepler's third law, we can estimate that 10^6 $M_\odot$ is needed to prevent this gas from flying off into interstellar space. Townes's observations thus suggest that an object with the mass of a million Suns is concentrated at Sagittarius A*. This object must be extremely compact—much smaller than the few light years Townes's observations could resolve. Many

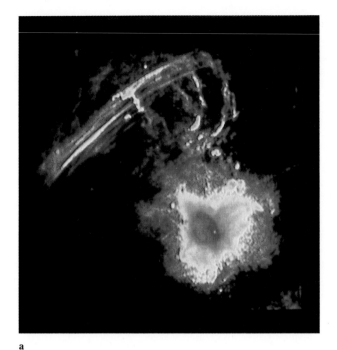

a

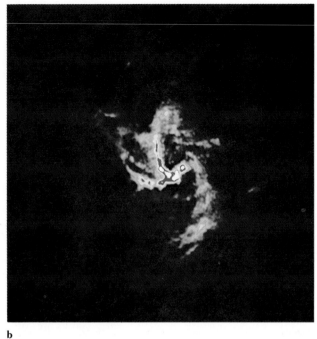

b

Figure 16-19 Two radio views of the galactic nucleus These two pictures show the appearance of the center of our Galaxy at radio wavelengths. The strongest radio emission is shown in red, weaker emission being colored green through blue. **(a)** This view covers an area of the sky about the same size as the full moon, corresponding to a distance of 250 ly across. The parallel filaments may be associated with a magnetic field. The galactic nucleus is toward the lower right, at the center of the strongest emission. **(b)** This high-resolution view shows details of the galactic center covering an area 30 ly across. The pinwheellike structure is centered on Sagittarius A*. (VLA, NRAO)

astronomers, such as Martin Ryle of Cambridge University, argue that an object this massive and this compact could only be a black hole. Because of its enormous mass, it is called a **supermassive black hole.** As we shall see in Chapter 18, astronomers find extraordinary activity also occurring at the nuclei of many other galaxies, which indicates the possibility of supermassive black holes at their centers.

Many astronomers disagree with the idea of a supermassive black hole at the center of our Galaxy. For instance, the high-resolution infrared view of the galactic center in Figure 16-18 shows nothing to suggest the presence of a supermassive black hole.

Astronomers are still groping for a better understanding of the galactic center. During the coming years, observations from Earth-orbiting satellites as well as from radio telescopes on the ground will certainly add to our knowledge and perhaps elucidate the nature of the core of the Milky Way.

Summary

· Our Galaxy has a disk about 80,000 light years in diameter and about 2000 light years thick, with a high concentration of interstellar dust and gas in the disk.

· The galactic nucleus is surrounded by a spherical distribution of stars called the central bulge; the entire Galaxy is surrounded by a spherical distribution of globular clusters called the halo of the Galaxy.

· OB associations, H II regions, and molecular clouds in the galactic disk outline huge spiral arms.

· From studies of the rotation of the Galaxy, astronomers estimate that the total mass of the Galaxy is about 6×10^{11} M$_\odot$, with much of this mass being in some nonvisible, unknown form.

- The Sun is located about 25,000 light years from the galactic nucleus, between two major spiral arms.

  The Sun moves in its orbit at a speed of about $\frac{1}{2}$ million miles per hour and takes about 200 million years to complete one orbit about the center of the Galaxy.

- Interstellar dust obscures our view at visual wavelengths along lines of sight that lie in the plane of the galactic disk.

  Hydrogen clouds can be detected despite the intervening interstellar dust by the 21-cm radio waves emitted in the spin-flip transition.

  The galactic nucleus has been studied at infrared and radio wavelengths, which also pass readily through intervening interstellar dust.

- According to the self-propagating star formation theory, spiral arms are caused by the birth of stars over an extended region of the Galaxy. Differential rotation of the Galaxy stretches the star-forming region into an elongated arch of stars and nebulae.

- According to the density wave theory, spiral arms are caused by density waves that sweep around the Galaxy.

- Each star moves in a precessing ellipse about the galactic nucleus; because the orbits are correlated, a spiral pattern is created.

  The gravitational field of this spiral pattern compresses the interstellar clouds through which it passes, thereby triggering the formation of the OB associations and H II regions that illuminate the spiral arms.

- Although infrared and radio observations have revealed many details of the galactic nucleus, astronomers are still puzzled by the processes occurring there.

  A supermassive black hole with a mass of about 5×10^6 $M_\odot$ possibly exists at the galactic center.

Review questions

1 Why do you suppose that the Milky Way is far more prominent in July than in December?

2 How would the Milky Way appear to us if our solar system were located at the edge of the Galaxy?

3 Why don't astronomers detect 21-cm radiation from the hydrogen in giant molecular clouds?

4 Describe the Doppler shifts of the 21-cm line that you would observe if the Galaxy were rotating like a rigid body.

5 Explain why globular clusters spend most of their time in the galactic halo, even though their eccentric orbits take them very close to the galactic center.

6 Approximately how many times has the solar system orbited the center of the Galaxy since the Sun and planets were formed?

7 Compare the apparent distribution of open clusters, which contain young stars, with the distribution of globular clusters relative to the Milky Way. Can you think why open clusters were originally referred to as galactic clusters?

8 Why are there no massive O and B stars in globular clusters?

9 How would you estimate the total number of stars in the Galaxy?

10 If there is a supermassive black hole at the center of our Galaxy, why hasn't it swallowed all the stars in the Galaxy?

Advanced questions

11 What can you surmise about galactic evolution from the fact that the galactic halo is dominated by metal-poor stars while metal-rich stars are predominantly found in the galactic disk?

*12 The Galaxy is about 25,000 parsecs in diameter and 600 parsecs thick. If supernovae occur randomly in the Galaxy at the rate of about five each century, how often on the average would we expect to see a supernova within 300 parsecs (1000 light years) of the Sun?

13 Speculate on the reasons for the rapid rise in the Galaxy's rotation curve (see Figure 16-13) at distances close to the galactic center.

Discussion questions

14 From what you know about stellar evolution, the interstellar medium, and the density-wave theory, explain the appearance and structure of the spiral arms of grand-design spiral galaxies.

15 What observations would you propose in order to determine the nature of the hidden mass in our Galaxy's halo?

For further reading

Bok, B. "Our Bigger and Better Galaxy." *Mercury,* September/October 1981, p. 130.

Bok, B., and Bok, P. *The Milky Way,* 5th ed. Harvard University Press, 1981.

Chaisson, E. "Journey to the Center of the Galaxy." *Astronomy,* August 1980, p. 6.

Geballe, T. "The Central Parsec of the Galaxy." *Scientific American,* July 1979.

Scoville, N., and Young, J. "Molecular Clouds, Star Formation, and Galactic Structure." *Scientific American,* April 1984.

Weaver, H. "Steps Toward Understanding the Large-Scale Structure of the Milky Way." *Mercury,* September/October 1975, p. 18; November/December 1975, p. 18; January/February 1976, p. 19.

17

Galaxies

*A **spiral galaxy*** *This galaxy, called NGC 1365, is the largest and most impressive member of a cluster of galaxies about 60 million light years from Earth. NGC 1365 is classified as a barred spiral because of the "bar" crossing through its nucleus. From the ends of the bar two distinct and quite open spiral arms branch off. The yellow color of the nucleus and the bar shows that these parts of the galaxy are dominated by old, relatively cool stars. The blue color of the arms is caused by light from young, hot stars. (European Southern Observatory)*

We now turn to the universe beyond the Milky Way and find that galaxies exist in many shapes and sizes. We learn that galaxies occur in clusters rather than being scattered randomly through space, and we discuss some of the spectacular phenomena associated with collisions of galaxies. We find that there is a simple relationship between the distance to a galaxy and the speed with which it moves away from us as measured by its redshift. In our discussion of this relationship, we learn about the measurement of intergalactic distances.

William Parsons was the third earl of Rosse in Ireland. He was rich, he liked machines, and he was fascinated with astronomy. Accordingly, he set about building gigantic telescopes. In February 1845, his *pièce de résistance* was finished. This telescope's massive mirror measured 6 ft in diameter and was mounted at one end of a 60-ft tube controlled by cables, straps, pulleys, and cranes. For many years, this triumph of nineteenth century engineering enjoyed the reputation of being the largest telescope in the world.

With this new telescope, Lord Rosse examined many of the nebulae discovered and catalogued by William Herschel. Lord Rosse observed that some of these nebulae have a distinct spiral structure. Perhaps the best example is M51 (also called NGC 5194, the 5194th object in the *New General Catalogue*, which is a list of all the nebulae and star clusters observed by William Herschel, his son John Herschel, and others).

Because Lord Rosse did not have any photographic equipment, he had to make drawings of what he saw. His drawing of M51 is shown in Figure 17-1, and a modern photograph appears in Figure 17-2. Views such as these inspired Lord Rosse to echo the famous German philosopher Immanuel Kant, who in 1755 suggested that these objects might be "island universes"—vast collections of stars far beyond the confines of the Milky Way.

Many astronomers did not subscribe to this notion of island universes. Many of the objects listed in the NGC were in fact nebulae and star clusters scattered about the Milky Way, and it seemed just as likely that these intriguing spiral nebulae could also be members of our Galaxy.

The astronomical community became increasingly divided over the nature of spiral nebulae. A debate on the topic occurred in April 1920 at the National Academy of Sciences in Washington, D.C. On one side was Harlow Shapley, the young, brilliant astronomer renowned for his recent determination of the size of the Milky Way Galaxy. Shapley believed the spiral nebulae to be relatively small, nearby objects scattered around our Galaxy like the globular clusters he had studied. Opposing Shapley was Heber D. Curtis of the Lick Observatory near San Jose, California. Curtis championed the island-universe theory and argued that each of these spiral nebulae is a rotating system of stars much like our own Galaxy.

The Shapley–Curtis debate generated much heat but little light. Nothing could be decided because no one had any firm evidence to demonstrate

Figure 17-1 [left] Lord Rosse's sketch of M51 *Using a large telescope of his own design, Lord Rosse was able to distinguish the spiral structure of this "spiral nebula." (Courtesy of Lund Humphries)*

Figure 17-2 [right] The spiral galaxy M51 *This spiral galaxy in the constellation of Canes Vantici is often called the Whirlpool galaxy because of its distinctive appearance. Its distance from Earth is about 15 million light years. The "blob" at the end of one of the spiral arms is a companion galaxy. (Canada-France-Hawaii Telescope)*

exactly how far away the spiral nebulae are. Astronomy desperately needed definitive measurements of the distances to the spiral nebulae. Making such measurements was the first great achievement of a young lawyer who abandoned a Kentucky law practice and moved to Chicago to study astronomy. His name was Edwin Hubble.

Edwin Hubble discovered the distances to galaxies and devised a system for classifying them

Figure 17-3 [left]   The Andromeda Galaxy (called M31 or NGC 224)   This nearby galaxy covers an area of the sky roughly five times as large as the full moon. Under good observing conditions, the galaxy's bright central bulge can be glimpsed with the naked eye in the constellation of Andromeda. The distance to the galaxy is 2.2 million light years. The white rectangle outlines the area shown in Figure 17-4. (Palomar Observatory)

Figure 17-4 [right]   Cepheid variables in the Andromeda Galaxy   Two Cepheid variables are identified in this view of the outskirts of the Andromeda Galaxy. Because these stars appear so faint (although they are in fact quite luminous), Hubble successfully demonstrated that the "Andromeda nebula" is extremely far away. (Mount Wilson and Las Campanas Observatories)

Edwin Hubble joined the staff of the Mount Wilson Observatory in Pasadena, California and, in 1923, took an historic photograph of the so-called Andromeda nebula, one of the spiral nebulae around which so much controversy raged. A modern photograph appears in Figure 17-3. Careful examination of Hubble's photographic plate revealed what was at first thought to be a nova. Reference to previous plates of that region soon showed, however, that the object is actually a Cepheid variable. Scrutiny of additional plates over the next several months revealed many additional Cepheids, two of which are identified in Figure 17-4.

As we saw in the previous chapter, Cepheid variables help astronomers determine distances. From the period–luminosity relation (recall Figure 13-23), astronomers can determine the average absolute magnitude of a Cepheid variable. From both the absolute magnitude and the apparent magnitude seen in the sky, a star's distance can be deduced.

Cepheid variables are intrinsically very bright. They typically have luminosities of a few thousand Suns. Hubble realized that, in order for these luminous stars to appear as dim as they do on his photographs of the "Andromeda nebula," they must be extremely far away. Straightforward calculations using modern data about Cepheids demonstrate that M31 is 2.2 million light years away, proving it to be not a traditional nebula, but an enormous stellar system located far beyond the Milky Way.

Hubble's results were presented at a meeting of the American Astronomical Society on December 30, 1924, settling the Shapley–Curtis debate once and for all. The universe was recognized to be far larger and populated with far bigger objects than anyone had thus far seriously imagined. Hubble had discovered the realm of the galaxies.

There are millions of galaxies all across the sky, visible in every unobscured direction. A typical spiral galaxy contains 100 billion stars and measures 100,000 light years in diameter. Galaxies are the biggest individual objects in the universe.

Hubble found that galaxies can be classified into four broad categories: spirals, barred spirals, ellipticals, and irregulars. These categories form the basis for the **Hubble classification scheme.**

Both M51 and M31 (Figures 17-2 and 17-3) are examples of **spiral galaxies.** While studying spiral galaxies, Hubble noted that this class could be further subdivided according to the size of the central bulge and the winding of the spiral arms. Spirals with tightly wound spiral arms and a prominent, fat central bulge he called **Sa galaxies.** Those with moderately wound spiral arms and a moderate-sized central bulge are **Sb galaxies.** Finally, loosely wound spirals with a tiny central bulge are **Sc galaxies.** Both M31 and M51 are Sb spirals. Various types of spirals are shown in Figure 17-5.

Fortunately, the size of the central bulge is correlated with the degree of winding of the spiral arms, and so we can classify even those spiral galaxies that are viewed nearly edge on. For example, M104 (see Figure 17-6a) must be an Sa galaxy because of its huge central bulge. An Sb galaxy (recall Figure 16-6) has a smaller central bulge. The tiny central bulge of an Sc is hardly noticeable at all in an edge-on view.

In **barred spiral galaxies** the spiral arms originate at the ends of a bar running through the galaxy's nucleus rather than from the nucleus itself (see Figure 17-7). As with ordinary spirals, Hubble subdivided barred spirals according to the size of their central bulge and the winding of the spiral arms. An **SBa galaxy** has a large central bulge and tightly wound spiral arms. A barred spiral with a moderate central bulge and moderately wound spiral arms is an **SBb galaxy,** and an **SBc galaxy** has loosely wound spiral arms and a tiny central bulge.

Figure 17-5 Various spiral galaxies (face-on views) Edwin Hubble classified spiral galaxies according to the winding of the spiral arms and the size of the central bulge. Three examples are shown here. (Sa courtesy of R. Schild; Sb and Sc courtesy of P. Seiden)

Sa   Sb   Sc

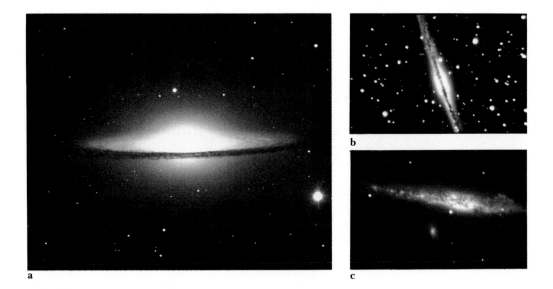

a

b

c

**Figure 17-6 Various spiral galaxies
(edge-on views)** *(a) Because of the large size
of its central bulge, this galaxy is classified as
an Sa. If we could see it face on, we would
find that the spiral arms are tightly wound
around a voluminous bulge. (b) Note the
smaller central bulge in this Sb galaxy.
(c) This galaxy is an Sc because of its
insignificant central bulge. (Sa from NOAO;
Sb and Sc courtesy of R. Schild)*

**Figure 17-7 Various barred spiral
galaxies** *As with spiral galaxies, Edwin
Hubble classified barred spirals according to
the winding of their spiral arms and the size
of the central bulge. Three examples are shown
here. (SBa and SBc courtesy of P. Seiden;
SBb courtesy of R. Schild)*

The development of a bar across a galaxy's nucleus is related to the motions of the stars within it. Computer simulations show that gravitational interactions among the stars in a huge rotating disk tend to make the stars "pile up" along a barlike structure. Indeed, it is difficult to explain why all spiral galaxies don't have bars. It seems as though the presence of a massive halo around a galaxy tends to inhibit the formation of a bar. So perhaps spiral galaxies possess such halos while barred spirals do not.

Elliptical galaxies, named for their distinctly elliptical shapes, have no spiral arms. Hubble subdivided elliptical galaxies according to how round or flattened they look. The roundest elliptical galaxies are called **E0 galaxies,** the flattest are **E7 galaxies.** Elliptical galaxies with intermediate amounts of flattening receive intermediate designations (see Figure 17-8).

The Hubble scheme classifies galaxies entirely by their appearance from our Earth-bound view. An E1 or E2 galaxy might actually be a very flattened disk of stars that just happens to be viewed face on, and a cigar-shaped E7 galaxy might look spherical when viewed end on. Statistical studies have shown, however, that elliptical galaxies of various actual degrees of ellipticity do exist.

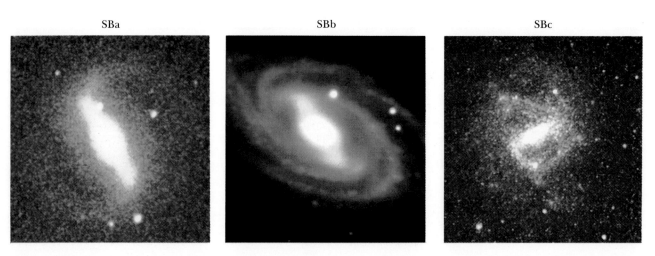

SBa

SBb

SBc

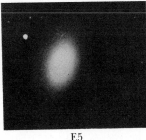

| E2 | E5 | E7 |

Figure 17-8 Various elliptical galaxies
Hubble classified elliptical galaxies according to how round or flattened they look. An E0 galaxy is round; the flattest elliptical are E7s. Three examples are shown here. (Yerkes Observatory)

Elliptical galaxies look far less dramatic than their spiral and barred spiral cousins because they are virtually devoid of interstellar gas and dust. In addition, there is no evidence of young stars in most elliptical galaxies because there is no material from which stars could have recently formed. Star formation in elliptical galaxies ended a long time ago.

Elliptical galaxies exist in an enormous range of sizes and masses. Both the biggest and the smallest galaxies in the universe are ellipticals. Figure 17-9 shows an example of a **giant elliptical galaxy.** This huge galaxy is about 20 times larger than a normal galaxy and is located near the middle of a large cluster of galaxies in the constellation of Coma Berenices.

Giant ellipticals are rather rare, but **dwarf elliptical galaxies** are quite common. Dwarf ellipticals are only a fraction the size of their normal counterparts, and each may contain only a few million stars. Some nearby dwarf ellipticals contain so few stars that these galaxies are transparent—you can actually see straight through the galaxy's nucleus and out the other side. Figure 17-10 shows a typical dwarf elliptical.

The relationships that might exist between spiral, barred spiral, and elliptical galaxies are current topics of research and debate. Hubble connected these three types of galaxies in his now-famous tuning-fork diagram (see Figure 17-11). According to this scheme, an S0 galaxy is a transition type between ellipticals and the two kinds of spirals.

Hubble found some galaxies that cannot be classified as either spirals, barred spirals, or ellipticals. He called these **irregular galaxies.** Examples include the Large Magellanic Cloud (LMC) and the Small Magellanic Cloud (SMC), both of which can be seen with the naked eye from southern latitudes.

Figure 17-9 [left] A giant elliptical galaxy
This huge elliptical galaxy sits near the center of a rich cluster of galaxies in the constellation of Coma Berenices. Many normal-sized galaxies surround this giant elliptical, which is about 2 million light years in diameter. (Courtesy of R. Schild)

Figure 17-10 [right] A dwarf elliptical galaxy *This nearby elliptical, called Leo I, is about 1 million light years from Earth. It is a satellite of the Milky Way and is thus a member of the cluster to which we belong. (Courtesy of R. Schild)*

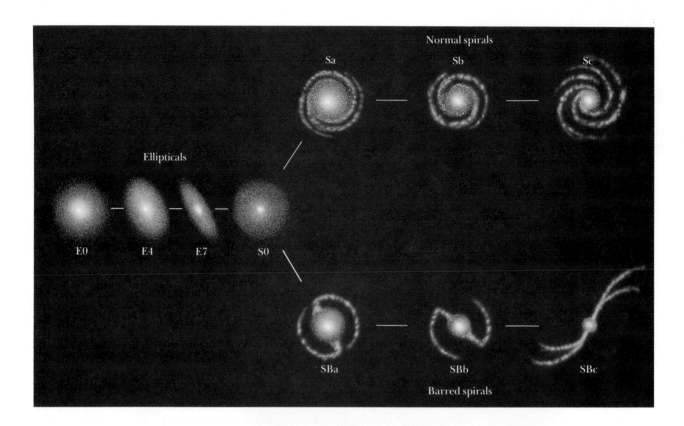

Normal spirals

Sa   Sb   Sc

Ellipticals

E0   E4   E7   S0

SBa   SBb   SBc

Barred spirals

Figure 17-11 [above]   Hubble's tuning-fork diagram   *Hubble summarized his classification scheme for regular galaxies with this tuning-fork diagram. An S0 galaxy is a transitional type between ellipticals and spirals.*

Figure 17-12 [top right]   The Large Magellanic Cloud (LMC)   *At a distance of only 160,000 light years, this galaxy is the nearest companion of our Milky Way Galaxy. Note the huge H II region (called the Tarantula Nebula or 30 Doradus) toward the left side of the photograph. Its diameter of 800 light years and mass of 5 million Suns makes it the largest known H II region. (Anglo-Australian Observatory)*

Figure 17-13 [bottom right]   The Small Magellanic Cloud (SMC)   *The SMC is only slightly farther away from us than the LMC is. Because of its sprawling, asymmetrical shape, the SMC is classified as an irregular galaxy. Note that the SMC is rich in young blue stars. (Anglo-Australian Observatory)*

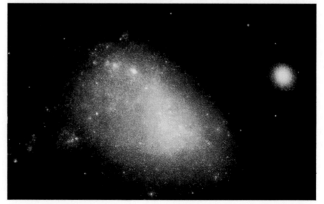

Telescopic views of the Magellanic clouds easily distinguish individual stars (see Figures 17-12 and 17-13). Note that the SMC does not exhibit any geometric symmetry characteristic of spirals or ellipticals and is therefore a true irregular. However, the LMC does apparently have a barlike structure. It is thus classified as a transition type between a barred spiral and an irregular. The Magellanic clouds are the nearest members of a cluster of galaxies to which the Milky Way and Andromeda galaxies belong.

Galaxies are grouped in clusters that are members of superclusters

Galaxies are not scattered randomly across the universe but are grouped in **clusters.** A typical cluster, called the Fornax cluster because it is located in the constellation of Fornax (the furnace), is seen in Figure 17-14.

A cluster is said to be either **poor** or **rich,** depending on how many galaxies it contains. For example, the Milky Way Galaxy, the Andromeda Galaxy, and the Large and Small Magellanic clouds belong to a poor cluster called the **Local Group.** At the present time, astronomers have identified about 30 galaxies in the Local Group, a dozen of which are dwarf ellipticals. A map of the Local Group is shown in Figure 17-15.

Figure 17-14   A cluster of galaxies
This cluster of galaxies, called the Fornax cluster, is about 60 million light years from Earth. Both elliptical and spiral galaxies are easily identified. An enlarged view of the barred spiral near the edge of this picture appears in the photograph on the first page of this chapter. (Anglo-Australian Observatory)

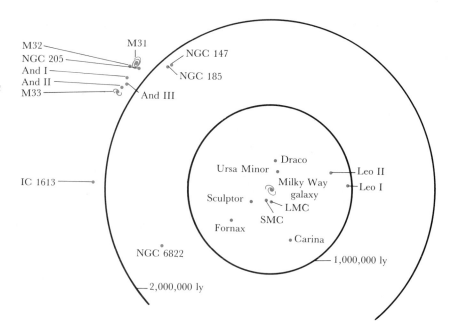

Figure 17-15 The Local Group
Our Galaxy belongs to a poor cluster consisting of about 30 galaxies, a dozen of which are dwarf ellipticals. The Andromeda Galaxy (M31) is the largest and most massive galaxy in the Local Group. The second largest is the Milky Way. Both M31 and the Milky Way Galaxy are each surrounded by a dozen satellite galaxies.

Figure 17-16 The center of the Virgo cluster *This fairly rich cluster lies about 50 million light years from us. Only the center of this huge cluster appears in this view. Note the two giant elliptical galaxies (M84 and M86). (Royal Observatory, Edinburgh)*

The nearest fairly rich cluster is the Virgo cluster. It is a sprawling collection of over 1000 galaxies covering a 10° × 12° area of the sky. The distance to the Virgo cluster is about 50 million light years, too far away for Cepheid variables to be seen from our Galaxy. Instead, the distance to the Virgo cluster has been determined by the apparent faintness of O and B supergiant stars and by the brightnesses of globular clusters surrounding some of the galaxies. The overall diameter of the Virgo cluster is about 7 million light years.

The center of the Virgo cluster is dominated by three giant elliptical galaxies. Two of them appear in Figure 17-16. These enormous galaxies are 2 million light years in diameter, 20 times as large as an ordinary elliptical or spiral. In other words, one giant elliptical is roughly the same size as the entire Local Group!

In addition to using the rich-versus-poor classification, astronomers often categorize clusters of galaxies as **regular** or **irregular,** depending on the overall shape of the cluster. The Virgo cluster, for example, is called irregular, because of the asymmetrical way its galaxies are scattered about a sprawling region of the sky. In contrast, a regular cluster has a distinctly spherical appearance, with a marked concentration of galaxies at its center. This distribution is what would be expected for a system of objects whose mutual gravitational interactions over the ages have equitably shared the system's energy among its members.

The nearest example of a rich, regular cluster is the Coma cluster, located 350 million light years from us toward the constellation of Coma Berenices. Despite the great distance to this cluster, more than 1000 bright galaxies within it are easily visible on photographic plates. Certainly, there must be many thousands of dwarf ellipticals that are too faint to be detected from our distance. The total membership of the Coma cluster may be as many as 10,000 galaxies. The core of the Coma cluster is dominated by two giant ellipticals surrounded by many normal-sized galaxies (see Figure 17-17).

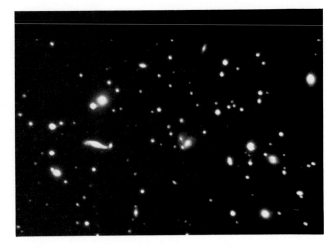

Figure 17-17 [left]   The center of the Coma cluster   *This rich, regular cluster is about 350 million light years from Earth. Only the cluster's center, dominated by two giant galaxies, appears in this view. Regular clusters are composed mostly of elliptical and S0 galaxies and are common sources of X rays. (Courtesy of R. Schild)*

Figure 17-18 [right]   The Hercules cluster   *This irregular cluster, which is about 700 million light years from Earth, contains a high proportion of spiral galaxies, often associated in pairs and small groups. (Courtesy of R. Schild)*

A correlation exists between the regular-versus-irregular classification scheme and the dominant types of galaxies in a cluster. Rich, regular clusters, like the Coma cluster, contain mostly elliptical and S0 galaxies. Only 15 percent of the Coma cluster's galaxies are spirals and irregulars. Irregular clusters, such as the Virgo cluster and the Hercules cluster (see Figure 17-18), have a more even mixture of galaxy types. For instance, of the 200 brightest galaxies in the Hercules cluster, 68 percent are spirals, 19 percent are ellipticals, and the rest are irregulars.

Clusters of galaxies are themselves grouped together, in huge associations called **superclusters.** A typical supercluster contains dozens of individual clusters spread over a volume up to 100 million light years across. Patterns in the distribution of galaxies can be seen in maps such as Figure

Figure 17-19   A million galaxies
This map displays nearly a million galaxies brighter than magnitude +19. The area shown here measures approximately 100° on each side. Although individual galaxies cannot be seen at this scale, clusters are recognizable as white blotches. Note that the clusters are not smoothly distributed across the sky but form a lacy, filamentary pattern. (Courtesy of E. J. Groth et al.)

Figure 17-20 A slice of the universe
This map shows the locations of 1099 galaxies in a thin slice of the universe. To produce this map, a team of astronomers painstakingly measured the positions of galaxies out to a distance of nearly a billion light years from Earth in a strip of the sky 6° wide and 153° long. Note that most galaxies surround voids that are roughly circular. This distribution of galaxies is like a slice cut through the suds in a kitchen sink. (Courtesy of V. de Lapparent, M. Geller, and J. Huchra)

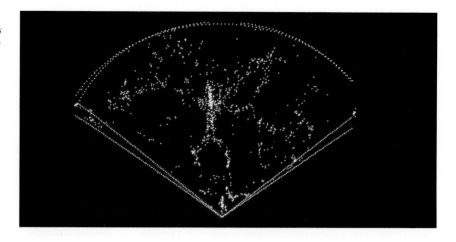

17-19, which covers many hundreds of square degrees. Note the delicate, filamentary structure spread across the sky.

The arrangement of superclusters in space was clarified in the early 1980s when astronomers began discovering enormous **voids** where exceptionally few galaxies are found. These voids are roughly spherical and measure 100 million to 400 million light years in diameter. Recent surveys reveal that galaxies are concentrated on the surfaces of these voids (see Figure 17-20). The distribution of clusters of galaxies is therefore said to be "sudsy" because it resembles a collection of giant soap bubbles. Galaxies surround voids in the same way a soap film is concentrated on the surface of bubbles. Many astronomers suspect that this sudsy pattern contains important clues about conditions shortly after the Big Bang that led to the formation of clusters of galaxies.

Most of the matter in the universe has yet to be observed

A cluster of galaxies must be a gravitationally bound system. In other words, there must be matter enough in the cluster to produce gravity sufficient to prevent the galaxies from wandering away. Nevertheless, careful examination of a rich cluster, like the Coma cluster, typically reveals that the mass of the visually luminous matter is not at all sufficient to bind the cluster gravitationally. The observed line-of-sight speeds of the cluster galaxies, measured by Doppler shifts, are so large that more mass than has been observed is needed to keep them bound in orbit about the center of the cluster. This dilemma is called the **missing-mass problem.** A lot of nonluminous matter must be scattered about each of the clusters, or else the galaxies would long ago have wandered away in random directions and the cluster would no longer exist today. Analyses demonstrate that the total mass needed to bind a typical rich cluster is 10 times greater than the mass of material that shows up on visual photographs.

Some of this mystery has been solved recently by X-ray astronomers. Satellite observations of rich clusters have revealed that X rays pour from the space between galaxies in rich clusters. This flow is evidence of substantial amounts of hot intergalactic gas at temperatures between 10 and 100 million kelvins. Analyses show that the mass of this hot intergalactic gas is typically as great as the combined mass of all the visible galaxies in the cluster.

Unfortunately, this discovery of hot intergalactic gas in rich clusters solves only part of the missing-mass problem. Most astronomers agree that a

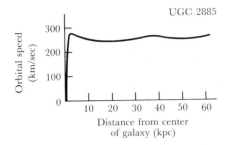

Figure 17-21 The rotation curve of a large spiral galaxy *This graph shows the orbital speed of material in the disk of the galaxy UGC 2885 (that is, the 2885th galaxy listed in the* Uppsala General Catalogue*). Many galaxies have flat rotation curves, indicating the presence of extended halos of dark matter. (Adapted from V. Rubin, K. Ford, and N. Thonnard)*

great deal of matter still remains to be discovered in rich clusters. One popular speculation is that these clusters may contain a lot of undetected dim stars. These faint stars could be located in extended halos surrounding individual galaxies and could be scattered throughout the spaces between the galaxies of a cluster.

Evidence to support the idea of extended halos comes from the rotation curves of galaxies. Many galaxies have rotation curves similar to that of our Milky Way Galaxy (recall Figure 16-13). These rotation curves remain remarkably flat out to surprisingly great distances from the galaxy's center. For example, Figure 17-21 shows the rotation curve for the Sc spiral galaxy UGC 2885. Note that the orbital speed is fairly constant out to 60 kpc (200,000 ly) from the galactic center. Beyond this distance, the galaxy's stars and H II regions are so dim and widely scattered that reliable measurements are not possible. Nevertheless, we have still not detected the true edge of this and many similar galaxies. In the outer portions of a galaxy we should see a decline in orbital speed, in accordance with Kepler's third law. Because this decline has not been observed, astronomers conclude that there must be a considerable amount of dark matter extending well beyond the visible portion of a galaxy's disk. Many proposals have been made to account for this dark matter. Some of the more exotic suggestions include black holes and various massive particles left over from the creation of the universe. The more mundane suggestions include dim stars and Jupiterlike planets. The nature of this unseen matter is one of the greatest mysteries in modern astronomy.

Colliding galaxies produce starbursts, spiral arms, and other spectacular phenomena

All the galaxies in a cluster are in orbit about their common center of mass. Occasionally two galaxies pass close enough to each other to collide. When they do collide, the stars within them pass by each other. There is so much space between the stars that the probability of two stars crashing into each other is quite small. However, huge clouds of interstellar gas and dust cannot interpenetrate. When two galaxies collide, interstellar clouds slam into each other, producing strong shock waves. The stars keep right on going, but the colliding interstellar clouds are stopped in their tracks. In this way, two colliding galaxies may be stripped of their interstellar gas and dust. The violence of the collision heats the gas stripped from these galaxies to extremely high temperatures. This process may be a major source of the hot intergalactic gas often observed in rich, regular clusters.

In a less violent collision or a near-miss between two galaxies, the compressed interstellar gas may have enough time to cool sufficiently to allow many protostars to form. Such collisions can thus stimulate prolific star formation, which may account for the **starburst galaxies** that blaze with the light of numerous newborn stars. These galaxies are characterized by bright centers surrounded by clouds of warm interstellar dust indicating a recent, vigorous episode of star birth (see Figure 17-22). The warm dust is so abundant that starburst galaxies are among the most luminous objects in the universe at infrared wavelengths.

Gravitational interaction between colliding galaxies can hurl thousands of stars out into intergalactic space, along huge, arching streams. These shapes have been dramatically illustrated in computer simulations like the one shown in Figure 17-23. Note the remarkable similarity with the colliding galaxies in Figure 17-24.

Figure 17-22 A starburst galaxy
Prolific star formation is occurring at the center of this galaxy, called M82 or NGC 3034, located about 120 million light years from Earth. This activity was probably triggered by a tidal interaction with a neighboring galaxy. (a) This wide-field view shows the extent of M82. (b) This image shows details of the turbulent interstellar gas and dust heated by numerous young stars around the galaxy's center. (Lick Observatory; R. Schild)

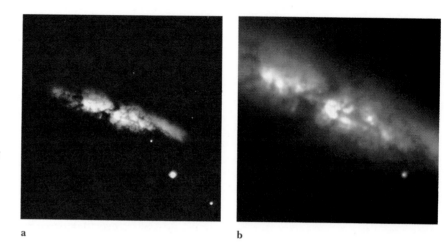

a   b

Spiral arms may be created during a collision that draws out long streamers of stars and gas. For instance, the spiral arms of M51 (examine Figure 17-2) were produced when a second galaxy pulled material out of the disk of M51. The disruptive galaxy is now located at the end of one of the spiral arms created by the collision. Some astronomers argue that the spiral arms of our Milky Way Galaxy were similarly produced by a close encounter with the Large Magellanic Cloud.

In a rich cluster there must be many near misses between galaxies. If galaxies are in fact surrounded by extended halos of dim stars, these near misses could strip the galaxies of their outlying stars. In this way, a loosely dispersed sea of dim stars might come to populate the space between galaxies in a cluster. Searching for these dim stars in extended halos and in intergalactic space is one of the main projects assigned to the Hubble Space Telescope in the 1990s.

Figure 17-23 A simulated collision between two galaxies *Two galaxies are simulated with 350 mass points, each point representing a star. As the two galaxies pass through each other, the gravitational forces between these masses spew out the stars along long streamers. The drawings represent the structure at intervals of 200,000 years. (Adapted from Alar and Juri Toomre)*

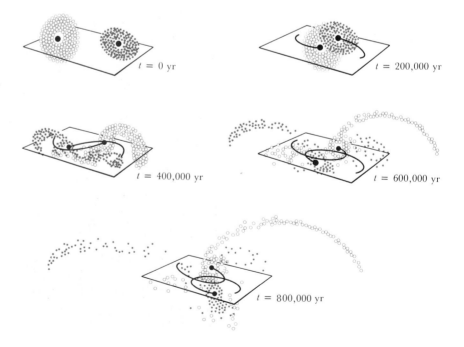

During any collision between galaxies, some of the stars are flung far and wide, scattering material into intergalactic space. However, other stars simply suffer a loss of energy and momentum. As these stars slow down, the galaxies may merge. Several dramatic examples of **galaxy mergers** have been discovered recently (see Figure 17-25).

When two galaxies merge, the result often is a bigger galaxy. If this new galaxy is located in a rich cluster, it may capture and devour additional galaxies and grow to enormous dimensions. This phenomenon is called **galactic cannibalism.** Cannibalism differs from mergers in that the dining galaxy is bigger than its dinner, whereas merging galaxies are about the same size.

Many astronomers suspect that galactic cannibalism explains why giant ellipticals are so huge. As we have seen, giant galaxies typically occupy the centers of rich clusters. In many cases, smaller galaxies are located around these giants (examine Figure 17-9). As they pass through the extended halo of a giant elliptical, these smaller galaxies slow down and are eventually devoured by the larger galaxy.

Computer simulations enable astronomers to study many details of galaxy collisions. For example, Figure 17-26 shows a simulated cannibalism in which a large, disk-shaped galaxy devours a small satellite galaxy. The large galaxy consists, by mass, of 90 percent stars (in blue) and 10 percent gas (in white). It is surrounded by a halo of dark matter having a mass about 3.3 times that of the disk. The satellite galaxy, which has a tenth of the mass of the large galaxy, contains only stars (in orange). Initially, the satellite is in circular orbit about the large galaxy. Note that spiral arms appear in the large galaxy as the collision proceeds. Two billion years elapse as the satellite spirals in toward the core of the large galaxy. Although much material is stripped from the satellite, most of its stars plunge into the nucleus of the large galaxy.

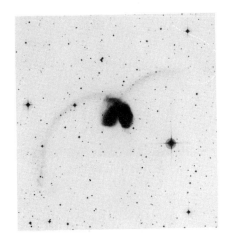

Figure 17-24 A colliding pair of galaxies with "antennae" *Many pairs of colliding galaxies exhibit long "antennae" of stars ejected by the collision. This particular system, called NGC 4038 and 4039, is located about 50 million light years from Earth. This negative print (white sky and black stars) shows faint details that are not easily seen on a positive print. The "antennae" extend about half a million light years from tip to tip—a distance equal to about five times the diameter of the Milky Way Galaxy. (Royal Observatory, Edinburgh)*

Figure 17-25 Merging galaxies
This contorted object in the constellation of Ophiuchus consists of two spiral galaxies in the process of merging. The collision between the two galaxies has triggered an immense burst of star formation. (Courtesy of W. C. Keel)

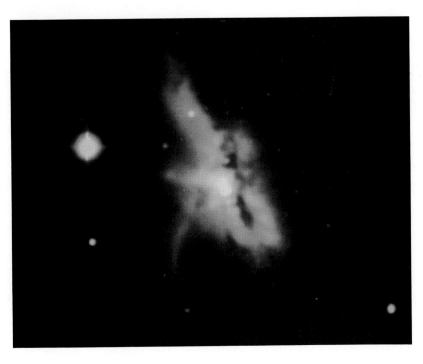

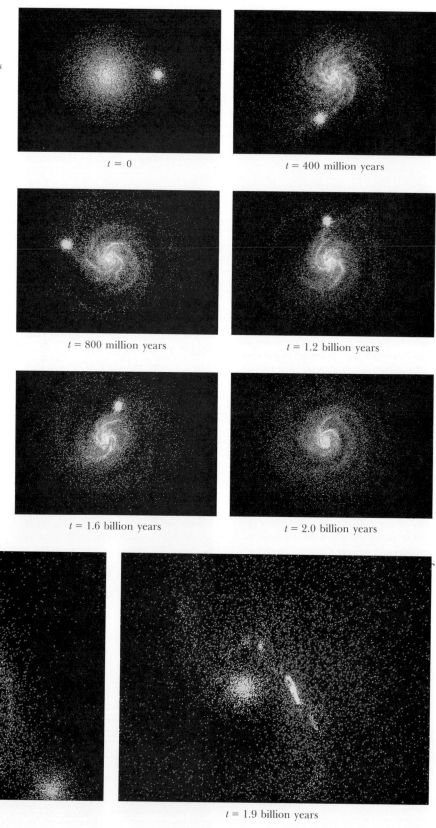

Figure 17-26 A simulated galactic cannibalism *This simulation, performed at the Pittsburgh Supercomputing Center, shows a small galaxy (stars in orange) being devoured by a larger, disk-shaped galaxy (stars in blue, gas in white). The upper six pictures display progress at 400-million-year intervals. Note how spiral arms are induced in the disk galaxy as a result of its interaction with the satellite galaxy. The lower two pictures, which cover the interval from 1.8 to 1.9 billion years, are closeup views of the satellite galaxy plunging into the core of the disk galaxy. As the satellite galaxy sweeps through the inner regions of the disk galaxy, a significant amount of gas becomes concentrated along one of the spiral arms. Vigorous star formation would be expected in these gas clouds. (Courtesy of L. Hernquist)*

$t = 0$

$t = 400$ million years

$t = 800$ million years

$t = 1.2$ billion years

$t = 1.6$ billion years

$t = 2.0$ billion years

$t = 1.8$ billion years

$t = 1.9$ billion years

As we shall see in the next chapter, some astronomers suspect very massive black holes may be located at the centers of certain peculiar galaxies that are unusually powerful sources of energy. A cannibalistic collision of the type shown in Figure 17-26 would "feed" the black hole at the core of such a galaxy. Astrophysicists calculate that the in-falling matter would release enormous amounts of energy, thereby explaining the galaxy's power output. Computer simulations are becoming an increasingly important technique for exploring these exotic processes.

Galaxies formed billions of years ago from the gravitational contraction of huge clouds of primordial gas

Astronomers can probe the past by looking deep into space to gain important clues about galactic evolution. The more distant a galaxy is, the longer its light takes to reach us. Consequently, as we examine galaxies at increasing distances from Earth, we are actually looking farther and farther back in time, seeing galaxies at increasingly earlier stages of evolution.

By observing remote galaxies, astronomers have discovered that galaxies were bluer and brighter in the past than they are today. These changes in color and brightness suggest that a newly formed galaxy has an abundance of young, bright, hot, massive stars. As the galaxy ages, these O and B stars become red supergiants and eventually die off. The galaxy therefore gradually becomes somewhat redder and dimmer.

By studying the rates at which the colors and luminosities of galaxies change, astronomers have discovered an important difference between spirals and ellipticals. Spiral galaxies, like our Milky Way, have been making stars steadily over the past 10 to 15 billion years. Indeed, there is still plenty of interstellar hydrogen in the disks of spiral galaxies to fuel star formation today. In contrast, star formation in elliptical galaxies ceased long ago. Elliptical galaxies formed nearly all their stars in one vigorous burst of activity that lasted for only about a billion years. Figure 17-27 compares the rates at which spiral and elliptical galaxies form stars.

It seems reasonable to suppose that galaxies formed from huge clouds of primordial hydrogen and helium roughly 15 billion years ago. This birth date for both spirals and ellipticals is deduced from the ages of the most ancient, metal-poor stars that both types of galaxies contain. Under the action of gravity, these pregalactic clouds of gas started to contract and form protogalaxies studded with the first generation of stars.

The rate of star formation in a protogalaxy determines whether it becomes a spiral or an elliptical. If the rate of star formation is low, then the gas has plenty of time to collapse to form a flattened disk. A flattened disk is the natural consequence of the overall rotation that the original gas cloud possessed. Star formation continues in this disk because it contains an abundance of hydrogen and thus a spiral galaxy is created. But if the stellar birthrate is high, then virtually all of the gas is used up in the formation of stars before a disk can form. In that case, an elliptical galaxy is created. These contrasting scenarios are depicted in Figure 17-28.

Galactic evolution is a difficult and often controversial subject in which many questions remain. For instance, our discussion of the birth of a galaxy began with a cloud of precisely the right mass and size to guarantee it would contract to become a galaxy. But where did this cloud come from? And what happened in the early universe to cause the primordial hydrogen and

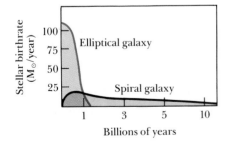

Figure 17-27 The stellar birth rate in galaxies *Most of the stars in an elliptical galaxy were created in a brief burst of star formation when the galaxy was very young. In spiral galaxies, star formation occurs at a more leisurely pace that extends over billions of years. (Adapted from J. Silk)*

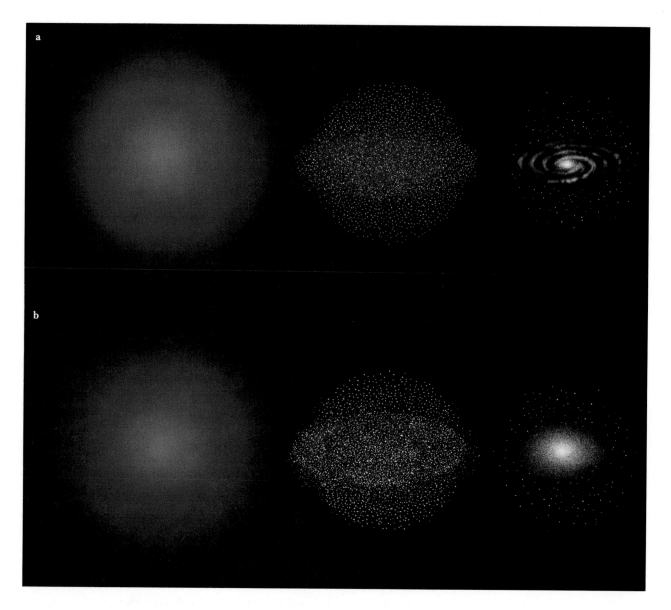

Figure 17-28 *The creation of spiral and elliptical galaxies* *A galaxy begins as a huge cloud of primordial gas that collapses gravitationally.* **(a)** *If the rate of star birth is low, then much of the gas collapses to form a disk and a spiral galaxy is created.* **(b)** *If the rate of starbirth is high, then the gas is converted into stars before a disk can form, resulting in an elliptical galaxy.*

helium to clump up in clouds destined to evolve into galaxies, instead of becoming objects a million times bigger or smaller? Even more troublesome is the issue of the unseen mass. The observable stars, gas, and dust in a galaxy account for only about 10 percent of the mass associated with the galaxy. We have little idea what the remaining 90 percent looks like, how it's distributed in space, or what it's made of. Some astronomers argue that this level of ignorance precludes any meaningful discussion of galactic evolution.

The Hubble law is a simple relationship between the redshifts of galaxies and their distances from Earth

Whenever an astronomer finds an object in the sky that can be seen or photographed, the natural inclination is to attach a spectrograph to a telescope and record the spectrum. As long ago as 1914, V. M. Slipher, working at the Lowell Observatory in Arizona, began taking spectra of "spiral nebulae." He was surprised to discover that, of the 15 spiral nebulae he studied,

Figure 17-29 *Five galaxies and their spectra* The photographs of these five elliptical galaxies all have the same magnification. The spectrum of each galaxy is the hazy band between the comparison spectra. In all five cases, the so-called H and K lines of calcium are seen. The recessional velocity (calculated from the Doppler shifts of the H and K lines) is given below each spectrum. Note that the more distant a galaxy is, the greater is its redshift. (Mt. Wilson and Las Campanas Observatories)

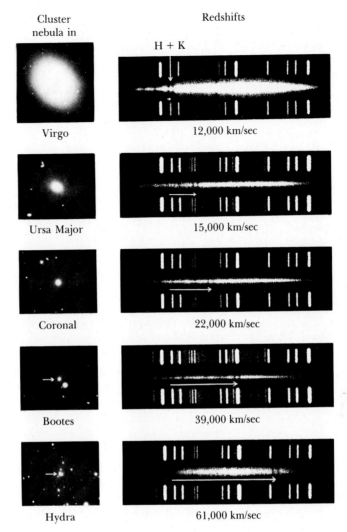

Cluster nebula in

Redshifts

H + K

Virgo — 12,000 km/sec

Ursa Major — 15,000 km/sec

Coronal — 22,000 km/sec

Bootes — 39,000 km/sec

Hydra — 61,000 km/sec

the spectral lines of 11 of them were shifted toward the red end of the spectrum, indicating substantial recessional velocities for those objects. This marked dominance of redshifts was presented by Curtis in the Shapley–Curtis debate as evidence that these spiral nebulae could not be ordinary nebulae in our Milky Way Galaxy.

During the 1920s, Edwin Hubble and Milton Humason photographed the spectra of many galaxies with the 100-in. telescope on Mount Wilson. Five representative elliptical galaxies and their spectra are shown in Figure 17-29. As indicated by this illustration, there seemed to be a direct correlation between the distance to a galaxy and the size of its redshift. In other words, nearby galaxies are moving away from us slowly, and more distant galaxies are rushing away from us much more rapidly. This recessional motion that pervades the universe is often called the **Hubble flow.**

Using various techniques, Hubble estimated the distances to a number of galaxies. Using the Doppler effect (recall Figure 5-18), Hubble calculated the speed with which each galaxy is receding from us. When he plotted the data on a graph of distance versus speed, he found that the points lie nearly along a straight line. Figure 17-30 is a modern version of Hubble's graph based on recent data.

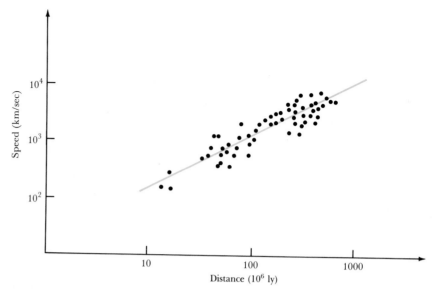

Figure 17-30 The Hubble law
The distances and recessional velocities of 60 Sc spiral galaxies are plotted on this graph. The straight line is the "best fit" for the data. This linear relationship between distance and speed is called the Hubble law. (Adapted from Sandage and Tammann)

This relationship between the distances to galaxies and their redshifts is one of the most important astronomical discoveries of the twentieth century. As we shall see in Chapter 19, this relationship tells us that we are living in an expanding universe. In 1929, Hubble published this discovery, which is now known as the **Hubble law.**

The Hubble law is most easily stated as the formula

$$v = H_0 r$$

where v is the recessional velocity, r is the distance, and H_0 is a constant commonly called the **Hubble constant.** This is the formula for the straight line displayed in Figure 17-30. The Hubble constant H_0 tells us the inclination of the line in Figure 17-30. From the data plotted on this graph we find that

$$H_0 = 15 \text{ km/sec/Mly}$$

where Mly stands for 10^6 light years (say "fifteen kilometers per second per million light years"). In other words, for each million light years to a galaxy, the galaxy's speed away from us increases by 15 km/sec. For example, a galaxy located 100 million light years from Earth should be rushing away from us with a speed of 1500 km/sec.

Incidentally, many astronomers prefer to speak of megaparsecs (Mpc) rather than millions of light years (Mly). Using that unit,

$$H_0 = 50 \text{ km/sec/Mpc}$$

(say "fifty kilometers per second per megaparsec").

The exact value of the Hubble constant is a topic of heated debate among astronomers today. The data plotted in Figure 17-30, as well as the values of H_0 given above, are all from recent meticulous work by Allan Sandage and Gustav Tammann, who did most of their observations with the 200-in. Palomar telescope. However, other prominent astronomers feel strongly that the

Sandage–Tammann value for H_0 is too low. They argue that the true value for H_0 is about 30 km/sec/Mly = 100 km/sec/Mpc. Many astronomers simply use a number between the two extremes, such as 75 km/sec/Mpc.

The Hubble constant is one of the most important numbers in all physical science. As we shall see in Chapter 19, this number expresses the rate at which the universe is expanding and thus tells us the age of the universe. Naturally, astronomers are very interested in an accurate determination of H_0.

In order to determine the Hubble constant, an astronomer must measure the redshifts and distances to many galaxies. Although redshift measurements can be quite precise, it is very difficult to determine the distances to remote galaxies accurately. Indeed, conflicting distance measurements are the main reason that the measurement of H_0 is so controversial.

Recall from our discussion in Chapter 12 that we can always find the distance to an object if we know both its apparent and absolute magnitudes. Astronomers use the term **standard candle** to denote any object whose absolute magnitude is known. Cepheid variables, the brightest supergiants, globular clusters, and supernovae are all useful as standard candles because astronomers know something about their absolute magnitudes. To determine the distance to a galaxy, an astronomer must measure the apparent magnitude of one or more of these standard candles in that galaxy. As soon as both the apparent and absolute magnitudes are known, the distance to the galaxy can be easily calculated.

Distances to nearby galaxies can be determined with the period–luminosity relation for Cepheid variables (recall Figure 13-23). The usefulness of Cepheids as standard candles is limited because they are not exceptionally bright. They can be seen only out to a distance of about 15 million light years from Earth. Furthermore, Cepheids are relatively massive stars and thus are only found in spiral galaxies. In elliptical galaxies, where star birth ceased long ago, all the Cepheids have died. Cepheid variables have been observed in about only 30 galaxies.

The most luminous stars are the supergiants that occupy the top of the H–R diagram. These stars have an absolute magnitude of about -8 and can be seen out to a distance of 80 million light years from Earth. Supergiants can therefore be used as standard candles out to that distance.

Beyond 80 Mly, individual stars are no longer discernible. Astronomers therefore turn to entire clusters and nebulae. The brightest globular clusters, which have a total absolute magnitude of about -10, can be seen out to 130 Mly from Earth. The brightest H II regions have absolute magnitudes of -12 and can be detected out to 300 Mly. From the faintness of these clusters and nebulae, distances to remote galaxies can be estimated.

To get beyond 300 Mly, astronomers must wait for supernova explosions. The brightest supernovae reach an absolute magnitude of -19 at the peak of their outbursts (see Figure 17-31). These brilliant outbursts can be seen out to distances of 8 billion light years from Earth.

Astronomers go to great lengths to check the reliability of their standard candles. After all, a tiny mistake in the absolute magnitude of a supergiant star or globular cluster can lead to an error of many millions of light years in calculating the distance to a remote galaxy.

The major obstacle in determining the Hubble constant is that the farther we look into space, the fewer standard candles we have. For example, the distance to a nearby galaxy can be cross-checked in many ways. The distance

Figure 17-31 A supernova in NGC 4303
In 1961, a supernova erupted in the spiral galaxy M61 (also called NGC 4303), which is a member of the Virgo cluster. Supernovae can be seen in extremely remote galaxies and are important "standard candles" used to determine the distances to these faraway galaxies. (Lick Observatory)

computed from the period–luminosity relation can be compared to the distance determined from the magnitudes of the most luminous supergiants. Then these results can be compared with the magnitudes of the galaxy's globular clusters and the angular sizes of its H II regions. The results from all these methods can then be averaged to obtain a relatively reliable distance to the galaxy.

As we turn to more distant galaxies, however, we can see fewer and fewer standard candles. For example, the nearest rich, regular cluster (the Coma cluster in Figure 17-17) is so far away that it is impossible to see individual stars. With fewer standard candles, fewer cross-checks can be made. The distance to these remote galaxies thus becomes less certain. Unfortunately, remote galaxies are precisely the objects whose distances we must determine to find the value of the Hubble constant. Uncertainty in determining distances is the cause of our uncertainty in the value of H_0.

Astronomers expect that the Hubble Space Telescope will solve many of the problems of determining H_0. This telescope, named after the father of extragalactic astronomy, will produce exceptional views of galaxies and thus will allow astronomers to identify standard candles that are undetectable from the Earth's surface.

Summary

- The Hubble classification system groups galaxies into four major categories: spirals, barred spirals, ellipticals, and irregulars.

   Spiral galaxies and barred spiral galaxies are sites of active star formation.

   Elliptical galaxies are virtually devoid of interstellar gas and dust; no star formation is occurring in these galaxies.

- Galaxies are found in clusters rather than scattered randomly through the universe.

   A rich cluster contains hundreds or even thousands of galaxies; a poor cluster may contain only a few dozen galaxies.

   A regular cluster has a nearly spherical shape with a central concentration of galaxies; an irregular cluster has an asymmetrical distribution of galaxies.

   Rich, regular clusters contain mostly elliptical and S0 galaxies; irregular clusters contain more spiral and irregular galaxies.

   Giant elliptical galaxies are often found near the centers of rich clusters.

   Our Galaxy is a member of a poor, irregular cluster called the Local Group.

- The observable mass of a cluster of galaxies is not large enough to account for the observed motions of the galaxies; a large amount of unobserved mass must be present between the galaxies.

   Hot intergalactic gases emit X rays in rich clusters. Extended halos of dim stars probably surround all galaxies.

- When two galaxies collide, their stars pass each other, but their interstellar media collide violently, either stripping the gas and dust from the galaxies or triggering prolific star formation.

   The gravitational effects of a galactic collision can throw stars out of their galaxies into intergalactic space.

Galactic mergers may occur; a large galaxy in a rich cluster may grow steadily through galactic cannibalism, perhaps producing in the process a giant elliptical galaxy.

· Galaxies probably formed 10 to 15 billion years ago by the gravitational contraction of huge clouds of hydrogen and helium.

The rate of star formation in a protogalaxy determines whether it will become a spiral galaxy or an elliptical galaxy.

A low rate of star birth results in a spiral galaxy because the interstellar gas has time enough to collapse and form a disk. A high rate of starbirth turns the gas into stars before a disk can form, resulting in an elliptical galaxy.

· There is a simple linear relationship between the distance from the Earth to a galaxy and the redshift of that galaxy (which is a measure of the speed with which it is receding from us); this relationship is the Hubble law, $v = H_0 r$.

Standard candles, such as Cepheid variables, the brightest supergiants, globular clusters, H II regions, and supernovae in a galaxy, are used in estimating intergalactic distances.

Because of difficulties in measuring the distances to galaxies, the value of the Hubble constant H_0 is not known with certainty; this leads to uncertainties about the rate at which the universe is expanding and about the age of the universe.

Review questions

1 In what types of galaxies are new stars most likely forming? Describe the observational evidence that supports your answer.

2 The Hubble classification scheme places the biggest galaxies into what category? Into what category do the smallest galaxies fall? Which type of galaxy is the most common?

3 How is it possible that galaxies in our Local Group still remain to be discovered? In what part of the sky would these galaxies be located? What sorts of observations might reveal these galaxies?

4 Are there any galaxies besides our own that can be seen with the naked eye? If so, which one(s) can you name?

5 How would you distinguish star images from unresolved images of remote galaxies on a photographic plate?

6 Explain why the "missing mass" in galaxy clusters could not be neutral hydrogen.

7 Outline a process that could have caused a protogalaxy to evolve into an elliptical galaxy. What would have had to occur differently for the protogalaxy to have become a spiral galaxy?

8 Why do some galaxies in the Local Group exhibit blueshifted spectral lines? Is this phenomenon a violation of the Hubble law? Explain.

9 Why are there so many differing values of H_0?

10 What kinds of stars would you expect to find populating space between galaxies in a cluster?

Advanced questions

*11 Suppose you were to take a spectrum of a distant galaxy and find that its redshift corresponds to a speed of 22,000 km/sec. How far away is the galaxy?

*12 A cluster of galaxies in the southern constellation of Pavo (the peacock) is located 100 Mpc from Earth. How fast is this cluster receding from us?

13 How might you determine what fraction of a galaxy's redshift is caused by the galaxy's orbital motion about the center of mass of its cluster?

Discussion questions

14 Discuss the advantages and disadvantages of using the various standard candles to determine extragalactic distances.

15 Discuss whether the various Hubble types of galaxies actually represent some sort of evolutionary sequence.

For further reading

Berendzen, R., et al. *Man Discovers the Galaxies.* Neale Watson, 1976.

DeVaucouleurs, G. "The Distance Scale of the Universe." *Sky & Telescope,* December 1983.

Field, G. "The Hidden Mass in Galaxies." *Mercury,* May/June 1982.

Gorenstein, P., and Tucker, W. "Rich Clusters of Galaxies." *Scientific American,* November 1978.

Hartley, K. "Elliptical Galaxies Forged by Collision." *Astronomy,* May 1989.

Hausman, M. "Galactic Cannibalism." *Mercury,* November/December 1979.

Hodge, P. *Galaxies.* Harvard University Press, 1986.

Kaufmann, W. *Galaxies and Quasars.* W. H. Freeman and Company, 1979.

Keel, W. "Crashing Galaxies, Cosmic Fireworks." *Sky & Telescope,* January 1989, p. 18.

Rubin, V. "Dark Matter in Spiral Galaxies." *Scientific American,* June 1983.

Silk, J. "Formation of the Galaxies." *Sky & Telescope,* December 1986.

Smith, R. *The Expanding Universe: Astronomy's Great Debate.* Cambridge University Press, 1982.

Tully, R. "Unscrambling the Local Supercluster." *Sky & Telescope,* June 1982.

18

Quasars and active galaxies

The core of a radio galaxy *This artist's rendition shows a scenario that many astronomers believe is responsible for double radio sources. A supermassive black hole at the center of a galaxy is surrounded by an accretion disk. In the inner regions of the accretion disk, matter crowding toward the hole is diverted outward along two oppositely directed beams. These beams deposit energy into two huge radio-emitting lobes located on either side of the galaxy.* (Astronomy)

Astronomers have discovered a large number of objects in the sky whose extreme redshifts indicate that they are at great distances from the Earth, according to the Hubble law. To be observable at all at such enormous distances, these quasi-stellar objects (quasars) and active galaxies must be extremely luminous. In this chapter, we learn of evidence indicating that the energy output of a typical quasar is equivalent to that of 100 galaxies and is emitted from a volume roughly the size of the solar system. We explore a variety of active galaxies and find that the centers of these objects also are powerful sources of energy. Finally, we learn about recent theories that supermassive black holes are involved in the intense energy production of these luminous objects.

The development of radio astronomy in the late 1940s ranks among the most important scientific accomplishments of the twentieth century. Prior to that time, everything known about the distant universe had to be gleaned from visual observations. Radio telescopes provided a view of the universe in a wavelength range far beyond that of visible light. Many surprising discoveries emerged from this new ability to examine the previously invisible universe.

The first radio telescope was built in 1937 by an amateur astronomer, Grote Reber, in his backyard in Illinois. By 1944, Reber had detected strong radio emission from Sagittarius, Cassiopeia, and Cygnus. Two of these sources (nicknamed Sgr A and Cas A) happen to be in our Galaxy, one the galactic nucleus and the other a supernova remnant. In 1951, Walter Baade and Rudolph Minkowski used the 200-in. optical telescope on Palomar Mountain to discover a strange-looking galaxy at the position of the third source, called Cygnus A (Cyg A). Figure 18-1 is a photograph of the optical counterpart of Cyg A.

The peculiar galaxy associated with Cygnus A is very dim. Nevertheless, Baade and Minkowski managed to photograph its spectrum, which shows a number of bright spectral lines, all shifted by 5.7 percent toward the red end of the spectrum. Astronomers were surprised by this large redshift because, according to the Hubble law, the distance to Cygnus A is roughly 1 billion light years!

The enormous distance to Cygnus A means that it has to be one of the brightest radio sources in the sky. Although Cyg A is barely visible through the 200-in. telescope at Palomar, its radio waves can be picked up by amateur astronomers with backyard equipment. The energy output in radio waves from Cyg A must therefore be enormous. Indeed, Cyg A shines with a

Figure 18-1   Cygnus A (also called 3C 405)   *This strange-looking galaxy was discovered at the location of the radio source Cygnus A. This galaxy has a redshift corresponding to a recessional speed of 6 percent of the speed of light. According to the Hubble law, this redshift corresponds to a distance of about 1 billion light years from Earth. Because Cyg A is one of the brightest radio sources in the sky, the energy output of this remote galaxy must be enormous. (Palomar Observatory)*

radio luminosity 10^7 times as bright as that of an ordinary galaxy such as M31 in Andromeda. The object corresponding to Cyg A obviously must be something quite extraordinary.

Quasars look like stars but have huge redshifts

During the late 1950s and early 1960s, radio astronomers were busy making long lists of all the radio sources they were finding across the sky. One of the most famous lists, published in 1959, is called the *Third Cambridge Catalogue* (the first two catalogues produced by the British team were filled with inaccuracies). It lists 471 radio sources. Even today, astronomers often refer to these sources by their "3C numbers." With the discovery of the extraordinary luminosity of Cyg A (also called 3C 405, because it is the 405th source on the Cambridge list), astronomers became eager to learn whether any other sources in the 3C catalog had similarly extraordinary properties.

One interesting case is 3C 48. In 1960, Allan Sandage used the 200-in. telescope to discover a "star" at the location of this radio source (see Figure 18-2). Because ordinary stars are not strong sources of radio emission, 3C 48 was recognized as something unusual. Its spectrum does in fact show a series of bright spectral lines that no one could identify. Although 3C 48 was clearly an oddball, many astronomers thought it was just another strange star in our Galaxy.

Another such "star," called 3C 273, was discovered in 1962. It was found to have a luminous "jet" protruding from one side, as shown in Figure 18-3. And, as was the case with 3C 48, this "star" contains a series of bright spectral lines that no one could identify.

Astronomers had difficulty identifying the emission lines in the spectra of 3C 48 and 3C 273 because they assumed that these starlike objects were just peculiar stars nearby in our Galaxy. A breakthrough finally came in 1963, when Maarten Schmidt at the California Institute of Technology found that four of the brightest spectral lines of 3C 273 are positioned relative to one other just like four familiar spectral lines of hydrogen. However, these

Figure 18-2 [left]   The quasar 3C 48
For several years, astronomers believed erroneously that this object is simply a nearby peculiar star that happens to emit radio waves. Actually, the redshift of this starlike object is so great that, according to the Hubble law, it must be roughly 5 billion light years away. (Palomar Observatory)

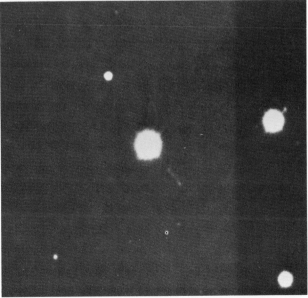

Figure 18-3 [right]   The quasar 3C 273
This greatly enlarged view shows the starlike object associated with the radio source 3C 273. Note the luminous jet to one side of this "star." By 1963, astronomers discovered that the redshift of this "star" is so great that its distance, according to the Hubble law, is nearly 3 billion light years from Earth. (NOAO)

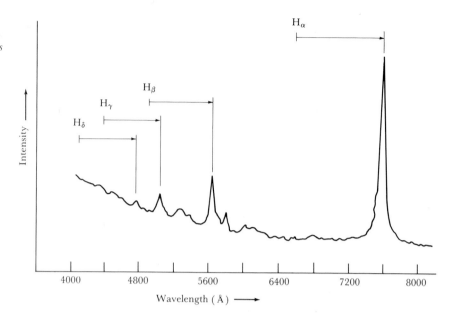

Figure 18-4 The spectrum of 3C 273
Four bright emission lines caused by hydrogen dominate the spectrum of 3C 273. The arrows indicate how far these spectral lines are redshifted from their usual wavelengths.

emission lines of 3C 273 are at much longer wavelengths than the usual positions of the Balmer lines. In other words, 3C 273 has a substantial redshift.

Stellar spectra exhibit comparatively small Doppler shifts, because a star in our galaxy cannot be moving extremely fast relative to the Sun, or it would soon escape from the Galaxy. Schmidt therefore conjectured that perhaps 3C 273 is not, after all, a nearby star. Pursuing this hunch, he promptly identified all four spectral lines as hydrogen lines that have suffered an enormous redshift corresponding to a speed of 15 percent of the speed of light. According to the Hubble law, this huge redshift implies the incredible distance to 3C 273 of roughly 3 billion light years.

Because of their starlike appearance, 3C 48 and 3C 273 were dubbed **quasi-stellar radio sources.** This term was soon shortened to **quasars.**

Instead of using photography to record a spectrum, many observatories use an electronic device at the focus of the spectrograph. The output of the machine is a graph of intensity versus wavelength, on which emission lines appear as peaks. The graph showing the spectrum of 3C 273 in Figure 18-4 was obtained in this way. Note the four hydrogen lines.

The spectral lines of 3C 273 are brighter than the intensity of the background radiation at other wavelengths. The background is called the **continuum,** and the bright lines are emission lines caused by excited atoms that are emitting radiation at specific wavelengths. Spectra of ordinary galaxies are dominated by dark absorption lines (recall Figure 17-29). Most quasars and many peculiar galaxies exhibit strong emission lines in their spectra, a sign that something unusual is going on.

Inspired by Schmidt's success, astronomers next identified the spectral lines of 3C 48 as having suffered a redshift corresponding to a velocity of nearly one-third the speed of light. According to the Hubble law, 3C 48 must therefore be twice as far away as 3C 273, or approximately 6 billion light years from Earth.

Incidentally, these distances to quasars assume that the Hubble constant (H_0) is 15 km/sec/Mly. As mentioned in the previous chapter, some astrono-

Figure 18-5 The quasar OH 471
This quasar has one of the largest redshifts ever discovered. This redshift corresponds to a speed slightly greater than 90 percent of the speed of light. According to the Hubble law, OH 471 must be 18 billion light years away. (Palomar Observatory)

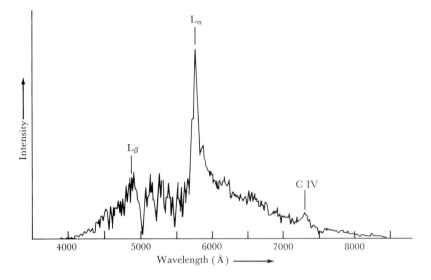

Figure 18-6 *The spectrum of a high-redshift quasar* The light from this quasar, known as PKS 2000-330, is so highly redshifted that spectral lines normally in the far-ultraviolet (L_α and L_β) can be seen at visible wavelengths. Note the large number of deep absorption lines on the short-wavelength side of L_α. These lines, collectively called the "Lyman-alpha forest," are probably caused by remote clouds of gas along our line of sight to the quasar. Hydrogen in these clouds absorbs photons from the quasar at wavelengths less redshifted than the quasar's L_α line.

mers believe that this value is too low and would prefer to use 30 km/sec/Mly for H_0. Doubling the Hubble constant would halve these quasar distances.

Hundreds of quasars have been discovered since the pioneering days of the early 1960s. All quasars look like stars, and all have enormous redshifts. There are even a few quasars, such as the one in Figure 18-5, with redshifts corresponding to speeds slightly greater than 90 percent of the speed of light. Figure 18-6 shows the spectrum of a quasar whose redshift corresponds to a speed of 92 percent of the speed of light. This is such an enormous redshift that the Lyman alpha line is shifted all the way from its usual wavelength of 1216 Å in the far-ultraviolet into the middle of the visible spectrum. From the Hubble law, it follows that the distance to these high-redshift quasars is about 18 billion light years. Their light has taken 18 billion years to reach us. When we look at these quasars, we are seeing objects as they existed when the universe was very young.

A quasar emits a huge amount of energy from a very small volume

The luminosities of quasars can best be appreciated in comparison with the luminosities of galaxies. Galaxies are big and bright. A typical large galaxy like our own or M31 in Andromeda contains several hundred billion stars and shines with a luminosity of 10 billion Suns. The most gigantic and most luminous galaxies (such as the giant ellipticals in Figure 17-16) are only 10 times brighter, shining with the brilliance of 100 billion Suns. Beyond 10 billion light years from Earth, even the brightest galaxies are too faint to be detected. Most ordinary galaxies are too dim to be detected at half that distance.

Although it is difficult to find high-redshift galaxies, high-redshift quasars are quite common. Quasars therefore must be incredibly luminous—far more so than galaxies. In fact, some quasars have an energy output 100 times that of the brightest galaxies we have ever seen.

In the mid-1960s, several astronomers discovered that some of the newly identified quasars had been inadvertently photographed in the past. For example, 3C 273 was found on numerous photographs, including one taken in 1887. By carefully examining the images of quasars on these old

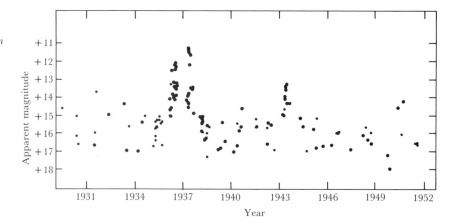

Figure 18-7   The brightness of 3C 279
This graph shows variations in magnitude of the quasar 3C 279. Note the large outburst in 1937. The data were obtained from careful examination of old photographic plates in the files of the Harvard College Observatory. (Adapted from L. Eachus and W. Liller)

photographs, astronomers found that quasars fluctuate in brightness, occasionally flaring up. For example, see the data from old photographs of 3C 279 plotted in Figure 18-7. Note particularly the prominent outbursts that occurred around 1937 and 1943. During these outbursts, the luminosity of 3C 279 increased by a factor of at least 25. Because of the enormous distance to this quasar, it must have been shining with a brilliance 10,000 times as great as that of the entire Andromeda galaxy at the peak of these outbursts.

Fluctuations in the brightness of quasars allow astronomers to place strict limits on the maximum sizes of quasars, because an object cannot be observed to vary in brightness faster than the time it takes light to travel across that object. For instance, an object that is 1 ly in diameter cannot vary in brightness with a period of less than 1 year.

To understand this limitation, imagine an object that measures 1 ly across, as shown in Figure 18-8. Suppose the entire object emits a brief flash of light. Photons from that part of the object nearest the Earth arrive at our telescopes first. Photons from the middle of the object arrive at Earth six months later. And finally, light from the far side of the object arrives a year after the first photons. Although the object emitted a sudden flash of light, we observe a gradual brightness variation that lasts a full year. In other words, the flash is stretched out over an interval equal to the difference in light travel time between the nearest and most remote observable regions of the object.

Figure 18-8   A limit in the speed of brightness variations   *The rapidity with which the brightness of an object can vary is limited by the light travel time across the object. Even if the object, shown here to be 1 ly in size, emits a sudden flash of light, photons from point A arrive at Earth one year before photons from point C. Thus, as seen from Earth, the sudden flash is observed over a full year.*

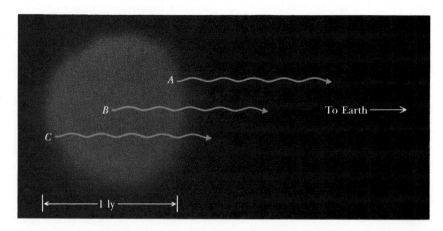

The brightness of many quasars may vary over periods of only a few weeks or months, and some quasars even fluctuate from night to night. Recent X-ray data from the Einstein Observatory reveal large variations in as few as three hours. Such rapid flickering would mean that a quasar is small. The energy-emitting region of a typical quasar—the "powerhouse" that blazes with the luminosity of 100 galaxies—is less than 1 light day in diameter. If quasars are indeed at the huge distances indicated by their redshifts, they must be producing the luminosity of 100 galaxies in a volume roughly the same size as our solar system!

Active galaxies bridge the gap in energy output between ordinary galaxies and quasars

The energy output of quasars is so huge that, in the 1960s, some astronomers preferred to question the Hubble law rather than believe that such highly luminous objects could exist. In recent years, however, astronomers have discovered various kinds of peculiar galaxies whose luminosities fall between those of ordinary galaxies and quasars. Some of these strange galaxies have unusually bright, starlike nuclei. Others have strong emission lines in their spectra. Still others are highly variable, changing their brightness every few weeks or months. Jets and beams of radiation are observed emanating from the cores of some of them. These various types of galaxies are called **active galaxies.**

The first active galaxies were discovered in 1943, during a survey of spiral galaxies by Carl Seyfert at the Mount Wilson Observatory. Called **Seyfert galaxies,** these luminous objects have bright, starlike nuclei and strong emission lines in their spectra. For example, NGC 4151 (see Figure 18-9) has an extremely rich emission spectrum with 28 percent of the galaxy's light concentrated in emission lines. These emission lines include Fe X and Fe XIV (iron atoms with 9 or 13 electrons, respectively, stripped away), indicating that NGC 4151 contains some extremely hot gas. Seyfert galaxies also commonly exhibit variability in brightness (the magnitude of NGC 4151 changes every few months).

Another example of a Seyfert galaxy is NGC 1068 shown in Figure 18-10. At infrared wavelengths, this galaxy shines with the brilliance of 10^{11} Suns. This extraordinary luminosity has been observed to vary by as much as 7×10^9 Suns over only a few weeks. In other words, the infrared power output

Figure 18-9 [left] The Seyfert galaxy NGC 4151 This is one of the best-studied Seyfert galaxies. Because of its bright, starlike nucleus and emission-line spectrum, NGC 4151 might be mistaken for a quasar if it were very far away. According to the Hubble law, the redshift of this galaxy gives it a distance of 70 million light years from Earth. (Palomar Observatory)

Figure 18-10 [right]   The Seyfert galaxy NGC 1068 This Seyfert galaxy, also called M77 or 3C 71, is renowned for its extraordinary infrared luminosity, which varies over intervals as short as a few weeks. Note how bright the inner spiral arms are, compared to the outer spiral arms. This galaxy is about the same distance from Earth as NGC 4151 in Figure 18-9. (Courtesy of R. Schild)

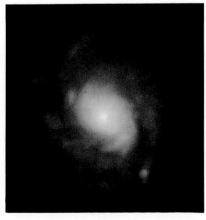

Figure 18-11 [left]   The active galaxy NGC 1275 (also called 3C 84)   This Seyfert galaxy, located in the Perseus cluster, is a strong source of X rays and radio radiation. According to the Hubble law, the galaxy's redshift gives a distance of nearly 400 million light years from Earth. Note the streamers of gas. Spectroscopic observations confirm significant mass ejection from the galaxy's center. (NOAO)

Figure 18-12 [right]   An X-ray image of NGC 1275   This picture from the Einstein Observatory shows the X-ray appearance of the Seyfert galaxy NGC 1275. Most of the X-ray emission comes from a point source at the galaxy's nucleus. (Harvard-Smithsonian Center for Astrophysics)

of the nucleus of NGC 1068 rises and falls by an amount equal to the total luminosity of our entire Galaxy.

Many more Seyfert galaxies have been discovered in recent years, largely through the efforts of Wallace Sargent at Caltech. Approximately 10 percent of the most luminous galaxies in the sky are Seyfert galaxies. Some of the brightest Seyfert galaxies shine as brightly as faint quasars, leading many astronomers to suspect that remote Seyfert galaxies could easily be interpreted as quasars.

Some Seyfert galaxies exhibit vestiges of violent, explosive phenomena in their nuclei. Examine the tangled galaxy NGC 1275 in Figure 18-11. Filaments of gas tens to thousands of light years long protrude from the galaxy in all directions. Spectroscopic studies indicate that this gas is being blasted away from the galaxy's nucleus at 3000 km/sec. The nucleus of this galaxy is a strong source of radio waves and X rays (see Figure 18-12). In 1977, Vera Rubin and her colleagues at the Carnegie Institution of Washington demonstrated that NGC 1275 actually consists of two galaxies. As we saw in the previous chapter, a collision or close encounter between two galaxies can result in the ejection of matter into intergalactic space.

The high-speed ejection of matter from active galaxies is often best seen at nonvisible wavelengths. For instance, Figure 18-13a is a photograph of the peculiar galaxy NGC 5128 in the southern constellation of Centaurus. An unusual broad dust lane stretching across the galaxy is the only obvious visible hint of the extraordinary activity occurring at the galaxy's center.

The galaxy NGC 5128 is one of the brightest sources of radio waves in the sky. It was one of the first sources discovered when radio telescopes were erected in Australia. As a radio source, it is called Centaurus A, because of its location in the sky. In part, the brightness of Cen A at radio wavelengths comes from its proximity to Earth, only 13 million light years away. As shown in Figure 18-13b, radio waves pour from two regions, called **radio lobes,** on either side of the galaxy's dust lane. Farther from the dust lane is a second set of radio lobes that spans a volume 2 Mly across. Recent X-ray pictures of NGC 5128 reveal an X-ray jet (see Figure 18-13c) sticking out of the galaxy's nucleus. This jet, which is perpendicular to the galaxy's dust lane, is aimed toward one of the radio lobes. These observations suggest that particles and energy stream out of the galaxy's nucleus toward the radio lobes.

By 1970, radio astronomers had discovered dozens of objects similar to Centaurus A that are now called **double radio sources.** An active galaxy, usually resembling a giant elliptical, is often found between the two radio

a

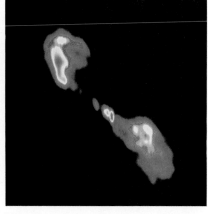

b

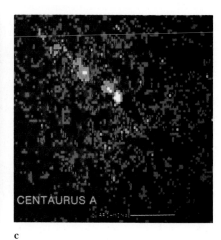

c

Figure 18-13 The peculiar galaxy NGC 5128 (also called Cen A) This extraordinary galaxy is located in the constellation of Centaurus, roughly 13 million light years from Earth. These three views show visible, radio, and X-ray images of this galaxy, all to the same scale. (a) This photograph at visible wavelengths shows a dust lane across the face of the galaxy. (b) This radio image shows that vast quantities of radio radiation pour from extended regions of the sky on either side of the dust lane. (c) This X-ray picture from the Einstein Observatory shows that NGC 5128 has a bright X-ray nucleus. An X-ray jet protrudes from the galaxy's nucleus along a direction perpendicular to the galaxy's dust lane. (NOAO; VLA, NRAO; Harvard-Smithsonian Center for Astrophysics)

lobes. For instance, the visible galaxy associated with Cygnus A (recall Figure 18-1) is located between two radio lobes shown in Figure 18-14.

Cygnus A is a fine example of a small double radio source that is comparatively far away. Generally, the nearer to us a double radio source is, the larger it is, sometimes spanning a volume as big as an entire cluster of galaxies. Because viewing objects at varying distances gives us glimpses of different "look-back times" into the past, we are able to track the probable evolutionary sequence of astronomical phenomena. Thus, since the larger double radio sources are closer to Earth, we conclude that double radio sources get bigger as they grow older.

All double radio sources seem to have some sort of central "engine" that squirts particles (probably electrons) and magnetic field outward along two oppositely directed jets at speeds very near the speed of light. After traveling many thousands or even millions of light years, this ejected material slows down, and interaction between the electrons and the magnetic field produces the radio radiation that we detect. A specific type of radio emission called **synchrotron radiation** occurs whenever electrons move in a spiral

Figure 18-14 A radio image of Cygnus A This color-coded radio picture was produced at the Very Large Array. Most of the radio emission from Cygnus A comes from the radio lobes located on either side of the peculiar galaxy seen in Figure 18-1. These two radio lobes are each about 160,000 light years from the optical galaxy, and each contains a brilliant, condensed region of radio emission. The rectangle indicates the area covered by Figure 18-1. (VLA, NRAO)

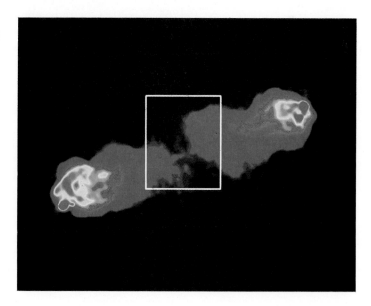

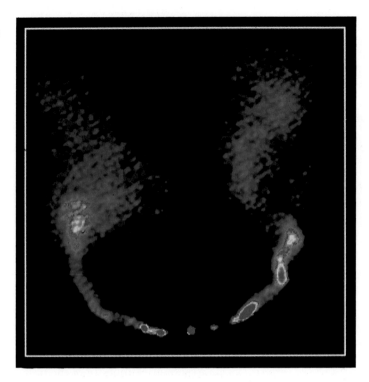

Figure 18-15 The head–tail source NGC 1265 The active elliptical galaxy NGC 1265 would probably be an ordinary double radio source except that the galaxy is moving at a high speed through the intergalactic medium. Because of this motion, the two jets trail the galaxy, giving this radio source a distinctly windswept appearance. (NRAO)

within a magnetic field. The radio waves that come from the lobes of a double radio source have all the characteristics of synchrotron radiation.

The idea that a double radio source involves powerful jets of particles traveling near the speed of light is supported by the existence of **head–tail sources,** so named because each such source appears to have a region of concentrated radio emission (its "head") followed by a weaker tail of emission. A good example is the active elliptical galaxy NGC 1265 in the Perseus cluster of galaxies. NGC 1265 is known to be moving at a high speed (2500 km/sec) relative to the cluster as a whole. Figure 18-15 is a radio map of NGC 1265. Note that the radio emission has a distinctly windswept appearance. Just as smoke pouring from a steam locomotive trails a rapidly moving train, particles ejected along two jets from this galaxy are deflected by the galaxy's passage through the sparse intergalactic medium.

At radio wavelengths, the double radio sources are among the brightest objects in the universe. The energy contained in the radio lobes of a typical double radio source roughly equals the energy released by 10 billion supernova explosions.

A final class of active galaxies is the **N galaxies,** so named because they have bright nuclei. Extreme examples are the **BL Lacertae objects,** named after their prototype, BL Lac, in the constellation of Lacerta (the Lizard).

BL Lacertae (see Figure 18-16) was first discovered in 1929, when it was mistaken for a variable star; its brightness varies by a factor of 15 in only a few months. BL Lac's most intriguing characteristic is a totally featureless spectrum that exhibits neither absorption nor emission lines.

Careful examination of BL Lac revealed some "fuzz" around its bright, starlike core. By blocking out the light from the bright center of BL Lac, astronomers have been able to obtain a spectrum of this "fuzz." This spec-

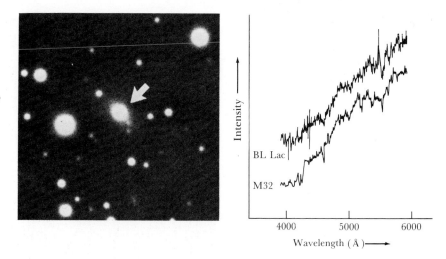

Figure 18-16 [left] BL Lacertae   This superb photograph shows "fuzz" around BL Lacertae. BL Lacertae objects appear to be giant elliptical galaxies with bright, starlike nuclei, much as Seyfert galaxies are spiral galaxies with quasarlike nuclei. BL Lacertae objects contain much less gas and dust than do Seyfert galaxies. (Courtesy of T. D. Kinman; NOAO)

Figure 18-17 [right]   The spectrum of BL Lacertae   The spectrum of the "fuzz" surrounding BL Lac is shown, along with the spectrum of M32, a small elliptical galaxy in the Local Group. The slight differences between these two spectra can be explained by assuming that BL Lac is a giant elliptical galaxy. The spectrum of BL Lac is redshifted by an amount corresponding to a distance of 1 billion light years from Earth. (Adapted from J. S. Miller, H. B. French, and S. A. Hawley)

trum contains many spectral lines and strongly resembles the spectrum of an elliptical galaxy (see Figure 18-17). In other words, a BL Lacertae object is an elliptical galaxy with a bright starlike center, much as a Seyfert galaxy is a spiral galaxy with a quasarlike center.

These discoveries have prompted many astronomers to suspect that quasars are the superluminous centers of very distant, very active galaxies. Painstaking observations have in fact revealed faint galaxylike "fuzz" around several quasars. Of course, quasars are extremely far away and are thus difficult to observe. Clues about the "engine" that powers a quasar therefore might best come from examining the cores of nearby galaxies.

Supermassive black holes may power quasars and active galaxies

How do quasars and active galaxies produce such enormous amounts of energy from such small volumes? As long ago as 1968, the British astronomer Donald Lynden-Bell suggested that an extremely massive black hole could be the "engine" that powers a quasar or active galaxy. At the center of a quasar or active galaxy, nature may be drawing upon the tremendous amounts of energy tied up in a black hole's powerful gravitational field.

As we saw in Chapter 15, finding black holes is a difficult business. At best, we can see only the effects of the hole's gravity and try to rule out all non–black-hole explanations of the data. This general approach has been applied to several nearby galaxies, including M31 and M32.

The Andromeda Galaxy (M31) is the largest, most massive galaxy in the local group. At a distance of only 2.2 million light years from Earth, M31 is so close to us that details in its core as small as 1 parsec across can be resolved under the best seeing conditions. A wide-field view of M31 is seen in Figure 17-3. A photograph of the core of M31 is shown in Figure 18-18.

In the mid-1980s, several astronomers made careful spectroscopic observations of the core of M31. By measuring the Doppler shifts of spectral lines at various locations in the core, they could determine the rotational velocities of stars about the galaxy's nucleus. The results show that stars within the central 15 pc of the galaxy's nucleus are orbiting the nucleus at exceptionally high speeds, which suggests that a massive object is located at the galaxy's center. Without the gravity of such an object to keep the stars in their high-

Figure 18-18 The central bulge of M31
The central bulge of the Andromeda Galaxy can be seen with the naked eye on a clear, moonless night. A photograph of the entire galaxy is shown in Figure 17-3. Spectroscopic observations suggest that a supermassive black hole is located at the center of this large, nearby galaxy. (NOAO)

Figure 18-19 The elliptical galaxy M32
This small galaxy is a satellite of M31, a portion of which is seen toward the left. Both galaxies are about 2.2 million light years from Earth. Spectroscopic observations suggest that an 8-million solar mass black hole may be located at the center of M32. (Palomar Observatory)

speed orbits, they would have escaped from the galaxy's core long ago. From such observations, astronomers estimate that the mass of the central object is about 50 million solar masses. That much matter confined to such a small volume strongly suggests the existence of a supermassive black hole.

Located near M31 is a small elliptical galaxy called M32 (see Figure 18-19). High-resolution spectroscopy of this galaxy also indicates that stars quite close to the galaxy's center are orbiting the galaxy's nucleus at unusually high speeds. In this case, the orbital motions suggest the presence of an 8-million solar mass black hole at the center of M32.

Astronomers have searched for evidence of supermassive black holes in other, more remote galaxies. For instance, John Kormendy used the 3.6-m Canada-France-Hawaii telescope on the top of Mauna Kea to examine the core of M104 spectroscopically (see Figure 18-20). Once again high-speed orbital motion around the galaxy's nucleus was found. Kormendy's observations suggest that the center of this galaxy is dominated by a supermassive black hole containing a billion solar masses.

Observations pointing to the existence of supermassive black holes have so far been successful only with rather ordinary galaxies. There are no nearby active galaxies that exhibit clearcut evidence of supermassive black holes at their centers. Nevertheless, the possibility that supermassive black holes might be quite commonplace has inspired astrophysicists to formulate mechanisms that could give rise to quasars and active galactic nuclei.

Various realistic scenarios about how to tap the gravitational energy of an extremely massive black hole have been worked out by Richard Lovelace at Cornell University and Roger Blandford at Caltech. In essence, these schemes are scaled-up versions of the explanation of Cygnus X-1 discussed in Chapter 15. A large black hole is needed to produce a large energy output.

Figure 18-20 [right]   The Sombrero Galaxy   This spiral galaxy in Virgo is nearly edge on to our Earth-based view. Spectroscopic observations suggest that a billion-solar mass black hole is located at the galaxy's center.

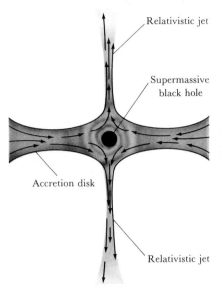

Relativistic jet

Supermassive black hole

Accretion disk

Relativistic jet

Figure 18-21 [above]   A supermassive black hole as the "central engine"   The energy output of active galaxies and quasars may involve extremely massive black holes that capture matter from their surroundings. In the scenario depicted here, the inflow of material through an accretion disk is redirected to produce two powerful jets of particles traveling near the speed of light. (Also see artist's rendition on the first page of this chapter.)

Imagine a billion-solar-mass black hole sitting at the center of a galaxy. The center of a galaxy is a very congested place, and we would expect this **supermassive black hole** to be surrounded by a huge rotating accretion disk of matter captured by the hole's gravity. Because of the constant inward crowding of matter in the accretion disk, the gas pressure surrounding the hole is tremendous. To relieve the pressure, matter is ejected perpendicular to the accretion disk, the direction along which the gases experience the least resistance. The result is two oppositely directed beams of particles. The overall scenario is sketched in Figure 18-21 and an artist's rendition is seen on the opening page of this chapter.

Supercomputer simulations help us explore and understand processes around black holes

The innermost region of the accretion disk around a supermassive black hole is "where the action is." If the scenario depicted in Figure 18-21 is correct, the matter spiraling in toward the black hole is somehow channeled into two oppositely directed beams. Because the inner accretion disk is heavily obscured by this infalling matter, astronomers have little hope of observing what goes on in the accretion disk around a supermassive black hole.

The fate of matter in the accretion disk around a black hole is governed by the laws of physics, which are expressed as equations written in the mathematical language of calculus. These laws include the equations of general relativity that describe the gravitational field of the black hole. In addition equations of hydrodynamics describe the behavior of the gas, including the pressure, density, and velocity at every point throughout the accretion disk. This massive set of equations is so overwhelmingly complicated that a person armed only with pencil and paper has no hope of solving them. In recent years, however, a new way of doing science has become possible with the development of supercomputers.

A **supercomputer** is the fastest, most powerful computer that can be built. Examples included the ETA-10 and the Cray X-MP, which can

Figure 18-22 A supercomputer
A supercomputer has a very high speed and a large internal memory that make it capable of rapidly performing elaborate, repetitive computations. With such a machine, scientists can simulate exotic physical circumstances—such as matter falling toward a black hole—that cannot otherwise be easily observed or studied. This photograph shows the main components of a Cray X-MP. Scientists use ordinary computers at remote locations to communicate with the supercomputer. (NCSA)

Figure 18-23 Standing shock waves in an accretion disk *These cross-sectional diagrams show (a) a schematic and (b) the color-coded results of a supercomputer simulation of an accretion disk around a black hole. Gas plunges at supersonic speed toward the black hole at the center. Centrifugal forces exclude matter from cone-shaped funnels surrounding the disk's axis of rotation. The black arrow-shaped features in the densest (red) part of the accretion disk are standing shock waves created where the infalling gas slows abruptly as it encounters the centrifugal barrier at the walls of the evacuated funnel. (Courtesy of J. F. Hawley and L. L. Smarr)*

perform billions of mathematical operations per second (see Figure 18-22). These machines will enjoy their exalted status for only a few years, because continuing technological advances will soon give rise to the next generation of supercomputers.

A supercomputer is a valuable research tool because the laws of physics can be written in a way that the supercomputer can solve them. In their basic form, the laws of physics are valid at every point in space and at every moment of time. Scientists seldom, however, need to apply the laws of physics on this fine a scale. Instead, they use a set of points in space that are separated by small distances compared to the size of the object under study. They use discrete intervals of time that are short compared to the duration of the process they want to explore. By programming a supercomputer with the laws of physics expressed only at selected points and time intervals, sci-

b

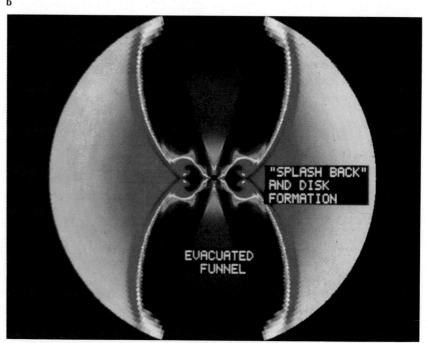

a

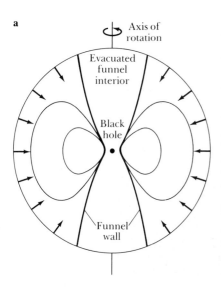

entists can reduce a complicated problem to a level that the machine can handle.

Using this technique with a Cray supercomputer, a team of astrophysicists has performed supercomputer simulations of an accretion disk orbiting a supermassive black hole. The setup for the computation is shown in Figure 18-23a. A black hole, at the center of the field of view, is surrounded by the inner regions of a rotating accretion disk (the axis of rotation is vertical). Matter flows inward toward the hole, as indicated by the arrows around the edge of the field of view. To account for the luminosity of a quasar, a billion-solar-mass black hole would have to capture or accrete 10 $M_\odot$ per year.

The program that runs the supercomputer divides space around the black hole into small boxes, some of which can be seen near the upper and lower edges of Figure 18-23b. The supercomputer goes from one box after the other, calculating such physical quantities as the pressure, temperature, density, and velocity of the infalling matter. The laws of physics also tell how these quantities change with time, and so the supercomputer can tell us how conditions around the black hole evolve.

Supercomputer simulations have demonstrated that infalling matter in an accretion disk plunges toward the black hole at supersonic speeds. Near the hole, however, this inward rush of gas is abruptly stopped by a "centrifugal barrier" caused by the rapid orbiting of the gas around the hole. This abrupt halt produces a **standing shock wave** that marks the inner edge of the accretion disk, as shown in Figure 18-23b. Much of the infalling matter "splashes back" off this shock wave, thereby keeping the funnel-shaped cavity surrounding the disk's axis of rotation empty.

The continuous deluge of matter crashing onto the standing shock wave propels gas along the walls of the evacuated funnel, as shown in Figure 18-24. This arrangement channels the gas into two oppositely directed

Figure 18-24 The inner edge of the accretion disk *This computer-generated picture shows the flow pattern of gas in an accretion disk surrounding a black hole. The black hole is located at the center of this cross-sectional diagram, where the broad-angle inner edges of the accretion disk point. Color displays gas pressure, ranging from red for high pressure across the spectrum to blue for low pressure. The flow pattern (indicated by the white arrows) shows how infalling gas is channeled into two oppositely directed jets. The centrifugal barrier created by the rapid rotation of the inner edge of the accretion disk prevents the gas from falling directly into the black hole. (Courtesy of J. F. Hawley and L. L. Smarr)*

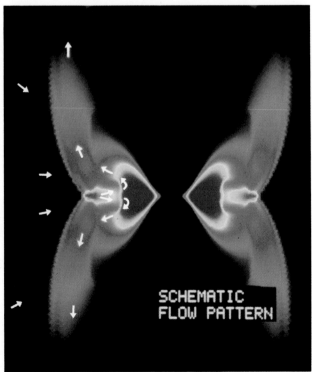

beams that emerge from either side of the accretion disk. Mechanisms of this type may explain double radio sources and some of the jets and beams we see protruding from active galaxies and from some quasars.

The possible existence of supermassive black holes in quasars and active galaxies is one of the most exciting topics in modern astronomy. In the 1990s, the Hubble Space Telescope will be used to probe the nuclei of galaxies with much higher resolution than is possible from the ground. These observations along with supercomputer simulations will undoubtedly give us a much better understanding of the powerful "engines" at the cores of active galaxies and quasars.

Summary

· A quasar (or quasi-stellar object) is an object that looks like a star but has a huge redshift corresponding to a great distance from the Earth.

A typical quasar is very luminous and shines with the energy output of roughly 100 ordinary galaxies.

Rapid fluctuations in the brightnesses of quasars indicate that they cannot be much larger than the diameter of our solar system.

· An active galaxy is an extremely luminous galaxy that has one or more unusual features: an unusually bright, starlike nucleus; strong emission lines in its spectrum; extreme variations in luminosity, and/or jets or beams of radiation emanating from its core. Seyfert galaxies, N galaxies, and BL Lacertae objects are examples of active galaxies.

A double radio source is an active galaxy, the bulk of whose radio emission comes from two regions, called radio lobes, on opposite sides of the galaxy.

Spectroscopic and photometric observations of certain active galaxies seems to indicate the presence of huge concentration of matter (perhaps a supermassive black hole) at their centers.

· The strong energy emission from quasars, active galaxies, and double radio sources may be produced as matter falls toward a supermassive black hole at the center of the object.

· Supercomputers are used to calculate the behavior of matter in an accretion disk around a supermassive black hole.

· Matter spiraling in toward a supermassive black hole can be channeled into two oppositely directed beams that propel particles and energy into intergalactic space.

Review questions

1 Suppose you suspected a certain object in the sky to be a quasar. What sort of observations might you perform to find out if it were indeed a quasar?

2 Explain why astronomers do not use any of the standard candles described in Chapter 17 to determine the distances to quasars.

3 In the 1960s, it was suggested that quasars might be compact objects ejected at high speeds from the centers of nearby ordinary galaxies. Why does the absence of blueshifted quasars disprove this hypothesis?

4 Why do astronomers believe that the energy-producing region of a quasar is very small?

5 In what ways are active galaxies similar to quasars?

6 Why do you suppose there are no quasars relatively near our galaxy?

7 How does SS433 (discussed in Chapter 14) compare with a typical double radio source?

8 Why do many astronomers believe that the "engine" at the center of a quasar is a supermassive black hole surrounded by an accretion disk?

Advanced questions

9 Some quasars show several sets of absorption lines whose redshifts are less than the redshift of the quasars' emission lines. For example, the quasar PKS 0237–23 has five sets of absorption lines, all with redshifts somewhat less than the redshift of the quasar's emission lines. Propose an explanation for these sets of absorption lines.

10 When quasars were first discovered, many astronomers were optimistic that these extremely luminous objects could be used to probe distant regions of the universe. For example, it was hoped that quasars would provide high-redshift data from which the Hubble constant could be accurately determined. Why do you suppose these hopes have not been realized?

Discussion question

11 Speculate on the possibility that quasars, double radio sources, giant elliptical galaxies, and so on form some sort of evolutionary sequence.

12 Speculate on the possibility that supercomputers offer a new way of doing science, perhaps as important and powerful as the observational and theoretical approaches that have been in use since the time of Galileo and Newton.

For further reading

Balick, B. "Quasars with Fuzz." *Mercury,* May/June 1983, p. 81.

Blandford, R., et al. "Cosmic Jets." *Scientific American,* May 1982.

Burns, J., and Price, R. "Centaurus A: The Nearest Active Galaxy." *Scientific American,* November 1983.

Ferris, T. "The Spectral Messenger: The Redshift Controversy." *Science 81,* October 1981, p. 66.

Kanipe, J. "M87: Describing the Indescribable." *Astronomy,* May 1987, p. 6.

Kaufmann, W. "Exploding Galaxies and Supermassive Black Holes." *Mercury,* September/October 1978, p. 97.

Tananbaum, H., and Lightman, A. "Cosmic Powerhouses: Quasars and Black Holes." In Cronell, J., and Lightman, A., eds. *Revealing the Universe.* MIT Press, 1982.

Wyckoff, S., and Wehinger, P. "Are Quasars Luminous Nuclei of Galaxies?" *Sky & Telescope,* March 1981, p. 200.

19

Cosmology and the creation of the universe

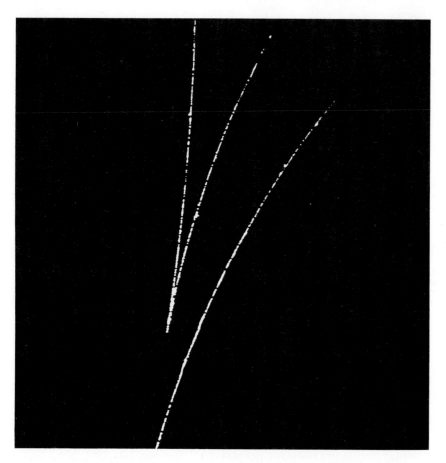

The creation of matter from energy *This photograph shows the conversion of a gamma ray into matter inside a bubble chamber, a device filled with liquid hydrogen that is designed to make the path of a charged particle visible as a long row of tiny bubbles. The path of the gamma ray is not visible because photons are electrically neutral. Near the bottom of the photograph, the energy carried by the gamma ray is converted into an electron and an antielectron. Because of a magnetic field surrounding the bubble chamber, the electron is deflected to the right while the antielectron veers toward the left. The path of a stray electron is also seen at the right. (Courtesy of Lawrence Berkeley Laboratory)*

We live in an expanding universe. This chapter describes how the expansion of the universe began with an explosion of space at the beginning of time. Evidence for this "Big Bang" exists in the form of microwave photons, which we interpret as a relics of the primordial fireball. Looking toward the future, we discover that the universe may either expand forever or collapse in a Big Crunch, depending on the density of matter throughout space. This analysis, coupled with the fact that space is almost perfectly flat, inspires us to probe the earliest moments of the universe. It appears that the universe suddenly "inflated" to many times its original size shortly after the Big Bang. To understand why the universe turned out the way it did, we explore the idea that all the fundamental forces had the same strength immediately after the Big Bang. We then see how the various familiar forces and particles that comprise ordinary matter "froze out" of the newborn universe as it expanded and cooled.

........**A**s foolish as it may seem, one of the most profound questions you can ask is, "Why is the sky dark at night?" This question apparently haunted Johannes Kepler as long ago as 1610, and it was popularized in the early 1800s by the German amateur astronomer Heinrich Olbers.

To appreciate the problem, we must begin by assuming that the universe is infinite and that stars are scattered more or less randomly across this infinite expanse of space. Isaac Newton argued that no other assumption makes sense. If the universe were not infinite or if stars were grouped in only one part of the universe, then the gravitational forces between the stars would soon cause all this matter to fall together into a compact blob. Obviously, this has not happened. Thus classical Newtonian mechanics describes our universe as infinite and static. This would mean each star feels a uniform gravitational pull from every part of the sky, from all the other stars in the universe. According to this model, the universe can exist forever without undergoing major changes in structure.

Imagine looking out into space in this static, infinite universe. Because space goes on forever with stars scattered throughout it, your line of sight must eventually hit a star. No matter where you look in the sky, you should see a star. The entire sky should therefore be as bright as an average star. Even at night, the entire sky should be blazing like the surface of the Sun. That this is not so is the dilemma called **Olbers's paradox.**

We live in an expanding universe

Olbers's paradox tells us that there is something wrong with Newton's idea of an infinite, static universe. According to the classical, Newtonian picture of reality, space is flat and inflexible. This rigid, flat space stretches on and on, totally independent of stars or galaxies or anything else. Similarly, a Newtonian clock ticks steadily and monotonously forever, never slowing down or speeding up.

Albert Einstein demonstrated that this view of space and time is wrong. In his special theory of relativity, he proved that measurements with clocks and rulers depend on the motion of the observer. As explained in Chapter 15, Einstein's general theory of relativity tells us that the shape of space is profoundly influenced by the masses occupying that space. Specifically, matter tells space how to curve, and curved space tells matter how to move.

Shortly after formulating the general theory of relativity in 1915, Albert Einstein applied his ideas to the structure of the universe. To his dismay, his calculations could not produce a truly static universe. His equations instead indicated that the universe must be either expanding or contracting. The prevailing Newtonian opinion was so strong that Einstein doubted the validity of his equations and missed the opportunity to propose that we live in an expanding universe.

Edwin Hubble is usually credited with discovering that we live in an expanding universe. As we saw in Chapter 17, Hubble discovered the simple linear relationship (now called the Hubble law) between the distances to remote galaxies and the redshifts of those galaxies' spectral lines. The greater the distance to a galaxy, the greater is its redshift. Thus, remote galaxies appear to be moving away from us with speeds proportional to their distances from us (recall Figure 17-30). Since the galaxies are getting

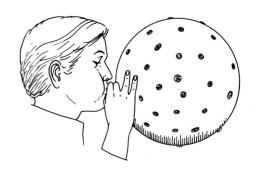

Figure 19-1 The expanding balloon analogy *The expanding universe can be compared to the expanding surface of an inflating balloon. All the coins on the balloon recede from one another as the balloon expands, just as all the galaxies recede from each other as the universe expands. (Adapted from C. Misner, K. Thorne, and J. Wheeler)*

farther and farther apart as time goes on, astronomers say that the universe is expanding.

What does it mean to say that "the universe is expanding?" According to relativity theory, space is not rigid. Instead, the amount of space between widely separated locations can change over time. A good analogy is that of a person blowing up a balloon, as sketched in Figure 19-1. Small coins, each representing a galaxy, are glued onto the surface of the balloon. As the balloon expands, the amount of space between the coins gets larger and larger. In the same way, as the universe expands, the amount of space between widely separated galaxies grows larger and larger. The expansion of the universe *is* the expansion of space.

The expanding-balloon analogy illustrates several important characteristics of the expanding universe. Imagine sitting on one of the coins in Figure 19-1. As the balloon expands, you see the other coins moving away from you. Specifically, you observe that nearby coins move away from you slowly, whereas more distant coins move away more rapidly. This Hubble-law-like relationship applies no matter which coin you call home. Your coin is not at the center of the balloon, of course. Indeed, the surface of the balloon does not have a center; you could walk around the surface forever without finding the center. And the surface of the balloon has no edge; you could explore every inch of the surface and never find an edge.

Just as the surface of the balloon has neither center nor edge, our universe has no center or edge. No matter what galaxy you call home, all the other galaxies are receding from you. No one is ever at the "center" of the universe. Questions like "What is beyond the edge of the universe?" or "What is the universe expanding into?" are as meaningless as asking "What is north of the Earth's north pole?"

The ongoing expansion of space explains how photons from remote galaxies are redshifted. Imagine a photon coming toward us from a distant galaxy. As the photon travels through space, space itself is expanding, and so the photon's wavelength becomes stretched. When the photon reaches our eyes, we see a drawn-out wavelength, which is the redshift. The longer the photon's journey, the more its wavelength will have been stretched. Thus, photons from distant galaxies have larger redshifts than those of photons from nearby galaxies, which is expressed as the Hubble law.

A redshift caused by the expansion of the universe is properly called a **cosmological redshift** to distinguish it from a Doppler shift. Doppler shifts are caused by an object's *motion through* space whereas a cosmological redshift is caused by the *expansion of* space.

It is important to realize that the expansion of space occurs primarily in the voids that separate clusters of galaxies. Just as the coins in Figure 19-1 do not expand as the balloon inflates, galaxies themselves do not expand. Einstein and others have established that a galaxy is always contained within a patch of nonexpanding space. The galaxy's gravitational field produces this nonexpanding patch, which is indistinguishable from the flat, rigid space of Newton. Thus, the Earth and your body are not getting bigger and bigger; the expansion of the universe occurs only in intergalactic space.

The Big Bang was an explosion of space at the beginning of time

The universe has been expanding for billions of years. Going back in time, we realize there must have been a point in the ancient past when all the matter in the universe was concentrated in a state of infinite density. Presumably some sort of colossal explosion occurred to start the expansion of the universe. This explosion, commonly called the **Big Bang,** marks the creation of the universe.

The Hubble constant, which gives the rate of expansion of the universe, can be used to estimate the time that has elapsed since the Big Bang as follows:

$$\frac{1}{H_0} = \frac{1}{15 \ \text{km/sec/Mly}} = 20 \ \text{billion years}$$

Astronomers believe, however, that the true age of the universe is probably between 15 and 18 billion years. Because of their mutual gravitational attraction, galaxies have not been flying away from each other with a constant velocity. Gravity has caused the speed of separation between galaxies to decrease gradually since the Big Bang. Thus, the expansion rate of the universe has been decreasing. An age of 20 billion years assumes no deceleration, and so this value would be valid only for an empty universe that contains no matter and thus no gravity to slow the expansion.

The finite age of the universe offers a resolution of Olbers's paradox. The entire sky is not as bright as the surface of the Sun because we cannot see any stars that are more than 20 billion light years away. The universe may indeed be infinite, but we can only observe a fraction of it. Because the universe is less than 20 billion years old, the light from stars more than 20 billion light years away has simply not had enough time to get here.

You can think of the Earth as being at the center of an enormous sphere with a radius of roughly 20 billion light years (see Figure 19-2). The surface of this sphere is called the **cosmic particle horizon.** The entire **observable universe** is located inside this sphere. We cannot see anything beyond the cosmic particle horizon because the travel time for light from these incredibly remote objects is greater than the age of the universe. Throughout the observable universe, the distribution of galaxies is sufficiently sparse that most of our lines of sight do not hit any stars, which explains why the night sky is dark.

The idea of a Big Bang origin of the universe is a straightforward, logical consequence of having an expanding universe. If we could look back far enough into the past, we would arrive at a time nearly 20 billion years ago when the density of matter throughout the universe was infinite. The entire universe was akin to the singularity at the center of a black hole. For this reason, perhaps a better name for the Big Bang is the **cosmic singularity.**

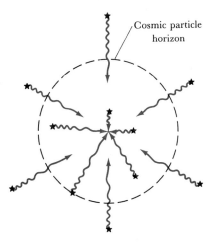

Cosmic particle horizon

Figure 19-2 The observable universe
The radius of the cosmic particle horizon is equal to the distance that light has traveled since the Big Bang. The Big Bang occurred about 20 billion years ago, and so the cosmic particle horizon is about 20 billion light years away. We cannot see stars beyond the cosmic particle horizon because their light has not had enough time to reach us.

It is easy to have misconceptions about the Big Bang. For one, it was not at all like an exploding bomb. When a bomb explodes, pieces of debris fly off *into space* from a central location. If you could trace all the pieces back to their origin, you could pinpoint the bomb's original location. You cannot do this with the Big Bang, however, because the cosmic singularity was not a point somewhere. When the density of matter throughout the universe was infinite, the cosmic singularity filled *all* space. The Big Bang therefore occurred everywhere throughout the entire universe. Indeed, we can think of the Big Bang as an explosion of space at the beginning of time.

Comparing the Big Bang to a black hole singularity helps us appreciate certain aspects of the creation of the universe. As we saw in Chapter 15, matter is crushed to infinite density at the center of a black hole. This location, called the singularity, is characterized by infinite curvature where space and time are all tangled up. Without a clear background of space and time, such concepts as *past, future, here,* and *now* cease to have any meaning.

At the moment of the Big Bang, a state of infinite density filled the universe. Space and time throughout the universe were completely jumbled up in a state of infinite curvature like that at the center of a black hole. Thus the laws of physics are useless in explaining what happened at the moment of the Big Bang. We certainly cannot use science to tell us what existed *before* the Big Bang. These things are fundamentally unknowable. The phrases "*before* the Big Bang" or "at the *moment* of the Big Bang" are meaningless, because time itself did not really exist.

In a very short time after the Big Bang, space and time came to exist in the way we think of them today. This interval, called the **Planck time,** equals 10^{-43} sec. From the moment of the Big Bang (at time $t = 0$) to the Planck time 10^{-43} sec later, all known science fails us. We do not know how space, time, and matter behaved under these extreme circumstances.

The explanation of the Big Bang as a space-time singularity rests on the validity of the general theory of relativity. That theory is, after all, the best description of gravity we have, and it clearly predicts the existence of singularities at the centers of black holes as well as throughout all space at the beginning of time. Because of these singularities, however, many physicists believe that general relativity may not give a correct picture of conditions earlier than the Planck time. These scientists look forward to the development of improved theories, sometimes called "quantized gravity" or "super-grand unified field theories," to shed new light on the nature of the Big Bang and perhaps do away with singularities. We will have more to say about these speculative developments near the end of this chapter.

....

Microwave radiation that fills all space is evidence for a hot Big Bang

One of the major successes of modern astronomy involves discoveries about the origin of the heavy elements. We know today that all the heavy elements are created in the fiery infernos at the centers of stars. As astronomers in the 1960s began understanding the details of thermonuclear synthesis, a new dilemma arose: there is too much helium around. For example, the Sun consists of about 73 percent hydrogen, 24 percent helium, and 3 percent for all the remaining heavier elements combined. This 3 percent can be understood as material produced at the centers of ancient stars that long ago cast these heavy elements into space. Certainly, some freshly made helium accompanied these heavy elements, but not nearly enough helium to account for one-quarter of the Sun's mass.

Shortly after World War II, George Gamow at George Washington University proposed that, immediately following the Big Bang, the universe must have been so incredibly hot that thermonuclear reactions could have occurred everywhere throughout space. Following up this idea in 1960, Princeton physicists Robert Dicke and P. J. E. Peebles confirmed that they could indeed account for today's high abundance of helium by assuming that the early universe was at least as hot as the Sun's center (where helium is currently being produced). The early universe must therefore have been filled with many high-energy, short-wavelength photons.

The universe has expanded considerably since those ancient times, and all those short-wavelength photons have become so stretched that they are now low-energy, long-wavelength photons. Thus, the temperature of this cosmic radiation field should now be quite low, perhaps only a few kelvins above absolute zero. According to Wien's law, radiation at this low temperature should have its peak intensity at microwave wavelengths of a few millimeters. This radiation had been predicted by Gamow and his colleagues in 1948. In the early 1960s, Dicke, Peebles, and their colleagues at Princeton began designing an antenna to detect this radiation.

Meanwhile, a few miles away from Princeton University, Arno Penzias and Robert Wilson of Bell Telephone Laboratories were working on a new microwave horn antenna (see Figure 19-3). This antenna was designed to relay telephone calls to Earth-orbiting communication satellites. Penzias and Wilson were deeply puzzled because no matter where they pointed their antenna in the sky, they detected a faint background noise. Thanks to a friend, they soon learned of the work of Dicke and Peebles. Penzias and Wilson realized that they had detected the cooled-down cosmic background radiation left over from the hot Big Bang.

Since those pioneering days, scientists have made many measurements of the intensity of this background radiation at a variety of wavelengths (see

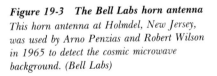

Figure 19-3   The Bell Labs horn antenna
This horn antenna at Holmdel, New Jersey, was used by Arno Penzias and Robert Wilson in 1965 to detect the cosmic microwave background. (Bell Labs)

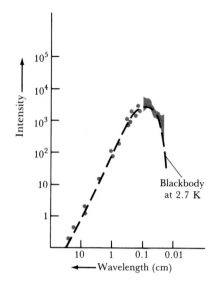

Figure 19-4 The spectrum of the cosmic microwave background Measurements of intensity at various wavelengths (indicated by the dots and the shaded area) demonstrate that the cosmic microwave background is blackbody radiation with a temperature of nearly 3 K.

Figure 19-4). The dashed curve that fits the data in Figure 19-4 is a blackbody spectrum with a temperature of nearly 3 K. Because of its various properties, this radiation field that fills all space is commonly called the **3-degree cosmic microwave background.**

An important feature of the microwave background is that its intensity is almost perfectly isotropic (meaning "the same in all directions"). In other words, we detect nearly the same intensity from all parts of the sky. However, extremely accurate measurements made in high-flying airplanes reveal a very slight temperature variation across the sky. The microwave background is slightly warmer than average in the direction of the constellation of Leo, and in the opposite direction (toward Aquarius) the microwave background is slightly cooler than average. Between the warm spot in Leo and the cool spot in Aquarius, the background temperature across the sky declines in a smooth fashion.

This temperature variation can be explained as a result of the Earth's overall motion through the cosmos. If we were at rest with respect to the microwave background, the radiation would be truly isotropic. Because we are moving through this radiation field, however, we see a Doppler shift. As sketched in Figure 19-5, we see shorter-than-average wavelengths in the direction toward which we are moving. A decrease in wavelength corresponds to an increase in photon energy and thus to an increase in temperature. The observed temperature excess corresponds to a speed of 390 km/sec.

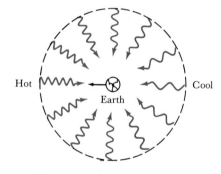

Figure 19-5 [above] Our motion through the microwave background Because of the Doppler effect, the microwave background is slightly warmer in that part of the sky toward which we are moving. Recent measurements indicate that our Galaxy, along with the rest of the Local Group, is moving in the general direction of the Virgo and Hydra clusters.

Figure 19-6 [right] The Cosmic Background Explorer (COBE) This satellite was launched in 1989 to study the spectrum and angular distribution of the cosmic background radiation over a wavelength range of 1 µm to 1 cm. COBE (pronounced CO-bee) will examine deviations from a perfect blackbody spectrum and from perfect isotropy. (Courtesy of John Mather, NASA)

Conversely, we see longer-than-average wavelengths in that part of the sky from which we are receding. An increase in wavelength corresponds to a decline in photon energy and hence to a decline in temperature. We are thus traveling from Aquarius toward Leo at a speed of 390 km/sec. Taking into account the known velocity of the Sun around the center of our Galaxy, we find that the entire Milky Way Galaxy is moving at 600 km/sec in the general direction of the Hydra–Centaurus supercluster, possibly because of the gravitational pull of an enormous mass in that direction dubbed the **Great Attractor.**

In 1989, the Cosmic Background Explorer satellite (COBE) was launched to observe the microwave background in detail (see Figure 19-6). Of particular interest is an energy excess discovered in 1987 on the short-wavelength side of the blackbody curve in Figure 19-4. This deviation from a perfect blackbody spectrum contains an amount of energy equivalent to 10 percent of the microwave background. Astronomers are intrigued by this excess because it suggests that a huge amount of energy was released in the universe by some unknown process shortly after the Big Bang. Also poorly understood are the kinds of primordial density fluctuation that eventually led to the formation of galaxies. These fluctuations should have left some sort of imprint on the microwave background detectable today as small-scale temperature variations across the sky. Astronomers believe that COBE's instruments are sensitive enough to discover these variations and thereby give us important insights into conditions that prevailed soon after the Big Bang.

The future of the universe is determined by the average density of matter in it

During the 1920s, general relativity theory was applied to cosmology by Alexandre Friedmann in Russia, Georges Lemaître in France, Willem de Sitter in the Netherlands, and, of course, by Einstein himself. The resulting picture of the structure and evolution of the universe, called **relativistic cosmology,** is in surprisingly good agreement with our intuitive notion that gravity should be slowing the cosmological expansion.

Imagine a rocket blasting off from the surface of the Earth. If the rocket's speed is less than the escape velocity (about 11 km/sec), the spacecraft will fall back to Earth. If the rocket's speed equals the escape velocity, the spacecraft will just barely manage not to fall back. If the rocket's speed exceeds the escape velocity, the spacecraft will easily leave the Earth and never fall back, despite the relentless pull of gravity.

The universe is subject to a similar set of circumstances. If the average density of matter throughout space is small, the gravity associated with this matter is weak, and the expansion of the universe will continue forever. Even infinitely far into the future, galaxies will continue to rush away from each other. In such circumstances, we say that the universe is **unbound.**

Conversely, if the average density of matter across space is large, the resulting gravity would be strong enough to halt the expansion of the universe eventually. The universe will reach a maximum size, then begin to contract as gravity starts to pull the galaxies back toward each other. In this case, we say that the universe is **bound.**

In between these two scenarios is the situation in which the galaxies just barely manage to keep moving away from each other. In this case, the universe is **marginally bound** and the average density of matter across space is called the **critical density** (ρ_c). This case is analogous to the rocket leaving the Earth at a speed exactly equal to the escape velocity.

Estimates of the critical density depend on the Hubble constant. Using a Hubble constant of 15 km/sec/Mly, we find that $\rho_c = 4.5 \times 10^{-30}$ g/cm^3, which is equivalent to a density of about three hydrogen atoms per cubic meter of space.

The average density of luminous matter that we see in the sky seems to be about 3×10^{-31} g/cm^3, clearly less than the critical density. However, as discussed in Chapter 17, the missing-mass problem for clusters of galaxies strongly suggests that we may be seeing only a tenth of the matter in the universe. Most of the mass of the universe seems to be dark and remains to be discovered. In short, present-day observations and estimates of density are not accurate enough to tell us whether the universe is bound or unbound.

The deceleration of the universe can be determined from observations of extremely distant galaxies

Figure 19-7 Deceleration and the Hubble diagram   These two graphs compare **(a)** *the appearance of the Hubble diagram, and* **(b)** *the evolution of the universe. The case* $\Omega_0 = 0$ *is an empty universe. The case* $\Omega_0 = 1$ *is a marginally bound universe. If* $0 < \Omega_0 < 1$, *then the universe is unbound and will expand forever. If* $\Omega_0 > 1$, *the universe is bound and will someday collapse.*

Another way of determining the future of the universe is to measure the rate at which cosmological expansion is slowing down. The deceleration of the universe shows up as a deviation from the straight-line relationship predicted by the Hubble law. Suppose that you measure the redshifts of galaxies several billion light years from Earth. Light from these galaxies has taken billions of years to get to your telescope, so your measurements reveal how fast the universe was expanding billions of years ago. Because the universe was expanding faster in the past than it is today, your data will deviate slightly from the straight-line Hubble law.

The graphs in Figure 19-7 display the relationship between the deceleration of the universe and the Hubble law, extended to great distances. Astronomers denote the amount of deceleration by the **density parameter** Ω_0, which is the density of matter in the universe divided by the critical

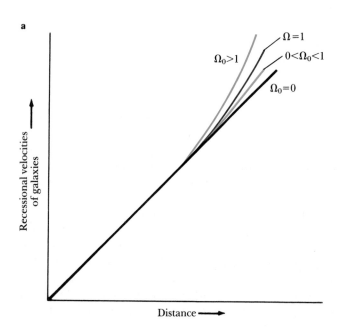

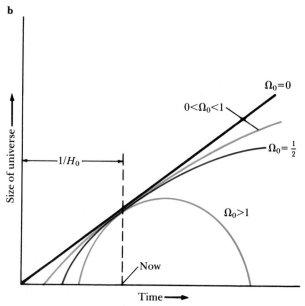

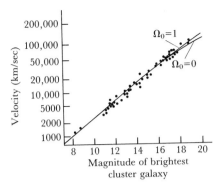

Figure 19-8 The Hubble diagram
This graph shows the Hubble diagram extended to include extremely remote galaxies. The magnitude of the brightest galaxy in a cluster is directly correlated with the distance to the cluster. If the data points fall between the curves marked $\Omega_0 = 0$ and $\Omega_0 = 1$, then the universe is unbound. If the data points fall above the curve marked $\Omega_0 = 1$, then the universe is bound. (Adapted from A. Sandage)

density. If the universe is empty, $\Omega_0 = 0$ and there is no deceleration, because there is no gravity to slow down the expansion. As sketched in Figure 19-7a, the $\Omega_0 = 0$ universe expands forever at a constant rate. As explained earlier in this chapter, the present age of this empty universe is $1/H_0$, or about 20 billion years.

The case $\Omega_0 = 1$ corresponds to a marginally bound universe. Such a universe just barely manages to expand forever, because it contains matter at the critical density ρ_c. In this case, the present age of the universe equals $\frac{2}{3}(1/H_0)$, or about 13 billion light years.

If Ω_0 is in the range between 0 and 1, the universe is unbound and will continue to expand forever. Such a universe contains matter at less than the critical density, and has a present age of 13 to 20 billion years.

If Ω_0 is greater than 1, the universe is bound and is filled with matter of a density greater than ρ_c. This universe, whose age is less than 13 billion years, is doomed to collapse in upon itself and ultimately end in a **Big Crunch** in the extremely distant future.

In principle, it should be possible to determine Ω_0 by measuring the redshifts and distances of many remote galaxies, then plotting the data on a Hubble diagram like the one in Figure 19-7b. If the data points fall above the $\Omega_0 = 1$ line, the universe is bound. If the data points fall between the $\Omega_0 = 0$ and $\Omega_0 = 1$ lines, the universe is unbound.

Unfortunately, such observations are extremely difficult to make. Galaxies nearer than a billion light years are of no help in determining Ω_0. Beyond a billion light years, galaxies are so faint and our understanding of galactic evolution is so poor that uncertainties cloud the distance determinations. For example, Figure 19-8 shows data obtained by Allan Sandage of the Mount Wilson and Las Campanas Observatories. Note how the data points are scattered about the $\Omega_0 = 1$ line. This scattering, a result of observational uncertainties, prevents us from determining conclusively whether the universe is bound or unbound. Nevertheless, because the data lie close to the $\Omega_0 = 1$ line, we may conclude that the universe is not far from being marginally bound. Many astronomers expect that, during the 1990s, the Hubble Space Telescope will enable them to do a better job of determining the value of Ω_0.

The shape of the universe is related to the average density of matter in the universe

There is another way of investigating the future of the universe. Our understanding of the universe is based on Einstein's general theory of relativity, which explains that gravity curves the fabric of space. The gravity that is slowing the expansion of the universe must also give space an overall "shape." This shape or geometry is directly related to the average density of matter throughout space, which is expressed by the density parameter. By measuring the shape of the universe, we might discover its ultimate fate.

To see what astronomers mean by the geometry of the universe, imagine shining two powerful laser beams out into space. Suppose that these two beams are aligned so that they are perfectly parallel as they leave the Earth. Finally, suppose that nothing gets in the way of these two beams, so that we can follow them for billions of light years across the universe, across the space whose curvature we wish to detect.

There are only three possibilities for what we might discover. First, we might find that our two beams of light remain perfectly parallel, even after traversing billions of light years. In this case, we would conclude that space is not curved: the universe has **zero curvature** and space is **flat.**

Alternatively, we might find that our two beams of light gradually get closer and closer together as they move across the universe, so that they eventually intersect at some enormous distance from Earth. In this case, space is not flat. Recall that lines of longitude on the Earth's surface are parallel at the equator but intersect at the poles. The shape of the universe must be analogous to the surface of a sphere. We therefore say that space is **spherical** and the universe has **positive curvature.**

The third and final possibility is that the two parallel beams of light will gradually diverge, becoming farther and farther apart as they move across the universe. In this case, we say that the universe has **negative curvature.** A sphere is a positively curved surface, and a saddle is a good example of a negatively curved surface. Parallel lines drawn on a saddle always diverge. Mathematicians say that saddle-shaped surfaces are hyperbolic. In a negatively curved universe, space is **hyperbolic.**

These three cases describing the geometry of the universe are sketched in Figure 19-9. To illustrate them, three surfaces are drawn: a plane, a sphere,

Figure 19-9   The geometry of the universe *The "shape" of space is determined by the matter contained in the universe. The curvature is either positive (**a**), zero (**b**), or negative (**c**), depending on whether the average density throughout space is greater than, equal to, or less than the critical density.*

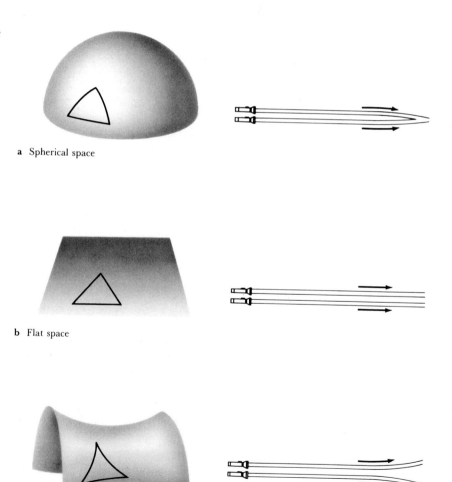

a Spherical space

b Flat space

c Hyperbolic space

Table 19-1   The geometry and fate of the universe

Geometry of space	Curvature of space	Average density throughout space	Density parameter (Ω_0)	Type of universe	Ultimate future of the universe
Spherical	Positive	Greater than the critical density	Greater than 1	Closed	Eventual collapse
Flat	Zero	Exactly equal to the critical density	Exactly equal to 1	Flat	Perpetual expansion (just barely)
Hyperbolic	Negative	Less than the critical density	Between 0 and 1	Open	Perpetual expansion

and a saddle. Real space is of course three dimensional, but it is much easier to visualize a two-dimensional surface. Thus, as you examine the drawings in Figure 19-9, remember that the real universe has one more dimension. For example, if the universe is hyperbolic, the geometry of space is the three-dimensional analogue of the two-dimensional surface of a saddle.

Each of the three possible geometries corresponds to a different behavior and fate for the universe. Flat space corresponds to the marginally bound case ($\Omega_0 = 1$) in which the galaxies just barely manage to keep receding from each other. This flat-space scenario divides the positive-curvature cases from the negative-curvature cases. If the density across space is greater than the critical density, then $\Omega_0 > 1$, and space is positively curved. Conversely, if the density across space is less than the critical density, then $0 < \Omega_0 < 1$, and space is negatively curved. These relationships are summarized in Table 19-1.

Note that both the flat and the hyperbolic universes are infinite. They extend forever in all directions, so that they have neither an "edge" nor a "center." In contrast, the spherical universe is finite. However, it also lacks a center or an edge. A good analogy is the Earth. Even though the Earth has a finite surface area (511 million km^2), you could walk forever around it without ever finding the center or an edge. Likewise, relativistic cosmology strictly rules out any possibility of a center or edge to the universe.

The flatness of the universe and the isotropy of the microwave background suggest that a period of vigorous inflation followed the Big Bang

Ever since Edwin Hubble discovered that the universe is expanding, astronomers have struggled to determine the density parameter. During the 1960s and 1970s, various teams of astronomers reported various values for Ω_0, some slightly larger than $\Omega_0 = 1$ and some smaller. Because $\Omega_0 = 1$ is the special case of a flat universe separating a bound cosmological model from an unbound model, the predicted fate of the universe swung back and forth. According to some data, the universe seems to be just barely open and infinite, whereas other data indicate that the universe is just barely closed and doomed to collapse.

Motivated by the fact that Ω_0 may be nearly equal to 1, physicists began looking for special conditions associated with a flat universe in which the average density of matter (ρ) is exactly equal to the critical density ρ_c. Specifically, suppose the average density of matter during the Big Bang to be slightly larger or smaller than the critical density. How would this deviation grow or decrease as the universe evolves?

We saw that the earliest understandable moment in the universe was the Planck time, about 10^{-43} sec after the Big Bang. Between $t = 0$ and $t = 10^{-43}$ sec, the universe was so dense and particles were interacting so violently that no known theory can properly describe what happened. Immediately after the Planck time, however, the universe became very sensitive to the density of matter. Calculations demonstrate that the slightest deviation from the precise critical density would have mushroomed very rapidly, multiplying itself every 10^{-43} sec. If the density were slightly less than ρ_c, the universe would soon have become wide open and virtually empty. If, on the other hand, the density were slightly greater than ρ_c, the universe would soon have become tightly closed and so packed with matter that the entire cosmos would have rapidly collapsed into a black hole. In other words, immediately after the Big Bang, the fate of the universe hung in the balance so that the tiniest deviation from the precise equality $\rho = \rho_c$ would have rapidly propelled the universe away from the special case of $\Omega_0 = 1$.

Observations reveal that Ω_0 is today approximately 1. Consequently, the density of the universe immediately after the Big Bang must have been equal to the critical density to an incredibly precise degree. Calculations demonstrate that, in order for Ω_0 to be roughly 1 today, ρ must have been equal to ρ_c to more than 50 decimal places!

What could have happened immediately after the Planck time to ensure that $\rho = \rho_c$ to such an astounding degree of accuracy? Because $\rho = \rho_c$ means that space is flat, this enigma is called the **flatness problem.**

A second enigma closely related to the flatness problem is the isotropy of the 3-K cosmic microwave background. We saw that the microwave background is so incredibly uniform across the sky that sensitive temperature measurements can reveal our motion through this radiation field. Subtracting the effects of our motion, we find that the temperature of the microwave background is the same in all parts of the sky to an accuracy of 1 part in 10,000.

To appreciate this isotropy dilemma, think about microwave radiation coming at us from two opposite parts of the sky. This radiation left over from the primordial fireball has been traveling toward us for nearly 20 billion years. The total distance between opposite sides of the observable universe is roughly 40 billion light years, and so these widely separated regions have absolutely no connection with each other. Why then do these unrelated parts of the universe have the same temperature?

In the early 1980s, Alan Guth, working at Stanford University, offered a remarkable solution to the problems of the flatness of the universe and the isotropy of the microwave background. Guth analyzed the suggestion that the universe had experienced a brief period of extremely rapid expansion shortly after the Planck time (see Figure 19-10). During this **inflationary epoch,** as the period is called, the universe ballooned outward in all directions. In a tiny fraction of a second, the universe expanded by a factor of perhaps 10^{30} times. This inflation placed much of the material that was originally near our location far *beyond* the edge of the observable universe today. Now the observable universe is now expanding into space containing matter and radiation that was once in close contact with our location.

Inflation accounts for the flatness of the universe. To see why, think about a small portion of the Earth's surface, such as your backyard. For all

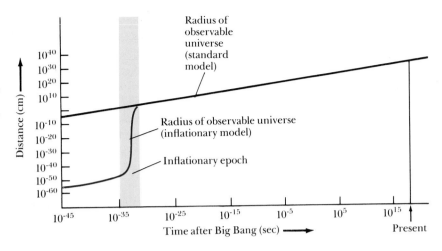

Figure 19-10 **The size of universe with and without inflation** *According to the inflationary model, the universe expanded by a factor of about 10^{30} shortly after the Big Bang. This sudden growth in the size of the observable universe probably occurred during a very brief interval, as indicated by the vertical shaded area on the graph. For comparison, the projected size of the universe without inflation is also shown. (Adapted from A. Guth)*

practical purposes, your backyard looks very flat and it is impossible to detect the Earth's curvature over such a small area. Similarly, the observable universe is such a tiny fraction of the inflated universe that any overall curvature is virtually undetectable. Like your backyard, our observable segment of space looks very flat.

The inflationary epoch also accounts for the isotropy of the microwave background. When we look at microwaves that are from opposite parts of the sky, we are seeing radiation from regions of the universe that were originally in intimate contact with each other. That is why they have the same temperature.

Finally, it is important to note that the concept of inflation does not violate Einstein's dictum that nothing can travel faster than the speed of light. Remember that the expansion of the universe is the expansion of space and does not involve the motion of objects through space. Inflation was a brief moment during which the distances between particles suddenly increased by an enormous amount. This expansion was accomplished entirely by a sudden vigorous expansion of space. Particles did not move; space inflated.

Cosmic background radiation dominated the universe during the first million years

Everything in the universe falls into one of two categories: matter or energy. The matter is contained in such objects as stars, planets, and galaxies, all of which are composed of particles like electrons, protons, and neutrons. The radiant energy in the universe consists of photons. There are, of course, many starlight photons traveling across space, but the vast majority of photons in the universe are members of the 3-degree microwave background.

Matter prevails over radiation today because the energy carried by microwave photons is so small. Using Einstein's famous equation $E = mc^2$, which expresses a fundamental relationship between mass and energy, astronomers can compare the matter density of the universe with the energy density of the microwave radiation that fills all space. The average density of matter throughout space today is much larger than the comparable energy density associated with the microwave background. Astronomers therefore say that we live in a **matter-dominated universe.**

Although the universe is dominated by matter, the number of photons in the microwave background is astounding. From the physics of blackbody radiation it can be demonstrated that there are today 550 million photons in every cubic meter of space. In contrast, if all the matter in the universe were uniformly spread throughout space, there would be roughly one hydrogen atom per cubic meter. In other words, photons outnumber atoms by roughly a billion to one. This radiation field no longer has much "clout," though, because its photons have been redshifted to long wavelengths and low energies, after nearly 20 billion years of being stretched by the expansion of the universe.

The universe was not always dominated by matter. To see why, think back toward the Big Bang. As we do so, we find the universe increasingly compressed, so that the density of matter was greater. The photons in the background radiation also were crowded more closely together then, but an additional effect must be considered. In those times, the photons were less redshifted and thus had shorter wavelengths and higher energy than they do today. Because of this added energy, there was a time in the ancient past when radiation held sway over matter. During that time, the energy density of the radiation field was greater than the average density of matter. Astronomers call this state a **radiation-dominated universe.**

The transition from a radiation-dominated universe to a matter-dominated universe occurred about 1 million years after the Big Bang. Since that time, the wavelengths of photons have been stretched by a factor of 1000. Today these microwave photons typically have wavelengths of about 1 mm, but when the universe was one million years old they had wavelengths of about 0.001 mm.

We can use Wien's law to calculate the temperature of the cosmic background radiation at the time of this transition from a radiation-dominated universe to a matter-dominated one. Just as a peak wavelength λ_{max} of

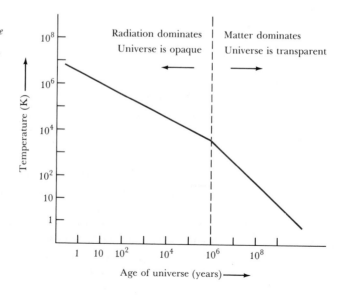

Figure 19-11   The declining temperature of the universe   *As the universe expands, the photons in the radiation background becomes increasingly redshifted and the temperature of the radiation field falls. Roughly 1 million years after the Big Bang, when the temperature fell below 3000 K, hydrogen atoms formed and the universe became transparent.*

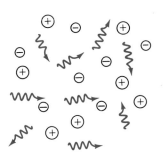

a Before recombination

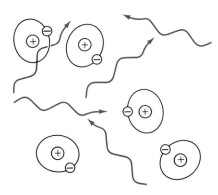

b After recombination

Figure 19-12 The era of recombination
(a) *Before recombination, photons in the cosmic background prevented protons and electrons from forming hydrogen atoms.*
(b) *As soon as hydrogen atoms could survive, the universe became transparent. The transition from opaque to transparent occurred roughly 1 million years after the Big Bang.*

1 mm corresponds to a blackbody temperature of 3 K, a peak wavelength of 0.001 mm corresponds to 3000 K. (The temperature history of the universe is graphed in Figure 19-11.)

The nature of the universe changed in a fundamental way when the temperature of the radiation field fell to 3000 K. To understand this change, recall that hydrogen is by far the most abundant element in the universe, and consists of a single proton orbited by a single electron. It takes only a little energy to knock the proton and electron apart. In fact, a radiation field warmer than about 3000 K easily ionizes hydrogen. Thus, hydrogen atoms could not exist earlier than about 1 million years after the Big Bang. As sketched in Figure 19-12*a*, the background photons prior to *t* = 1 million years had energies great enough to prevent electrons and protons from getting together to form hydrogen atoms. Only since *t* = 1 million years have these photons been redshifted enough to permit hydrogen atoms to exist (see Figure 19-12*b*).

Prior to *t* = 1 million years, the universe was completely filled with a shimmering expanse of high-energy photons colliding vigorously with protons and electrons. This state of matter, called a **plasma,** is opaque, just as the glowing gases inside a fluorescent light bulb are opaque. The term **primordial fireball** is used to describe the universe during this time.

After *t* = 1 million years, the photons no longer had enough energy to keep the protons and electrons apart. As soon as the temperature of the radiation field fell below 3000 K, protons and electrons everywhere began combining to form hydrogen atoms. Because hydrogen is transparent, the universe suddenly became transparent! The photons that just a few seconds earlier had been vigorously colliding with charged particles could now stream unimpeded across space. Today we see these same photons in the microwave background.

This dramatic moment, when the universe went from being opaque to being transparent, is referred to as the **era of recombination.** Because the universe was opaque prior to *t* = 1 million years, we cannot see any farther into the past than the era of recombination. The microwave background contains the most ancient photons we shall ever be able to observe. This microwave background is today only a ghostly relic of its former dazzling splendor.

During the first second, most of the matter and antimatter in the universe annihilated each other

When the universe was young, it was dominated by a hot radiation field that filled all space. We have also seen that the photons of this radiation field are a billion times more plentiful than ordinary particles of matter. Where did all these photons and particles come from? To search for clues, we must delve still further into the past and examine conditions within a few seconds of the Big Bang.

As we move back in time toward the Big Bang, we find that the energy of the photons in the universe increases. If we go far enough into the past, we come to the time when these photons possessed enough energy to create matter, according to Einstein's equation $E = mc^2$. This equation explains that matter and energy can be converted one into the other. To make a particle of mass *m*, you need an amount of energy *E* that is at least as great as mc^2, where *c* is the speed of light.

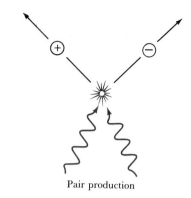

Pair production

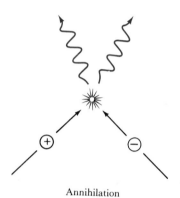

Annihilation

Figure 19-13 Pair production and annihilation *A particle and an antiparticle can be created during the collision of high-energy photons. Conversely, a particle and an antiparticle can annihilate each other by giving up energy in the form of gamma rays.*

The creation of matter from energy is routinely observed in laboratory experiments involving high energy gamma rays. When a highly energetic gamma ray collides with a second photon, both vanish and matter appears in their place, as sketched in Figure 19-13. (Also see the photograph on the first page of this chapter.) This process, called **pair production,** always creates one ordinary particle and one antiparticle. For instance, if one of the particles is an electron, the other is an antielectron.

There is nothing mysterious about antimatter. An antiparticle is like an ordinary particle, except that it has an opposite electric charge. For example, an antielectron (e^+) has a positive charge, whereas an ordinary electron (e^-) has a negative charge. If a particle and its antiparticle collide, they annihilate each other and their mass is converted back into high-energy photons. Particles and antiparticles are always created or destroyed in equal numbers, thus ensuring that the total electric charge in the universe remains constant. Physicists often refer to this balance between matter and antimatter as a **symmetry.**

Close to the time of the Big Bang, photon energy was high enough to initiate pair creation throughout space. When the universe was 1 second old, the temperature of the radiation field was about 6 billion K. This was so hot that colliding photons produced electrons and antielectrons. Thus at times earlier than $t = 1$ second, the universe contained vast numbers of these particles. When the universe was 0.0001 second old, the temperature was 10 trillion K. That was hot enough to create protons and antiprotons or neutrons and antineutrons. Thus at times earlier than $t = 0.0001$ second, the universe was teaming with these particles also. In other words, shortly after the Big Bang all space was chock full of particles and antiparticles immersed in an inconceivably hot bath of high-energy photons.

Now let us think forward in time from these early moments when the universe was filled with particles and antiparticles. As the universe expanded, all the gamma-ray photons became increasingly redshifted. The temperature soon declined to a level at which the gamma rays were no longer energetic enough to create pairs of particles and antiparticles. Collisions between existing particles and antiparticles did add photons to the cosmic radiation background, but collisions between photons no longer replenished the supply of particles and antiparticles. The first second of the universe was thus dominated by the wholesale annihilation of vast amounts matter and antimatter.

In laboratory experiments, physicists observe a strict equality in numbers of particles and antiparticles. Particles and antiparticles are always created or destroyed in equal numbers. If this symmetry is truly valid, then for every proton there should have been an antiproton. For every electron there should have been an antielectron. This poses a dilemma because by the time the universe was 1 second old, every particle should have been annihilated by an antiparticle, leaving no matter at all in the universe.

Obviously this did not happen. Physicists therefore say that a **symmetry breaking** occurred during the earliest moments of the universe, which caused the number of particles to be slightly greater than the number of antiparticles. For every billion antiprotons, a billion plus one protons must have existed. For every billion antielectrons, a billion plus one electrons existed. This particular slight excess is suggested by the fact that there are roughly a billion photons in the microwave background today for each proton and neutron in the universe.

Grand unified theories explain that all four forces had the same strength immediately after the Big Bang

All phenomena in the universe can be explained as the result of four physical forces: gravity, electromagnetism, and the strong and weak nuclear forces. These four forces differ sharply from one other. For instance, their relative strengths are quite dissimilar. Gravity, by far the weakest force, has a strength only 6×10^{-39} that of the strong nuclear force, which is the most powerful.

We are all familiar with the force of gravity as a long-range force that dominates the universe over astronomical distances. Electromagnetism is also a long-range force (in principle, its influence extends to infinity, as gravity's does), but it is much stronger than the gravitational force. Just as the force of gravity holds the Moon in orbit about the Earth, the electromagnetic force holds electrons in orbit about the nuclei in atoms. We do not generally observe the long-distance effects of the electromagnetic force, however, because in most cases there is a negative electric charge for every positive charge and a south magnetic pole for every north magnetic pole. Over large volumes of space, the net effects of electromagnetism effectively cancel each other. A similar canceling does not occur with gravity because there is no "negative mass."

The strong and the weak nuclear forces are both said to be short-range because their influence extends only over distances less than about 10^{-13} cm. The **strong nuclear force** holds protons and neutrons together inside the nuclei of atoms. Without this force, nuclei would disintegrate because of the electromagnetic repulsion of the positively charged protons. Thus, the strong nuclear force overpowers the electromagnetic force inside nuclei. The **weak nuclear force** is so weak that it does not hold anything together. Instead, it is at work in certain kinds of radioactive decay, such as the transformation of a neutron into a proton.

Numerous experiments in nuclear physics strongly suggest that protons and neutrons are composed of more basic particles called **quarks,** the most common varieties being "up" quarks and "down" quarks. A proton is composed of two up quarks and one down quark, whereas a neutron is made of two down quarks and one up quark.

In the 1970s, the concept of quarks gave rise to a more fundamental description of the strong and weak nuclear forces. In its most basic form, the strong nuclear force is the force that holds quarks together. Similarly, the weak nuclear force is at work whenever a quark changes from one variety to another. For example, when a neutron decays into a proton, one of the neutron's down quarks changes into an up quark.

A highly successful description of forces, called **quantum field theory,** explains that two particles exert a force on each other by exchanging a third particle. The exchanged particle is the carrier of the force. For instance, the gravitational force occurs when particles exchange **gravitons,** and the weak nuclear force exists when particles exchange **weakons.** The carrier of the electromagnetic force is the photon, and quarks stick together by exchanging **gluons.** Thus we may summarize the four forces as indicated in Table 19-2.

To examine details of the forces, scientists use particle accelerators that hurl high-speed electrons and protons at targets. In such experiments, physicists find that the different forces begin to look the same as the particles' speeds approach the speed of light. In fact, during experiments at the CERN accelerator in Europe in the 1980s, particles were slammed together

Table 19-2 The four forces

Force	Relative strength	Particles exchanged	Particles acted upon	Range	Example
Strong	1	Gluons	Quarks	10^{-13} cm	Holds nuclei together
Electromagnetic	1/137	Photons	Charged particles	Infinite	Holds atoms together
Weak	1/10,000	Weakons	Quarks, electrons neutrinos	$<10^{-14}$ cm	Radioactive decay
Gravity	6×10^{-39}	Gravitons	Everything	Infinite	Holds the solar system together

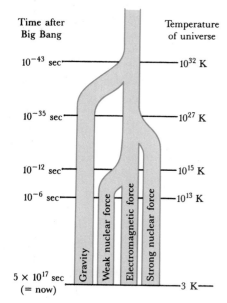

Figure 19-14 Unification of the four forces *The strength of the four physical forces depends on the speed or energy with which these particles interact. As shown in this diagram, the higher the temperature of the universe, the more the forces resemble each other. Also included here is the age of the universe at which the various forces were equal.*

with such violence that the electromagnetic force and the weak nuclear force had equal strength—they were "unified."

In recent years, physicists have labored to produce a comprehensive theory that would completely describe these forces. Some of the most promising attempts, such as the supergrand unified theory, predict that all four forces would have the same strength if only we could slam particles together with energies trillions of times greater than that in the CERN accelerator. Physicists have no hope of building accelerators that powerful. However, the universe was so hot and particles were moving with such high speeds immediately after the Big Bang that all four forces were unified into a single "superforce." The earliest moments of the universe thus provide a laboratory wherein scientists can explore some of the most elegant ideas in physics. Figure 19-14 shows the age and temperature of the universe when the various forces become unified.

Many of the ideas connecting particle physics with cosmology are very new and still quite speculative. Nevertheless, Figure 19-15 attempts to summarize these ideas. During the Planck time (from $t = 0$ to $t = 10^{-43}$ sec), particles collided with such high energies that all four forces were unified. By the end of the Planck time, however, the energy of particles in the universe had declined to the extent that gravity was no longer unified with the other three forces. We can therefore say that, at $t = 10^{-43}$ sec, gravity "froze out" of the otherwise unified hot soup that filled all space. The temperature of the universe was 10^{32} K when gravity emerged as a separate force.

By $t = 10^{-35}$ sec, the temperature of the universe had fallen to 10^{27} K and the energy of particles had declined to the extent that the strong nuclear force was no longer unified with the electromagnetic and weak nuclear forces. Thus, at $t = 10^{-35}$ sec, the strong nuclear force made its appearance, freezing out of an otherwise unified hot soup. Calculations suggest that the inflationary epoch lasted from $t = 10^{-35}$ sec to about $t = 10^{-24}$ sec, during which time the universe increased its size by a factor of between 10^{20} and 10^{30}.

At $t = 10^{-12}$ sec, when the temperature of the universe had dropped to 10^{15} K, a final "freeze out" separated the electromagnetic force from the weak nuclear force. From that moment on, all four forces interacted with particles essentially as they do today.

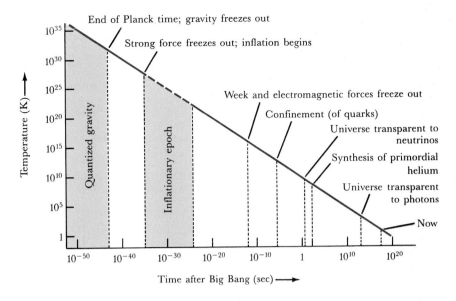

Figure 19-15 The early history of the universe *As the universe cooled, the four forces "froze out" of their initial unified state. The inflationary epoch lasted from 10^{-35} sec to 10^{-24} sec after the Big Bang. Neutrons and protons "froze out" of the hot "quark soup" one millionth of a second after the Big Bang. The universe became transparent to light (that is, photons "froze out") when the universe was a million years old.*

The next significant event occurred at $t = 10^{-6}$ sec, when the universe's temperature was 10^{13} K. Prior to this moment, particles collided so violently that individual protons and neutrons could not exist, because they were constantly being fragmented into quarks. After this moment, appropriately called **confinement,** quarks could finally stick together to form individual protons and neutrons.

As noted on Figure 19-15, all the primordial helium was produced by $t = 3$ min, and the universe became transparent to photons at $t = 1$ million years. These photons are today observed as the 3-K microwave background, the ghostly reminder of the first few moments of the universe.

Summary

· The Hubble law describes the ongoing expansion of the universe; the space between widely separated galaxies is growing ever larger.

  The observable universe extends about 20 billion light years in every direction from the Earth. We cannot see objects beyond the cosmic particle horizon at the distance of 20 billion light years because light from these objects has not had enough time to reach us.

· The universe began as an infinitely dense cosmic singularity, called the Big Bang, throughout all space at the beginning of time.

  During the Planck time (which lasted until about 10^{-43} sec after the Big Bang), the universe was so dense that known laws of physics do not properly describe the behavior of space, time, and matter.

· The 3-K cosmic microwave background radiation is the greatly redshifted remnant of the very hot universe that existed about 1 million years after the Big Bang.

· The average density of matter in the universe determines the curvature of space and the ultimate fate of the universe.

  If the average density of matter in the universe is greater than the critical density ρ_c, then space is spherical (with positive curvature), the density parameter Ω_0 has a value greater than 1, and the universe is closed (bounded) and ultimately will collapse.

If the average density of matter in the universe is less than ρ_c, then space is hyperbolic (with negative curvature), Ω_0 has a value less than 1, and the universe is open (unbounded) and will continue to expand forever.

If the average density of matter in the universe is exactly equal to ρ_c, then space is flat (with zero curvature), Ω_0 is exactly equal to 1, the universe is marginally bounded, and expansion will just barely continue forever.

· That the universe is nearly flat and that the 3-K microwave is almost perfectly isotropic may be explained as the result of a brief period of very rapid expansion (the inflationary epoch).

During the inflationary period, much of the material originally near our location moved far beyond the limits of our observable universe. The observable universe thus is today expanding into space containing matter and radiation that was in close contact with our matter and radiation during the first instant after the Big Bang.

· For the first million years, the universe was opaque and dominated by radiation. During the era of recombination, roughly a million years after the Big Bang, the universe became transparent and matter-dominated, as it is today.

· Four basic forces—gravity, electromagnetism, the strong nuclear force, and the weak nuclear force—explain all of the interactions observed in the universe.

· A grand unified theory is an attempt to explain two or more forces in terms of a single consistent set of physical laws. Such theories suggest that all four forces were identical just after the Big Bang.

At the end of the Planck time, gravity "froze out" to become a distinctive force. A short time later, the strong nuclear force became a distinctive force. A final "freeze out" separated the electromagnetic force from the weak nuclear force.

Review questions

1 What is meant by the expansion of the universe?

2 Why is it incorrect to think of the redshifts of remote galaxies and quasars as being a result of the Doppler effect?

3 How does the expansion of the universe offer a resolution to Olbers's paradox?

4 In what ways are the fate of the universe, the geometry of the universe, and the average density of the universe related?

5 How does modern cosmology preclude the possibility of a "center" or an "edge" to the universe?

6 Suppose that the universe will expand forever. What will eventually become of the microwave background radiation?

7 Describe an example of each of the four basic forces in the physical universe.

8 What is the observational evidence for **(a)** the Big Bang, **(b)** the inflationary epoch, and **(c)** the confinement of quarks?

Advanced questions

9 Prior to the discovery of the 3-K cosmic microwave background, it seemed possible that we might be living in a "steady state" universe whose overall properties do not change with time. The steady-state model, like the Big Bang model, assumes an expanding universe, but it does not assume a "creation event." Instead, in the steady-state theory matter is assumed to be created continuously everywhere in space to ensure that the average density of the universe remains constant. Explain why the cosmic microwave background is a major blow to the steady-state theory.

10 With a diagram show how you would expect the observed number of galaxies to depend on the distance from Earth for various values of q_0. Discuss some of the problems of using such a diagram to determine q_0.

11 The COBE satellite was launched about the same time this textbook was published. Read up on the COBE mission in recent issues of such science magazines as *Sky & Telescope* and *Science News*. Is the satellite operating properly? Have any observations or results been reported?

Discussion questions

12 Discuss the theological implications of the idea that we cannot use science to tell us what existed before the Big Bang.

13 Do you think that there can be "other universes," regions of space and time that are not connected to our universe? Should astronomers be concerned with such possibilities? Why or why not?

For further reading

Barrow, J., and Silk, J. *The Left Hand of Creation: Origin and Evolution of the Universe.* Basic Books, 1983.

Bartusiak, M. "Before the Big Bang: The Big Foam." *Sky & Telescope,* September 1987, p. 76.

Davies, P. *Superforce.* Simon & Schuster, 1984.

Dicus, D., et al. "The Future of the Universe." *Scientific American,* March 1983.

Guth, A., and Steinhardt, P. "The Inflationary Universe." *Scientific American,* May 1984.

Harrison, E. "The Paradox of the Dark Night Sky." *Mercury,* July/August 1980, p. 83.

Kaufmann, W. *Black Holes and Warped Spacetime.* W. H. Freeman and Company, 1979.

Overbye, D. "The Universe According to Guth." *Discover,* June 1983. p. 92.

Shu, F. "The Expanding Universe and the Large-Scale Geometry of Space-Time." *Mercury,* November/December 1983, p. 162.

Silk, J. *The Big Bang.* W. H. Freeman and Company, 1989.

Silk, J., et al. "The Large-Scale Structure of the Universe." *Scientific American,* October 1983.

Wagoner, R., and Goldsmith, D. "Quarks, Leptons, and Bosons." *Mercury,* July/August 1983, p. 98.

Webster, A. "The Cosmic Background Radiation." *Scientific American,* August 1974.

Afterword: The search for extraterrestrial life

The prevalence of life *This painting, entitled* DNA Embraces the Planets, *artistically expresses the suspicion of many scientists that carbon-based life may be a common phenomenon in the universe. Other scientists argue, however, that we may be unique and no intelligent alien civilizations exist. In either case, humanity has the clear mandate to preserve and protect the abundance of life-forms with which we share our planet. (Courtesy of J. Lomberg)*

The heavens inspire us to contemplate a variety of profound topics, including the creation of the universe, the nature of the stars, and the formation of the Earth. Of all the fascinating subjects we might explore, perhaps none is as compelling as the question of extraterrestrial life. Are we alone? Does life exist elsewhere in the universe? What are the chances that we might someday make contact with an alien civilization?

There are no firm answers to such questions. Earth is the only planet on which life is known to exist and our probes to other worlds have so far failed to detect any life-forms on them. This lack of data does not, however, undermine the possibility of extraterrestrial biology.

As you have seen throughout this book, one of the great lessons of modern astronomy is that our circumstances are quite ordinary. Contrary to the beliefs of our ancestors, we do not occupy a special location, like the "center of the universe." Over the past four centuries it has become clear that we inhabit one of nine planets orbiting an unremarkable star—just one of billions in an undistinguished galaxy. Is it possible that we are also biologically commonplace? The answer to this question is sought by scientists involved in SETI, the search for extraterrestrial intelligence.

Although the possibility cannot be ruled out that an alternative form of biochemistry exists, scientists for now confine their search to life as we now know it. All terrestrial life is based on the unique properties of the carbon atom. Carbon is an extremely versatile element whose atoms are capable of forming chemical bonds that result in especially long and complex molecules. Among these carbon-based compounds called **organic molecules,** are the molecules of which living organisms are made.

Organic molecules can be linked together to form elaborate structures such as chains, lattices, and fibers. Some of these structures are capable of complex, self-regulating chemical reactions. Furthermore, the primary constituents of organic molecules—carbon, hydrogen, nitrogen, oxygen, sulfur, and phosphorus—are among the most abundant elements in the universe. Indeed, the versatility and abundance of carbon suggest that extraterrestrial biology may also be based on organic chemistry.

Organic molecules are scattered abundantly throughout the Galaxy. In interstellar clouds, carbon atoms have combined with other elements to produce an impressive variety of organic compounds. Beginning in the 1960s, radio astronomers have detected telltale microwave emission lines from interstellar clouds that help identify dozens of these carbon-based chemicals. Examples include ethyl alcohol (CH_3CH_2OH), formaldehyde (H_2CO), methyl cyanoacetylene (CH_3C_3N), and acetaldehyde (CH_3CHO) to name just a few.

Further evidence of extraterrestrial organic molecules comes from newly fallen meteorites called carbonaceous chondrites (see Figure A-1), which are often found to contain a variety of organic substances. As noted in Chapter 10, carbonaceous chondrites are ancient meteorites dating from the formation of the solar system. So it seems reasonable to conclude that, even from their earliest days, the planets have been continually bombarded with organic compounds.

Interstellar space is not the only source of organic material. In a classic experiment performed in 1952, American chemists Stanley Miller and Harold Urey demonstrated that simple chemicals can combine to form prebiological compounds under supposedly primitive Earthlike conditions. In a closed container, they subjected a mixture of hydrogen, ammonia, methane, and water vapor to an electric arc (to simulate lightning bolts) for a week. At the end of this period, the inside of the container had become coated with a reddish-brown substance rich in compounds essential to life.

Scientists today tend to believe that Earth's primordial atmosphere probably was composed of carbon dioxide, nitrogen, and water vapor outgassed from volcanoes along with some hydrogen. Modern versions of the Miller-Urey experiment (see Figure A-2) using these common gases also succeed in producing a wide variety of organic compounds.

It is important to emphasize that scientists have not created life in a test tube. Biologists have yet to figure out, among other things, how these

Figure A-1 A carbonaceous chondrite
Carbonaceous chondrites are ancient meteorites that date back to the formation of the solar system. Chemical analysis of newly fallen specimens disclose that they are rich in organic molecules, many of which are the chemical building blocks of life. This sample is a piece of a large carbonaceous chondrite that fell in Mexico in 1969. (From the collection of Ronald A. Oriti)

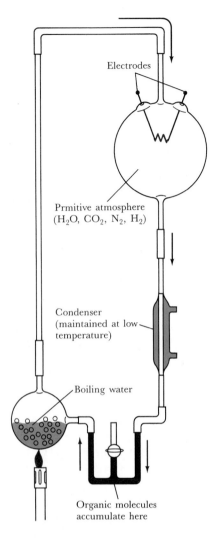

Electrodes

Prmitive atmosphere
(H_2O, CO_2, N_2, H_2)

Condenser
(maintained at low
temperature)

Boiling water

Organic molecules
accumulate here

Figure A-2 The Miller-Urey experiment (updated) Modern versions of this classic experiment prove that numerous organic compounds important to life can be synthesized from gases that were present in Earth's primordial atmosphere. This experiment supports the hypothesis that life on Earth arose as a result of ordinary chemical reactions.

organic molecules gathered themselves into cell-like arrangements and managed to develop systems for self-replication. Nevertheless, since so many chemical components of life are so easily synthesized under conditions that simulate the primordial Earth, it seems reasonable to suppose that life could have originated as the result of chemical processes involving materials from the Earth or from space.

A widespread abundance of organic precursors does not guarantee that life is commonplace throughout the universe. If a planet's environment is hostile, life may never get started or quickly becomes extinct. It seems quite possible, however, that there are planets orbiting other stars. Perhaps conditions on some of these worlds are sufficiently suitable for life as we know it.

The development of life on Earth seems to suggest that extraterrestrial life, including intelligent species, might evolve on habitable planets, given sufficient time and appropriately hospitable conditions. How might we ascertain whether such worlds exist, given the tremendous distances that indubitably separate us from them? Many astronomers hope to learn about extraterrestrial civilizations by detecting radio transmissions from them. As we have seen in previous chapters, radio waves can travel immense distances without being significantly degraded by the gas and dust through which they pass. Because of this ability to penetrate the interstellar medium, radio waves are a logical choice for interstellar communication.

Over the past several decades, astronomers have proposed various ways to search for alien radio transmission and several limited searches have been undertaken. One of the first searches occurred in 1960, when Frank Drake used a radio telescope at the National Radio Astronomy Observatory in West Virginia to "listen" to two Sunlike stars, τ Ceti and ϵ Eridani, without success. About forty similar unsuccessful searches have taken place since then using radio telescopes in both the United States and the Soviet Union. For instance, in 1973, astronomers "listened" to 600 nearby solar-type stars for $\frac{1}{2}$ hour each, but no unusual signals were detected. Since 1983, a radio telescope belonging to Harvard University has been used along with a sophisticated computer program to scan a wide range of frequencies over a large portion of the sky. So far, nothing has turned up.

Should we be discouraged by this lack of success? What are the chances that a radio astronomer might someday detect radio signals from an extraterrestrial civilization? The first person to tackle such issues was Frank Drake, now at the University of California at Santa Cruz. Drake proposed that the number of technologically advanced civilizations in the Galaxy (designated by the letter N) could be estimated with the equation:

$$N = R_* f_p n_e f_l f_i f_c L$$

where

R_* = the rate at which solar-type stars form in the Galaxy

f_p = the fraction of stars that have planets

n_e = the number of planets per solar system suitable for life

f_l = the fraction of those habitable planets on which life actually arises

f_i = the fraction of those life-forms that evolve into intelligent species

f_c = the fraction of those species that develop adequate technology and then choose to send messages out into space

and

L = the lifetime of that technologically advanced civilization

The Drake equation is enlightening because it expresses the number of extraterrestrial civilizations in a series of terms, some of which can be estimated from what we know about stars and stellar evolution. For instance, the first two factors, R_* and f_p, could be determined by observation. In estimating R_*, we should probably exclude massive stars (those larger than about $1.5_\odot$), because they have main-sequence lifetimes shorter than the time it took to develop intelligent life here on Earth. Life on Earth originated some 3.5 to 4.0 billion years ago. If that is typical of the time needed to evolve higher life forms, then a massive star probably becomes a red giant or a supernova before intelligent creatures appear on any of its planets. While low-mass stars have much longer lifetimes, they, too, seem unsuited for life because they are so cool. Only planets very near a low-mass star would be sufficiently warm for life as we know it, and a planet that close can become tidally coupled to the star, with one side continually facing the star, while the other is in perpetual frigid darkness. This leaves us with main-sequence stars like the Sun, those with spectral types between F5 and M0. Based on statistical studies of star formation in the Milky Way, some astronomers estimate that roughly one of these Sunlike stars forms in the Galaxy each year, thus setting R_* at 1 per year.

We learned in Chapter 6 that the planets in our solar system formed as a natural consequence of the birth of the Sun, and have seen evidence suggesting that similar processes of planetary formation may be commonplace around single stars. Yet, so far, no planet outside our solar system has been discovered. Nevertheless, many astronomers give f_p a value of 1, meaning they believe it likely that most Sunlike stars have planets.

Unfortunately, the rest of the terms in the Drake equation are very uncertain. Let's play with some hypothetical values. The chances that a planetary system has an Earthlike world are not known. Were we to consider our own solar system as representative, we could put n_e at 1. Let's be more conservative, however, and suppose that 1 in 10 solar-type stars is orbited by a habitable planet, making $n_e = 0.1$. From what we know about the evolution of life on Earth, we might assume that, given appropriate conditions, the development of life is a certainty, which would make $f_l = 1$. This is, of course, an area of intense interest to biologists. For the sake of argument, we might also assume that evolution might naturally lead to the development of intelligence (a conjecture that is hotly debated) and also make $f_i = 1$. It's anyone's guess as to whether these intelligent extraterrestrial beings would attempt communication with other civilizations in the Galaxy, but were we to assume they would, f_c would be put at 1 also. The last variable, L, involving the longevity of civilization, is the most uncertain of all, and certainly cannot be subjected to testing! Looking at our own example, we see a planet whose atmosphere and oceans are increasingly polluted by creatures that possess nuclear weapons. If we are typical, perhaps L is as short as 100 years. Putting all these numbers together, we arrive at

$$N = 1/\text{year} \times 1 \times 0.1 \times 1 \times 1 \times 1 \times 100 \text{ years} = 10$$

In other words, out of the hundreds of billions of stars in the Galaxy, we would estimate that there are only ten technologically advanced civilizations from which we might receive communications.

A wide range of values has been proposed for the terms in the Drake equation, and these various guesses produce vastly different estimates of N. Some scientists argue that there is exactly one advanced civilization in the Galaxy and that we are it. Others speculate that there may be hundreds or thousands of planets inhabited by intelligent creatures.

If extraterrestrial beings were purposefully sending messages into space, it seems reasonable that they might choose a frequency that is fairly free of emissions from extraneous sources. SETI pioneer Bernard Oliver has pointed out that a range of relatively noise-free frequencies exists in the neighborhood of the microwave emission lines of hydrogen and OH (see Figure A-3). This region of the microwave spectrum is called "the water hole," a humorous reference to the H and OH lines being so close together. Or, perhaps, they would choose to transmit at a wavelength of 21 cm, because astronomers studying the distribution of hydrogen around the Galaxy would already have their radio telescopes tuned to that wavelength (recall Figure 16-7).

Even if there are only a few alien civilizations scattered across the Galaxy, we have the technology to detect radio transmission from them. The most ambitious plan ever proposed was Project Cyclops, which would consist of 1000 to 2500 radio antennae, each 100 feet in diameter (see Figure A-4). This colossal array would be so sensitive that it could detect signals from virtually anywhere in the Galaxy. Unfortunately, it seems unlikely that such a project will ever be funded.

On a more modest scale, NASA has funded a project called SETI. From 1992 through 1998, existing radio telescopes will be used along with sophisticated computers to examine all of nearly 800 known stars like the Sun

Figure A-3 The "water hole"
The so-called water hole is a range of radio frequencies (from about 1000 to 10,000 megahertz) that happens to have relatively little cosmic noise. Some scientists suggest that this noise-free region would be well suited for interstellar communication. (Adapted from C. Sagan and F. Drake)

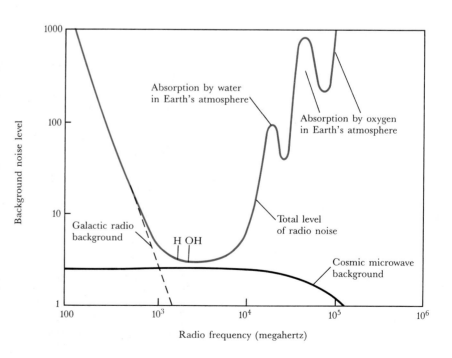

Figure A-4 *Project Cyclops* *Project Cyclops was the grandest plan ever seriously considered for the purpose of searching for extraterrestrial civilizations. It would have consisted of 1000 to 2500 radio antennae spread over an area 16 kilometers in diameter. (NASA)*

(spectral types F, G, and K, luminosity class V) within a distance of 25 parsecs (82 light years) from Earth. In addition, the plan calls for a less sensitive survey of the entire sky.

The detection of a message from an alien civilization would be one of the greatest events in human history. Such a message could dramatically change the course of civilization, whether it were to share scientific information or bring about some sort of social or humanistic enlightenment. In a few short years, our technology, industry, and social structure might advance the equivalent of centuries into the future. Such changes would touch every person on Earth. Mindful of these profound implications, scientists push ahead with the search for extraterrestrial communication.

For further reading

Beatty, J. K. "The New, Improved SETI." *Sky & Telescope,* May 1983, p. 411.

Davies, P. *The Cosmic Blueprint.* Simon and Schuster, 1988.

Dawkins, R. *The Blind Watchmaker.* W. W. Norton & Company, 1987.

Goldsmith, D., and Owen, T. *The Search for Life in the Universe.* Benjamin/Cummings, 1980.

Kutter, G. S. *The Universe and Life.* Jones and Bartlett, 1987.

Papagiannies, M. D. "Bioastronomy: The Search for Extraterrestrial Life." *Sky & Telescope,* June 1984, p. 508.

Rood, R., and Trefil, J. *Are We Alone?* Scribner, 1981.

Sagan, C., and Drake, F. "The Search for Extraterrestrial Intelligence." *Scientific American,* May 1975.

Schorn, R. A. "Extraterrestrial Beings Don't Exist." *Sky & Telescope,* September 1981, p. 207.

Shklovskii, I. S., and Sagan, C. *Intelligent Life in the Universe.* Holden-Day, 1966.

Wilson, A. C. "The Molecular Basis of Evolution." *Scientific American,* October 1985.

Appendixes

The terrestrial worlds *This montage of photographs taken by various spacecraft shows the terrestrial planets and six large moons at the same scale. (Prepared for NASA by S. P. Meszaros)*

1 The planets: Orbital data

Planet	Semimajor axis (AU)	Semimajor axis (10⁶ km)	Sidereal period (tropical years)	Sidereal period (days)	Synodic period (days)	Mean orbital speed (km/sec)	Orbital eccentricity	Inclination of orbit to ecliptic (°)
Mercury	0.3871	57.9	0.2408	87.97	115.88	47.9	0.206	7.00
Venus	0.7233	108.2	0.0615	224.70	583.96	35.0	0.007	3.39
Earth	1.0000	149.6	1.0000	365.26	—	29.8	0.017	0.0
Mars	1.5237	227.9	1.8809	686.98	779.87	24.1	0.093	1.85
(Ceres)	2.7671	414	4.603		466.6	17.9	0.077	10.6
Jupiter	5.2028	778	11.86		399	13.1	0.048	1.31
Saturn	9.588	1427	29.46		378	9.6	0.056	2.49
Uranus	19.191	2871	84.07		370	6.8	0.046	0.77
Neptune	30.061	4497	164.82		367	5.4	0.010	1.77
Pluto	39.529	5913	248.6		367	4.7	0.248	17.15

2 The planets: Physical data

Planet	Diameter (km)	Diameter (Earth = 1)	Mass (Earth = 1)	Mean density (g/cm³)	Rotation period (days)	Inclination of equator to orbit	Surface gravity (Earth = 1)	Albedo	Brightest visual magnitude	Escape velocity (km/sec)
Mercury	4,878	0.38	0.055	5.43	58.6	0°0	0.38	0.106	−1.9	4.3
Venus	12,104	0.95	0.82	5.24	−243.0	177.4	0.91	0.65	−4.4	10.4
Earth	12,756	1.00	1.00	5.52	0.997	23.4	1.00	0.37	—	11.2
Mars	6,794	0.53	0.107	3.9	1.026	25.2	0.38	0.15	−2.0	5.0
Jupiter	142,796	11.2	317.8	1.3	0.41	3.1	2.53	0.52	−2.7	60
Saturn	120,000	9.41	94.3	0.7	0.43	26.7	1.07	0.47	+0.7	36
Uranus	50,800	3.98	14.6	1.3	−0.65	97.9	0.92	0.50	+5.5	21
Neptune	50,450	3.81	17.2	1.6	0.77	29	1.18	0.5	+7.8	24
Pluto	3,400	0.27	0.0023	0.5 (?)	6.387	90	0.03	0.5	+15.1	1

3 Satellites of the planets

Planet	Satellite	Discovered by	Mean distance from planet (km)	Sidereal period (days)	Orbital eccentricity	Diameter of satellite* (km)	Approximate magnitude at opposition
Earth	Moon	—	384,404	27.322	0.055	3476	−12.5
Mars	Phobos	A. Hall (1877)	9,380	0.319	0.021	25	+12
	Diemos	A. Hall (1877)	23,500	1.262	0.003	13	13
Jupiter	Almalthea	Barnard (1892)	181,300	0.498	0.003	240	13
	Io	Galileo (1610)	421,600	1.769	0.000	3640	5
	Europa	Galileo (1610)	670,900	3.551	0.000	3130	6
	Ganymede	Galileo (1610)	1,070,000	7.155	0.002	5270	5
	Callisto	Galileo (1610)	1,880,000	16.689	0.008	4840	6
	Himalia	Perrine (1904)	11,470,000	250.57	0.158	(170)	14
	Elara	Perrine (1905)	11,800,000	259.65	0.207	(40)	18
	Lysithea	Nicholson (1938)	11,850,000	263.55	0.130	(10)	19
	Leda	Kowal (1974)	11,110,000	239.2	0.147	(8)	20
	Aranke	Nicholson (1951)	21,200,000	631.1	0.169	(10)	18
	Carme	Nicholson (1938)	22,600,000	692.5	0.207	(15)	19
	Pasiphae	Melotte (1908)	23,500,000	738.9	0.378	(25)	17
	Sinope	Nicholson (1914)	23,700,000	758	0.275	(15)	18
Saturn	Mimas	Herschel (1789)	185,500	0.942	0.020	390	13
	Enceladus	Herschel (1789)	237,900	1.370	0.004	500	12
	Tethys	Cassini (1684)	294,700	1.888	0.000	1050	10
	Dione	Cassini (1684)	377,400	2.737	0.002	1120	10
	Rhea	Cassini (1672)	526,700	4.518	0.001	1530	10
	Titan	Huygens (1655)	1,222,000	15.945	0.029	5120	8
	Hyperion	Bond (1848)	1,481,000	21.277	0.104	310	14
	Iapetus	Cassini (1671)	3,560,000	79.331	0.028	1440	11
	Phoebe	Pickering (1898)	12,930,000	550.45	0.163	40	16
Uranus	Miranda	Kuiper (1948)	129,900	1.414	0	480	17
	Ariel	Lassell (1851)	190,900	2.520	0.003	1160	14
	Umbriel	Lassell (1851)	266,000	4.144	0.004	1190	15
	Titania	Herschel (1787)	436,300	8.706	0.002	1610	14
	Oberon	Herschel (1787)	583,400	13.463	0.001	1550	14
Neptune	Triton	Lassell (1846)	353,400	5.877	0.000	2720	13
	Nereid	Kuiper (1949)	5,560,000	359.881	0.749	170	19
Pluto	Charon	Christy (1978)	17,000	6.387	0	(1200)	17

NOTE: This table does not include several very small satellites about Jupiter, Saturn, Uranus, and Neptune that were discovered by *Voyagers 1* and *2*.
*A diameter of a satellite given in parentheses is estimated from the amount of sunlight it reflects.

Name	Parallax (arc sec)	Distance (ly)	Spectral type	Radial velocity (km/sec)	Proper motion (arc sec/yr)	Apparent visual magnitude	Luminosity (Sun = 1.0)
Sun			G2 V			−26.7	1.0
α Cen A	0.750	4.3	G2 V	−22	3.68	−0.01	1.6
B			K0 V			1.3	0.45
C	0.772	4.2	M5e			11.0	0.00006
Barnard's star	0.552	5.9	M5 V	−108	10.30	9.5	0.00045
Wolf 359	0.431	7.6	M8e	+13	4.84	13.5	0.00002
Lalande 21185	0.402	8.1	M2 V	−84	4.78	7.5	0.0055
Luyten 726-8A	0.1387	8.4	M6e	+30	3.35	12.5	0.00006
B(UV Ceti)			M6e			13.0	0.00004
Sirius A	0.377	8.6	A1 V	−8	1.32	−1.5	23.5
B			wd			8.7	0.003
Ross 154	0.345	9.4	M5e	−4	0.74	10.6	0.00048
Ross 248	0.314	10.3	M6e	−81	1.82	12.3	0.00011
ε Eri	0.303	10.7	K2 V	+16	0.97	3.7	0.30
Luyten 789-6	0.302	10.8	M7e	−60	3.27	12.2	0.00014
Ross 128	0.301	10.8	M5	−13	1.40	11.1	0.00036
61 Cyg A	0.292	11.2	K5 V	−64	5.22	5.2	0.083
B			K7 V			6.0	0.040
ε Ind	0.291	11.2	K5 V	−40	4.67	4.7	0.13
Procyon A	0.287	11.4	F5 IV–V	−3	1.25	0.4	7.65
B			wd			10.7	0.00055
Σ 2398 A	0.284	11.5	M3.5 V		2.29	8.9	0.0028
B			M4 V			9.7	0.0013
Groombridge 34 A	0.282	11.6	M1 V		2.91	8.1	0.0058
B			M6 V			11.0	0.00040
Lacaille 9352	0.279	11.7	M2 V		6.87	7.4	0.013
τ Ceti	0.273	11.9	G8 V	−16	1.92	3.5	0.45
BD + 5° 1668	0.266	12.2	M5	+26	3.73	9.8	0.0015
L725-32 (YZ Ceti)	0.262	12.4	M5e		1.31	11.6	0.0002
Lacaille 8760	0.260	12.5	M1		3.46	6.7	0.028
Kapteyn's star	0.256	12.7	M0 V	+245	8.79	8.8	0.0040
Kruger 60 A	0.254	12.8	M4	−26	0.87	9.8	0.0017
B			M5e			11.3	0.00044

5 The brightest stars

Star	Name	Apparent visual magnitude	Spectral type	Absolute magnitude	Distance (ly)	Radial velocity (km/sec)	Proper motion (arc sec/yr)
α CMa A	Sirius	−1.46	A1 V	+1.42	8.7	−8	1.324
α Car	Canopus	−0.72	F0 I–II	−3.1	98	+21	0.025
α Boo	Arcturus	−0.06	K2 III	−0.3	36	−5	2.284
α Cen A	Rigil Kentaurus	0.01	G2 V	+4.39	4.3	−25	3.676
α Lyr	Vega	0.04	A0 V	+0.5	26.5	−14	0.345
α Aur	Capella	0.05	G8 III(?)	−0.6	45	+30	0.435
β Ori A	Rigel	0.14	B8 Ia	−7.1	900	+21	0.001
α CMi A	Procyon	0.37	F5 IV–V	+2.7	11.3	−3	1.250
α Ori	Betelgeuse	0.41	M2 Iab	−5.6	520	+21	0.028
α Eri	Achernar	0.51	B3 V	−2.3	118	+19	0.098
β Cen AB	Hadar	0.63	B1 III	−5.2	490	−12	0.035
α Aql	Altair	0.77	A7 IV–V	+2.2	16.5	−26	0.658
α Tau A	Aldebaran	0.86	K5 III	−0.7	68	+54	0.202
α Vir	Spica	0.91	B1 V	−3.3	220	+1	0.054
α Sco A	Antares	0.92	M1 Ib	−5.1	520	−3	0.029
α PsA	Fomalhaut	1.15	A3 V	+2.0	22.6	+7	0.367
β Gem	Pollux	1.16	K0 III	+1.0	35	+3	0.625
α Cyg	Deneb	1.26	A2 Ia	−7.1	1600	−5	0.003
β Cru	Beta Crucis	1.28	B0.5 III	−4.6	490	+20	0.049
α Leo A	Regulus	1.36	B7 V	−0.7	87	+4	0.248

6 Some important astronomical quantities

Astronomical unit	1 AU = 1.496×10^{13} cm
Parsec	1 pc = 3.086×10^{18} cm = 3.262 ly
Light year	1 ly = 9.460×10^{17} cm = 63,240 AU
Solar mass	1 $M_\odot$ = 1.989×10^{33} g
Solar radius	1 $R_\odot$ = 6.960×10^{10} cm
Solar luminosity	1 $L_\odot$ = 3.90×10^{33} erg/sec

Glossary

absolute magnitude   The apparent magnitude that a star would have at a distance of 10 parsecs. (page 234)

absolute zero   A temperature of $-273°C$ (or 0 K) where all molecular motion stops; the lowest possible temperature. (pages 72–73)

absorption line   A dark line at a specific wavelength in a continuous spectrum. (pages 82–83)

absorption line spectrum   Dark lines superimposed on a continuous spectrum. (pages 82–83)

acceleration   A change in velocity. (page 42)

accretion   The gradual accumulation of matter in one location, usually from the action of gravity. (page 102)

active galactic nucleus   The center of a galaxy that is emitting exceptionally large amounts of energy; the center of a Seyfert galaxy or quasar. (page 361ff)

active galaxy   A very luminous galaxy, often containing an active galactic nucleus. (pages 361–362)

albedo   The fraction of sunlight that a planet, asteroid, or satellite reflects.

Angstrom (Å)   A unit of length equal to 10^{-8} cm. (page 52)

angular diameter (angular size)   The angle subtended by the diameter of an object. (page 2)

angular momentum   A measure of the momentum associated with rotation. (page 101)

annihilation   The process by which the mass of a particle and antiparticle is converted into energy. (page 388)

annular eclipse   An eclipse of the Sun in which the Moon is too distant to completely cover the Sun so that a ring of sunlight is seen around the Moon at mid-eclipse. (page 28)

antimatter   Matter consisting of antiparticles such as antiprotons, antielectrons (positrons), and antineutrons. (page 388)

aperture   The diameter of an opening; the diameter of the primary lens or mirror of a telescope.

aphelion   The point in its orbit where a planet is farthest from the Sun. (page 38)

apogee   The point in its orbit where a satellite or the Moon is farthest from the Earth.

Apollo asteroid   An asteroid whose orbit brings it closer to the Sun than to the Earth.

apparent magnitude   A measure of the brightness of light from a star or other object at Earth. (page 232)

asteroid (minor planet)   One of tens of thousands of small, rocky planetlike objects in orbit about the Sun. (page 196)

asteroid belt   A $1\frac{1}{2}$-AU-wide region between the orbits of Mars and Jupiter in which most asteroids are found. (page 197)

asthenosphere   A warm, plastic layer of the mantle beneath the lithosphere of the Earth. (page 122)

astronomical unit (AU)   The semimajor axis of the Earth's orbit; the average distance between the Earth and the Sun. (page 6)

astronomy   The branch of science dealing with objects and phenomena that lie beyond the Earth's atmosphere.

astrophysics   That part of astronomy dealing with the physics of astronomical objects and phenomena.

atom   The smallest particle of an element that has the properties that characterize that element. (page 72)

atomic number   The number of protons in the nucleus of an atom. (pages 78, 84)

aurora   Light radiated by atoms and ions in the Earth's upper atmosphere, mostly in the polar regions. (page 135)

autumnal equinox   The intersection of the ecliptic and the celestial equator where the Sun crosses the equator from north to south. (page 18)

average density The mass of an object divided by its volume. (page 95)

Balmer lines Emission or absorption lines in the hydrogen spectrum involving electron transitions between the second and higher energy levels. (page 85)

Balmer series All of the Balmer lines. (page 85)

belt asteroid An asteroid whose orbit lies largely in the asteroid belt. (page 197)

barred spiral galaxy A spiral galaxy in which the spiral arms begin from the ends of a "bar" running through the nucleus rather than from the nucleus itself. (pages 335–336)

Big Bang An explosion throughout all space roughly 20 billion years ago from which the universe emerged. (page 375)

Big Crunch The gravitational collapse of the universe; the ultimate fate of a bound universe. (page 381)

binary asteroid Two asteroids revolving about each other. (page 200)

binary star Two stars revolving about each other; a double star. (page 241)

BL Lacertae object A type of active galaxy. (pages 364–365)

blackbody A hypothetical perfect radiator that absorbs and reemits all radiation falling upon it. (page 74)

black hole An object whose gravity is so strong that the escape velocity exceeds the speed of light. (page 305)

blueshift A shift toward shorter wavelengths; the Doppler shift of light from an approaching source. (pages 87–88)

Bode's law A numerical sequence that gives the approximate distances of the planets from the Sun in astronomical units. (page 195)

Bohr atom A model of the atom, described by Niels Bohr, in which electrons revolve about the nucleus in circular orbits. (pages 85–86)

breccia A rock formed by the sudden amalgamation of various rock fragments under pressure. (page 176)

burster A nonperiodic X-ray source that emits powerful bursts of X rays. (page 298)

caldera The crater at the summit of a volcano. (page 124)

Callisto One of the four Galilean satellites. (pages 186–187)

carbon burning The thermonuclear fusion of carbon nuclei. (page 280)

Cassegrain focus An optical arrangement in a reflecting telescope in which light rays are reflected by a secondary mirror to a focus behind the primary mirror. (page 57–58)

celestial equator A great circle on the celestial sphere 90° from the celestial poles. (pages 16–17)

celestial mechanics The branch of astronomy dealing with the motions and gravitational interactions of objects in the solar system.

celestial poles Points about which the celestial sphere appears to rotate. (pages 16–17)

celestial sphere A sphere of very large radius centered on the observer; the apparent sphere of the sky. (pages 16–17)

Celsius temperature scale *See* temperature (Celsius). (pages 72–73)

center of mass That point in an isolated system that moves at a constant velocity in accordance with Newton's first law. (page 242)

Cepheid variable One of two types of yellow, supergiant pulsating stars. (pages 271–272)

Ceres The largest asteroid and the first to be discovered. (pages 196–197)

Chandrasekhar limit The maximum mass of a white dwarf. (page 279)

Charge-coupled device (CCD) A type of solid-state silicon wafer designed for the detection of photons. (page 60)

chromatic aberration An optical defect whereby different colors of light passing through a lens are focused at different locations. (pages 55–56)

chromosphere The middle layer in the solar atmosphere, between the photosphere and the corona. (pages 218–220)

Clouds of Magellan Two nearby galaxies visible to the naked eye from southern latitudes. (page 338)

cluster of galaxies A collection of galaxies containing a few to several thousand member galaxies. (pages 339–341)

color index The difference in the magnitudes of a star measured in two separate wavelength bands. (page 236)

coma (of a comet) The diffuse gaseous component of the head of a comet. (pages 205–206)

comet A small body of ice and dust in orbit about the Sun. While passing near the Sun, a comet's vaporized ices give rise to a "coma" and "tail." (pages 205–208)

condensation temperature The temperature at which a substance turns from gas to solid at low pressure. (page 100)

conduction The transfer of heat by passing energy directly from atom to atom. (page 215)

configuration (of a planet) A particular geometric arrangement of the Earth, a planet, and the Sun. (page 36)

conic section The curve of intersection between a circular cone and a plane. This curve can be a circle, ellipse, parabola, or hyperbola. (page 43)

conservation of angular momentum   The law of physics that says that the total amount of angular momentum in an isolated system remains constant. (page 290)

constellation   A configuration of stars often named after an object, person, or animal. (pages 14–15)

continental drift   The gradual movement of the continents over the surface of the Earth from plate tectonics. (page 120)

continuous spectrum   A spectrum of light over a range of wavelengths without any spectral lines. (pages 82–83)

convection   The transfer of energy by moving currents of fluid or gas containing that energy. (page 120)

convective zone (convective envelope)   A layer in a star where energy is transported outward by means of convection. (page 217)

corona   The Sun's outer atmosphere. (pages 220–221)

cosmic singularity   The Big Bang. (page 375)

cosmological model   A specific theory about the organization and evolution of the universe.

cosmological redshift   An increase in wavelength caused by the expansion of the universe. (page 374)

cosmology   The study of the organization and evolution of the universe. (page 372)

coudé focus   A reflecting telescope in which a series of mirrors direct light to a remote focus away from the moving parts of the telescope. (pages 57–58)

crater   A circular depression on a planet or satellite caused by the impact of a meteoroid. (pages 172, 174)

crescent moon   One of the phases of the Moon in which it appears less than half full. (pages 21, 22)

critical density   The average density throughout the universe at which space is flat and galaxies just barely continue receding from each other infinitely far into the future. (pages 380–383)

cyclonic motion   Circular wind motion (counterclockwise in the Earth's northern hemisphere) in a planet's atmosphere. (page 144)

dark nebula   A cloud of interstellar gas and dust that obscures the light of more distant stars. (page 254)

deferent   A stationary circle in the Ptolemaic system along which another circle (an epicycle) moves carrying a planet, the Sun or Moon. (pages 33–34)

degenerate gas   A gas in which all the allowed states for particles (electrons or neutrons) have been filled, thereby causing the gas to behave different from ordinary gases. (page 267)

density   The ratio of the mass of an object to its volume. (page 95)

density-wave theory   An explanation of spiral arms in galaxies proposed by C. C. Lin and his colleagues. (pages 324–326)

deuterium   An isotope of hydrogen whose nuclei each contain one proton and one neutron; heavy hydrogen.

differential rotation   The rotation of a nonrigid object in which parts adjacent to each other do not always stay close together. (pages 142, 222, 322)

differentiation (geological)   The separation of different kinds of material in different layers inside a planet. (page 132)

diffraction grating   A system of closely spaced slits or reflecting strips (usually on a piece of glass) used to produce a spectrum. (page 81)

direct motion   The gradual, apparent eastward motion of a planet as seen from Earth. (page 33)

diurnal   Daily.

diurnal motion   Motion in one day.

Doppler effect   The apparent change in wavelength of radiation due to relative motion between the source and the observer along the line of sight. (pages 87–88)

double radio source   An extragalactic radio source characterized by two large regions of radio emission, often located on either side of an active galaxy. (pages 362–363)

dust tail (of a comet)   That part of a comet's tail caused by dust particles escaping from its nucleus. (page 206)

dynamo effect   The generation of a magnetic field by circulating electric charges. (page 133)

eclipse   The cutting off of part or all the light from one celestial object by another. (pages 25–29)

eclipse path   The track of the tip of the Moon's shadow along the Earth's surface during a total or annular solar eclipse. (page 28)

eclipse season   A period during the year when a solar or lunar eclipse is possible.

eclipsing binary star   A double star system in which the stars periodically pass in front of each other, as seen from Earth. (pages 246–248)

ecliptic   The apparent annual path of the Sun on the celestial sphere. (pages 17–18)

electromagnetic radiation   Radiation consisting of oscillating electric and magnetic fields such as gamma rays, X rays, visible light, ultraviolet and infrared radiation, radio waves, and microwaves. (pages 52–53)

electromagnetic spectrum   The entire array or family of electromagnetic radiation. (pages 53–54)

electromagnetic theory A comprehensive description of electricity and magnetism first formulated by J. C. Maxwell. (page 52)

electron A negatively charged subatomic particle usually found in orbit about the nuclei of atoms. (page 84)

electron volt (eV) The energy acquired by an electron accelerated through an electric potential of one volt. (page 76)

element A substance that cannot be decomposed by chemical means into simpler substances. (pages 78–79)

ellipse A conic section obtained by cutting completely through a circular cone with a plane. (pages 37, 43)

elliptical galaxy A galaxy with an elliptical shape and no conspicuous interstellar material. (pages 336–338)

elongation The angular distance between a planet and the Sun. (pages 35–36)

emission line A bright spectral line. (pages 82–83)

emission nebula A glowing gaseous nebula whose light comes from fluorescence caused by a nearby star. (page 257)

emission line spectrum A spectrum that contains emission lines. (pages 82–83)

energy The ability to do work.

energy level (in an atom) A particular amount of energy possessed by an atom above the atom's least energetic state. (pages 86–87)

epicycle In the Ptolemaic system, a moving circle about which planets revolve. (pages 33–34)

equinox One of the intersections of the ecliptic and the celestial equator. (page 18)

ergosphere The region of space immediately outside the event horizon of a rotating black hole where it is impossible to remain at rest. (page 307)

escape velocity The speed needed by one object to achieve a parabolic orbit away from a second object and thereby permanently move away from the second object.

Europa One of the Galilean satellites. (pages 183–184)

event horizon The location around a black hole where the escape velocity equals the speed of light; the surface of a black hole. (pages 305–306)

evolutionary track On the H–R diagram, the path followed by a point representing an evolving star. (pages 254–255)

excitation The process of imparting energy to an atom or ion.

eyepiece A magnifying lens used to view the image produced at the focus of a telescope. (page 55)

extragalactic Beyond the galaxy.

Fahrenheit scale *See* temperature (Fahrenheit). (pages 72–73)

flare A sudden, temporary outburst of light from an extended region of the solar surface. (page 224)

focal length The distance from a lens or mirror where converging light rays meet. (pages 55, 57)

focus (pl., foci) The point where light rays converged by a lens or mirror meet. (pages 55, 57)

force That which can change the momentum of an object. (pages 41–43)

frequency The number of waves that cross a given point per unit time; the number of vibrations per unit time.

full moon A phase of the Moon during which its full daylight hemisphere can be seen from Earth. (pages 21–22)

galactic cannibalism A collision between two galaxies of unequal mass and size in which the smaller galaxy seems to be absorbed into the larger galaxy. (page 345)

galactic cluster A loose association of young stars in the disk of the galaxy. (pages 327–329)

galactic equator The intersection of the principal plane of the Milky Way with the celestial sphere.

galaxy A large assemblage of stars; the galaxy of which the Sun and nearby stars are members. (page 332ff)

Galilean satellite Any one of the four large moons of Jupiter. (pages 178–179)

gamma rays The most energetic form of electromagnetic radiation. (pages 53–54)

Ganymede One of the Galilean satellites. (pages 184–185)

general theory of relativity A description of gravity formulated by Einstein explaining that gravity affects the geometry of space and the flow of time. (pages 303–304)

geomagnetic The Earth's magnetic field.

giant molecular cloud A large cloud of cool gas in a galaxy. (pages 259–261)

gibbous moon A phase of the Moon in which more than half, but not all, of the Moon's daylight hemisphere is visible from Earth. (pages 21–22)

globular cluster A large spherical cluster of stars usually found in the outlying regions of a galaxy. (pages 268–269)

globule A small, dense, dark nebula.

gluon A particle that is exchanged between quarks. (page 389)

grand unified theory A theory that attempts to explain the physical forces. (pages 389–390)

granulation   The rice-grainlike structure of the solar photosphere. (page 218)

gravitation (gravity)   The tendency of matter to attract matter. (page 43)

graviton   A particle that carried the gravitational force. (pages 389–390)

gravitational radiation (gravitational waves)   Ripples in the overall geometry of space produced by moving objects. (page 307)

greatest elongation   The largest possible angle between the Sun and an inferior planet. (pages 35–36)

greenhouse effect   The trapping of infrared radiation near a planet's surface by the planet's atmosphere. (pages 114–115)

H I region   A region of neutral hydrogen in interstellar space.

H II region   A region of ionized hydrogen in interstellar space. (page 257)

helio-   A prefix referring to the Sun.

heliocentric   Centered on the Sun.

helium flash   The nearly explosive ignition of helium burning in the dense core of a red giant star. (page 267)

Hertzsprung–Russell (H–R) diagram   A plot of the absolute magnitude or luminosity of stars against their surface temperatures. (pages 238–239)

horizontal branch   A group of stars on the Hertzsprung–Russell diagram of a typical globular cluster near the main sequence and having roughly constant absolute magnitude. (pages 268–269)

Hubble constant   The constant of proportionality in the relation between the velocities of remote galaxies and their distances. (page 350)

Hubble law   The relationship which says that the redshifts of remote galaxies are directly proportional to their distances from Earth. (page 350)

hydrogen burning   The thermonuclear fusion of hydrogen nuclei to produce helium. (page 214)

hydrostatic equilibrium   A balance between the weight of a layer in a star and the pressure that supports it. (page 214)

hyperbola   A conic section formed by cutting a circular cone with a plane at an angle steeper than the sides of the cone. (page 43)

image   The optical representation of an object produced by the focusing of light rays by lenses or mirrors.

inertia   The property of matter that requires a force to act on it to change its state of motion.

inferior conjunction   The configuration when Mercury or Venus is between the Sun and the Earth. (pages 35–36)

inflationary epoch   A brief period shortly after the Big Bang during which the scale of the universe increased very rapidly. (page 384)

infrared radiation   Electromagnetic radiation of a wavelength longer than visible light yet shorter than radio waves. (pages 53–54)

interstellar dust   Microscopic solid grains of various compounds in interstellar space. (pages 315–316)

interstellar gas   Sparse gas in interstellar space. (page 259)

interstellar medium   Interstellar gas and dust. (page 259)

Io   One of the Galilean satellites. (pages 180–182)

ion   An atom that has become electrically charged due to the addition or loss of one or more electrons. (page 87)

ionization   The process by which an atom loses electrons. (page 87)

ion tail (of a comet)   The relatively straight tail of a comet produced by the solar wind acting on ions. (pages 206–207)

irregular galaxy   An unsymmetrical galaxy having neither spiral arms nor an elliptical shape. (page 337)

isotope   Any of several forms for the same chemical element whose nuclei all have the same number of protons, but different numbers of neutrons. (page 85)

isotropic   The same in all directions. (page 378)

Jovian planet   Any of the four largest planets: Jupiter, Saturn, Uranus, and Neptune. (page 96)

kelvin   *See* temperature (Kelvin). (pages 72–73)

Kepler's laws   Three statements, discovered by Johannes Kepler, that describe the motions of the planets. (page 38)

kiloparsec   One thousand parsecs; about 3260 light years. (page 7)

Kirchhoff's laws   Three statements formulated by Gustav Kirchhoff describing spectra and spectral analysis. (page 82)

Kirkwood gaps   Gaps in the spacing of asteroid orbits discovered by Daniel Kirkwood. (page 198)

light   Electromagnetic radiation that is visible to the eye. (page 51)

light curve   A graph that displays variations in the brightness of a star or other astronomical object. (page 246)

light year   The distance light travels in a vacuum in one year. (page 6)

limb (of Sun or Moon)   The apparent edge of the Sun or Moon as seen in the sky.

limb darkening The phenomenon whereby the Sun is darker near its limb than near the center of its disk. (page 218)

limiting magnitude The faintest magnitude that can be observed with a given telescope under certain conditions.

line of nodes A line connecting the nodes of an orbit. (pages 25–26)

liquid metallic hydrogen A metal-like form of hydrogen that can be produced under extreme pressure. (pages 157–158)

lithosphere The solid, upper layer of the Earth; essentially the Earth's crust. (page 122)

Local Group The cluster of galaxies of which our own galaxy is a member. (pages 399–340)

LMC The Large Magellanic Cloud, a companion galaxy to the Milky Way. (pages 337–338)

luminosity The rate at which electromagnetic radiation is emitted from a star or other object. (page 235)

luminosity class The classification of a star of a given spectral type according to its luminosity. (pages 240–241)

lunar Referring to the Moon.

lunar eclipse An eclipse of the Moon. (pages 25–27)

Lyman lines A series of spectral lines of hydrogen produced by electron transitions to and from the lowest energy state of the hydrogen atom. (pages 86–87)

magnetic field A region of space near a magnetized body within which magnetic forces can be detected.

magnetosphere The region around a planet occupied by its magnetic field. (page 134)

magnification (magnifying power) The number of times larger in angular diameter an object appears through a telescope than with the naked eye. (page 55)

magnitude A measure of the amount of light received from a star or other luminous object. (page 232)

main sequence A grouping of stars on the Hertzsprung–Russell diagram extending diagonally across the graph from the hottest, brightest stars to the dimmest, coolest stars. (page 239)

major axis (of an ellipse) The longest diameter of an ellipse. (page 37)

mantle (of a planet) That portion of a terrestrial planet located between its crust and core. (page 132)

mare (pl., maria) Latin for sea; a large relatively crater-free plain on the Moon. (pages 172–173)

mare basalt Dark solidified lava that covers the lunar maria. (pages 175–176)

mass A measure of the total amount of material in an object. (page 43)

mass–luminosity relation The relationship between the masses and luminosities of main sequence stars. (page 243)

mechanics The branch of physics dealing with the behavior and motions of objects acted upon by forces.

megaparsec (Mpc) One million parsecs. (page 7)

mesosphere The layer in a planet's atmosphere above the stratosphere. (pages 112–113)

meteor The luminous phenomenon seen when a meteoroid enters the Earth's atmosphere; a "shooting star." (page 202)

meteorite A fragment of a meteoroid that has survived passage through the Earth's atmosphere. (page 202)

meteoroid A small rock in interplanetary space. (page 202)

meteor shower Many meteors that seem to radiate from a common point in the sky. (page 209)

microwaves Short wavelength radio waves. (page 54)

Milky Way Galaxy Our galaxy. (page 314ff)

minor axis (of an ellipse) The smallest diameter of an ellipse. (page 37)

minor planet *See* asteroid. (pages 196–197)

molecule A combination of two or more atoms.

momentum A measure of the inertia of an object; an object's mass multiplied by its velocity.

monochromatic Of one wavelength or color.

N galaxy A galaxy with a bright starlike nucleus. (page 364)

nebula (pl., nebulae) A cloud of interstellar gas and dust. (page 7)

neutrino A subatomic particle with no electric charge and little or no mass, yet which is important in many nuclear reactions. (page 283)

neutron A subatomic particle with no electric charge and with a mass nearly equal to that of the proton. (page 84)

neutron star A very compact, dense star composed almost entirely of neutrons. (page 288)

New General Catalogue (NGC) A catalogue of star clusters, nebulae, and galaxies first published in 1888. (page 333)

new moon A phase of the Moon when it is nearest the Sun in the sky. (pages 21–22)

Newtonian reflector An optical arrangement in a reflecting telescope in which a small mirror reflects converging light rays to a focus on one side of the telescope tube. (pages 56–57)

Newton's laws   The laws of mechanics and gravitation formulated by Isaac Newton. (pages 41–43)

node   The intersection of an orbit with a reference plane such as the plane of the celestial equator or the ecliptic. (page 25)

nonthermal radiation   Radiation emitted by charged particles moving through a magnetic field; synchrotron radiation.

nova   A star that experiences a sudden outburst of radiant energy, temporarily increasing its luminosity roughly a thousandfold. (pages 297–298)

nuclear   Referring to the nucleus of an atom.

nuclear density   The density of matter in the nucleus of an atom; about 10^{14} g/cm^3. (page 289)

nucleus (of an atom)   The massive part of an atom, composed of protons and neutrons, about which electrons revolve. (page 84)

nucleus (of a comet)   A collection of ices and dust that constitute the solid part of a comet. (pages 205–206)

nucleus (of a galaxy)   The concentration of stars and dust at the center of a galaxy. (pages 317–318)

OB association   An association of very young, massive stars predominantly of spectral types O and B. (page 257)

objective   The principal lens or mirror of a telescope. (page 55)

oblateness   A measure of how much a flattened sphere (or spheroid) differs from a perfect sphere. (page 156)

Obler's paradox   An argument based on Newtonian cosmology that the night sky should be as bright as the Sun's surface. (page 373)

Oort cloud   A region surrounding the solar system where comets spend most of their time. (page 208)

occultation   The eclipsing of an astronomical object by the Moon or a planet.

open cluster   A loose association of young stars in the disk of the galaxy; a galactic cluster. (page 259)

opposition   The configuration of a planet when it is at an elongation of 180° and thus appears opposite the Sun in the sky. (page 36)

optics   The branch of physics dealing with the behavior and properties of light. (page 53ff)

orbit   The path of an object that is moving about a second object or point.

outgassing   Volcanic processes by which gases escape from a planet's crust into its atmosphere. (page 128)

Pallas   The second asteroid to be discovered. (page 196)

parabola   A conic section formed by cutting a circular cone at an angle parallel to the sides of the cone. (page 43)

parallax   The apparent displacement of an object caused by the motion of the observer. (page 233)

parsec   A unit of distance; 3.26 light years. (page 7)

partial eclipse   A lunar or solar eclipse in which the eclipsed object does not appear completely covered. (page 27)

Pauli exclusion principle   A principle of quantum mechanics that says that two identical particles cannot have the same position and momentum. (page 266)

penumbra   The portion of a shadow in which only part of the light source is covered by an opaque body. (page 26)

penumbral eclipse   A lunar eclipse in which the Moon passes only through the Earth's penumbra. (page 27)

perigee   The point in its orbit where a satellite or the Moon is nearest the Earth.

perihelion   The point in its orbit where a planet is nearest the Sun. (page 38)

period   The interval of time between successive repetitions of a periodic phenomenon.

periodic table   A listing of the chemical elements according to their properties invented by D. Mendeleev. (pages 78–79)

period–luminosity relation   A relationship between the period and average density of a pulsating star. (page 272)

perturbation   A small, disturbing effect. (page 325)

phases (of the Moon)   The appearance of the Moon as it orbits the Earth. (pages 21–23)

photometry   The measurement of light intensities. (page 235)

photon   A discrete unit of electromagnetic energy. (page 52)

photosphere   The region in the solar atmosphere from which most of the visible light escapes into space. (pages 217–218)

Planck time   The brief interval of time immediately after the Big Bang when all four forces had the same strength; about 10^{-43} sec. (page 376)

planetary nebula   A luminous shell of gas ejected from an old, low-mass star. (pages 277–278)

planetesimals   Primordial asteroidlike objects from which the planets accreted. (pages 102–103)

plasma   A hot ionized gas. (page 223)

plate tectonics The motions of large segments (plates) of the Earth's surface over the underlying mantle. (pages 120–123)

Population I star A star whose spectrum exhibits spectral lines of many elements heavier than helium; a metal-rich star.

Population II star A star whose spectrum exhibits comparatively few spectral lines of elements heavier than helium; a metal-poor star.

positron An electron with a positive rather than negative electric charge; an antielectron. (page 388)

precession (of the Earth) A slow, conical motion of the Earth's axis of rotation caused by the gravitational pull of the Moon and Sun on the Earth's equatorial bulge. (page 20)

precession (of the equinoxes) The slow westward motion of the equinoxes along the ecliptic because of precession. (page 20)

prime focus The point in a telescope where the objective focuses light. (page 57)

primordial fireball The extremely hot gas that filled the universe immediately following the Big Bang. (page 387)

principle of equivalence A principle of general relativity that states that, in a small volume, it is impossible to distinguish between the effects of gravitation and acceleration. (page 303)

prism A wedge-shaped piece of glass used to disperse white light into a spectrum. (pages 51–52)

prominences Flamelike protrusions seen near the limb of the Sun and extending into the solar corona. (page 224)

proper motion The annual change in the location of a star on the celestial sphere. (page 88)

proton A heavy, positively charged subatomic particle that is one of two principal constituents of atomic nuclei. (page 84)

proto- A prefix referring to the embryonic stage of a young astronomical object (planet, star, etc.) that is still in the process of formation.

pulsar A pulsating radio source believed to be associated with a rapidly rotating neutron star. (page 289)

pulsating variable A star that pulsates in size and luminosity. (page 271)

quantum mechanics The branch of physics dealing with the structure and behavior of atoms and their interaction with light. (page 86)

quark One of several hypothetical particles presumed to be the internal constituents of certain heavy subatomic particles such as protons and neutrons. (page 389)

quarter moon A phase of the Moon when it is located 90° from the Sun in the sky. (pages 21–22)

quasar A starlike object with a very large redshift; a quasi-stellar object or quasi-stellar radio source. (pages 357–358)

RR Lyrae variable One class of pulsating star with periods less than one day. (page 271)

radial velocity That portion of an object's velocity parallel to the line of sight. (page 88)

radial velocity curve A plot showing the variation of radial velocity with time for a binary star or variable star. (pages 245–246)

radiation Electromagnetic energy; photons.

radiation pressure The transfer of momentum carried by radiation to an object on which the radiation falls. (page 208)

radiative diffusion The movement of photons through a gas from a hot location to a cooler one. (page 215)

radiative zone A region inside a star where energy is transported outward by radiative diffusion. (page 217)

radio astronomy That branch of astronomy dealing with observations at radio wavelengths. (pages 60, 356)

radio galaxy A galaxy that emits an unusually large amount of radio waves. (page 356)

radio telescope A telescope designed to detect radio waves. (pages 60–61)

radio wave Long wavelength electromagnetic radiation. (pages 53–54)

radioactivity The process whereby certain atomic nuclei naturally decompose by spontaneously emitting particles. (page 82)

red giant A large, cool star of high luminosity. (pages 240–241)

redshift The shifting to longer wavelengths of the light from remote galaxies and quasars; The Doppler shift of light from a receding source. (pages 87–88)

reflecting telescope A telescope in which the principal optical component is a concave mirror. (pages 57–59)

reflection The return of light rays by a surface. (page 57)

reflection nebula A comparatively dense cloud of dust in interstellar space that is illuminated by a star. (pages 258–259)

refracting telescope A telescope in which the principal optical component is a lens. (pages 55–56)

refraction The bending of light rays passing from one transparent medium to another. (page 55)

regolith The rocky material of the surface of a lifeless planet like the Moon or Mars. (page 175)

resolution The degree to which fine details in an optical image can be distinguished. (page 58)

resolving power A measure of the ability of an optical system to distinguish or resolve fine details in the image it produces. (page 58)

retrograde motion The apparent westward motion of a planet with respect to background stars. (pages 33–34)

revolution The motion of one body about another.

Roche lobe The smallest distance from a planet or other object at which a second object can be held together by purely gravitational forces. (page 247)

rotation The turning of a body about an axis passing through through the body.

rotation curve (of a galaxy) A graph showing how the orbital speed of material in a galaxy depends on the distance from the galaxy's center. (pages 322–323)

satellite A body that revolves about a large one.

Schwarzschild radius The distance from the singularity to the event horizon in a nonrotating black hole. (page 305)

seismic waves Vibrations traveling through a terrestrial planet, usually associated with earthquakelike phenomena. (page 132)

seismograph A device used to record and measure seismic waves, such as those produced by earthquakes. (page 132)

seismology The study of earthquakes and related phenomena.

semimajor axis Half of the major axis of an ellipse. (page 37)

Seyfert galaxy A spiral galaxy with a bright nucleus whose spectrum exhibits emission lines. (pages 361–362)

shepherd satellite A small satellite whose gravity is responsible for maintaining a sharply defined ring of matter around a planet such as Saturn or Uranus. (page 152)

shock wave An abrupt, localized region of compressed gas caused by an object traveling through the gas at a speed greater than the speed of sound. (page 143)

shooting star *See* meteor. (page 202)

sidereal month The period of the Moon's revolution about the Earth with respect to the stars. (page 21)

sidereal period The orbital period of one object about another with respect to the stars. (page 36)

SMC The Small Magellanic Cloud, a companion galaxy to the Milky Way. (pages 337–338)

solar activity Phenomena that occur in the solar atmosphere, such as sunspots, plages, flares, and so forth.

solar cycle A 22-year cycle during which the Sun's magnetic field reverses its polarity. (page 226)

solar flare A violent eruption of the Sun's surface. (page 224)

solar nebula The cloud of gas and dust from which the Sun and solar system formed. (pages 100–101)

solar seismology The study of the Sun's interior from observations of vibrations of its surface. (page 227)

solar system The Sun, planets, their satellites, asteroids, comets and related objects that orbit the Sun. (pages 92–93)

solar wind A radial flow of particles (mostly electrons and protons) from the Sun. (page 105)

solstice Either of two points along the ecliptic at which the Sun reaches its maximum distance north or south of the celestial equator. (pages 18–19)

special theory of relativity A description of mechanics and electromagnetic theory formulated by Einstein that explains that measurements of distance, time, and mass are affected by the observer's motion. (page 303)

spectral analysis The identification of chemicals by the appearance of their spectra. (page 78)

spectral type A classification of stars according to the appearance of their spectra. (page 236)

spectrogram The photograph of a spectrum. (page 80)

spectrograph A device for photographing a spectrum. (pages 80–81)

spectroscope A device for directly viewing a spectrum. (page 80)

spectroscopic binary star A double star whose binary nature can be deduced from the periodic Doppler shifting of lines in its spectrum. (pages 244–246)

spectroscopy The study of spectra.

speed The rate at which an object moves.

spicule A narrow jet of rising gas in the solar chromosphere. (pages 219–220)

spin A small amount of angular momentum possessed by certain particles such as electrons, protons, and neutrons. (page 318)

spiral arms Lanes of interstellar gas, dust, and young stars that wind outward in a plane from the central regions of a galaxy. (page 260)

spiral galaxy A flattened, rotating galaxy with pinwheel-like spiral arms winding outward from the galaxy's nucleus. (pages 232–324)

standard candle An object whose known luminosity can be used to deduce the distance to a galaxy. (page 351)

star A self-luminous sphere of gas.

Stefan-Boltzmann law A relationship between the temperature of a blackbody and the rate at which it radiates energy. (page 74)

Stefan's law *See* Stefan-Boltzmann law. (page 74)

stellar evolution The changes in size, luminosity, temperature, and so forth, that occur as a star ages. (page 253)

stellar model The result of theoretical calculations that give details of physical conditions inside a star. (page 214)

stratosphere A layer in the atmosphere of a planet directly above the troposphere. (pages 112–113)

strong nuclear force The force that binds protons and neutrons together in nuclei. (page 389)

subduction zone A location where colliding tectonic plates cause the Earth's crust to be pulled down into the mantle. (page 122)

subgiant A star whose luminosity is between that of main sequence stars and normal giants of the same spectral type. (page 240)

summer solstice The point on the ecliptic where the Sun is farthest north of the celestial equator. (pages 18–19)

Sun The star about which the Earth and other planets revolve. (page 212ff)

sunspot A temporary cool region in the solar photosphere. (page 222)

sunspot cycle The semiregular 11-year period with which the number of sunspots fluctuates. (page 223)

supergiant A star of very high luminosity. (page 239)

superior conjunction The configuration of a planet being behind the Sun. (pages 35–36)

supermassive black hole A black hole whose mass exceeds a thousand solar masses. (page 329)

supernova A stellar outburst during which a star suddenly increases its brightness roughly a millionfold. (page 282)

synchrotron radiation The radiation emitted by charged particles moving through a magnetic field; nonthermal radiation. (page 328)

synodic month The period of revolution of the Moon with respect to the Sun; the length of one cycle of lunar phases. (pages 21, 23)

synodic period The interval between successive occurrences of the same configuration of a planet. (page 36)

T Tauri stars Young variable stars associated with interstellar matter that show erratic changes in luminosity. (page 258)

T Tauri wind A flow of particles away from a T Tauri star. (page 104)

tail (of a comet) Gas and dust particles from a comet's nucleus that have been swept away from the comet's head by the radiation pressure of sunlight and the solar wind. (pages 205–206)

telescope An instrument for viewing remote objects. (page 50ff)

temperature (Celsius) Temperature measured on a scale where water freezes at 0° and boils at 100°. (pages 72–73)

temperature (Fahrenheit) Temperature measured on a scale where water freezes at 32° and boils at 212°. (pages 72–73)

temperature (Kelvin) Absolute temperature measured in centigrade degrees. (pages 72–73)

terminator The line dividing day and night on the surface of the Moon or a planet; the line of sunset or sunrise.

terrae Cratered regions of the Moon; lunar highlands. (page 174)

terrestrial planet Any of the planets Mercury, Venus, Earth, or Mars, and sometime also the Galilean satellites and Pluto. (page 95)

thermal energy The energy associated with the motions of atoms or molecules in a substance.

thermal equilibrium A balance between the input and outflow of heat in a system. (page 215)

thermal radiation The radiation naturally emitted by any object not at absolute zero. (pages 73–74)

thermonuclear fusion A reaction in which the nuclei of atoms are fused together at a high temperature. (pages 213–214)

thermonuclear reaction A reaction resulting from a high-speed collision of nuclear particles moving rapidly because they are at a high temperature. (page 213)

thermosphere A region in the Earth's atmosphere between the mesosphere and the exosphere. (pages 112–113)

3-K cosmic microwave background Photons from every part of the sky with a blackbody spectrum at nearly 3 K; the cooled-off radiation from the primordial fireball that filled all space. (page 378)

tidal force A gravitational force whose strength and/or direction varies over a body and thus tends to deform the body. (page 180)

total eclipse A solar eclipse during which the Sun is completely hidden by the Moon, or a lunar eclipse during which the Moon is completely immersed in the Earth's umbra. (page 27)

transit The passage of a celestial body across the meridian; the passage of a small object in front of a larger object. (page 168)

triple point The pressure and temperature at which a substance can simultaneously exist as a solid, liquid, and gas. (page 189)

Trojan asteroid One of several asteroids that share Jupiter's orbit about the Sun. (pages 198–199)

troposphere The lowest level in the Earth's atmosphere. (page 112)

turbulence Random motions in a gas or liquid.

UBV system A system of stellar magnitudes involving measurements of starlight intensity in the ultraviolet, blue, and visible spectral regions. (page 235)

ultraviolet radiation Electromagnetic radiation of wavelengths shorter than those of visible light but longer than those of X rays. (pages 53–54)

umbra The central, completely dark portion of a shadow. (page 26)

universal constant of gravitation The constant of proportionality in Newton's law of gravitation. (page 43)

universe All space along with all the matter and radiation in space.

Van Allen belts Two doughnut-shaped regions around the Earth where many charged particles (protons and electrons) are trapped by the earth's magnetic field. (pages 134–135)

variable star A star whose luminosity varies. (page 271)

vernal equinox The point on the ecliptic where the Sun crosses the celestial equator from south to north. (page 18)

visual binary star A double star in which the two components can be resolved through a telescope. (page 241)

VLBI Very long baseline interferometry; a method of connecting widely separated radio telescopes to make observations of very high resolution. (page 62)

wavelength The distance between two successive peaks in a wave. (pages 52–53)

white dwarf A low-mass star that has exhausted all its thermonuclear fuel and contracted to a size roughly equal to the size of the Earth. (pages 239–240)

Widmanstätten patterns Crystalline structure seen in certain types of meteorites. (pages 203–204)

Wien's law A relationship between the temperature of a blackbody and the wavelength at which it emits the greatest intensity of radiation. (page 74)

winter solstice The point on the ecliptic where the Sun reaches its greatest distance south of the celestial equator. (pages 18–19)

X ray Electromagnetic radiation whose wavelength is between that of ultraviolet light and gamma rays. (pages 53–54)

year The period of revolution of the Earth about the Sun.

Zeeman effect A splitting or broadening of spectral lines because of a magnetic field. (pages 223–224)

zenith The point on the celestial sphere opposite to the direction of gravity. (page 19)

zero-age main sequence (ZAMS) The main sequence of young stars that have just begun to burn hydrogen at their cores. (page 267)

zodiac A band of twelve constellations around the sky centered on the ecliptic. (page 19)

Answers to selected exercises

Chapter 1

3 (a) 10^7 (b) 4×10^5 (c) 6×10^{-2} (d) 7×10^9
7 8.3 minutes
8 8700 km

Chapter 2

2 The equator
4 The south pole; $23\frac{1}{2}°$; December 21
5 Due east

Chapter 3

10 5.2 square AUs in 1984; 26 square AUs in five years
11 Average distance = 100 AU; farthest distance = 200 AU
12 Yes; sidereal period = synodic period = 2 years; Mars (sidereal period = 1.88 yr; synodic period = 2.14 yr)

Chapter 5

8 $7\frac{1}{2}$
9 2500 Å; 16 times brighter than the Sun
10 The Sun is sixteen times brighter
12 To shift photons from 7000 to 5000 Å requires a speed of $\frac{2}{7}$ the speed of light

Chapter 6

6 They are the simplest stable chemical combinations of hydrogen (the most abundant element) and the next three most abundant elements (excluding helium, which is inert)
11 The planet would be made mostly of ice; density = 1 g/cm^3; the planet's diameter is 1.77 times lager than the Earth's diameter

Chapter 7

12 Core = 30%; mantle = 69%; crust = 1%
13 About 220 million years ago

Chapter 11

11 During 5 billion years, 5 percent of the Sun's hydrogen is converted into helium

Chapter 12

11 $10 M_\odot$; $10^{-2} L_\odot$

Chapter 16

12 One supernova about every 55,000 years

Chapter 17

11 1500 light years
12 5000 km/sec (for these answers, H_0 = 15 km/sec/Mly = 50 km/sec/Mpc)

Illustration credits

Chapter 1 p. 1: Photography by David F. Malin of the Anglo-Australian Observatory from original negatives by the U.K. 1.2-m Schmidt telescope, copyright © 1987 Royal Observatory, Edinburgh; Fig. 1-2: Courtesy of Scientific American, NASA, and AAO; Figs. 1-3, 1-4, 1-5, 1-11: NASA; Fig. 1-7: USNO; Fig. 1-8: Lick Observatory; Fig. 1-9: copyright © R. J. Dufour, Rice University; Fig. 1-10: Palomar Observatory.

Chapter 2 p. 13: Copyright © Anglo-Australian Telescope Board; Fig. 2-1: British crown copyright, reproduced with permission of the Controller of the Brittanic Majesty's Stationary Office; Fig. 2-2: (left) Drawing from Elijah Burritt's *Atlas,* Janus Publications, Wichita Omnisphere Earth-Space Center, 220 S. Main, Wichita, Kansas, (right) Courtesy of Robert C. Mitchell, Central Washington University; Figs. 2-10, 2-21, 2-23: NASA; Fig. 2-12: Lick Observatory; Fig. 2-20: Courtesy of Mike Harms, Fig. 2-24: Courtesy of Dennis di Cicco.

Chapter 3 p. 32: NASA; Fig. 3-8: New Mexico State University Observatory; Fig. 3-10: Courtesy of Clifford Holmes; Figs. 3-11, 3-12: Yerkes Observatory; Fig. 3-14: Copyright © 1986 Jack B. Marling, Lumicon; Fig. 3-15; Lick Observatory.

Chapter 4 p. 50: Copyright © Anglo-Australian Telescope Board; Fig. 4-8: Yerkes Observatory; Fig. 4-12: NOAO; Fig. 4-14: Courtesy of the Multiple-Mirror Telescope Observatory; Fig. 4-15: California Institute of Technology; Fig. 4-16: Smithsonian Institution Astrophysical Observatory; Fig. 4-17: Courtesy of Patrick Seitzer, NOAO; Figs. 4-18, 4-19: NRAO; Figs. 4-20a, 4-21, 4-22, 4-23b, 4-24, 4-25: NASA; Fig. 4-20b: Copyright © 1982 Associated Universities, Inc., under contract with the National Science Foundation (VLA observations by Imke de Pater, J. R. Dickel); Fig. 4-23: (a) George R. Carruthers, NRL, (c) Robert C. Mitchell, Central Washington University; Fig. 4-26: TRW; Fig. 4-27: Donald A. Kniffen, NASA and TRW; Fig. 4-28: (a) Max Planck Institut für Radioastronomie, (b) Jet Propulsion Laboratory, (c) Griffith Observatory, (d) Royal Observatory, Edinburgh, (e) Elihu A. Boldt, NASA.

Chapter 5 p. 71: Courtesy of Bausch and Lomb; Figs. 5-5, 5-20: NOAO; Figs. 5-8, 5-16: Mount Wilson and Las Campanas Observatories; Fig. 5-10: Palomar Observatory.

Chapter 6 p. 91: Photography by David F. Malin of the Anglo-Australian Observatory from Original negatives by the U.K. 1.2-m Schmidt telescope, copyright © 1890 Royal Observatory, Edinburgh; Figs. 6-3, 6-4, 6-5, 6-14, 6-15: NASA; Fig. 6-6: Stephen P. Meszaros, NASA; Figs. 6-7, 6-8, 6-13: copyright © Anglo-Australian Telescope Board; Fig. 6-9: University of Arizona and JPL; Fig. 6-16: Courtesy of Marlin E. Kipp, Sandia National Laboratories.

Chapter 7 p. 110: NASA; Figs. 7-1, 7-3, 7-8, 7-10, 7-11, 7-18, 7-19, 7-20, 7-21, 7-22, 7-23, 7-24, 7-28, 7-29, 7-30, 7-31: NASA; Fig. 7-7: Courtesy of Stephen M. Larson; Fig. 7-13: W. Kaufmann; Fig. 7-15: Courtesy of Marie Tharp and Bruce Heezen © 1977; Fig. 7-25: Stephen P. Meszaros, NASA; Fig. 7-26: A. Post, Project Office Glaciology, U.S. Geological Survey; Fig. 7-27: Courtesy of Carle M. Pieters and the U.S.S.R. Academy of Sciences; Fig. 7-34: Courtesy of Syun-Ichi Akasofu, Geophysical Institute, University of Alaska, Fairbanks; Fig. 7-36: Courtesy of Michael Carroll.

Chapter 8 p. 140: Stephen P. Meszaros, NASA; Fig. 8-1: Copyright © The University of Texas McDonald Observatory; Figs. 8-2, 8-3, 8-6, 8-7, 8-8, 8-11, 8-12, 8-14, 8-15, 8-16, 8-17, 8-18, 8-19, 8-20, 8-21, 8-23, 8-24, 8-26, 8-32, 8-33, 8-34, 8-35, 8-36, 8-37: NASA; Fig. 8-13: Lowell Observatory; Figs. 8-29, 8-30: New Mexico State University Observatory.

Chapter 9 p. 167: NASA; Fig. 9-1: Yerkes Observatory; Fig. 9-2: New Mexico State University Observatory; Fig. 9-3: (left) NASA. (right) Lick Observatory; Figs. 9-4, 9-8, 9-9, 9-11, 9-12, 9-13, 9-14, 9-15, 9-16, 9-17, 9-18, 9-19, 9-20, 9-21, 9-22, 9-23, 9-24, 9-25, 9-26, 9-27, 9-29, 9-30: NASA; Fig. 9-7: Lick Observatory; Fig. 9-31: Courtesy of James Christy and Robert Harrington.

Chapter 10 p. 194: Copyright © Anglo-Australian Telescope Board; Figs. 10-1, 10-5: Yerkes Observatory; Fig. 10-6: Meteor Crater Enterprises, Arizona; Fig. 10-7: Courtesy of

Walter Alvarez; Fig. 10-8: Courtesy of Ronald A. Oriti; Figs. 10-9, 10-10, 10-11, 10-12, 10-13: Courtesy of Ronald A. Oriti; Fig. 10-14: Courtesy of Hans Vehrenberg; Fig. 10-15: Max Planck Institut für Aeronomie; Fig. 10-17: Johns Hopkins University and the Naval Research Laboratory; Figs. 10-18, 10-19: Lick Observatory; Fig. 10-21: Palomar Observatory; Fig. 10-22: New Mexico State University Observatory; Fig. 10-23: TASS, Sovfoto.

Chapter 11 p. 212: High Altitude Observatory, NCAR, University of Colorado; Fig. 11-1: Celestron International; Fig. 11-4: San Fernando Observatory; Fig 11-5, 11-14, 11-16, 11-17: NOAO; Fig. 11-7: C. Keller, Los Alamos Scientific Laboratory; Fig. 11-8: NASA and the High Altitude Observatory; Fig. 11-9: NASA and Harvard College Observatory; Fig. 11-10: Project Stratoscope, Princeton University; Fig. 11-11: Mount Wilson and Las Campanas Observatories; Fig. 11-13: NASA; Fig. 11-15: Naval Research Laboratory; Fig. 11-20: National Solar Observatory.

Chapter 12 p. 231: Copyright © Anglo-Australian Telescope Board; Fig. 12-8: Courtesy of Nancy Houk, Nelson Irvine, and David Rosenbush; Fig. 12-13: Yerkes Observatory; Fig. 12-17: Lick Observatory;

Chapter 13 p. 252, Figs. 13-2, 13-5, 13-6, 13-8, 13-9, 13-11, 13-13, 13-20: Copyright © Anglo-Australian Telescope Board; Fig. 13-1: Photography by David F. Malin of the Anglo-Australian Observatory from original negatives by the U.K. 1.2-m Schmidt telescope, copyright © 1980 Royal Observatory, Edinburgh; Fig. 13-4: (a) IRAS photograph produced by NRL/ROG, using spline fitted survey data [IRAS was developed and operated by the Netherlands Agency for Aerospace Programs (NIVR), NASA, and the U.K. Science and Engineering Research Council (SERC)], (b) optical photography by David F. Malin of the Anglo-Australian Observatory from original negatives by the U.K. 1.2-m Schmidt telescope, copyright © 1980 Royal Observatory, Edinburgh; Figs. 13-7, 13-16: U.S. Naval Observatory; Fig. 13-12: Courtesy of Hans Vehrenberg; Fig. 13-19: Lick Observatory; Fig. 13-21: NOAO.

Chapter 14 pp. 275, Figs. 14-19, 14-25: Lick Observatory photographs; Figs. 14-2, 14-3, 14-8: copyright © Anglo-Australian Telescope Board; Fig. 14-5: Courtesy of R. B. Minton; Fig. 14-7: European Southern Observatory; Fig. 14-10: Courtesy of K. S. Luttrell; Fig. 14-11: Mount Wilson and Las Campanas Observatories; Figs. 14-13, 14-17: Palomar Observatory; Fig. 14-14: Photography by David A. Malin of the Anglo-Australian Observatory from original negatives by the U.K. 1,2-m Schmidt telescope, copyright © 1980 Royal Observatory, Edinburgh; Fig. 14-15: S. S. Murray, Harvard-Smithsonian Center for Astrophysics, (b) NRAO/AUI (VLA observations by A. Angerhofer, R. A. Perley, B. Balick, D. K. Milne); Fig. 14-20: NASA; Fig. 14-24: NRAO/AUI (VLA observations by R. M. Hjellming and K. J. Johnson)

Chapter 15 pp. 302: Courtesy of D. Norton, Science Graphics; Fig. 15-6: National Radio Astronomy Observatory, operated by Associated Universities, Inc., under contract with the National Science Foundation (VLA observations by P. E. Greenfield, D. H. Roberts, B. F. Burke); Fig. 15-7: National Radio Astronomy Observatory, operated by Associated Universities, Inc., under contract with the National Science Foundation (VLA observations by J. N. Hewitt and E. L. Turner); Fig. 15-8: Courtesy of J. Kristian, Mt. Wilson and Las Campanas Observatories.

Chapter 16 p. 314: Courtesy of Dennis di Cicco; Fig. 16-1: Steward Observatory, University of Arizona; Fig. 16-3: Harvard Observatory; Fig. 16-4: NOAO; Fig. 16-6: U.S. Naval Observatory; Fig. 16-9: Courtesy of G. Westerhout; Fig. 16-10: (a) Copyright © The Anglo-Australian Telescope Board; (b) National Radio Astronomy Observatory, operated by Associated Universities, Inc. under contract with the National Science Foundation; Fig. 16-14: Courtesy of P. Seiden, D. Elmegreen, B. Elmegreen, and A. Mobarak, IBM; Fig. 16-17: (a) Courtesy of Dennis di Cicco, (b and c) NASA; Fig. 16-18: Copyright © The Anglo-Australian Telescope Board; Fig. 16-19: National Radio Astronomy Observatory, operated by Associated Universities, Inc., under contract with the National Science Foundation [VLA observations by (a) F. Yusef-Zadeh, M. R. Morris, D. R. Chance, (b) K-Y Lo, M. J. Claussen].

Chapter 17 p. 332: European Southern Observatory; Fig. 17-1: Courtesy of Lund Humphries; Fig. 17-2: Copyright © 1984 Regents, University of Hawaii, Canada-France-Hawaii Telescope; Fig. 17-3: Palomar Observatory, copyright © California Institute of Technology; Fig. 17-4: Mount Wilson and Las Campanas Observatories; Fig. 17-5: (Sa) Courtesy of Rudolph Schild, Harvard-Smithsonian Center for Astrophysics, (Sb, Sc) Courtesy of Philip E. Seiden, IBM; Fig. 17-6: (Sa) NOAO, (Sb, Sc) Courtesy of Rudolph Schild, Harvard-Smithsonian Center for Astrophysics; Fig. 17-7: (SBa, SBc) Courtesy of Philip E. Seiden, IBM; (SBb) courtesy of Rudolph Schild, Harvard-Smithsonian Center for Astrophysics; Fig. 17-8: Yerkes Observatory; Figs. 17-9, 17-10, 17-17, 17-18, 17-22b: Courtesy of Rudolph Schild, Harvard-Smithsonian Center for Astrophysics: Figs. 17-12, 17-13, 17-14: Copyright © Anglo-Australian Telescope Board; Fig. 17-16: Photography by David A. Malin of the Anglo-Australian Observatory from original negatives by the U.K. 1,2-m Schmidt telescope, copyright © 1987 Royal Observatory, Edinburgh; Fig. 17-19: Courtesy of E. J. Groth, Princeton University; Fig. 17-20: Courtesy of V. de Lapparent, M. Geller, and J. Huchra; Figure 17-22a, 17-31: Lick Observatory photographs; Fig. 17-24: Royal Observatory, Edinburgh; Fig. 17-25: Courtesy of W. C. Keel, University of Alabama; Fig. 17-26; Courtesy of Lars Hernquist, Institute for Advanced Study with simulations performed at the Pittsburgh Supercomputing Center; Fig. 17-29: Mt. Wilson and Las Campanas Observatories.

Index

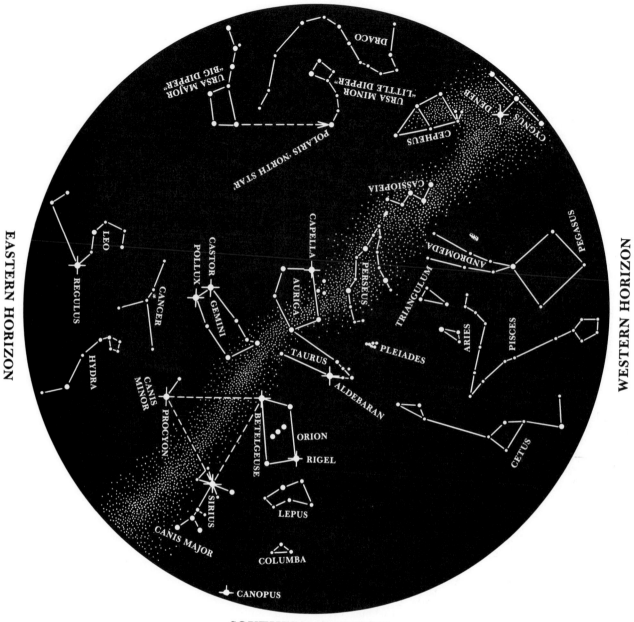

THE NIGHT SKY IN WINTER

Latitude of chart is 34°N, but it is practical throughout the continental United States.

To use: Hold chart vertically and turn it so the direction you are facing shows at the bottom.

Chart time (local standard time):
Mid-December.. 11 pm
Mid-January...... 9 pm
Mid-February.... 7 pm

Star Chart from *GRIFFITH OBSERVER*, Griffith Observatory, Los Angeles

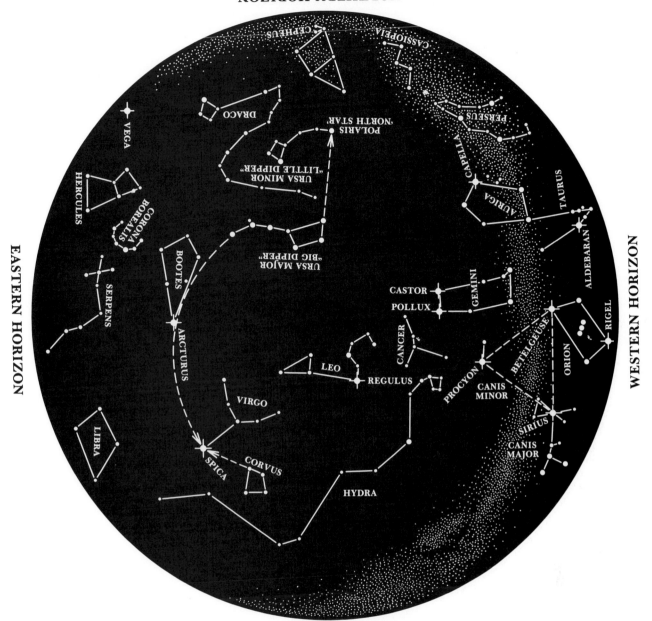

NORTHERN HORIZON

EASTERN HORIZON

WESTERN HORIZON

SOUTHERN HORIZON

THE NIGHT SKY IN SPRING

Latitude of chart is 34°N, but it is practical throughout the continental United States.

To use: Hold chart vertically and turn it so the direction you are facing shows at the bottom.

Chart time (local standard time):
Mid–March....... 11 pm
Mid–April.......... 9 pm
Mid–May.......... 7 pm

Star Chart from *GRIFFITH OBSERVER*, Griffith Observatory, Los Angeles

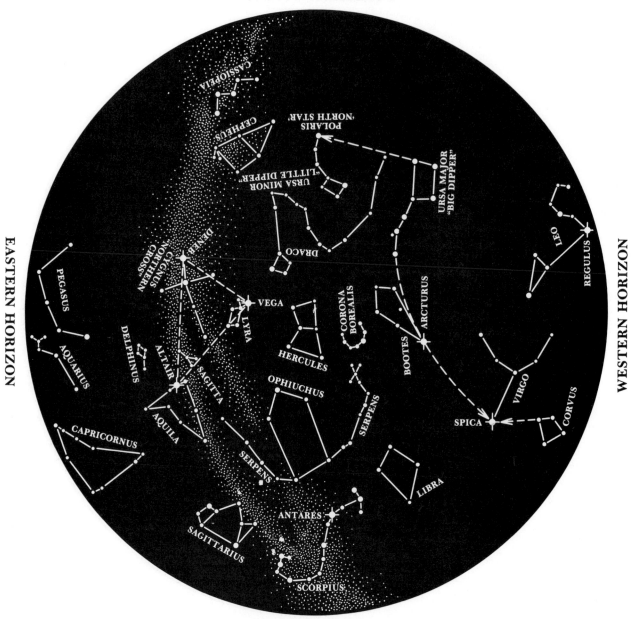

NORTHERN HORIZON

CASSIOPEIA

CEPHEUS

POLARIS 'NORTH STAR'

URSA MINOR "LITTLE DIPPER"

URSA MAJOR "BIG DIPPER"

DENEB

CYGNUS NORTHERN CROSS

DRACO

LEO

REGULUS

EASTERN HORIZON

PEGASUS

VEGA

LYRA

CORONA BOREALIS

ARCTURUS

AQUARIUS

DELPHINUS

ALTAIR

SAGITTA

HERCULES

OPHIUCHUS

BOOTES

VIRGO

WESTERN HORIZON

AQUILA

SERPENS

SERPENS

SPICA

CORVUS

CAPRICORNUS

LIBRA

SAGITTARIUS

ANTARES

SCORPIUS

SOUTHERN HORIZON

THE NIGHT SKY IN SUMMER

Latitude of chart is 34°N, but it is practical throughout the continental United States.

To use: Hold chart vertically and turn it so the direction you are facing shows at the bottom.

Chart time (daylight saving time):

Mid–June .... Midnight
Mid–July .... 10 pm
Mid–August......... 8 pm

Star Chart from *GRIFFITH OBSERVER*, Griffith Observatory, Los Angeles

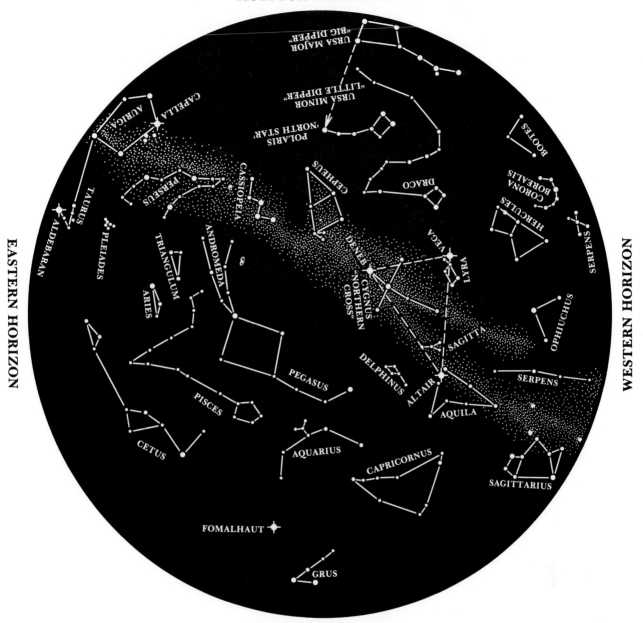

NORTHERN HORIZON

NORTHERN HORIZON

URSA MAJOR "BIG DIPPER"

URSA MINOR "LITTLE DIPPER"

POLARIS "NORTH STAR"

BOÖTES

CASSIOPEIA

CEPHEUS

DRACO

CORONA BOREALIS

HERCULES

CAPELLA

AURIGA

PERSEUS

TAURUS

ALDEBARAN

PLEIADES

TRIANGULUM

ANDROMEDA

ARIES

DENEB

CYGNUS "NORTHERN CROSS"

VEGA

LYRA

SERPENS

OPHIUCHUS

SAGITTA

PEGASUS

DELPHINUS

ALTAIR

SERPENS

PISCES

AQUILA

CETUS

AQUARIUS

CAPRICORNUS

SAGITTARIUS

FOMALHAUT

GRUS

EASTERN HORIZON

WESTERN HORIZON

SOUTHERN HORIZON

THE NIGHT SKY IN AUTUMN

Latitude of chart is 34°N, but it is practical throughout the continental United States.

To use: Hold chart vertically and turn it so the direction you are facing shows at the bottom.

Chart time (local standard time):
Mid–September..... 11 pm
Mid–October.......... 9 pm
Mid–November....... 7 pm

Star Chart from *GRIFFITH OBSERVER*, Griffith Observatory, Los Angeles